Lecture Notes in Computer Science 2051

Edited by G. Goos, J. Hartmanis and J. van Leeuwen

Springer
Berlin
Heidelberg
New York
Barcelona
Hong Kong
London
Milan
Paris
Singapore
Tokyo

Aart Middeldorp (Ed.)

Rewriting Techniques and Applications

12th International Conference, RTA 2001
Utrecht, The Netherlands, May 22-24, 2001
Proceedings

Springer

Series Editors

Gerhard Goos, Karlsruhe University, Germany
Juris Hartmanis, Cornell University, NY, USA
Jan van Leeuwen, Utrecht University, The Netherlands

Volume Editor

Aart Middeldorp
University of Tsukuba
Institute of Information Sciences and Electronics
Tsukuba 305-8573, Japan
E-mail: ami@is.tsukuba.ac.jp

Cataloging-in-Publication Data applied for

Die Deutsche Bibliothek - CIP-Einheitsaufnahme

Rewriting techniques and applications : 12th international conference ; proceedings / RTA 2001, Utrecht, The Netherlands, May 22 - 24, 2001. Aart Middeldorp (ed.). - Berlin ; Heidelberg ; New York ; Barcelona ; Hong Kong ; London ; Milan ; Paris ; Singapore ; Tokyo : Springer, 2001
(Lecture notes in computer science ; Vol. 2051)
ISBN 3-540-42117-3

CR Subject Classification (1998): F.4, F.3.2, D.3, I.2.2-3, I.1

ISSN 0302-9743
ISBN 3-540-42117-3 Springer-Verlag Berlin Heidelberg New York

Springer-Verlag Berlin Heidelberg New York
a member of BertelsmannSpringer Science+Business Media GmbH

http://www.springer.de

Typesetting: Camera-ready by author, data conversion by PTP Berlin, Stefan Sossna
Printed on acid-free paper SPIN 10781501 06/3142 5 4 3 2 1 0

Preface

This volume contains the proceedings of the *12th International Conference on Rewriting Techniques and Applications* (RTA 2001), which was held May 22-24, 2001 at Utrecht University in The Netherlands. RTA is the major forum for the presentation of research on all aspects of rewriting. Previous RTA conferences were held in Dijon (1985), Bordeaux (1987), Chapel Hill (1989), Como (1991), Montreal (1993), Kaiserslautern (1995), Rutgers (1996), Sitges (1997), Tsukuba (1998), Trento (1999), and Norwich (2000).

There were 55 submissions from Argentina ($\frac{2}{3}$), Australia (1), France ($12\frac{2}{3}$), Germany ($11\frac{2}{3}$), Israel ($1\frac{1}{3}$), Italy (2), Japan ($8\frac{1}{2}$), The Netherlands (6), Slovakia ($\frac{1}{3}$), Spain (4), UK ($2\frac{5}{6}$), USA (3), and Venezuela (1), of which the program committee selected 23 regular papers and 2 system descriptions for presentation. In addition, there were invited talks by Arvind (Rewriting the Rules for Chip Design), Henk Barendregt (Computing and Proving), and Michael Rusinowitch (Rewriting for Deduction and Verification).

The program committee awarded the *best paper* prize to Jens R. Woinowski for his paper *A Normal Form for Church-Rosser Language Systems.* In this paper the surprising and important result is shown that all Church-Rosser languages can be defined by string rewrite rules of the form $uvw \to uxw$ with v being nonempty and x having a maximum length of one.

Many people helped to make RTA 2001 a success. I am grateful to the members of the program committee and the external referees for reviewing the submissions and maintaining the high standards of the RTA conferences. It is a particular pleasure to thank Vincent van Oostrom and the other members of the local organizing committee for organizing an excellent conference in a rather short period. Finally, I thank the organizers of the four events that collocated with RTA 2001 for making the conference even more attractive:

- 4th International Workshop on Explicit Substitutions: Theory and Applications to Programs and Proofs (Pierre Lescanne),
- 5th International Workshop on Termination (Nachum Dershowitz),
- International Workshop on Reduction Strategies in Rewriting and Programming (Bernhard Gramlich and Salvador Lucas),
- IFIP Working Group 1.6 on Term Rewriting (Claude Kirchner).

March 2000 — Aart Middeldorp

Conference Organization

Program Chair

Aart Middeldorp — *University of Tsukuba*

Conference Chair

Vincent van Oostrom — *Utrecht University*

Program Committee

Zena Ariola	*Eugene*
David Basin	*Freiburg*
Mariangiola Dezani-Ciancaglini	*Torino*
Philippe de Groote	*Nancy*
Christine Paulin-Mohring	*Orsay*
Ian Mackie	*Palaiseau*
José Meseguer	*Menlo Park*
Aart Middeldorp	*Tsukuba*
Robert Nieuwenhuis	*Barcelona*
Enno Ohlebusch	*Bielefeld*
Friedrich Otto	*Kassel*
Sándor Vágvölgyi	*Szeged*
Joe Wells	*Edinburgh*

Local Organizing Committee

Jan Bergstra	*Utrecht*
Hans Zantema	*Eindhoven*
Vincent van Oostrom	*Utrecht*

RTA Steering Committee

Leo Bachmair	*Stony Brook*	
Nachum Dershowitz	*Tel Aviv*	(chair)
Hélène Kirchner	*Nancy*	
José Meseguer	*Menlo Park*	(publicity chair)
Tobias Nipkow	*Munich*	
Michael Rusinowitch	*Nancy*	

Sponsors

Centrum voor Wiskunde en Informatica (CWI)
Instituut voor Programmatuurkunde en Algoritmiek (IPA)
International Federation for Information Processing (IFIP)
University of Amsterdam, Informatics Institute
University of Tsukuba, Institute of Information Sciences and Electronics
Utrecht University, Department of Philosophy
Utrecht University, Lustrum Committee
ZENO Institute for Philosophy, The Leiden-Utrecht Research Institute

List of Referees

Yohji Akama
Thomas Arts
Jürgen Avenhaus
Steffen van Bakel
Franco Barbanera
László Bernátsky
Gavin Bierman
Stefan Blom
Roel Bloo
Gerhard Buntrock
Wei Ngan Chin
Adam Cichon
Manuel Clavel
Evelyne Contejean
Mario Coppo
Andrea Corradini
Roberto Di Cosmo
Dan Dougherty
Frank Drewes
Catherine Dubois
Irène Durand
Steven Eker
Maribel Fernández
Wan Fokkink
Zoltán Fülöp
Philippa Gardner
Simon Gay
Neil Ghani
Robert Giegerich
Jürgen Giesl
Isabelle Gnaedig
Guillem Godoy
Eric Goubault
Jean Goubault-Larrecq
Bernhard Gramlich
Stefano Guerrini
Pál Gyenizse
Chris Hankin
Michael Hanus
Thérèse Hardin
Dieter Hofbauer
Markus Holzer
Maria Huber
Benedetto Intrigila
Jean-Pierre Jouannaud
Fairouz Kamareddine
Yoshinobu Kawabe
Richard Kennaway
Delia Kesner
Felix Klaedtke
Yves Lafont
François Lamarche
Ingo Lepper
Pierre Lescanne
Jordi Levy
Ugo de'Liguoro
Luigi Liquori
Markus Lohrey
Salvador Lucas
Klaus Madlener
Luis Mandel
Claude Marché
Maurice Margenstern
Gundula Niemann
Yoshikatsu Ohta
Peter Ølveczky
Vincent van Oostrom
Adriano Peron
Jorge Sousa Pinto
Adolfo Piperno
François Pottier
Femke van Raamsdonk
Christophe Ringeissen
Albert Rubio
Michael Rusinowitch
Kai Salomaa
Andrea Sattler-Klein
Manfred Schmidt-Schauß
Marco Schorlemmer
Aleksy Schubert
Carsten Schürmann
Helmut Seidl
Paula Severi
Mark-Oliver Stehr
Magnus Steinby
Georg Struth
Taro Suzuki
Ralf Treinen
Marisa Venturini Zilli
Laurent Vigneron
Fer-Jan de Vries
Uwe Waldmann
Andreas Weiermann
Benjamin Werner
Hans Zantema

Table of Contents

Invited Talks

Regular Papers

System Descriptions

Computing and Proving

Henk Barendregt

University of Nijmegen
henk@cs.kun.nl

Abstract. Computer mathematics is the enterprise to represent substantial parts of mathematics on a computer. This is possible also for arbitrary structures (with non-computable predicates and functions), as long as one also represents proofs of known properties of these. In this way one can construct a 'Mathematical Assistant' that verifies the well-formedness of definitions and statements, helps the human user to develop theories and proofs.
An essential part of the enterprise consists of a reliable representation of computations $f(a) = b$, say for a, b in some concrete set A. We will discuss why this is so and present two reliable ways to do this. One consists of following the trace of the computation in the formal system used to represent the mathematics 'from the outside'. The other way consist of doing this 'from the inside', building the assistant around a term rewrite system. The two ways will be compared.
Other choices in the design of a Mathematical Assistant are concerned with the following qualities of the system

1. reliability;
2. choice of ontology;
3. choice of quantification strength;
4. constructive or classical logic;
5. aspects of the user interface.

These topics have been addressed by a number of 'competing' projects, each in a different way. From many of these systems one can learn, but a system that is satisfactory on all points has not yet been built. Enough experience through case studies has been obtained to assert that now time is ripe for building a satisfactory system.

A. Middeldorp (Ed.): RTA 2001, LNCS 2051, p. 1, 2001.

Rewriting for Deduction and Verification

Michael Rusinowitch

LORIA — INRIA
615, rue du Jardin Botanique
BP 101, 54602 Villers-lès-Nancy Cedex France
rusi@loria.fr

Abstract. Rewriting was employed even in early theorem-proving systems to improve efficiency. Theoretical justifications for its completeness were developed through work on Knuth-Bendix completion and convergent reduction systems, where the key-notions of termination and confluence were identified. A first difficulty with reduction in automated theorem-proving is to ensure termination of simplification steps when they are interleaved with deduction. Term orderings have been widely investigated and provide us with good practical solutions to termination problems. A second difficulty is to keep completeness with unidirectional use of equations, both for deduction and simplification, which amounts to restoring a confluence property. This is obtained by extending the superposition rule of Knuth-Bendix procedure to first-order clauses.
Rewrite-based deduction has found several applications in formal verification. We shall outline some of them in the presentation. In computer-assisted verification, *decision procedures* are typically applied for eliminating trivial subgoals (represented, for example, as sequents modulo a background theory). The computation of limit sets of formulas by iterating superposition rules generalizes Knuth-Bendix completion and permits the uniform design of these decision procedures. Rewriting combined with a controlled instantiation mechanism is also a powerful *induction* tactic for verifying safety properties of infinite-state systems. A nice feature of the so-called rewriting induction approach is that it allows for the refutation of false conjectures too. A more recent application of rewriting to verification concerns *security protocols.* These protocols can be compiled to rewrite systems, since rewriting nicely simulates the actions of participants and malicious environments.

A. Middeldorp (Ed.): RTA 2001, LNCS 2051, p. 2, 2001.

Universal Interaction Systems with Only Two Agents

Denis Bechet

LIPN - UPRES-A 7030
Institut Galilée
Université Paris 13 & CNRS
99, avenue J.-B. Clément
93430 VILLETANEUSE
France
Denis.Bechet@lipn.univ-paris13.fr

Abstract. In the framework of interaction nets [6], Yves Lafont has proved [8] that every interaction system can be simulated by a system composed of 3 symbols named γ, δ and ϵ. One may wonder if it is possible to find a similar universal system with less symbols. In this paper, we show a way to simulate every interaction system with a specific interaction system constituted of only 2 symbols. By transitivity, we prove that we can find a universal interaction system with only 2 agents. Moreover, we show how to find such a system where agents have no more than 3 auxiliary ports.

1 Introduction

In [6], Yves Lafont introduces *interaction nets*, a programming paradigm inspired by Girard's proof nets for *linear logic* [3]. Some translations from λ-calculus into interaction nets [9,4,5] or from proof nets [7,10,2,1,11] show that universal interaction systems are interesting for computation. We can explain this interest for these translations by the fact that computation with interaction nets is purely local and naturally confluent. Reductions can be made in parallel. Moreover, the number of steps that are necessary to reduce completely a net is independent of the way one may choose. From the point of view of λ-calculus, translations used in [4,5] captures optimal reduction.

In [8], Lafont introduces a universal interaction system with only three different symbols γ, δ and ϵ. δ and ϵ are respectively a duplicator and an eraser and γ is a constructor. This system preserves the complexity of computation for a particular system. The number of steps that are necessary to reduce a simulated interaction net is just (at most) multiplied by a constant (which depends only on the simulated system and not on the size of the simulated net).

One may wonder if it is possible to find a simpler universal interaction system with only 2 symbols. This paper answers yes to this question. In fact, we prove that we can simulate a particular interaction system with only two symbols. By simulating a universal system, we prove that a universal system constituted of only two symbols exists.

A. Middeldorp (Ed.): RTA 2001, LNCS 2051, pp. 3–14, 2001.

Using the universal system in [8], the resulting universal system has two symbols, one is an eraser and the second is a constructor/rule encoder. The eraser has no auxiliary port but the second one has 16 auxiliary ports! In fact, we show a way to reduce this number to only three auxiliary ports.

This paper is organized as follows: after an introduction to interaction nets and interaction systems, the notions of translations, simulations and universal interaction systems are presented. Section 4 is the heart of this article. It shows how to reduce a system to a system with only two agents. Section 5 reduces the number of auxiliary ports of the agents in this system to 0 and 3.

2 Interaction Systems

This model of computing is introduced in [6]. We briefly recall what interaction nets and interaction systems are.

2.1 Agents and Nets

An interaction net is a set of agents linked together through their ports. An individual agent is an instance of a particular symbol which is characterized by its *name* α and its *arity* $n \geq 0$. The arity defines the number of auxiliary ports associated to each agent. In addition to auxiliary ports, an agent owns a *principal port*. Graphically, an agent is represented by a triangle:

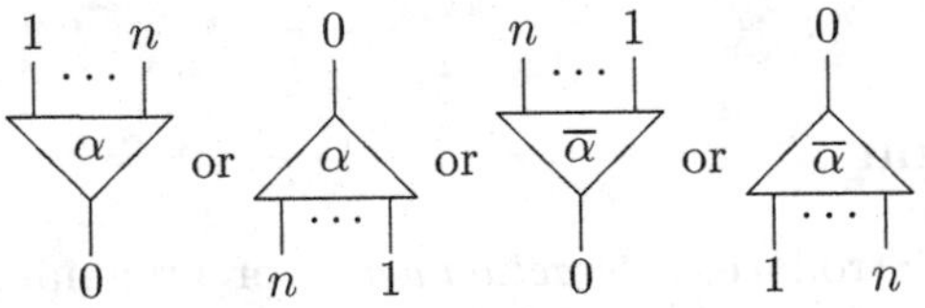

With α, auxiliary ports go clockwise from 1 to n but with $\overline{\alpha}$ it goes in the other direction (an agent is obtained by symmetry from α to $\overline{\alpha}$). Here, the principal port is noted 0, the auxiliary ports $1 \ldots n$.

An interaction net is a set of agents where the ports are connected two by two. The ports that are not connected to another one are the *free ports* of the net and are distinguished by a unique symbol. The set of the symbols of the free ports of a net consists the *interface* of this net. Below, the interface is $\{y, x\}$. α has one auxiliary port, β has two and ϵ has none.

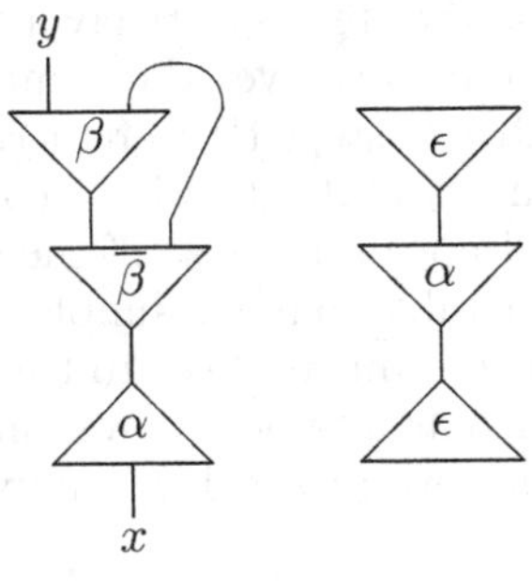

2.2 Interaction Rules and Interaction Systems

An interaction net can evolve when two agents are connected through their principal port. An *interaction rule* is a rewriting rule where the left member is constituted of only two agents connected through their principal ports and the right member is any interaction net with the same interface.

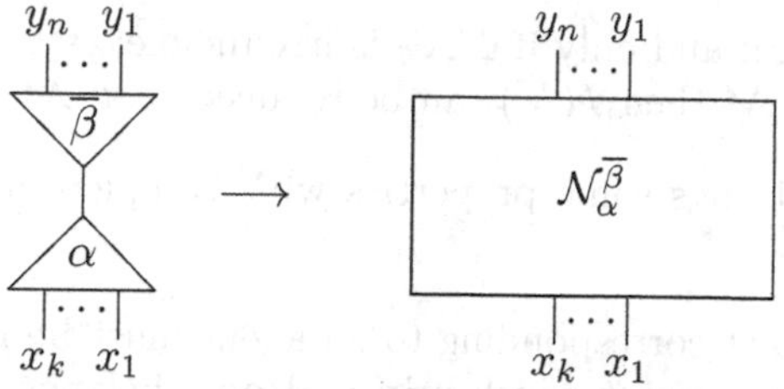

An interaction net that does not contain two agents connected by their principal port is *irreducible* (we say also *reduced*). A net is reduced by applying an interaction rule to a couple of agents connected through their principal port. This step substitutes the couple by the right member of the rule. A reduction can be repeated several times.

An *interaction system* $\mathcal{I} = (\Sigma, \mathcal{R})$ is a set of symbols Σ and a set of interaction rules $\mathcal{R}$ where agents in the left and right members are instances of the symbols of Σ.

An interaction system $\mathcal{I}$ is *deterministic* when (1) there exists at most one interaction rule for each couple of different agent and (2) there exists at most one interaction rule for the interaction of an agent with itself. In this case, the right member of this rule must be symmetric from a center point. An interaction system $\mathcal{I}$ is *complete* when there is at least one rule for each couple of agent. In this a paper we consider deterministic and complete systems. With these systems, we can prove that reduction is strongly confluent. In fact, this property is true whenever the system is deterministic. Moreover, it is assumed that right member of every rule has no *deadlock* and do not introduce an infinite recursive computation or a computation that creates a deadlock. Thus, we can always erase every part of the right member of a rule with eraser agents noted ϵ.

3 Universal Interaction Systems

Universality means that every interaction system can be *simulated* by a universal interaction system. Here, we use a very simple notion of simulation that is based on *translation*.

3.1 Translation

Let Σ and Σ' be two sets of symbols. A *translation* Φ from Σ to Σ' is a map that associates to each symbol in Σ an interaction net of agents of Σ' with the same interface. This translation is naturally extended to interaction nets of agents of Σ.

3.2 Simulation

We say that a translation Φ from Σ to Σ' defines a *simulation* of an interaction system $\mathcal{I} = (\Sigma, \mathcal{R})$ by an interaction system $\mathcal{I}' = (\Sigma', \mathcal{R}')$ if the reduction mechanism on interaction nets of $\mathcal{I}$ and $\mathcal{I}'$ are *compatible.* More precisely, that means that, if $\mathcal{N}$ is an interaction net of $\mathcal{I}$ then:

1. $\mathcal{N}$ is irreducible if and only if $\Phi(\mathcal{N})$ is irreducible;
2. if $\mathcal{N}$ reduces to $\mathcal{M}$ then $\Phi(\mathcal{N})$ can be reduced to $\Phi(\mathcal{M})$.

This definition brings some properties with complete and deterministic interaction systems:

- the interaction net corresponding to an agent must be reduced;
- it has at most one agent which principal port belongs to the interface and the symbol of this interface is the same as the symbol of the principal port of the initial agent;
- a translation is a simulation if and only if each rule of $\mathcal{R}$ is compatible with Φ;
- the simulation relation is transitive and symmetric.

In this paper, we have an approach that is not exactly the same as in [8]. Here, we work only with complete interaction systems but right members of interaction rules need not be reduced. They just need to be erasable by ϵ agents. However, it has a very small influence on the properties studied here.

3.3 Universal Interaction System

An interaction system $\mathcal{U}$ is said to be *universal* if for any interaction system $\mathcal{I}$, there exists a simulation $\Phi^{\mathcal{I}}$ of $\mathcal{I}$ by $\mathcal{U}$. In [8], Lafont introduces a system of 3 combinators γ, δ and ϵ defined by 6 rules and he proves that this system is universal:

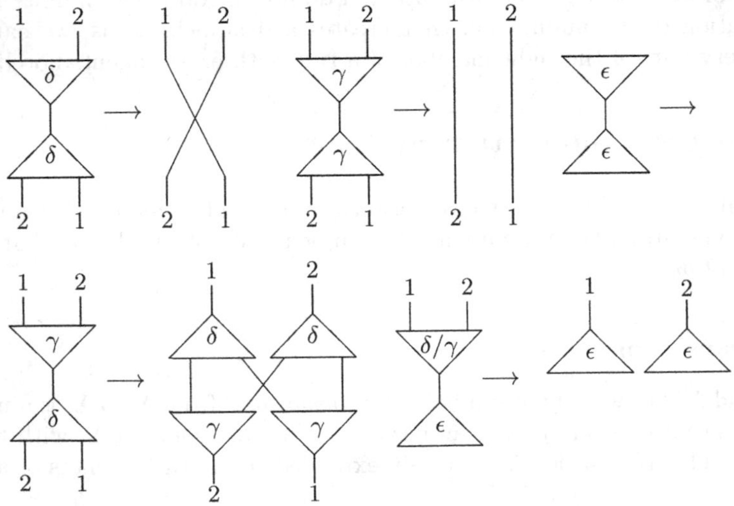

4 Universal System with Only 2 Agents

In this section, we show how to simulate a particular interaction system $\mathcal{I}$ with an interaction system composed of only two symbols ϵ and $\Pi_{\mathcal{I}}$. This system has exactly 3 rules.

ϵ agents have no auxiliary port. They erase everything even another ϵ agent (you can see for instance the above ϵ-rules). The other symbol $\Pi_{\mathcal{I}}$ has more auxiliary ports depending on the complexity of the system which is simulating. The rule between ϵ and itself gives the empty net. The rule between ϵ and $\Pi_{\mathcal{I}}$ gives m ϵ agents where m is the number of auxiliary ports of $\Pi_{\mathcal{I}}$. Finally, the rule between $\Pi_{\mathcal{I}}$ and itself is complex and depends of $\mathcal{I}$. This rule must be symmetric because it is a rule between an agent and itself.

4.1 Normalizing the Number of Auxiliary Ports in $\mathcal{I}$

Before giving directly the simulation of a system $\mathcal{I}$, we simulate it by a system $\mathcal{I}'$ where all the agents have the same number of auxiliary ports except one ϵ that has a null arity.

For $\mathcal{I} = (\Sigma, \mathcal{R})$, let $n \geq 0$ be the maximum arity of the symbols. If $n = 0$, we set $n = 1$. We define $\Sigma' = \{(\alpha', n), (\alpha, i) \in \Sigma\} \cup \{(\epsilon, 0)\}$. ϵ has a null arity and erases everything that it meets. The other ones have the same arity n.

We translate an agent α of arity i in $\mathcal{I}$ by an agent α' where the $n - i$ ports number $i + 1, \ldots, n$ are connected to $n - i$ agents ϵ. The rule between agents α' and β' is derived from the rule between α and β by substituting in the right member each agent by its translation and by adding ϵ agents to the symbols in the interface that do not correspond to the interface of the rule between α and β. If the arity of α is i and the arity of β is j, there are $(n-i)+(n-j)$ ϵ agents added to complete the interface of the rule between α' and β'

A short proof shows that $\mathcal{I}$ is simulated by $\mathcal{I}'$.

4.2 Simulation with ϵ and $\Pi_{\mathcal{I}}$

We can assume that our interaction system $\mathcal{I}$ is composed of k symbols all of arity n and ϵ agents (which erase everything). This system has $\frac{k\times(k+1)}{2}$ proper rules between the k agents of arity n, n rules between these agents and ϵ which are the same (except for the symbol of the agent) and create n ϵ agents and a rule for ϵ and itself which gives the empty net.

$\Pi_{\mathcal{I}}$ agents. The interaction system $\mathcal{I}$ has k symbols which arity is n and the ϵ symbol. It is simulated by a system composed of two symbols: ϵ and $\Pi_{\mathcal{I}}$. $\Pi_{\mathcal{I}}$ has exactly $n \times k \times (k+2)$ auxiliary ports.

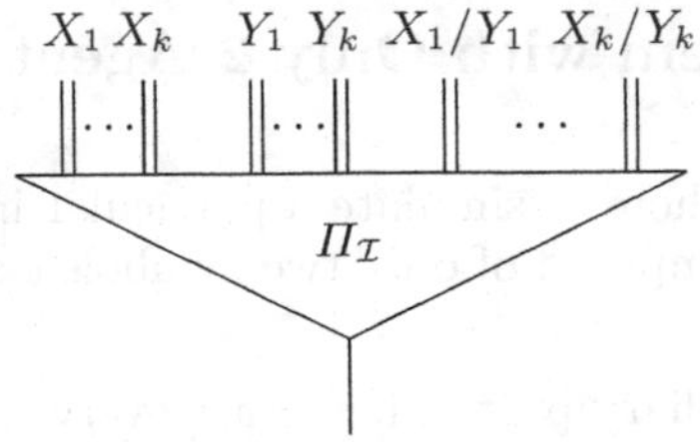

The auxiliary ports are grouped together by n. In the picture, they correspond to two vertical lines. Each group of n auxiliary ports are put in three different partitions. $X_1, \ldots, X_k$ are the inputs of the agent. One of this group corresponds to the n auxiliary ports of the initial agent of $\mathcal{I}$. The other input ports are connected to ϵ agents. The second group of auxiliary ports $Y_1, \ldots, Y_k$ are the inputs of the other agent when this agent interacts with another $\Pi_\mathcal{I}$ agent. The agent connects all the auxiliary ports in this partition to the auxiliary ports of the third group of ports. Finally, $X_1/Y_1, \ldots, X_k/Y_k$ are the interface of the right members of the rules generated by the interaction of this agent and another $\Pi_\mathcal{I}$ agent:

- $n \times k$ input ports X_i^t, $1 \leq t \leq n$ and $1 \leq i \leq k$;
- $n \times k$ intermediate ports Y_j^t, $1 \leq t \leq n$ and $1 \leq j \leq k$;
- $n^2 \times k$ rule ports $(X_i/Y_j)^t$, $1 \leq t \leq n$ and $1 \leq i, j \leq k$.

The ϵ agent of $\mathcal{I}$ is directly translated into itself. An agent α of $\mathcal{I}$, different from ϵ is simulated by an agent $\Pi_\mathcal{I}$, $n \times k$ links between auxiliary ports of this agent and $n \times (k^2 - 1)$ ϵ agents. In fact, the symbols of $\mathcal{I}$ that are different from ϵ are numbered from 1 to k. Thus, if the number associated to α is i, its translation is the net:

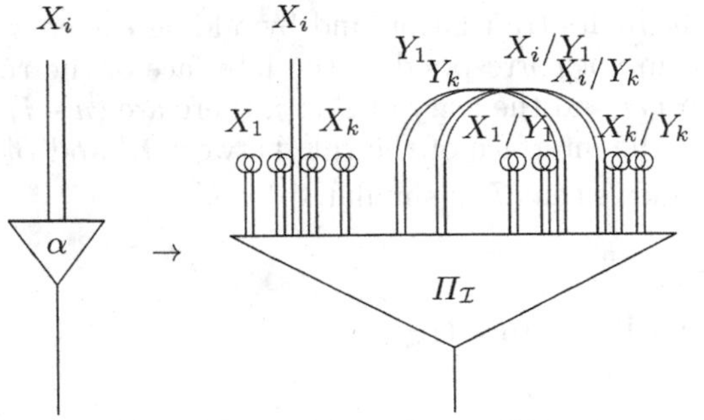

In this figure, the small circles are ϵ agents. The n input ports $X_i^1, \ldots, X_i^n$ correspond to the auxiliary ports of α. The other input ports are connected to ϵ agents. The $n \times k$ intermediate ports are connected to rule ports: for $1 \leq t \leq n$ and $1 \leq j \leq k$, Y_j^t is connected to $(X_i/Y_j)^t$. The other rule ports are connected to ϵ agents.

The rule between $\Pi_\mathcal{I}$ and itself. We need to define the interaction rule between two agents $\Pi_\mathcal{I}$. To simplify this presentation, we use the symmetric agent $\overline{\Pi_\mathcal{I}}$ for bottom agent. The rule is as follows:

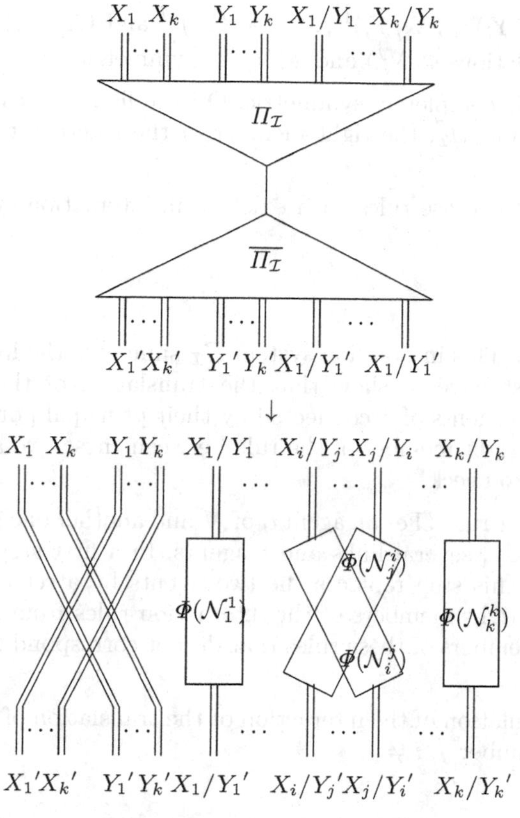

In this rule, variables without $'$ are from top $\Pi_\mathcal{I}$ agent and variables with $'$ are from bottom $\overline{\Pi_\mathcal{I}}$ agent. In the right member of the rule, input port X_i^t of top agent is connected to the intermediate port $Y_i^{t'}$ of bottom agent. Symmetrically, $X_i^{t'}$ is connected to Y_i^t.

The groups of n rule ports $(X_i/Y_i)^1, \ldots, (X_i/Y_i)^n$ and $(X_i/Y_i)^{1'}, \ldots, (X_i/Y_i)^{n'}$ are connected to the translation of the rule between the agent number i of $\mathcal{I}$ and itself. This interaction net $\mathcal{N}_i^i$ is by hypothesis symmetric, thus $\Phi(\mathcal{N}_i^i)$ is also symmetric and this part of the rule between $\Pi_\mathcal{I}$ and itself is symmetric.

For $i \neq j$, the two groups $(X_i/Y_j)^1, \ldots, (X_i/Y_j)^n$ and $(X_j/Y_i)^{1'}, \ldots, (X_j/Y_i)^{n'}$ are connected to the translation $\Phi(\mathcal{N}_i^j)$ of the rule between the agent number i of $\mathcal{I}$ and the agent number j. Because there is an inversion of ports, the auxiliary ports corresponding to the side of the agent number i (the top agent on the figure) are connected to $(X_i/Y_j)^1, \ldots, (X_i/Y_j)^n$ and the auxiliary ports corresponding to the side of the agent number j (the bottom agent on the figure) are connected to $(X_j/Y_i)^{1'}, \ldots, (X_j/Y_i)^{n'}$. In the same way, the two groups of n rule ports $(X_j/Y_i)^1, \ldots, (X_j/Y_i)^n$ and $(X_i/Y_j)^{1'}, \ldots, (X_i/Y_j)^{n'}$ are connected to the translation $\Phi(\mathcal{N}_j^i)$ of the rule between the agent number i of $\mathcal{I}$ and the agent number j but in the opposite direction (upside-down). The 4 groups of rule ports $(X_i/Y_j)^1, \ldots, (X_i/Y_j)^n$,

$(X_j/Y_i)^1, \ldots, (X_j/Y_i)^n$, $(X_i/Y_j)^{1'}, \ldots, (X_i/Y_j)^{n'}$ and $(X_j/Y_i)^{1'}, \ldots, (X_j/Y_i)^{n'}$ and the two translations $\Phi(\mathcal{N}_i^j)$ and $\Phi(\mathcal{N}_j^i)$ is symmetric.

Thus the rule is completely symmetric. On the figure, because we write the rule between $\Pi_{\mathcal{I}}$ and $\overline{\Pi_{\mathcal{I}}}$, the right member of the rule has to be horizontally symmetric.

This rule and the erase rules with ϵ define an interaction system $\Pi_{\mathcal{I}}$ with 2 agents and 3 rules.

***$\Pi_{\mathcal{I}}$* simulates *$\mathcal{I}$*.** This interaction system $\Pi_{\mathcal{I}}$ simulates the interaction system $\mathcal{I}$. In fact, we just have to show that the translation of the interaction net constituted by two agents of $\mathcal{I}$ connected by their principal ports reduces to the translation of the right member of the rule between these two agents. This task is not so difficult to check.

It is obvious for ϵ rules. For an agent α of $\mathcal{N}$ and another one β, the translation gives two agents $\Pi_{\mathcal{I}}$, several links and ϵ agents. In a first step, the two agents $\Pi_{\mathcal{I}}$ are reduced. This step replaces the two agents by a set of links and a set of translations of right members of the interaction rules from $\mathcal{I}$. A second step erases the right members of these rules that do not correspond to the interaction between α and β.

Below is a simulation of the interaction of the translation of an agent number i and an agent number j, $i \neq j$:

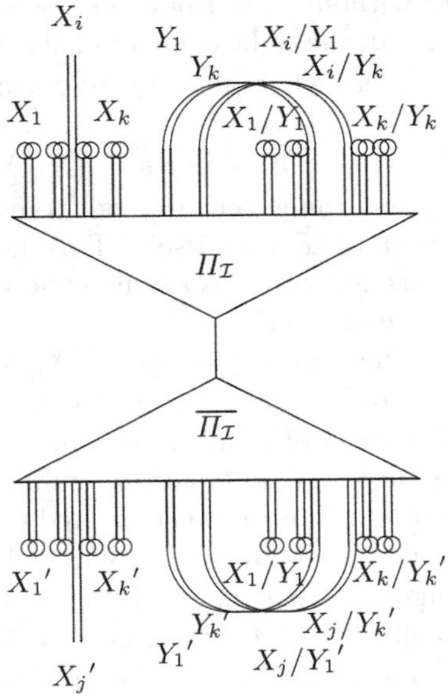

↓

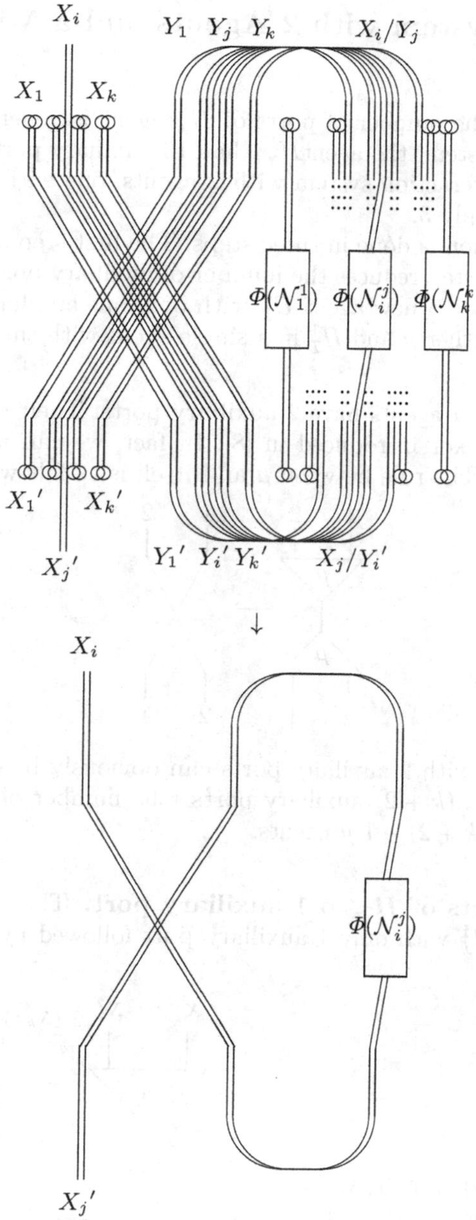

Universal system with 2 agents: first version. Starting with a universal system $\mathcal{I}$, we obtain a universal system composed of only two agents $\Pi_{\mathcal{I}}$ and ϵ. For instance, with Lafont's combinators γ, δ and ϵ, we obtain a universal system with ϵ and an agent $\Pi_{\gamma,\delta}$ which has $2 \times 2 \times (2+2) = 16$ auxiliary ports.

5 Universal System with 2 Agents and a Minimum of Ports

We have seen that the number of ports of $\Pi_{\mathcal{I}}$ agent is generally very big. For Lafont's universal system, the agent$\Pi_{\gamma,\delta}$ has 16 auxiliary ports. We can reduce this system to an interaction system with 2 agents, one with no auxiliary port and the other with only 3.

This transformation is done in three steps. The first step adds a multiplexor agent μ. The second step reduces the number of auxiliary ports of $\Pi_{\mathcal{I}}$ using the multiplexor. It leads to a new agent $\Pi_{\mathcal{I}}^1$ with only one auxiliary port. Finally, a last step merges together μ and $\Pi_{\mathcal{I}}^1$ in a single agent with three auxiliary ports.

Adding μ agents. μ agents have 2 auxiliary ports. Their construction is the same as the multiplexor introduced in [8]. In fact, we can use either Lafont's combinators δ or γ. The rule between μ and itself is as follows (δ version):

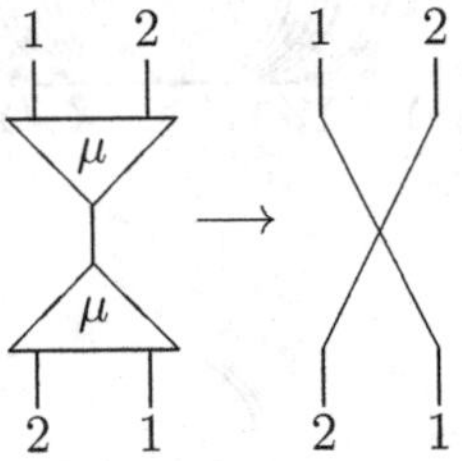

The multiplexor with 2 auxiliary ports can obviously be extended to multiplexors with $n \times k \times (k+2)$ auxiliary ports (the number of auxiliary ports of $\Pi_{\mathcal{I}}$) using $n \times k \times (k+2) - 1$ μ agents.

Reducing the ports of $\Pi_{\mathcal{I}}$ to 1 auxiliary port. Then, we can transform $\Pi_{\mathcal{I}}$ into an agent $\Pi_{\mathcal{I}}^1$ with only 1 auxiliary port followed by a $n \times k \times (k+2)$ multiplexer:

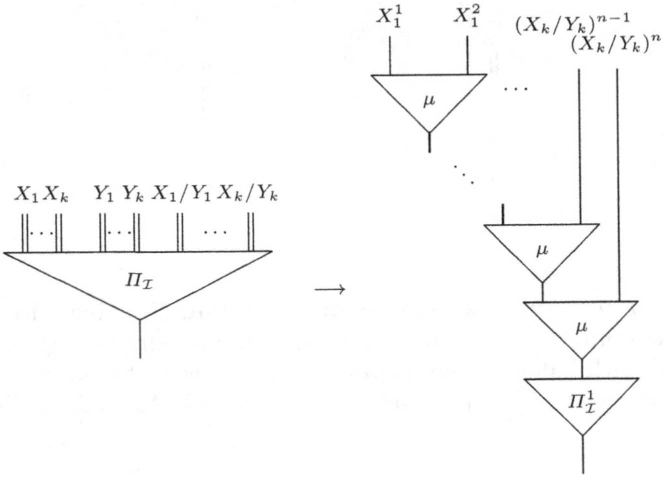

The rule between $\Pi^1_{\mathcal{I}}$ and itself is deduced from the rule between $\Pi_{\mathcal{I}}$ and itself by replacing in the right member of this rule each occurrence of $\Pi_{\mathcal{I}}$ by its translation and by folding both group of $n \times k \times (k+2)$ auxiliary ports with a $n \times k \times (k+2)$ multiplexor.

Merging μ and $\Pi^1_{\mathcal{I}}$. The functionalities of μ and $\Pi^1_{\mathcal{I}}$ do not overlap. In fact we only use the rule between μ and itself and $\Pi^1_{\mathcal{I}}$ and itself but never μ with $\Pi^1_{\mathcal{I}}$. As a consequence, we can put together these two agents into a single agent $\mu \times \Pi^1_{\mathcal{I}}$. The two translations from μ to $\mu \times \Pi^1_{\mathcal{I}}$ and from $\Pi^1_{\mathcal{I}}$ to $\mu \times \Pi^1_{\mathcal{I}}$ are as follows:

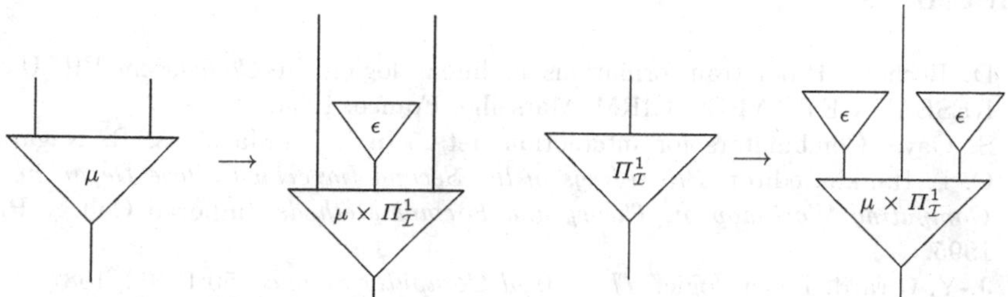

The rule between two $\mu \times \Pi^1_{\mathcal{I}}$ is as follows:

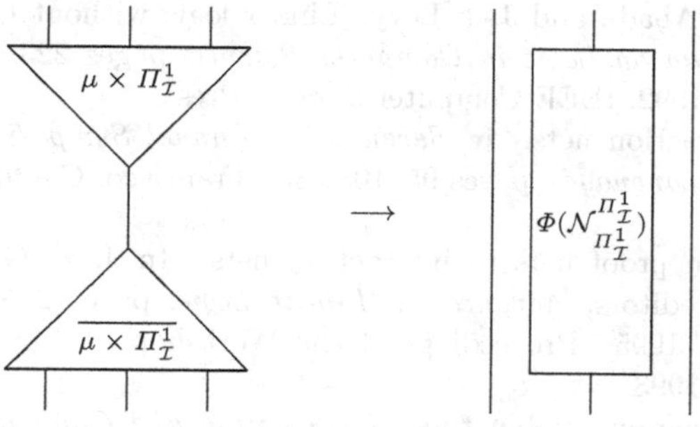

Universal system with 2 agents: second version. Starting with a universal system $\mathcal{I}$, we obtain a universal system composed of only two agents $\mu \times \Pi^1_{\mathcal{I}}$ and ϵ. The first one has three auxiliary ports and the second one has no auxiliary port.

6 Conclusion

Three results are given in this paper. The first one gives a way to simulate every interaction system with a system composed of only two symbols. A corollary is that there exists a universal interaction system with only two symbols (in [8] the universal system has 3 symbols). In the last part of this article, it is shown how to reduce the number of auxiliary ports of one of the symbols to only 3. This leads to a universal system with two symbols, one without auxiliary port and the other with only 3.

We have succeeded in finding a simpler universal system than Lafont's one. However, the price is that the right member of the rule between $\Pi^1_{\mathcal{I}}$ or $\mu \times \Pi^1_{\mathcal{I}}$ and itself is big and not very pleasant (like Lafont's system). One may wonder if it is possible to find a system that has simpler right members for the rules. Moreover, an open question remains: is it possible to find a universal system with 2 agents one with 0 auxiliary ports and the other with 2 (less is obviously impossible)?

References

[1] D. Bechet. Proof transformations in linear logic. In *Colloquium* PREUVES, RÉSEAUX ET TYPES. CIRM, Marseille, France, 1994.

[2] S. Gay. Combinators for interaction nets. In I. C. Mackie & R. Nagarajan C. L. Hankin, editor, *Proceedings of the Second Imperial College Department of Computing Workshop on Theory and Formal Methods.* Imperial College Press, 1995.

[3] J.-Y. Girard. Linear logic. *Theoretical Computer Science*, 50:1–102, 1987.

[4] G. Gonthier, M. Abadi, and J.-J. Levy. The geometry of optimal lambda reduction. In *Proceedings of the Nineteenth Annual Symposium on Principles of Programming Languages (POPL '90)*, pages 15–26, Albuquerque, New Mexico, January 1992. ACM Press.

[5] G. Gonthier, M. Abadi, and J.-J. Levy. Linear logic without boxes. In *Seventh Annual Symposium on Logic in Computer Science*, pages 223–234, Santa Cruz, California, June 1992. IEEE Computer Society Press.

[6] Y. Lafont. Interaction nets. In *Seventeenth Annual Symposium on Principles of Programming Languages*, pages 95–108, San Francisco, California, 1990. ACM Press.

[7] Y. Lafont. From proof nets to interaction nets. In J.-Y. Girard, Y. Lafont, and L. Regnier, editors, *Advances in Linear Logic*, pages 225–247. Cambridge University Press, 1995. Proceedings of the Workshop on Linear Logic, Ithaca, New York, June 1993.

[8] Y. Lafont. Interaction combinators. *Information and Computation*, 137(1):69–101, 1997.

[9] J. Lamping. An algorithm for optimal lambda calculus reduction. In *Seventeenth Annual Symposium on Principles of Programming Languages (POPL '90)*, pages 16–46, San Francisco, California, 1990. ACM Press.

[10] I. Mackie. *The Geometry of Implementation (an investigation into using the Geometry of Interaction for language implementation).* PhD thesis, Departement of Computing, Imperial College of Science, Technology and Medecine, 1994.

[11] I. Mackie. Interaction nets for linear logic. *Theoretical Computer Science*, 247:83–140, 2000.

General Recursion on Second Order Term Algebras*

Alessandro Berarducci[1] and Corrado Böhm[2]

[1] Università di Pisa, Dipartimento di Matematica,
Via Buonarroti 2, 56127 Pisa, Italy,
berardu@dm.unipi.it,
http://fibonacci.dm.unipi.it/~berardu/
[2] Università di Roma "La Sapienza",
Dipartimento di Scienze dell'Informazione,
Via Salaria 113, 00198 Roma, Italy,
boehm@dsi.uniroma1.it

Abstract. Extensions of the simply typed lambda calculus have been used as a metalanguage to represent "higher order term algebras", such as, for instance, formulas of the predicate calculus. In this representation bound variables of the object language are represented by bound variables of the metalanguage. This choice has various advantages but makes the notion of "recursive definition" on higher order term algebras more subtle than the corresponding notion on first order term algebras. Despeyroux, Pfenning and Schürmann pointed out the problems that arise in the proof of a canonical form theorem when one combines higher order representations with primitive recursion.
In this paper we consider a stronger scheme of recursion and we prove that it captures all partial recursive functions on second order term algebras. We illustrate the system by considering typed programs to reduce to normal form terms of the untyped lambda calculus, encoded as elements of a second order term algebra. First order encodings based on de Bruijn indexes are also considered. The examples also show that a version of the intersection type disciplines can be helpful in some cases to prove the existence of a canonical form. Finally we consider interpretations of our typed systems in the pure lambda calculus and a new gödelization of the pure lambda calculus.

1 Introduction

Despite some limitations, first order term rewriting systems can serve as a framework for functional programming. In this approach one has a first order signature whose symbols are partitioned in two sets: constructor symbols and program symbols. A data structure is identified with the set of normal forms of a given type built up from a given set of constructor symbols. To each program symbol is associated a set of recursive equations which, when interpreted as rewrite rules, define the operational semantics of the program.

* Dedicated to Nicolaas G. de Bruijn

A. Middeldorp (Ed.): RTA 2001, LNCS 2051, pp. 15–30, 2001.

This approach however is not adequate to handle programs on syntactic data structures involving variable-binding operators. Consider for instance a program to find the prenex normal form of a formula of the predicate calculus, or a program to reduce lambda terms. In principle one can encode the relevant data structures as first order term algebras, and represent the corresponding programs by first order term rewriting systems. It is however more convenient to assume that the programming environment (namely the "metatheory") is based on some version of the lambda calculus. In this approach bound variables of the object language are represented by bound variables of the metalanguage (rather than by strings of characters or de Bruijn indexes). This means that one can make use of the built-in procedures to handle renaming of bound variables and substitutions, which otherwise must be implemented separately in each case. Along these lines [14] shows that many important syntactic data structures involving variable-binding mechanisms can be represented in a version of the typed lambda calculus with additional constant symbols (higher order constructors). In this representation scheme the elements of a given data structure correspond to the "canonical forms" of a certain type built up from a certain set of higher order constructor symbols. We call "higher order term algebra" the set of canonical forms representing the elements of a given data structure.

In order to do some functional programming on such algebras we must introduce a notion of "recursion". A difficulty is that an element of a higher order algebra is a normal form of atomic type, and yet it can have subterms of functional type; so it is not clear, at least semantically, in which sense these algebras can be considered to be inductively defined. The problem is considered in [13, 12] under different perspectives. In the presence of recursion the existence of canonical forms becomes a rather delicate issue. Essentially this depends on the fact that, a recursively defined function applied to a formal parameter (i.e. a bound variable), cannot reduce to a canonical form. In other words, to be able to evaluate a recursive function, or more generally a function defined by cases, it is necessary that its recursive argument is "constructed" rather than parametric. Thus, to prove that a given term has a canonical form, one must ensure that occurrences of recursively defined programs applied to formal parameters do not arise dynamically during the computation. The solution proposed in [12], in the case of *primitive* recursion on higher order algebras, is a typed lambda calculus enriched with modalities whose purpose is to make a type distinction between parametric and constructed objects. This is reminiscent of the notion of "safe" recursion of [5] or of the "tiers" of [16]. Following a different line of research, Constable [11] introduced an extension of the typed lambda calculus with *general* recursion, based on fixed points operators. Here one is only interested in first order data structures (booleans, integers), and the issues are different: the main type distinction to be made concerns partial versus total functions rather then parametric versus constructed objects.

Inspired by the work of these authors and continuing the work done in [6, 9] in an untyped setting, we propose an extension of the simply typed lambda calculus with a scheme of recursion based on the distinction between program symbols and constructor symbols. We show that our scheme is able to capture all partial recursive functions on first order or second order term algebras.

Much of the emphasis of this paper is on the examples. We begin with a program to compute the prenex normal form of a formula of the predicate calculus (represented as in [14]). Then, as an example of a partial recursive function on a higher order term algebra, we consider (typed) programs to reduce terms of the untyped lambda calculus to normal form (where inputs and outputs are encoded as elements of a higher order term algebra). Similar programs have been discussed in [17,6,3] in an untyped setting.

In the case of partial functions the canonical form theorem takes the following conditional form: if a computation terminates, then it terminates in a canonical form. To prove that a given program has this property, a typing system based on the simple types may not be sufficient and a more refined typing based on a variant of the intersection type disciplines of [4] may be useful. We illustrate this fact proving the canonical form theorem for our program to reduce lambda terms.

All the experiments have been computer-tested by a software called "CuCh machine". The CuCh machine is essentially an interpreter of the pure lambda calculus together with a macro to transform recursive definitions into lambda terms developed by Böhm and Piperno and explained in [6,9]. Quite remarkably the interpretation described in these papers does not make use of the fixed point operator.

2 Extensions of the Typed Lambda Calculus

Given a set of atomic types we generate a set of types as follows.

Definition 1. *The* **types** α *are generated by the grammar*

$$\alpha ::= \langle \textit{atomic type} \rangle \mid \alpha_1 \times \ldots \times \alpha_n \to \langle \textit{atomic type} \rangle$$

Note that we allow cartesian products only on the left of the arrow. So $\alpha_1 \times \alpha_2$ is not a type.

Definition 2. *A* **signature** Σ *is a set of "constant declarations" of the form* $c\colon \alpha$*, where* α *is a type (over a given set of atomic types). If* $c\colon \alpha \in \Sigma$ *we say that* c *is a constant symbol of type* α*.*

We allow a constant symbol to have more than one type in a given signature, even an infinite set of types. If the set is finite, this amounts to the possibility of assigning to a constant symbol the intersection of all the types of the set, in the sense of the intersection type disciplines of [4].

Definition 3. *Given a signature* Σ*, a* **basis** B *is a set of "parameter declarations" of the form* $y : \alpha$*, where* y *is a variable and* α *is a type. We stipulate that a basis cannot contain two different declarations* $y : \alpha$ *and* $y : \beta$ *for the same variable. The following type assignment system defines inductively the set of terms* t *of type* α *over a basis* B*.*

$$\text{(Const.)}\frac{c{:}\,\alpha \in \Sigma}{\vdash c{:}\,\alpha} \qquad \text{(Var.)}\frac{x{:}\,\alpha \in B}{B \vdash x{:}\,\alpha}$$

$$(\to \text{I}) \ \frac{B, x_1{:}\,\alpha_1, \ldots, x_n{:}\,\alpha_n \vdash t{:}\,\alpha}{B \vdash \lambda x_1 \ldots x_n.t{:}\,\alpha_1 \times \ldots \times \alpha_n \to \alpha}$$

$$(\to \text{E}) \ \frac{B \vdash t{:}\,\alpha_1 \times \ldots \times \alpha_n \to \alpha \quad B \vdash t_1{:}\,\alpha_1 \ \cdots \ B \vdash t_n{:}\,\alpha_n}{B \vdash tt_1 \ldots t_n{:}\,\alpha}$$

$$\text{(Weakening)}\frac{B \vdash t{:}\,\alpha \quad B' \supset B}{B' \vdash t{:}\,\alpha}$$

In rule $(\to \text{I})$, B *does not contain declarations for* $x_1, \ldots, x_n$. *If we can deduce* $B \vdash t{:}\,\alpha$ *by the above rules we say that* t *is a term of type* α *over the basis* B.

Note that $\lambda xy.t$ cannot be applied to a single argument: it necessarily requires two arguments (so it is not identified with $\lambda x.(\lambda y.t)$, which is not a legitimate term because we only allow atomic types on the left hand side of an arrow). This feature of our system enables to encode bijectively the elements of various (higher order) data structures as the closed normal forms of a given atomic type over a given signature (thus rendering superfluous the distinction between normal form and canonical form in [14]).

The behaviour of the constant symbols of the signature is dictated by a set of reduction rules.

Definition 4. *Given a signature* Σ, *a* **reduction rule** *is a pair* (t_1, t_2), *written* $t_1 := t_2$, *such that for every basis* B *and every type* α, *if* $B \vdash t_1{:}\,\alpha$, *then* $B \vdash t_2{:}\,\alpha$.

Definition 5. *A set* P *of reduction rules over a signature* Σ *determines an extension* $\Lambda(\Sigma, P)$ *of the simply typed lambda calculus as follows. The terms of this calculus are as defined in Definition 3. The* **reduction relation** *"*$\to$*" between typable terms of* $\Lambda(\Sigma, P)$ *is obtained by adding to the* β*-rule* $(\lambda x_1 \ldots x_n.t)t_1 \ldots t_n \to t[t_1/x_1, \ldots, t_n/x_n]$ *all the reductions* $t_1 \to t_2$ *for each reduction rule* $t_1 := t_2$ *of the program* P. *By definition, the reduction relation is transitive and closed under substitutions and contexts. We identify terms which differ only by a renaming of bound variables.*

The following "subject reduction" theorem holds.

Theorem 1. *If* $t_1 \to t_2$ *in* $\Lambda(\Sigma, P)$ *and* $B \vdash t_1{:}\,\alpha$, *then* $B \vdash t_2{:}\,\alpha$.

3 General Recursive Programs on Second Order Term Algebras

So far nothing forbids the presence of reduction rules for which the Church-Rosser property for $\Lambda(\Sigma, P)$ fails. We will now introduce further restrictions on the shape of the reduction rules which are always satisfied in all the examples, and which ensure that the Church-Rosser property holds and that $\Lambda(\Sigma, P)$ is interpretable in the untyped lambda calculus. This is done by distinguishing between constructor symbols and program symbols (or destructors), as in [15,6, 2,10].

Definition 6. *We say that a set of reduction rules P over a signature Σ is* **dichotomic** *if the constant symbols of Σ can be partitioned in two sets,* **program symbols** *and* **constructor symbols** *in such a way that each reduction rule of P has the form*

$$\langle program\rangle(\langle constructor\rangle x_1, \dots, x_n) y_1 \dots y_m := t$$

We assume moreover that the rules of P are mutually exclusive and exhaustive in the sense that for each closed term t whose outermost symbol is a program symbol, and whose first argument begins with a constructor symbol, there is one and only one rule $t_1 := t_2$ such that t is a substitution instance of t_1. We also require that each program symbol appears in as the left-most symbol of at least one equation of P. So if P is empty all the symbols of the signature are constructor symbols. Finally we will assume that the right-hand sides of the reduction rules of P have no β-redexes.

We now make some further assumptions concerning the complexity of the types of the signature.

Definition 7. *The* **level** *of a type is defined as the maximum number of nested occurrences of arrow symbols.*

So the level of an atomic type is zero and the level of a type of the form $\alpha_1 \times \dots \times \alpha_n \to \beta$ is the maximum of the levels of $\alpha_1, \dots, \alpha_n$ plus 1 (recall that β is atomic, so it has level zero).

Definition 8. *Given a dichotomic set of rules P over a signature Σ, we say that Σ and P are* **second order**, *if the types of the program symbols have level at most 1 and the types of all constructor symbols have level at most 2.*

All the examples we consider in this paper are based on dichotomic rules on a second order signature. The second order assumption will be used in section 8 to prove the representability of every partial recursive function.

4 The Prenex Normal Form Example

Our first example of a program on second order term algebras is a procedure **Pnf** to put a formula of the predicate calculus into prenex normal form.

Definition 9. *(Formulas of the predicate calculus) For simplicity we consider formulas of the predicate calculus whose only non-logical symbol is a binary relation symbol* `A`. *Following [14], to represent such formulas as typed terms of the extended lambda calculus we need an atomic type ι (individuals), an atomic type o (predicates), and the following constructor symbols:*

$$\begin{aligned}
&\mathtt{Fa} \;: (\iota \to o) \to o\\
&\mathtt{Ex} \;: (\iota \to o) \to o\\
&\mathtt{not} : o \to o\\
&\mathtt{imp} : o \times o \to o\\
&\mathtt{A} \quad : \iota \times \iota \to o
\end{aligned}$$

Universal quantification $\forall x P$ *will be represented as* $\mathtt{Fa}(\lambda x.P)$ *and existential quantification* $\exists x P$ *will be represented as* $\mathtt{Ex}(\lambda x.P)$. *So a quantifier transforms a function from individuals to predicates into a predicate. The constructors* `not` *and* `imp` *stand for the negation and the implication sign.*

For example, the formula $\forall x \forall y(\mathtt{A}\, x\, y \;\to\; \neg \mathtt{A}\, y\, x)$ is represented by the term $\mathtt{Fa}(\lambda x.(\mathtt{Fa}(\lambda y.\mathtt{imp}(\mathtt{A}\, x\, y)(\mathtt{not}(\mathtt{A}\, y\, x)))))$ of type o. It is easy to see that the representation function maps bijectively closed formulas of the predicate calculus into closed normal forms of type o.

Definition 10. *(Prenex normal form) The program* ***Pnf*** *defined below, computes the prenex normal form of a formula.*

Program symbols:

$$\begin{aligned}
&\boldsymbol{Pnf} \;\;: o \to o\\
&\boldsymbol{nPnf} : o \to o\\
&\boldsymbol{iPnf} \;: o \times o \to o\\
&\boldsymbol{iP2} \;\;: o \times o \to o
\end{aligned}$$

Reduction rules:

$$\boldsymbol{Pnf}(\mathtt{Fa}\, t) := \mathtt{Fa}(\lambda T.(\boldsymbol{Pnf}(\, t\, T))) \tag{1}$$
$$\boldsymbol{Pnf}(\mathtt{Ex}\, t) := \mathtt{Ex}(\lambda T.(\boldsymbol{Pnf}(\, t\, T))) \tag{2}$$
$$\boldsymbol{Pnf}(\mathtt{not}\, L) := \boldsymbol{nPnf}(\boldsymbol{Pnf}\, L) \tag{3}$$
$$\boldsymbol{Pnf}(\mathtt{imp}\, L\, M) := \boldsymbol{iPnf}(\boldsymbol{Pnf}\, L)(\boldsymbol{Pnf}\, M) \tag{4}$$
$$\boldsymbol{Pnf}(\mathtt{A}\, u\, v) := \mathtt{A}\, u\, v \tag{5}$$
$$\boldsymbol{nPnf}(\mathtt{Fa}\, t) := \mathtt{Ex}(\lambda T.(\boldsymbol{Pnf}(\mathtt{not}(\, t\, T)))) \tag{6}$$
$$\boldsymbol{nPnf}(\mathtt{Ex}\, t) := \mathtt{Fa}(\lambda T.(\boldsymbol{Pnf}(\mathtt{not}(\, t\, T)))) \tag{7}$$
$$\boldsymbol{nPnf}(\mathtt{not}\, L) := L \tag{8}$$
$$\boldsymbol{nPnf}(\mathtt{imp}\, L\, M) := \mathtt{not}(\mathtt{imp}\, L\, M) \tag{9}$$
$$\boldsymbol{nPnf}(\mathtt{A}\, u\, v) := \mathtt{not}(\mathtt{A}\, u\, v) \tag{10}$$
$$\boldsymbol{iPnf}(\mathtt{Fa}\, t)\, y := \mathtt{Ex}(\lambda T.(\boldsymbol{Pnf}(\mathtt{imp}(\, t\, T)\, y))) \tag{11}$$
$$\boldsymbol{iPnf}(\mathtt{Ex}\, t)\, y := \mathtt{Fa}(\lambda T.(\boldsymbol{Pnf}(\mathtt{imp}(\, t\, T)\, y))) \tag{12}$$
$$\boldsymbol{iPnf}(\mathtt{not}\, L)\, y := \boldsymbol{iP2}\, y(\mathtt{not}\, L) \tag{13}$$
$$\boldsymbol{iPnf}(\mathtt{imp}\, L\, M)\, y := \boldsymbol{iP2}\, y(\mathtt{imp}\, L\, M) \tag{14}$$
$$\boldsymbol{iPnf}(\mathtt{A}\, u\, v)\, y := \boldsymbol{iP2}\, y(\mathtt{A} u\, v) \tag{15}$$
$$\boldsymbol{iP2}\,(\mathtt{Fa}\, t)\, x := \mathtt{Fa}(\lambda T.(\boldsymbol{Pnf}(\mathtt{imp}\, x(t\, T)))) \tag{16}$$
$$\boldsymbol{iP2}\,(\mathtt{Ex}\, t)\, x := \mathtt{Ex}(\lambda T.(\boldsymbol{Pnf}(\mathtt{imp}\, x(t\, T)))) \tag{17}$$

$$\boldsymbol{iP2}(\texttt{not}\,L)\,x := \texttt{imp}\,x(\texttt{not}\,L) \tag{18}$$

$$\boldsymbol{iP2}(\texttt{imp}\,L\,M)\,x := \texttt{imp}\,x(\texttt{imp}\,L\,M) \tag{19}$$

$$\boldsymbol{iP2}(\texttt{A}\,u\,v)\,x := \texttt{imp}\,x(\texttt{A}\,u\,v) \tag{20}$$

The above set P of equations defines an extension $\Lambda(P, \Sigma)$ of the typed lambda calculus as in Definition 5 (where Σ is signature of P).

Remark 1. The role of the various equations should be clear. For instance equation 11 takes care of the fact that the prenex normal form of a formula of the form $(\forall x A) \to B$ is equal to the prenex normal form of $\exists x(A \to B)$, provided x does not occur free in B. Note that the proviso is implicitly taken into account by the built-in rules of the λ-calculus with renaming of bound variables to avoid unwanted capture of variables.

5 A Partial Recursive Example

The program for the prenex normal form is total recursive, namely it always terminates. As an example of a partial recursive second order program we will define a program **Nf** (typable within our system) to reduce closed terms of the untyped lambda calculus to normal form. The program **Nf** is an improvement of the one we presented in [6] (in an untyped metatheory). The main novelty is the use of the auxiliary data structure "list" which allows for a considerable gain in efficiency. Our program should also be compared with the one in [17], which is very elegant but conceptually rather complex, as it requires, recursively, a duplication of terms into a "functional part" and an "argument part".

Definition 11. *To represent untyped lambda terms in our typed metatheory we use the following signature Σ_0 of second order constructors.*

$$\begin{aligned}&\texttt{App} : \texttt{exp} \times \texttt{exp} \to \texttt{exp};\\&\texttt{Abs} : (\texttt{exp} \to \texttt{exp}) \to \texttt{exp}\end{aligned}$$

Definition 12. *Given a term t of the untyped lambda calculus, define $\lceil t \rceil$ inductively as follows:*

$$\begin{aligned}\lceil x \rceil &:= x;\\ \lceil MN \rceil &:= \texttt{App}\lceil M \rceil \lceil N \rceil;\\ \lceil \lambda x.M \rceil &:= \texttt{Abs}(\lambda x.\lceil M \rceil)\end{aligned}$$

So for instance

$$\lceil (\lambda x.xx)(\lambda x.xx) \rceil = \texttt{App}(\texttt{Abs}(\lambda x.\texttt{App}\,x\,x))(\texttt{Abs}(\lambda x.\texttt{App}\,x\,x))$$

Remark 2. The above encoding is adequate: the map $t \mapsto \lceil t \rceil$ is a bijection from terms of the untyped lambda calculus, to normal forms of type exp *over Σ_0.*

The above representation is a variant of the one in [17] where the author defines $\lceil x \rceil = \mathtt{Var}\, x$, with $\mathtt{Var}$ a new constructor. We will use the following characterization of normal forms:

Remark 3.

$$\mathrm{nf}(xA_1 \ldots A_n) = x\ \mathrm{nf}(A_1) \ldots \mathrm{nf}(A_n); \tag{1}$$

$$\mathrm{nf}(\lambda x.B) = \lambda x.\mathrm{nf}(B); \tag{2}$$

$$\mathrm{nf}((\lambda x.B)A_1 \ldots A_n) = \mathrm{nf}(B[A_1/x]A_2 \ldots A_n) \tag{3}$$

We will define a program **Nf** such that $\mathbf{Nf}\lceil t \rceil = \lceil \mathrm{nf}(t) \rceil$ for each *closed* term t having a normal form $\mathrm{nf}(t)$. Although we are not interested in the behaviour of **Nf** on open terms, the above equations suggest that to carry out the recursion we are forced to take into account not only closed terms, but also open terms. However we cannot hope to have $\mathbf{Nf}\lceil t \rceil = \lceil \mathrm{nf}(t) \rceil$ even for open terms, because otherwise taking $n = 0$ in the first equation above we get $\mathbf{Nf}\lceil x \rceil = \lceil x \rceil$, which implies $\mathbf{Nf}\, x = x$ (as $\lceil x \rceil = x$), namely **Nf** is the identity function. So we must decide what is the relevant equation for **Nf** on open terms. The solution is $\mathbf{Nf}\lceil t \rceil^{\mathtt{Box}} = \lceil \mathrm{nf}(t) \rceil$, where $\lceil t \rceil^{\mathtt{Box}}$ is obtained from $\lceil t \rceil$ by substituting each free variable x by $\mathtt{Box}\, x$. Here $\mathtt{Box}$ is an auxiliary constructor of type $\mathtt{exp} \to \mathtt{exp}$. So for instance $\lceil \lambda y.xy \rceil^{\mathtt{Box}} = \mathtt{Abs}(\lambda y.\mathtt{App}(\mathtt{Box}\, x)\, y)$. Note that for closed terms $\lceil t \rceil^{\mathtt{Box}} = \lceil t \rceil$.

We are finally ready to give the reduction rules defining **Nf**. The idea is that **Nf** will introduce a $\mathtt{Box}$ each time an abstraction is passed over, and will eliminate it when the corresponding variable is reached. The dots "..." in the equations of Remark 3 suggest the use of the data structure "list". In other words it is convenient to generalize the program $\mathbf{Nf} : \mathtt{exp} \to \mathtt{exp}$ to a program $\mathbf{Reduce} : \mathtt{exp} \times \mathtt{list} \to \mathtt{exp}$ which compute the normal form of a term applied to a list of terms.

Definition 13. *(A program to compute the normal form of a closed lambda term) Besides the constructors* $\mathtt{App}, \mathtt{Abs}$ *we use the following auxiliary constructor symbols:*

$$\begin{aligned} &\mathtt{Box} &&: \mathtt{exp} \to \mathtt{exp};\\ &\mathtt{nil} &&: \mathtt{list};\\ &\mathtt{cons} &&: \mathtt{exp} \times \mathtt{list} \to \mathtt{list} \end{aligned}$$

where $\mathtt{list}$ *is an atomic type to represent lists of objects of type* $\mathtt{exp}$.

Program symbols:

$$\begin{aligned} &\boldsymbol{Reduce} &&: \mathtt{exp} \times \mathtt{list} \to \mathtt{exp};\\ &\boldsymbol{RBox} &&: \mathtt{list} \times \mathtt{exp} \to \mathtt{exp};\\ &\boldsymbol{RAbs} &&: \mathtt{list} \times (\mathtt{exp} \to \mathtt{exp}) \to \mathtt{exp} \end{aligned}$$

Reduction rules:

$$\boldsymbol{Reduce}\,(\mathtt{App}\, x\, y)\, L := \boldsymbol{Reduce}\, x(\mathtt{cons}\, y\, L); \tag{1}$$

$$\textit{\textbf{Reduce}}\,(\texttt{Abs}\,f)\,L := \textit{\textbf{RAbs}}\,L\,f; \quad (2)$$
$$\textit{\textbf{Reduce}}\,(\texttt{Box}\,u)\,L := \textit{\textbf{RBox}}\,L\,u; \quad (3)$$
$$\textit{\textbf{RAbs}}\,\texttt{nil}\,f := \texttt{Abs}(\lambda\,u.\textit{\textbf{Reduce}}\,(\,f(\texttt{Box}\,u))\texttt{nil}); \quad (4)$$
$$\textit{\textbf{RAbs}}\,(\texttt{cons}\,y\,L)\,f := \textit{\textbf{Reduce}}\,(\,f\,y)\,L; \quad (5)$$
$$\textit{\textbf{RBox}}\,\texttt{nil}\,u := u; \quad (6)$$
$$\textit{\textbf{RBox}}\,(\texttt{cons}\,y\,L)\,u := \textit{\textbf{RBox}}\,L(\texttt{App}\,u(\textit{\textbf{Reduce}}\,y\,\texttt{nil})) \quad (7)$$

The above rules define a system $\Lambda(P,\Sigma)$, and within that system we define $\textit{\textbf{Nf}} = \lambda x.\textit{\textbf{Reduce}}\, x\, \texttt{nil}$.

Note that an occurrence of the auxiliary constructor Box is introduced in step 4 and is eliminated in step 3.

Example 1. (Example of the computation of a normal form)
Let us compute the normal form of the term $(\lambda xy.xyy)\lambda x.x$. We will use the notation $t_1 \longrightarrow_i t_2$ to express that the term t_1 has been rewritten as t_2 applying the rule i to a subterm of t_1. For better readability we will underline the relevant subterm (when it is not the whole term).

$$\begin{aligned}
&\textbf{Reduce}\,(\texttt{App}(\texttt{Abs}(\lambda\,x.\texttt{Abs}(\lambda\,y.\texttt{App}(\texttt{App}\,x\,y)\,y)))(\texttt{Abs}(\lambda\,x.\,x)))\texttt{nil}\\
\longrightarrow_1\;&\textbf{Reduce}\,(\texttt{Abs}(\lambda\,x.\texttt{Abs}(\lambda\,y.\texttt{App}(\texttt{App}\,x\,y)\,y)))(\texttt{cons}(\texttt{Abs}(\lambda\,x.\,x))\texttt{nil})\\
\longrightarrow_2\;&\textbf{RAbs}\,(\texttt{cons}(\texttt{Abs}(\lambda\,x.\,x))\texttt{nil})(\lambda\,x.\texttt{Abs}(\lambda\,y.\texttt{App}(\texttt{App}\,x\,y)\,y))\\
\longrightarrow_5\;&\textbf{Reduce}\,(\underline{(\lambda\,x.\texttt{Abs}(\lambda\,y.\texttt{App}(\texttt{App}\,x\,y)\,y))(\texttt{Abs}(\lambda\,x.\,x))})\texttt{nil}\\
\longrightarrow_\beta\;&\textbf{Reduce}\,(\texttt{Abs}(\lambda\,y.\texttt{App}(\texttt{App}(\texttt{Abs}(\lambda\,x.\,x))\,y)\,y))\texttt{nil}\\
\longrightarrow_2\;&\textbf{RAbs}\,\texttt{nil}(\lambda\,y.\texttt{App}(\texttt{App}(\texttt{Abs}(\lambda\,x.\,x))\,y)\,y)\\
\longrightarrow_4\;&\texttt{Abs}(\lambda\,u.\textbf{Reduce}\,(\underline{(\lambda\,y.\texttt{App}(\texttt{App}(\texttt{Abs}(\lambda\,x.\,x))\,y)\,y)(\texttt{Box}\,u)})\texttt{nil})\\
\longrightarrow_\beta\;&\texttt{Abs}(\lambda\,u.\underline{\textbf{Reduce}\,(\texttt{App}(\texttt{App}(\texttt{Abs}(\lambda\,x.\,x))(\texttt{Box}\,u))(\texttt{Box}\,u))\texttt{nil}})\\
\longrightarrow_1\;&\texttt{Abs}(\lambda\,u.\underline{\textbf{Reduce}\,(\texttt{App}(\texttt{Abs}(\lambda\,x.\,x))(\texttt{Box}\,u))(\texttt{cons}(\texttt{Box}\,u)\texttt{nil})})\\
\longrightarrow_1\;&\texttt{Abs}(\lambda\,u.\underline{\textbf{Reduce}\,(\texttt{Abs}(\lambda\,x.\,x))(\texttt{cons}(\texttt{Box}\,u)(\texttt{cons}(\texttt{Box}\,u)\texttt{nil}))})\\
\longrightarrow_2\;&\texttt{Abs}(\lambda\,u.\textbf{RAbs}\,(\texttt{cons}(\texttt{Box}\,u)(\texttt{cons}(\texttt{Box}\,u)\texttt{nil}))(\lambda\,x.\,x))\\
\longrightarrow_5\;&\texttt{Abs}(\lambda\,u.\textbf{Reduce}\,(\underline{(\lambda\,x.\,x)(\texttt{Box}\,u)})(\texttt{cons}(\texttt{Box}\,u)\texttt{nil})\\
\longrightarrow_\beta\;&\texttt{Abs}(\lambda\,u.\underline{\textbf{Reduce}\,(\texttt{Box}\,u)(\texttt{cons}(\texttt{Box}\,u)\texttt{nil})})\\
\longrightarrow_3\;&\texttt{Abs}(\lambda\,u.\textbf{RBox}\,(\texttt{cons}(\texttt{Box}\,u)\texttt{nil})u)\\
\longrightarrow_7\;&\texttt{Abs}(\lambda\,u.\underline{\textbf{RBox}\,\texttt{nil}(\texttt{App}\,u(\textbf{Reduce}\,(\texttt{Box}\,u)\texttt{nil}))})\\
\longrightarrow_6\;&\texttt{Abs}(\lambda\,u.\texttt{App}\,u(\underline{\textbf{Reduce}\,(\texttt{Box}\,u)\texttt{nil}}))\\
\longrightarrow_3\;&\texttt{Abs}(\lambda\,u.\texttt{App}\,u(\underline{\textbf{RBox}\,\texttt{nil}\,u}))\\
\longrightarrow_6\;&\texttt{Abs}(\lambda\,u.\texttt{App}\,u\,u)
\end{aligned}$$

A formal proof of correctness of the program **Nf** is long and tedious: it is based on the observation that our reduction rules simulate the equations in Remark 3.

Remark 4. The program in Definition 13, unlike the program in [17], has the property that $\textbf{Nf}\,\lceil t \rceil$ is strongly normalizing (every reduction path terminates) whenever t is strongly normalizing.

6 Canonical Forms

One of the main difficulties which must be overcome when programming on higher order term algebras is that the typing information may not be sufficient to ensure that a given closed normal form is canonical, namely it represents an element of a given data structure.

Example 2. Consider the system $\Lambda(P, \Sigma)$ of Definition 13. By Remark 2, the closed normal forms of type exp over the signature $\{\texttt{App}, \texttt{Abs}\} \subset \Sigma$ represent the untyped lambda terms. Unfortunately within the system $\Lambda(P, \Sigma)$ there are "exotic" closed normal forms of type exp (over Σ) which do not represent any lambda term. Examples of such terms are $\texttt{Abs}(\lambda x.\mathbf{Nf}\ x)$ and $\texttt{Box}(\lambda x. x)$. The first term is particularly bad since it is a normal form containing an occurrence of a program symbol: this difficulty arises with second order programming and has no analogue in the first order case (i.e. when all constructors have level at most 1).

The presence of exotic terms shows that the typing information may be too weak to prove that a program has the correct range. To solve the problem we use a feature of our system that we have not yet exploited: the fact that we allow constructor symbols to have more than one type. The following result illustrates the technique.

Theorem 2. *(Canonical form theorem for* **Nf***) Given a closed term t of the untyped lambda calculus, if* $\mathbf{Nf}\lceil t \rceil$ *has a normal form in the system* $\Lambda(P, \Sigma)$ *of Definition 13, then its normal form has the shape* $\lceil t' \rceil$ *for some* t'.

Proof. By the adequacy of the representation (Remark 2), it suffices to show that the normal form of $\mathbf{Nf}\lceil t \rceil$ is a term over the signature $\{\texttt{App}, \texttt{Abs}\} \subset \Sigma$ (necessarily of type exp by the subject reduction theorem). To this aim we redefine the signature used in Definition 13 as follows, using two new atomic types $\Box\texttt{exp}$ and $\Box\texttt{list}$ (*warning*: we are not taking $\Box$ as a new type constructor):

$$
\begin{array}{ll}
\texttt{App} & : \texttt{exp} \times \texttt{exp} \to \texttt{exp},\ \ \Box\texttt{exp} \times \Box\texttt{exp} \to \Box\texttt{exp};\\
\texttt{Abs} & : (\texttt{exp} \to \texttt{exp}) \to \texttt{exp},\ \ (\Box\texttt{exp} \to \Box\texttt{exp}) \to \Box\texttt{exp};\\
\texttt{Box} & : \texttt{exp} \to \Box\texttt{exp};\\
\texttt{nil} & : \Box\texttt{list};\\
\texttt{cons} & : \Box\texttt{exp} \times \Box\texttt{list} \to \Box\texttt{list};\\
\mathbf{Reduce} & : \Box\texttt{exp} \times \Box\texttt{list} \to \texttt{exp};\\
\mathbf{RBox} & : \Box\texttt{list} \times \texttt{exp} \to \texttt{exp};\\
\mathbf{RAbs} & : \Box\texttt{list} \times (\Box\texttt{exp} \to \Box\texttt{exp}) \to \texttt{exp}
\end{array}
$$

So App and Abs have two types each. The reduction rules of Definition 13 are still correctly typed in this new signature and **Nf** has type $\Box\text{exp} \to \text{exp}$. The idea is that $\Box\texttt{exp}$ represents the objects of the form $\lceil t \rceil^{\texttt{Box}}$ and exp represents the objects of the form $\lceil t \rceil$, as in the discussion following Remark 3. Over the new signature the representation $\lceil t \rceil$ of a closed term t, has both type exp and $\Box\texttt{exp}$, so $\mathbf{Nf}\lceil t \rceil$ is correctly typed and has type exp. By the subject reduction theorem if the computation of $\mathbf{Nf}\lceil t\rceil$ terminates, the result has type exp. An easy induction shows that a closed normal form of type exp is necessarily a term over the signature $\{\mathtt{App}, \mathtt{Abs}\} \subset \Sigma$, hence it is a term of the form $\lceil t'\rceil$.

Remark 5. The signature Σ of Definition 13 is obtained from the signature Σ' in the proof of Theorem 2 by identifying $\Box$exp with exp and $\Box$list with list. This suggests the following definition.

Definition 14. *We say that a signature Σ'* **refines** *a signature Σ, if Σ and Σ' have the same number of constant symbols, with the same names but possibly different types, and Σ can obtained from Σ' by substituting, in the constant declarations, some atomic types with other types (so in particular by identifying some atomic types).*

Theorem 2 shows that to prove that a program of a given system $\Lambda(P, \Sigma)$ has the correct range (or more generally has some desired property), it may be convenient to try to refine the signature, in such a way that the new signature still respects P. Note that when we refine a signature we make less terms typable: in fact, if Σ' refines Σ, and a closed term t has type α over Σ', then t has type α^s over Σ, where α^s is a substitution instance of α (with the same substitution as in the definition of refinement). The notion of refinement is clearly related with the notion of principal type scheme in the intersection type disciplines [19].

Example 3. Another interesting refinement of the signature of Definition 13, which respects the corresponding reduction rules, uses infinitely many types for each constant symbol parameterized by an index i ranging over the integers $\mathbf{Z}$:

$$\begin{array}{ll}
\mathtt{Abs} & : (\mathtt{exp_{i+1}} \to \mathtt{exp_{i+1}}) \to \mathtt{exp_{i+1}};\\
\mathtt{App} & : \mathtt{exp_{i+1}} \times \mathtt{exp_{i+1}} \to \mathtt{exp_{i+1}};\\
\mathtt{Box} & : \mathtt{exp_{i}} \to \mathtt{exp_{i+1}};\\
\mathtt{nil} & : \mathtt{list_{i}};\\
\mathtt{cons} & : \mathtt{exp_{i+1}} \times \mathtt{list_{i+1}} \to \mathtt{list_{i+1}};\\
R & : \mathtt{exp_{i+1}} \times \mathtt{list_{i+1}} \to \mathtt{exp_{i}};\\
\mathbf{RBox} & : \mathtt{list_{i+1}} \times \mathtt{exp_{i}} \to \mathtt{exp_{i}};\\
\mathbf{RAbs} & : \mathtt{list_{i+1}} \times (\mathtt{exp_{i+1}} \to \mathtt{exp_{i+1}}) \to \mathtt{exp_{i}}
\end{array}$$

Now consider the terms $\lambda x.\mathbf{Nf}(\mathbf{Nf}\, x)$ and $\mathtt{Abs}(\lambda x.\mathbf{Nf}\, x)$. Both terms can be typed in the original signature of Definition 13 in which $\mathbf{Nf}$ has type exp $\to$ exp. Neither of them can be typed in the refined signature in the proof of Theorem 2 in which $\mathbf{Nf}$ has type $\Box$exp $\to$ exp. Only the first one can be typed in the refined signature of Example 3. Collecting information coming from different signatures we gain insight on the program $\mathbf{Nf}$.

7 From Second Order to First Order Representations

Using de Bruijn indexes we can pass from a second order to a first order representation of lambda terms (or formulas, or any other second order syntactic structure). Once a second order program is found, it is easy to transform it into a first order program by implementing the relevant procedures to handle variable substitutions. To illustrate this idea we modify the program of Definition 13 to obtain a program to reduce a lambda term to normal form in de Bruijn notation.

Definition 15. *(De Bruijn notation) We recall the* **de Bruijn** *notation for lambda-terms. In this notation variable occurrences are replaced by positive integers. For example* $\lambda x.\lambda y.xy(\lambda z.x)$ *becomes* $\lambda\lambda 21(\lambda 3)$. *The positive integer* n *indicates a variable which is bounded by the* n*-th occurrence of* λ *going upward in the parsing tree of the term. If such an occurrence of* λ *does not exist, then the integer indicates a free variable.*

Note that closed terms which differ only for a renaming of bound variables have the same de Bruijn notation.

Definition 16. *To represent lambda terms in de Bruijn notation we use the following signature:*

Constructor symbols:

$$\begin{array}{ll} \mathtt{1} & : \mathtt{nat}; \\ \mathtt{S} & : \mathtt{nat} \to \mathtt{nat}; \\ \mathtt{var} & : \mathtt{nat} \to \mathtt{term}; \\ \mathtt{abs} & : \mathtt{term} \to \mathtt{term}; \\ \mathtt{app} & : \mathtt{term} \times \mathtt{term} \to \mathtt{term} \end{array}$$

For example $\lambda\lambda 21(\lambda 3)$ becomes $\mathtt{abs}(\mathtt{abs}(\mathtt{app}\, v2\, v1)(\mathtt{abs}\, v3))$, where $2 = S(1)$, $3 = S(2)$, $vn = (\mathtt{var}\, n)$. The de Bruijn terms correspond bijectively to the closed normal forms of type $\mathtt{term}$.

The program **nf** : $\mathtt{term} \to \mathtt{term}$ defined below reduces a de Bruijn term to normal form. Unlike the higher order program **Nf** of Definition 13, it works well even when applied to representations of open terms (the free variables are represented by de Bruijn indexes which point "above" the root of the term). The reduction rules in the definition of **nf** are almost identical to those of **Nf**. The main difference is in equation (5) below, where the auxiliary program **sub** is used to simulate the single β-reduction which in the higher order program is built-in.

Definition 17. *(A program to reduce de Bruijn terms to normal form) We set* ***nf*** $= \lambda x.$***reduce*** $x\, \mathtt{nil}$ *where* ***reduce*** *is defined as follows:*

Auxiliary constructor symbols:

$$\begin{array}{ll} \mathtt{nil} & : \mathtt{list}; \\ \mathtt{cons} & : \mathtt{term} \times \mathtt{list} \to \mathtt{list} \end{array}$$

Program symbols:

$$\begin{array}{ll} \textbf{\textit{reduce}} & : \mathtt{term} \times \mathtt{list} \to \mathtt{term}; \\ \textbf{\textit{Rabs}} & : \mathtt{list} \times \mathtt{term} \to \mathtt{term}; \\ \textbf{\textit{Rvar}} & : \mathtt{list} \times \mathtt{term} \to \mathtt{term}; \\ \textbf{\textit{sub}} & : \mathtt{term} \times \mathtt{term} \to \mathtt{term}; \\ \textbf{\textit{subs}} & : \mathtt{term} \times \mathtt{term} \times \mathtt{nat} \to \mathtt{term} \\ \textbf{\textit{update}} & : \mathtt{term} \times \mathtt{nat} \times \mathtt{nat} \to \mathtt{term} \end{array}$$

Reduction rules:

$$\begin{aligned}
\textit{\textbf{reduce}}\,(\texttt{app}\,x\,y)\,L &:= \textit{\textbf{reduce}}\,x\,(\texttt{cons}\,y\,L); && (1)\\
\textit{\textbf{reduce}}\,(\texttt{abs}\,f)\,L &:= \textit{\textbf{Rabs}}\,L\,f; && (2)\\
\textit{\textbf{reduce}}\,(\texttt{var}\,n)\,L &:= \textit{\textbf{Rvar}}\,L\,(\texttt{var}\,n); && (3)\\
\textit{\textbf{Rabs}}\,\texttt{nil}\,f &:= \texttt{abs}(\textit{\textbf{reduce}}\,f\,\texttt{nil}); && (4)\\
\textit{\textbf{Rabs}}\,(\texttt{cons}\,y\,L)\,f &:= \textit{\textbf{reduce}}\,(\textit{\textbf{sub}}\,f\,y)\,L; && (5)\\
\textit{\textbf{Rvar}}\texttt{nil}\,u &:= u; && (6)\\
\textit{\textbf{Rvar}}\,(\texttt{cons}\,y\,L)\,u &:= \textit{\textbf{Rvar}}\,L(\texttt{app}\,u(\textit{\textbf{reduce}}\,y\,\texttt{nil})) && (7)
\end{aligned}$$

The purpose of ***sub*** $f\ y$ *in equation (5) is to find the* β*-reduct of* abs f *applied to* y. *We set* ***sub*** $= \lambda f\,y.$***subs*** $f\,y\,0$ *where* ***subs*** *is defined as follows:*

$$\begin{aligned}
\textit{\textbf{subs}}\,(\texttt{app}\,u\,x)\,x\,m &:= \texttt{app}(\textit{\textbf{subs}}\,u\,x\,m)(\textit{\textbf{subs}}\,x\,x\,m); && (8)\\
\textit{\textbf{subs}}\,(\texttt{abs}\,u)\,x\,m &:= \texttt{abs}(\textit{\textbf{subs}}\,u\,x\,(m+1)); && (9)\\
\textit{\textbf{subs}}\,(\texttt{var}\,n)\,x\,m &:= \begin{cases} (\textit{\textbf{update}}\,x\,n\,1) & \textit{if } m = n-1,\\ \texttt{var}(n-1) & \textit{if } m < n-1,\\ (\texttt{var}\,n) & \textit{otherwise.}\end{cases} && (10)\\
\textit{\textbf{update}}\,(\texttt{app}\,x\,y)\,m\,j &:= \texttt{app}(\textit{\textbf{update}}\,x\,m\,j)(\textit{\textbf{update}}\,y\,m\,j); && (11)\\
\textit{\textbf{update}}\,(\texttt{abs}\,x)\,m\,j &:= \texttt{abs}(\textit{\textbf{update}}\,x\,m\,(j+1)); && (12)\\
\textit{\textbf{update}}\,(\texttt{var}\,n)\,m\,j &:= \begin{cases} \texttt{var}(n+m-1) & \textit{if } n \geq j,\\ (\texttt{var}\,n) & \textit{otherwise.}\end{cases} && (13)
\end{aligned}$$

The program ***update*** *takes care of updating the de Bruijn indexes of the free variables after the substitution performed by* ***subs***.

We do not enter in the details of the equations for the updating of the de Bruijn indexes since similar equations have already been used by various people, see for instance [1] and [18]. In these papers the authors consider a "lambda calculus with explicit substitutions" using a suitable notation based on de Bruijn indexes. What we do is different: we are not defining a lambda calculus, but rather a program to reduce lambda terms. So in our approach along with the normalization program, we can also define a wealth of other programs on lambda terms, for instance a program to count the variables. The lambda terms, represented in de Bruijn notation, do not reduce by themselves to normal form: they must be given as inputs to the normalization program.

8 Computability of All Partial Recursive Functions

We define a **second order term algebra** as the set of closed normal forms of a given atomic type α over a given signature Σ of second order constructors. So such an algebra can be denoted by the pair (α, Σ). We have seen that many interesting data structures can be represented by second order term algebras.

Theorem 3. *For every partial recursive function f between two second order term algebras (α_1, Σ_1) and (α_2, Σ_2), there is a dichotomic set of reduction rules over a second order signature $\Sigma \supset \Sigma_1 \cup \Sigma_2$ which computes f. A similar result holds for functions of several arguments.*

Proof. (Sketch) We take for granted that the result is true for first order algebras ("folklore theorem"), namely when the constant symbols of the signatures have level 1. In the general case the idea is to show that to every second order term algebra we can associate a first order term algebra and a bijection between the two algebras, such that the bijection and its inverse are computable by a dichotomic set of reduction rules over a second order signature. For instance the second order term algebra of type `exp` that we have used to represent untyped lambda terms (Definition 12) can be associated bijectively to the first order term algebra of type `term` which represents lambda terms in de Bruijn notation (Definition 16). For the details of how to translate between the two representations using a dichotomic set of reduction rules see the appendix.

9 Interpretation in the Pure Untyped Lambda Calculus

Definition 18. *An* **interpretation** *of $\Lambda(\Sigma, P)$ into the untyped lambda calculus is a map ϕ which assigns to every closed term t of $\Lambda(\Sigma, P)$ a closed term t^ϕ of the untyped lambda calculus and has the following properties:*

1. $(tt_1 \ldots t_n)^\phi \equiv t^\phi t_1^\phi \ldots t_n^\phi$, $(\lambda x_1 \ldots x_n.t)^\phi \equiv \lambda x_1 \ldots x_n.(t^\phi)$. *So ϕ is uniquely determined by its restriction $\phi_{|\Sigma}$ to the symbols of the signature.*
2. *If $t_1 \to t_2$ in $\Lambda(\Sigma, P)$, then $t_1^\phi \to t_2^\phi$ in the untyped lambda calculus.*

The above definition admits many variants. A weaker notion is obtained by replacing the reduction relation by the convertibility relation in clause 2. A stronger version is obtained by requiring that the converse implication of clause 2 also holds. A reasonable compromise is to require that the interpretation is injective on the data structures. This can be formalized as follows:

Definition 19. *An interpretation ϕ of $\Lambda(\Sigma, P)$ into the untyped lambda calculus is injective on the data structures if whenever t_1 and t_2 are distinct normal forms of $\Lambda(\Sigma, P)$ having atomic type and not containing program symbols, then t_1^ϕ and t_2^ϕ have distinct normal forms.*

Using a technique introduced by Böhm and Piperno and studied in [9,6] one can prove the following theorem:

Theorem 4. *$\Lambda(\Sigma, P)$ can be interpreted in the untyped lambda calculus by an interpretation that is injective on the data structures.*

Quite remarkably the interpretation described in [9,6] does not make use of the fixed point combinator. Using this fact it is shown in [6] that in the first order case the interpretation "preserves strong normalization".

A different interpretation of an higher order term algebra into the pure lambda calculus is obtained by replacing the constructors by variables and abstracting them, as in the definition of the Church numerals (see [7,8,17] for similar proposals).

Consider for instance the higher order algebra of type $\mathtt{exp}$ in Definition 12. The term $\ulcorner(\lambda x.xx)(\lambda x.xx)\urcorner = \mathtt{App}(\mathtt{Abs}(\lambda\, x.\mathtt{App}\, x\, x))(\mathtt{Abs}(\lambda\, x.\mathtt{App}\, x\, x))$ is an element of that algebra. By replacing the constructors $\mathtt{App}, \mathtt{Abs}$ by variables a, b and abstracting them, we obtain the lambda term $\lambda ab.a(b(\lambda\, x.a\, x\, x))(b(\lambda\, x.a\, x\, x))$ (which is a normal form).

In this way we have defined an embedding $t \mapsto \theta(t)$ of the pure lambda calculus into itself (e.g. θ sends $(\lambda x.xx)(\lambda x.xx)$ into $\lambda ab.a(b(\lambda\, x.a\, x\, x))(b(\lambda\, x.a\, x\, x)))$ which is probably the simplest "gödelization" of the lambda calculus which has ever been considered (compare with [17,6,3]). The name "gödelization" is justified by the fact that there is a combinator which defines a bijection from the image of θ onto the Church numerals. This can be easily proved applying the interpretation of [9,6] to the translations between first order and higher order term algebras given in the appendix. We also need the fact that all infinite first order algebras (suitably embedded in the lambda calculus) admit a lambda definable bijection onto the Church numerals. Note that the image of θ consists of typable terms of type $(\mathtt{exp} \times \mathtt{exp} \to \mathtt{exp}) \to ((\mathtt{exp} \to \mathtt{exp}) \to \mathtt{exp}) \to \mathtt{exp}$.

10 Appendix

The program $\mathbf{M}: \mathtt{term} \to \mathtt{exp}$ below, translates from de Bruijn notation as in Definition 16, to lambda terms represented as in in Definition 12. We need an auxiliary program $\mathbf{ch} : \mathtt{term} \times \mathtt{term} \times \mathtt{nat} \to \mathtt{term}$ and an auxiliary constructor $\mathtt{Bx} : \mathtt{exp} \to \mathtt{term}$.

$$\mathbf{M}\,(\mathtt{abs}\, t) := \mathtt{Abs}(\lambda\, u.\mathbf{M}\,(\mathbf{ch}\; t(\mathtt{Bx}\, u)1) \quad (1)$$

$$\mathbf{M}\,(\mathtt{app}\, x\, y) := \mathtt{App}(\mathbf{M}\; x)(\mathbf{M}\; y) \quad (2)$$

$$\mathbf{M}\,(\mathtt{Bx}\, u) := u \quad (3)$$

$$\mathbf{ch}\,(\mathtt{app}\, x\, y)\, u\, j := \mathtt{App}(\mathbf{ch}\; x\, u\, j)(\mathbf{ch}\; y\, u\, j) \quad (4)$$

$$\mathbf{ch}\,(\mathtt{abs}\, x)\, u\, j := \mathtt{Abs}(\mathbf{ch}\; x\, u(1+j)) \quad (5)$$

$$\mathbf{ch}\,(\mathtt{var}\, m)\, u\, j := \text{ if } m = j \text{ then } u \text{ else } (\mathtt{Var}\, m) \quad (6)$$

$$\mathbf{ch}\,(\mathtt{Bx}\, x)\, u\, j := \mathtt{Bx}\, x \quad (7)$$

We now define a term $\lambda\, x.\mathbf{db}\; x\, 0 : \mathtt{exp} \to \mathtt{term}$ which performs the inverse translation. The program $\mathbf{db} : \mathtt{exp} \times \mathtt{nat} \to \mathtt{term}$ uses the auxiliary constructor $\mathtt{Var}: \mathtt{nat} \to \mathtt{exp}$ (not to be confused with $\mathtt{var}: \mathtt{nat} \to \mathtt{term}$).

$$\mathbf{db}\,(\mathtt{Abs}\, t)\, n := \mathbf{abs}(\mathbf{db}\,(\, t(\mathtt{Var}\, n))(1+n)) \quad (8)$$

$$\mathbf{db}\,(\mathtt{App}\, x\, y)\, n := \mathbf{app}(\mathbf{db}\; x\, n)(\mathbf{db}\; y\, n) \quad (9)$$

$$\mathbf{db}\,(\mathtt{Var}\, m)\, n := \mathbf{var}(n - m) \quad (10)$$

References

1. M. Abadi, L. Cardelli, P.-L. Curien, J.-J. Lévy. Explicit Substitutions. *Journal of Functional Programming*, 1(4): 375–416, 1991
2. A. Asperti and C. Laneve. Interaction Systems I, The Theory of Optimal Reductions. *Mathematical Structures in Computer Science*, 4(4): 457–504, 1995
3. H. Barendregt. Self-interpretation in lambda calculus. *Journal of Functional Programming*, 1(2): 229–233, 1991
4. H. Barendregt, M. Coppo, and M. Dezani-Ciancaglini. A filter lambda model and the completeness of type assignment. *J. Symbolic Logic*, 48: 931-940, 1983
5. S. Bellantoni and S. Cook. New recursion-theoretic characterization of the polytime functions. *Computational Complexity*, 2: 97–110, 1992
6. A. Berarducci and C. Böhm. A self-interpreter of lambda calculus having a normal form. 6th Workshop, CSL '92,San Miniato, Italy, *E. Börger & al. eds. LNCS* 702: 85–99, Springer-Verlag, 1992
7. C. Böhm and A. Berarducci. Automatic synthesis of typed Λ-programs on term algebras. *Theoretical Computer Science* 39: 135–154,1985
8. C. Böhm. Fixed Point Equations Inside the Algebra of Normal Form. *Fundamenta Informaticae*, 37(4): 329–342, 1999
9. C. Böhm, A. Piperno and S. Guerrini. Lambda-definition of function(al)s by normal forms. In: *ESOP'94, LNCS* 788: 135–149, Springer-Verlag, 1994
10. S. Byun, R. Kennaway, R. Sleep. Lambda-definable term rewriting systems. Second Asian Computing Science Conference, ASIAN '96, Singapore, December 2-5, 1996, *LNCS* 1179:105–115, Springer-Verlag, 1996
11. R. L. Constable and S. F. Smith. Computational foundations of basic recursive function theory. *Theoretical Computer Science*, 121: 89–112, 1993
12. J. Despeyroux, F. Pfenning, C. Schürmann. Primitive Recursion for Higher-Order Abstract Syntax. In: *R.Hindley, ed., Proc. TLCA'97 Conf., LNCS* 1210: 147–163, Springer-Verlag, 1997
13. M. Gabbay, A. Pitts. A New Approach to Abstract Syntax Involving Binders. In: *Proc. 14th Symp. Logic in Comp. Sci. (LICS) Trento, Italy*: 214–224, IEEE, Washington, 1999.
14. R. Harper, F. Honsell, G. Plotkin. A framework for defining logics. Journal of the ACM, 40(1): 143-184, 1993
15. Y. Lafont. Interaction Combinators. *Information and Computation* 137 (1): 69–101, 1997
16. D. Leivant, J.-Y. Marion. Lambda calculus characterizations of polytime. *Fundamenta Informaticae*, 19: 167–184, 1993
17. T. Æ. Mogensen. Efficient Self-Interpretation in Lambda Calculus. *Journal of Functional Programming*, 2(3): 354–364, 1992
18. M. Ayala-Rincón, F. Kamareddine. Unification via λs_e-Style of Explicit Substitution. In: *International Conference on Principles and Practice of Declarative Programming, PPDP'00, ACM Publications*:163–174, 2000
19. S. Ronchi della Rocca and B. Venneri. Principal type schemes for an extended type theory. *Theoretical Computer Science*, 28:151–169, 1984

Beta Reduction Constraints

Manuel Bodirsky, Katrin Erk*, Alexander Koller[1]**, and Joachim Niehren**

Programming Systems Lab [1] Dept. of Computational Linguistics
Universität des Saarlandes, Saarbrücken, Germany
www.ps.uni-sb.de/~bodirsky,erk,koller,niehren

Abstract. The constraint language for lambda structures (CLLS) can model lambda terms that are known only partially. In this paper, we introduce beta reduction constraints to describe beta reduction steps between partially known lambda terms. We show that beta reduction constraints can be expressed in an extension of CLLS by group parallelism. We then extend a known semi-decision procedure for CLLS to also deal with group parallelism and thus with beta-reduction constraints.

1 Introduction

The constraint language for lambda structures (CLLS) [7,6,8] can model λ-terms that are known only partially. The idea is to see a λ-term as a λ-structure: a tree decorated with binding edges. One can then describe a λ-term partially as one would describe a standard tree structure. CLLS provides dominance [13,2, 5], parallelism [9] and binding constraints for this purpose.

This paper shows how to lift β-reduction to partial descriptions of λ-terms in CLLS. We define *beta reduction constraints*, which allow a declarative description of the result of a single β-reduction step. At first, this description is very implicit; it is made explicit by *solving* the constraints. To this end, we show how beta reduction constraints can be expressed as *group parallelism constraints.* Then we adapt a known semi-decision procedure for CLLS to also deal with group parallelism and thus with beta-reduction constraints.

Beta-reduction constraints lay the foundation for *underspecified beta reduction,* which is needed in the application of CLLS to *semantic underspecification* of natural language [15,17,14]. Given a CLLS constraint describing many lambda terms, the aim is to compute a compact description of all corresponding beta normal forms efficiently. In particular, we want to avoid enumerating and individually beta-reducing the described lambda terms. (Enumerating is neither efficient, nor is its result compact.) A recent proposal towards *underspecified beta reduction* is described by the authors in a follow-up paper [4].

Solving beta reduction constraints is very much different from higher-order unification [10] in that CLLS constraints express α-equality rather than $\alpha\beta\eta$-equality. CLLS is closely linked to context unification [12,16], and it can express

* Supported by the DFG through the Graduiertenkolleg Kognition in Saarbrücken.
** Supported by the Collaborative Research Center (SFB) 378 of the DFG and the Procope project of the DAAD.

A. Middeldorp (Ed.): RTA 2001, LNCS 2051, pp. 31–46, 2001.

sharing as in optimal lambda reduction [11] or calculi with explicit substitutions [1] but can also describe several lambda terms at once.

Plan. We first recall the definition of the CLLS restricted to dominance and λ-binding constraints (Sec. 2); then we go through two examples to give an idea of how one might lift β-reduction to partial descriptions (Sec. 3). We next define β-reduction constraints (Sec. 4). Then we define group parallelism constraints and show how they can express β-reduction constraints (Sec. 5). Finally, we present a sound and complete semi-decision procedure for CLLS with group parallelism (Sec. 6) and illustrate it with an example (Sec. 7).

2 CLLS with Dominance and Lambda Binding Constraints

We first introduce λ-structures and then a fragment of CLLS for their description. This fragment contains dominance and λ-binding constraints, but not parallelism and anaphoric binding constraints.

We assume a signature $\Sigma = \{f, g, \dots\}$ of function symbols, each equipped with an arity $\mathsf{ar}(f) \geq 0$. Symbols of arity 0 are constants, written as $a, b, \dots$

A tree θ is a ground term over Σ, e.g. $g(f(a,b))$. A *node* of a tree is identified with a *path* from the root to this node, expressed by a word over the naturals (excluding 0). ϵ is the empty path, and $\pi_1\pi_2$ the concatenation of π_1 and π_2. π is a *prefix* of a path π' if there is a (possibly empty) π'' s.t. $\pi\pi'' = \pi'$. The set of all nodes of a tree θ is defined as

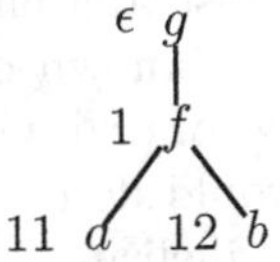

Fig. 1. Tree structure for $g(f(a,b))$

$$D_\theta(f(\theta_1, \dots, \theta_n)) \;=\; \epsilon \cup \{i\pi \mid \pi \in D_\theta(\theta_i),\ 1 \leq i \leq n\}$$

A tree θ can be characterized uniquely by the set D_θ of its nodes and a labeling function $L_\theta : D_\theta \to \Sigma$.

Now we can consider λ-terms as pairs of a tree and a *binding function* that encodes variable binding. We assume that Σ contains the symbols var (arity 0, for variables), lam (arity 1, for abstraction), and @ (arity 2, for application), and quantifier $\exists$ and $\forall$ (arity 1). The tree uses these symbols to reflect the structure of the λ-term and first-order connectives. The binding function λ explicitly maps var-labeled nodes to binders. For example, Fig. 2 shows a representation of the term $\lambda x.f(x)$. Here $\lambda(12) = \epsilon$. Such a pair of a tree and a binding function is called a λ-structure.

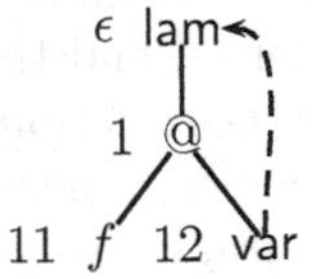

Fig. 2. The λ-structure of $\lambda x.f(x)$

Definition 1. *A* λ-structure τ *is a pair* (θ, λ) *of a tree* θ *and a total binding function* $\lambda : L_\theta^{-1}(\mathsf{var}) \to L_\theta^{-1}(\{\mathsf{lam}, \exists, \forall\})$ *such that* $\lambda(\pi)$ *is always a prefix of* π.

A λ-structure corresponds uniquely to a closed λ-term modulo α-renaming. We freely consider λ-structures as first-order structures with domain D_θ. As such, they define relations of labeling, binding, inverse binding, dominance, disjointness, and inequality of nodes. (Later we will add group parallelism and β-reduction relations.) The labeling relation $\pi{:}f(\pi_1,\ldots,\pi_n)$ holds in a λ-structure τ if $L_\theta(\pi) = f$, $\mathsf{ar}(f) = n$ and $\pi_i = \pi i$ for all $1 \le i \le n$. *Dominance* $\lhd^*$ is the prefix relation between paths of D_θ; inequality $\neq$ is simply inequality of paths; *disjointness* $\pi\perp\pi'$ holds if neither $\pi\lhd^*\pi'$ nor $\pi'\lhd^*\pi$. We will also consider intersections, unions, and complements of these relations; for instance, *proper dominance* $\lhd^+$ is $\lhd^*\cap\neq$, and *equality* $=$ is $\lhd^*\cap\rhd^*$. The relation $\lambda^{-1}(\pi_0)=\{\pi_1,\ldots,\pi_n\}$ states that $\pi_1,\ldots,\pi_n$, and only those nodes, are λ-bound by π_0. Note that an element of a set can be mentioned multiply, i.e. $\{\pi,\pi\} = \{\pi\}$.

Now we can define *dominance and binding* constraints to talk about λ-structures as follows; X, Y, Z are variables that denote nodes.

$$\begin{array}{rll} \varphi, \psi & ::= XRY \mid X{:}f(X_1,\ldots,X_n) \mid \varphi \wedge \psi \mid \textbf{false} & (\mathsf{ar}(f) = n) \\ & \mid \lambda(X){=}Y \mid \lambda^{-1}(X_0){=}\{X_1,\ldots,X_n\} & \\ R, R' & ::= \lhd^* \mid \rhd^* \mid \perp \mid \neq \mid R{\cup}R' \mid R{\cap}R' & \end{array}$$

A constraint φ is a conjunction of *literals* (for dominance, labeling, etc). Set operators in relation descriptors R [5] are mainly needed for processing purposes. As above we also use $\lhd^+, =$ to abbreviate set operators. The one literal that has not appeared in the literature before is the *inverse binding literal* $\lambda^{-1}(X){=}\{X_1,\ldots,X_n\}$, which matches the inverse binding relation.

We will also use first-order formulas Φ built over constraints. We write $\mathcal{V}(\Phi)$ for the set of variables occurring in Φ. Given a pair (τ,σ) of a λ-structure τ and a variable assignment $\sigma : \mathcal{G} \to D_\tau$, for some set $\mathcal{G} \supseteq \mathcal{V}(\varphi)$, we can associate a truth value to Φ in the usual Tarskian sense. We say that (τ,σ) *satisfies* Φ iff Φ evaluates to true under (τ,σ). In this case, we write $(\tau,\sigma) \models \Phi$ and say that (τ,σ) is a *solution* of Φ. Φ is *satisfiable* iff it has a solution. Entailment $\Phi \models \Phi'$ means that all solutions of Φ are also solutions of Φ', equivalence $\Phi \models\!\!\!\!\!\!= \Phi'$ is mutual entailment.

We draw constraints as graphs with the nodes representing variables. E.g. Fig. 3 is the graph of $\lambda^{-1}(X){=}\{X_1,X_2\} \;\wedge\; X\lhd^*X_1 \;\wedge\; X\lhd^*X_2$. Labels and solid lines indicate labeling literals, while dotted lines represent dominance. Dashed arrows indicate the binding relation; disjointness and inequality literals are not represented.

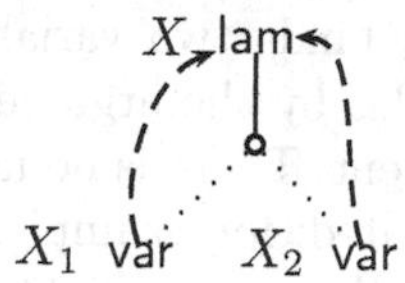

Fig. 3. A constraint graph

3 Examples

Before we begin with the formal investigation of beta reduction constraints, we first go through two examples which illustrate how beta-reduction can be lifted to descriptions of λ-structures in CLLS, and why the problem is nontrivial.

First, consider the left constraint in Fig. 4. The constraint contains just one redex, and it is easy to see how to obtain a description of the reduced formulas.

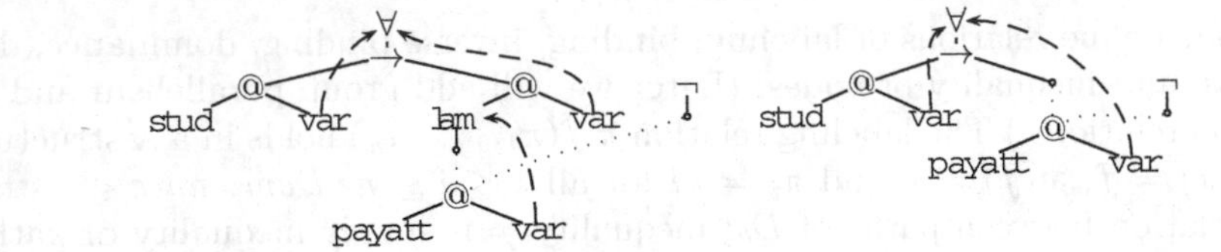

Fig. 4. Underspecified representations of 'Every student did not pay attention' before and after beta reduction

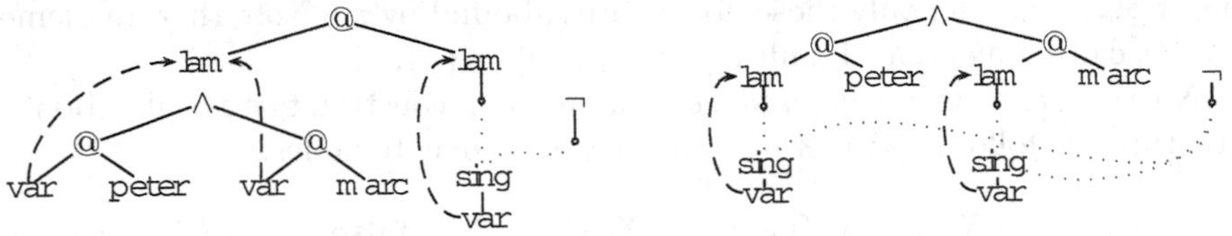

Fig. 5. Representation of 'Peter and Marc do not sing', wrong description of the reduct

We can essentially replace the bound variables with the argument description; the result is shown on the right-hand side of Fig. 4.

Incidentally, the left constraint in Fig. 4 is an underspecified description of the ambiguous sentence *Every student didn't pay attention*. Its two readings are given by the HOL formulas:

$$\forall x\ (\text{stud}\ x \rightarrow (\lambda y \neg(\text{payatt}\ y))\ x), \qquad \neg\forall x\ (\text{stud}\ x \rightarrow (\lambda y\ \text{payatt}\ y)\ x).$$

(These are the only models of the constraint that do not contain additional material not mentioned in the constraint. We ignore this aspect of "solution minimality" in this paper and always consider all solutions.)

The naive replacement approach, however, does not work in general. Fig. 5 shows an example (which describes the ambiguous sentence 'Peter and Marc do not sing.') This constraint also describes a β-redex, this time one where the binder binds two variables. Here it is no longer trivial to replace the bound variables by the argument description, as we do not know what belongs to the argument. There is no useful choice for the part of the constraint that should be duplicated; for example, if we decide not to duplicate the negation, we get the description on the right-hand side of Fig. 5, which lacks one solution. Describing the reduct using β-reduction constraints solves this problem; the description is correct even if it is not yet known which variables belong to the body and the argument of the redex.

4 Beta Reduction Constraints

In this section, we add the *β-reduction relation* to lambda structures and extend the constraint language with *β-reduction constraints* to talk about it. The β-reduction relation on nodes of a lambda structure corresponds exactly to traditional beta reduction on lambda terms.

Stated in the unfolded notation for λ-terms we use to build the λ-structures (with application as an internal label @, etc.), β-reduction looks as follows:

$$C(@(\lambda x.B, A)) \quad \rightarrow_\beta \quad C(B[x/A]) \qquad\qquad x \text{ free for } A$$

We call the left-hand side the *reducing tree*, the right-hand side *the reduct* of the β-reduction. We call C the context, B the body, and A the argument of the reduction step.

Now an important notion throughout the paper are *tree segments.* Intuitively, a tree segment is subtree which may itself be missing some subtrees (see Fig. 6). The context, body, and argument of a beta reduction step will all be tree segments.

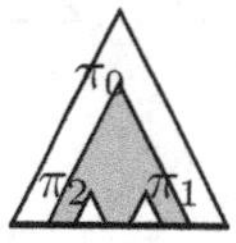

Fig. 6. The tree segment $\pi_0/\pi_1, \pi_2$

Definition 2. *A* tree segment α *of a λ-structure* τ *is given by a tuple* $\pi_0/\pi_1, \ldots, \pi_n$ *of nodes in* D_τ*, such that* $\tau \models \pi_0 \triangleleft^* \pi_i$ *and* $\tau \models \pi_i (\perp \cup =) \pi_j$ *for* $1 \leq i < j \leq n$. *The node* $r(\alpha) = \pi_0$ *is called the* root, *and* $hs(\alpha) = \pi_1, \ldots, \pi_n$ *is the sequence of* holes *of* α. *If* $n = 0$ *we write* $\alpha = \pi_0/$. *The nodes* between *the root* $r(\alpha)$ *and the holes* $hs(\alpha)$ *are defined as*

$$\mathsf{b}(\alpha) =_{\mathsf{df}} \{\pi \in D_\tau \mid r(\alpha) \triangleleft^* \pi \wedge \bigwedge_{\pi' \in \{hs(\alpha)\}} \pi' \neg \triangleleft^+ \pi\}$$

To exempt the holes of the segment, we define $\mathsf{b}^-(\alpha) =_{\mathsf{df}} \mathsf{b}(\alpha) - \{hs(\alpha)\}$.

Definition 3. *A* correspondence function *between tree segments* α, β *in a lambda structure* τ *is a bijective mapping* $c : \mathsf{b}(\alpha) \rightarrow \mathsf{b}(\beta)$ *which satisfies for all nodes* $\pi_1, \ldots, \pi_n$ *of* τ*:*

1. *The roots correspond:* $c(r(\alpha)) = r(\beta)$
2. *The sequences of holes correspond:*

$$hs(\alpha) = \pi_1, \ldots, \pi_n \Leftrightarrow hs(\beta) = c(\pi_1), \ldots, c(\pi_n)$$

3. *Labels and children correspond within the proper segments. For* $\pi \in \mathsf{b}^-(\alpha)$ *and label* f*:*

$$\pi{:}f(\pi_1, \ldots, \pi_n) \Leftrightarrow c(\pi){:}f(c(\pi_1), \ldots c(\pi_n)).$$

We next define the β-reduction relation on λ-structures to be a relation between nodes in the *same* λ-structure. This allows us to see the β-reduction relation as a conservative extension of the existing λ-structures. The representations both of the reducing and reduced term are part of same big λ-structure—in Fig. 7, these are the trees rooted by $r(\gamma)$ and $r(\gamma')$ respectively.

A *redex* in a lambda structure is a sequence of segments (γ, β, α) of that λ-structure that are connected by nodes π_0, π_1 with the following properties.

$$hs(\gamma) = \pi_0,\ \pi_0{:}@(\pi_1, r(\alpha)),\ \pi_1{:}\mathsf{lam}(r(\beta)), \text{ and } \lambda^{-1}(\pi_1) = \{hs(\beta)\}$$

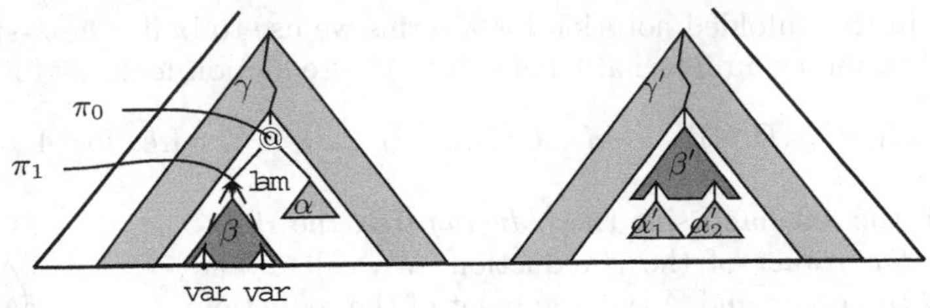

Fig. 7. The beta reduction relation for a binary redex

The lambda structure in Fig. 7 contains a redex (γ, β, α) and also its reduct $(\gamma', \beta', \alpha'_1, \alpha'_2)$. There, corresponding segments (γ to γ', β to β', α to both α'_1 and α'_2) have the same structure.

Definition 4 (Beta Reduction Relation). Let τ be a λ-structure. Then

$$(\gamma, \beta, \alpha) \rightarrow^{\beta} (\gamma', \beta', \alpha'_1, \ldots, \alpha'_n)$$

holds in τ iff first, (γ, β, α) form a redex. Second, there are correspondence functions c_γ between γ, γ', c_β between β, β' and c^i_α between α, α'_i (for $1 \leq i \leq n$), such that for each δ, δ' among these segment pairs with correspondence function c between them and for each $\pi \in \mathsf{b}^-(\delta)$, the following conditions hold:

1. for a var-labeled node bound in the same segment, the correspondent is bound by the c-corresponding binder node.

$$\lambda(\pi) \in \mathsf{b}^-(\delta) \Rightarrow \lambda(c(\pi)) = c(\lambda(\pi))$$

2. for a var-labeled node bound in the context γ, the correspondent is bound by the c_γ-corresponding binder node.

$$\lambda(\pi) \in \mathsf{b}^-(\gamma) \Rightarrow \lambda(c(\pi)) = c_\gamma(\lambda(\pi))$$

3. for a var-labeled node bound above the reducing tree, the corresponding node is bound at the same place:

$$\lambda(\pi) \notin \mathsf{b}(r(\gamma)/) \Rightarrow \lambda(c(\pi)) = \lambda(\pi)$$

The β-reduction relation on λ-structures models β-reduction on λ-terms faithfully. This even holds for λ-terms with global variables, although λ-structures can only model closed λ-terms. Global variables correspond to var-labeled nodes that are bound in the surrounding tree, i.e. above the root node of the context of the redex. Rule 3 of Def. 4 thus ensures a proper treatment of global variables.

Capturing in β-reduction is usually avoided by a freeness condition. For instance, one cannot simply β-reduce $(\lambda x.\lambda y.x)y$ without renaming the bound occurrence of y beforehand. Otherwise, the global variable y in the argument gets captured by the binder λy. The following proposition states that the analogous problem can never arise with the β-reduction relation on λ-structures.

Proposition 1 (No Capturing). *Global variables in the argument are never captured by a λ-binder in the body: with the notation of Def. 4, this means that no* var*-labeled node in* $\mathsf{b}(\alpha_i')$ *is bound by a* lam*-labeled node in* $\mathsf{b}^-(\beta')$.

Proof. Assume there exists a node π' in $\mathsf{b}(\alpha_i')$ such that $\lambda(\pi') \in \mathsf{b}^-(\beta')$. There must be a corresponding var-labeled node π with $c_\alpha^i(\pi) = \pi'$, which is bound either in α, in γ or outside the reducing tree. In the first case property (1) leads to a contradiction, in the second case property (2), and in the third case (3).

The β-reduction relation conservatively extends λ-structures. We extend our constraint syntax similarly, by *β-reduction literals*, which are interpreted by the β-reduction relation. Let a *segment term* A, B, C be given by the following abstract syntax:

$$A, B, C =_{\mathsf{df}} X_0/X_1, \ldots, X_n$$

Then β-reduction literals have the following form:

$$(C, B, A) \to^\beta (C', B', A_1', \ldots, A_n')$$

5 Group Parallelism

In this section, we extend dominance and binding constraints with *group parallelism constraints* (Def. 5), a generalization of the parallelism constraints found in CLLS [6,9]. Then we show that CLLS with group parallelism can express β-reduction constraints (Thm. 1).

Group parallelism relates two *groups*, i.e. sequences of tree segments. It requires that corresponding entries in the two sequences must have the same tree structures, and binding in the two groups must be parallel. The following definition makes this precise; all conditions but the last are illustrated in Fig. 8.

Definition 5. *The* group parallelism relation $\sim$ *of a λ-structure τ is the greatest symmetric relation between groups of the same size such that*

$$(\alpha_1, \ldots, \alpha_n) \sim (\alpha_1', \ldots, \alpha_n')$$

implies there are correspondence functions $c_k : \mathsf{b}(\alpha_k) \to \mathsf{b}(\alpha_k')$ *for all* $1 \leq k \leq n$ *that satisfy the following properties for all* $1 \leq i, j \leq n$ *and* $\pi \in \mathsf{b}^-(\alpha_i)$*:*

(same.seg) *for a* var*-labeled node bound in the same segment, the corresponding node is bound correspondingly:*

$$\lambda(\pi) \in \mathsf{b}^-(\alpha_i) \Rightarrow \lambda(c_i(\pi)) = c_i(\lambda(\pi))$$

(diff.seg) *for a* var*-labeled node bound outside α_i but inside α_j, the correspondent is bound at the corresponding place with respect to c_j:*

$$\lambda(\pi) \in \mathsf{b}^-(\alpha_j) \wedge \lambda(\pi) \notin \mathsf{b}^-(\alpha_i) \quad \Rightarrow \lambda(c_i(\pi)) = c_j(\lambda(\pi))$$

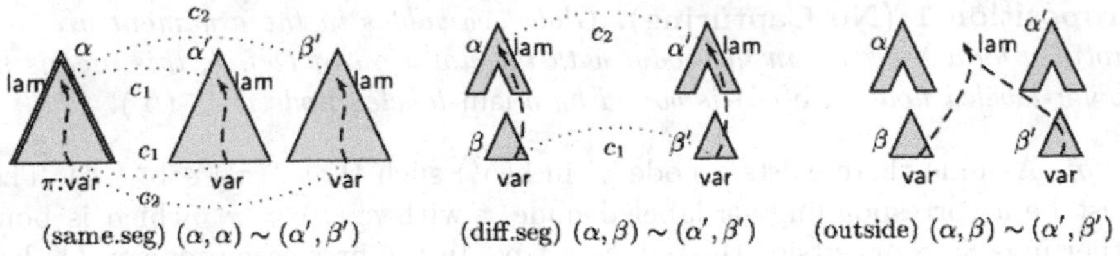

Fig. 8. Possible bindings in a group parallelism

(outside) *corresponding* var*-labeled nodes with binders outside the group segments are bound by the same binder node:*

$$\lambda(\pi) \notin \cup_{k=1}^{n} \mathsf{b}^{-}(\alpha_k) \Rightarrow \lambda(c_i(\pi)) = \lambda(\pi)$$

(hang) *there are no hanging binders:*

$$\lambda^{-1}(\pi) \subseteq \cup_{k=1}^{n} \mathsf{b}^{-}(\alpha_k)$$

On the syntactic side, we extend CLLS by *group parallelism literals* that are interpreted by the group parallelism relation. Let $A_1, \ldots A_m, A'_1, \ldots, A'_m$ be segment terms, then group parallelism literals have the form

$$(A_1, \ldots, A_m) \sim (A'_1, \ldots, A'_m)$$

Group parallelism extends ordinary parallelism constraints [6,9], which are simply the special case for groups of size one. This extension is proper; ordinary parallelism constraints cannot handle the case where a node is bound in a different segment of the group, as illustrated in the (diff.seg) part of Fig. 8. From the perspective of ordinary parallelism, the node is bound outside the parallel segment, and thus the (outside) condition applies, and the corresponding node must be bound by the *same* binder.

Another interesting observation in Fig. 8 is that the conditions (same.seg) and (diff.seg) must be mutually exclusive. If (diff.seg) was applicable in the leftmost case, it would enforce $\lambda(c_1(\pi)) = c_2(\lambda(\pi))$, which is clearly wrong.

Now we show how to encode beta reduction constraints in CLLS. First, we define the following abbreviation to express that the segment term $A = X_0/X_1, \ldots, X_n$ indeed denotes a tree segment:

$$\mathsf{seg}(A) =_{\mathsf{df}} \wedge_{i=1}^{n} X_0 \triangleleft^* X_i \;\wedge\; \wedge_{1 \le i < j \le n} X_i (\perp \cup =) X_j$$

Using this, we can axiomatize a redex in CLLS. For segment terms $A = X_2/$, $B = X_3/X_4, \ldots, X_n$, $C = X'/X_0$, we set:

$$\begin{aligned}\mathsf{redex}_{X_0,X_1}(C, B, A) \quad &=_{\mathsf{df}} \mathsf{seg}(A) \wedge \mathsf{seg}(B) \wedge \mathsf{seg}(C) \wedge X_0{:}@(X_1, X_2) \\ &\wedge \quad X_1{:}\mathsf{lam}(X_3) \wedge \lambda^{-1}(X_1) = \{X_4, \ldots, X_n\}\end{aligned}$$

Theorem 1. *Beta reduction constraints can be expressed in CLLS with group parallelism via the following equivalence:*

$$(C, B, A) \to^{\beta} (C', B', A'_1, \ldots, A'_n) \quad \models\!\!\!\mid \quad \exists X_0, X_1 : \mathsf{redex}_{X_0,X_1}(C, B, A) \\ \wedge\ (C, B, A, \ldots, A) \sim (C', B', A', \ldots, A')$$

Proof. We will check the two-side entailment separately, first from right to left. Let σ be a variable assignment into some λ-structure that solves the right hand side. Properties (same.seg), (diff.seg), and (outside) for group parallelism (Def. 5) then subsume the corresponding properties of β-reduction (Def. 4).

For the other direction, let (τ, σ') solve the beta-reduction literal on the left hand side. According to (Def. 4) there exists a redex (γ, β, α) in τ with nodes π_0, π_1 as in Sec. 4. Let σ be the variable assignment $\sigma'[\pi_0/X_0, \pi_1/X_1, \alpha/A, \beta/B, \gamma/C]$. It remains to check that (τ, σ) solves the group parallelism literal on the right hand side.

We consider the symmetric relation $\approx$ which relates the group $(\sigma(C), \sigma(B), \sigma(A), \ldots, \sigma(A))$ to $(\sigma(C'), \sigma(B'), \sigma(A'_1), \ldots, \sigma(A'_n))$ and conversely. We show that $\approx$ satisfies all conditions in the definition of group parallelism (Def. 5), which means that $\approx$ is subsumed by the group parallelism relation $\sim$.

First of all, both above groups satisfy condition (hang). This is clear for the group $(\sigma(C'), \sigma(B'), \sigma(A'_1), \ldots, \sigma(A'_n))$, which covers the complete subtree below $r(\sigma(C'))$. A similar argument applies to $(\sigma(C), \sigma(B), \sigma(A), \ldots, \sigma(A))$, which covers the whole tree below $r(\sigma(C))$ except the @-labeled node π_0, the lam-labeled node π_1 and the var-labeled nodes $hs(\sigma(B))$. But these var-labeled nodes are bound by π_1.

By Def. 4 there exist correspondence functions c_γ between segments $\sigma(C)$, $\sigma(C')$, c_β between $\sigma(B)$, $\sigma(B')$ and c^i_α between $\sigma(A), \sigma(A'_i)$ for $1 \leq i \leq n$. Since $\approx$ is symmetric, we have to check properties (same.seg), (diff.seg), and (outside) for group parallelism (Def. 5) for the correspondence functions and their inverse functions.

We only show the particularly interesting property (diff.seg) for a correspondence function $(c^i_\alpha)^{-1}$ with $1 \leq i \leq n$. Let π' be a var-labeled node in $\mathsf{b}^-(\sigma(A'_i))$, and $\lambda(\pi') \notin \mathsf{b}^-(\sigma(A'_i))$. There are three cases: $\lambda(\pi') \in \mathsf{b}^-(\sigma(C'))$, or $\lambda(\pi') \in \mathsf{b}^-(\sigma(B'))$, or $\lambda(\pi') \in \mathsf{b}^-(\sigma(A'_j))$ for some $1 \leq j \leq n$. The second case is impossible by Proposition 1. The third case is impossible as the holes of the segment $\sigma(B')$ are disjoint or equal (Def. 2). We can thus concentrate on the first case. Let π be the corresponding node of π'. i.e. $c^i_\alpha(\pi) = \pi'$. The node π has to be var-labeled by Def 3. Properties (same.seg) and (outside) of Def. 4 yield $\lambda(\pi) \in \sigma(C)$ (some computation is needed here). Thus, Property 2 of Def. 4 can be applied. It implies $\lambda(\pi') = c_\gamma(\lambda(\pi))$, i.e. $c_\gamma^{-1}(\lambda(\pi')) = \lambda(c_\gamma^{-1}(\pi'))$ as required.

6 Solving Group Parallelism Constraints

We now turn to a sound and complete semi-decision procedure for CLLS with group parallelism, which thus solves β-reduction constraints. To keep the pre-

(D.clash.ineq)	$X{=}Y \wedge X{\neq}Y \rightarrow \mathbf{false}$	
(D.dom.trans)	$X{\triangleleft^*}Y \wedge Y{\triangleleft^*}Z \rightarrow X{\triangleleft^*}Z$	
(D.lab.ineq)	$X{:}f(\dots) \wedge Y{:}g(\dots) \rightarrow X{\neq}Y$	where $f \neq g$
(D.lab.dom)	$X{:}f(\dots, Y, \dots) \rightarrow X{\triangleleft^+}Y$	
(D.distr.notDisj)	$X{\triangleleft^*}Z \wedge Y{\triangleleft^*}Z \rightarrow X{\triangleleft^*}Y \vee Y{\triangleleft^*}X$	
(D.distr.child)	$X{\triangleleft^*}Y \wedge X{:}f(X_1, \dots, X_m) \rightarrow Y{=}X \vee \bigvee_{i=1}^m X_i{\triangleleft^*}Y$	

Fig. 9. Saturation rules for dominance constraints

sentation readable, we focus on the most relevant rules. The full procedure is given in [3]. An illustrative example follows in Section 7.

The procedure is obtained by extending an existing semi-decision procedure for CLLS [8] that is based on *saturation*. A constraint is freely identified with the *set* of its literals. Starting with a set of literals, more literals are added according to some *saturation rules*. Our saturation rules are implications of the form $\varphi_0 \rightarrow \vee_{i=1}^n \varphi_i$ for some $n \geq 1$. To write down rules more compactly, we will also use arbitrary positive existential formulas on the left hand side. These can be eliminated in a preprocessing step: $\exists$-quantified variables can be replaced by arbitrary variables, and disjunction is eliminated by explosion into several rules.

A saturation rule of the above form is applicable to a constraint φ if φ_0 is contained in φ, but none of the φ_i is. A rule $\varphi \rightarrow \Phi$ is *sound* if $\varphi \models \Phi$. Apart from that, we have saturation rules of the form $\varphi_0 \rightarrow \exists X \varphi_1$, which introduce fresh variables. Such a rule is applicable to φ if φ_0 is in φ, but φ_1, modulo renaming of X, is not. Given a set S of saturation rules, we call a constraint *saturated* (under S) if no further rule of S applies to it. We say that a constraint is in *S-solved form* if it is saturated under S and clash-free (i.e. it does not contain **false**).

Fig. 9 contains an (incomplete) set of saturation rules for dealing with dominance constraints (the constraints of Sec. 2 without binding). A more complete collection including a treatment of set operators can be found in [5,8,3]. To deal with parallelism, we first introduce some formulas that describe membership in (proper) segments and groups.

$$
\begin{aligned}
X{\in}\mathsf{b}(A) &=_{\mathsf{df}} X_0{\triangleleft^*}X \ \wedge\ \textstyle\bigwedge_{i=1}^n X({\triangleleft^*}{\cup}{\perp})X_i \\
X{\in}\mathsf{b}^-(A) &=_{\mathsf{df}} X{\in}\mathsf{b}(A) \wedge \textstyle\bigwedge_{i=1}^n X{\neq}X_i \\
X{\notin}\mathsf{b}^-(A) &=_{\mathsf{df}} X({\triangleleft^+}{\cup}{\perp})X_0 \vee \textstyle\bigvee_{i=1}^n X_i{\triangleleft^*}X \\
X{\in}\mathsf{b}(A_1, \dots, A_m) &=_{\mathsf{df}} \textstyle\bigvee_{i=1}^m X{\in}\mathsf{b}(A_i) \\
X{\in}\mathsf{b}^-(A_1, \dots, A_m) &=_{\mathsf{df}} \textstyle\bigvee_{i=1}^m X{\in}\mathsf{b}^-(A_i)
\end{aligned}
$$

Note that the terms $\mathsf{b}(A)$, $\mathsf{b}^-(A)$, $\mathsf{b}(A_1, \dots, A_m)$ are not given any formal meaning, even though it would be correct to interpret them as the corresponding sets of nodes.

(P.symm)	$\overline{A} \sim \overline{B} \to \overline{B} \sim \overline{A}$
(P.init)	$\overline{A} \sim \overline{B} \to \mathsf{seg}(A_i) \wedge \mathrm{co}(A_i, B_i)(X_i^j)=Y_i^j$ where $1 \leq i \leq n$, $A_i = X_i^0/X_i^1, \ldots, X_i^{m_i}$, $B_i = X_i^0/Y_i^1, \ldots, Y_i^{m_i}$, and $0 \leq j \leq m_i$
(P.new)	$\overline{A} \sim \overline{B} \wedge U \in \mathsf{b}(A_i) \to \exists V\ \mathrm{co}(A_i, B_i)(U)=V$ where V fresh, $1 \leq i \leq n$
(P.copy.lab)	$\bigwedge_{i=0}^{m} \mathrm{co}(A, B)(X_i)=Y_i \wedge X_0{:}f(X_1, \ldots, X_m) \wedge X_0 \in \mathsf{b}^-(A) \to Y_0{:}f(Y_1, \ldots, Y_m)$
(P.copy.dom)	$U_1\ R\ U_2 \wedge \bigwedge_{i=1}^{2} \mathrm{co}(A, B)(U_i)=V_i \to V_1\ R\ V_2$
(P.distr.eq)	$\varphi \to X=Y \vee X \neq Y$ for $X, Y \in \mathcal{V}(\varphi)$

Fig. 10. Saturation rules, where $\overline{A} = A_1, \ldots, A_n$ and $\overline{B} = B_1, \ldots, B_n$

We also want to be able to speak about correspondence functions. So we extend our constraint language by auxiliary literals

$$\varphi \quad ::= \quad \ldots \mid \mathrm{co}(A, B)(X)=Y$$

where A and B are segment terms for segments with the same number of holes. Such a literal states that A and B are parallel within some group parallelism, that $X \in \mathsf{b}(A)$ and $Y \in \mathsf{b}(B)$, and that X corresponds to Y with respect to the correspondence function for A and B. We introduce two more formulas. Let $\overline{A} = (A_1, \ldots, A_n)$, $\overline{B} = (B_1, \ldots, B_n)$, and $1 \leq k \leq n$.

$$\mathrm{co}^-(A, B)(X)=Y =_{\mathrm{df}} \mathrm{co}(A, B)(X)=Y \wedge X \in \mathsf{b}^-(A)$$
$$\mathrm{co}_k^-(\overline{A}, \overline{B})(X)=Y =_{\mathrm{df}} \overline{A} \sim \overline{B} \wedge \mathrm{co}^-(A_k, B_k)(X)=Y$$

The second lets us talk about correspondence functions for a group parallelism, picking out the k-th correspondence function. In that respect, $\mathrm{co}_k^-(\overline{A}, \overline{B})$ matches the c_k of Def. 5 (except that $\mathrm{co}_k^-(\overline{A}, \overline{B})(X)=Y$ additionally demands $X \in \mathsf{b}^-(A_k)$ for convenience).

The main rules for handling parallelism are given in Fig. 10. A complete set can be found in [9,8,3]. The rules (P.init) and (P.new) introduce correspondence literals; between them, they state that each node in a parallel segment needs to have a correspondent. (P.init) states that in a correspondence function, root corresponds to root, and hole to hole, while (P.new) is responsible for all other nodes. (P.copy.dom) and (P.copy.lab) between them ascertain the structural isomorphism that Def. 3 demands for a correspondence function.

Fig. 11 shows saturation rules for the interaction of group parallelism and lambda binding. The first four rule schemata directly express the conditions of Def. 5. The rules (L.distr.gr.1) and (L.distr.gr.2) decide, loosely speaking, whether variables occurring in a labeling literal belong to some segment of a group or not. This is necessary because we need to know which of the schemata

(L.same.seg), (L.diff.seg), (L.outside) and (L.hang) is applicable. This is expressed by using the following formula, where $\overline{A} = (A_1, \ldots, A_n)$:

$$\mathsf{distr}_{\overline{A}}(U) =_{\mathsf{df}} \bigwedge_{i=1}^{n} (U \in \mathsf{b}^-(A_i) \vee U \notin \mathsf{b}^-(A_i).$$

Finally, (L.inverse) deals with the copying of λ^{-1} literals. This is necessary if we want to perform a second beta reduction step, where we need the λ^{-1} information again. The schema uses two more formulas. The first one is simple:

$$\lambda(X) \neq Y =_{\mathsf{df}} \exists Z(\lambda(X) = Z \wedge Z \neq Y)$$

The second formula collects, for a finite set S_1 of variables, all correspondents with respect to $\overline{A} \sim \overline{B}$. Let S_1, S_2 stand for finite sets of variables, and let $\overline{A} = A_1 \ldots, A_n$.

$$\begin{aligned} \mathrm{co}^-(\overline{A}, \overline{B})(S_1) = S_2 =_{\mathsf{df}} & \textstyle\bigwedge_{i=1}^{n} \bigwedge_{X \in S_1} (X \notin \mathsf{b}^-(A_i) \vee \bigvee_{Y \in S_2} \mathrm{co}_i^-(\overline{A}, \overline{B})(X) = Y) \\ & \wedge \textstyle\bigwedge_{Y \in S_2} \bigvee_{X \in S_1} \bigvee_{i=1}^{n} \mathrm{co}_i^-(\overline{A}, \overline{B})(X) = Y \end{aligned}$$

So (L.inverse) collects all correspondents of all variables bound by X; for each of these correspondents it must be known whether it is bound by Y or definitely bound by something else. Then we can determine $\lambda^{-1}(Y)$. The soundness of this rule is not obvious: is it really sufficient to look among the correspondents of $\lambda^{-1}(X)$ to compute $\lambda^{-1}(Y)$? The following proposition shows that it is.

Proposition 2 (Inverse lambda binding). *Suppose* $(\alpha_1, \ldots, \alpha_n) \sim (\alpha'_1, \ldots, \alpha'_n)$ *holds with correspondence functions* $c_1, \ldots, c_n$. *Then for all* $1 \leq k \leq n$ *and all* $\pi \in \mathsf{b}^-(\alpha_k)$,

$$\lambda^{-1}(c_k(\pi)) \subseteq \bigcup_{i=1}^{n} \{c_i(\pi') \mid \pi' \in \lambda^{-1}(\pi) \cap \mathsf{b}^-(\alpha_i)\}$$

Proof. Let $\omega \in \lambda^{-1}(c_k(\pi))$. The "no hanging binders" condition (hang) of Def. 5 is critical here: it enforces $\omega \in \bigcup_{i=1}^{n} \mathsf{b}^-(\alpha'_i)$. If $\omega \in \mathsf{b}^-(\alpha'_k)$, then there exists some $\pi' \in \mathsf{b}^-(\alpha_k)$ with $c_k(\pi') = \omega$. π' is var-labeled by Def. 3 and has a binder since λ is total. So we must have $\lambda(c_k(\pi')) = c_k(\lambda(\pi'))$ by condition (same.seg) of Def. 5. Now $\lambda(c_k(\pi')) = c_k(\pi)$ and c_k is a bijection, so $\pi' \in \lambda^{-1}(\pi)$. If, on the other hand, $\omega \notin \mathsf{b}^-(\alpha'_k)$ but $\omega \in \mathsf{b}^-(\alpha'_j)$, there is again a π' with $c_j(\pi') = \omega$, and $\lambda(c_j(\pi')) = c_k(\lambda(\pi'))$ by condition (diff.seg), so again $\pi' \in \lambda^{-1}(\pi)$.

The rules we have presented are part of a sound and complete semi-decision procedure for group parallelism constraints given in [3]. The omitted rules state additional properties of dominance constraints, ensure that correspondence functions are indeed bijective functions, and regulate the interaction between different correspondence functions.

(L.same.seg) $\lambda(U_1)=U_2 \wedge \bigwedge_{i=1}^{2} \mathrm{co}_k^-(\overline{A},\overline{B})(U_i)=V_i \rightarrow \lambda(V_1)=V_2$

(L.diff.seg) $\lambda(U_1)=U_2 \wedge \bigwedge_{i=1}^{2} \mathrm{co}_{k_i}^-(\overline{A},\overline{B})(U_i)=V_i \wedge U_2 \notin \mathrm{b}^-(A_{k_1}) \rightarrow \lambda(V_1)=V_2$

(L.outside) $\lambda(U)=Y \wedge \mathrm{co}_k^-(\overline{A},\overline{B})(U)=V \wedge Y \notin \mathrm{b}^-(\overline{A}) \rightarrow \lambda(V)=Y$

(L.hang) $\lambda(U_1)=U_2 \wedge \overline{A} \sim \overline{B} \wedge U_2 \in \mathrm{b}^-(\overline{A}) \rightarrow U_1 \in \mathrm{b}^-(\overline{A})$

(L.distr.1) $\lambda(U_1)=U_2 \wedge \overline{A} \sim \overline{B} \wedge U_1 \in \mathrm{b}^-(\overline{A}) \rightarrow \mathrm{distr}_{\overline{A}}(U_2)$

(L.distr.2) $\lambda(U_1)=U_2 \wedge \overline{A} \sim \overline{B} \wedge U_2 \in \mathrm{b}^-(\overline{A}) \rightarrow \mathrm{distr}_{\overline{A}}(U_1)$

(L.equal) $\lambda(X_1)=X_2 \wedge \bigwedge_{i=1}^{2} X_i=Y_i \rightarrow \lambda(Y_1)=Y_2$

(L.inverse) $\lambda^{-1}(X)=S_1 \wedge \mathrm{co}_k^-(\overline{A},\overline{B})(X)=Y \wedge \mathrm{co}^-(\overline{A},\overline{B})(S_1)=S_2 \cup S_3 \wedge$
$\bigwedge_{V \in S_2} \lambda(V)=Y \wedge \bigwedge_{V \in S_3} \lambda(V) \neq Y \rightarrow \lambda^{-1}(Y)=S_2$

Fig. 11. Lambda binding rules for group parallelism

$(C,B,A,A) \sim (C',B',A',A'')$
with $C = X/X_0$,
$C' = X'/X_0'$,
$B = X_t/X_1, X_2$,
$B' = X_0'/X_1', X_2'$,
$A = X_a/$,
$A' = X_1'/$, and $A'' = X_2'/$.
$\lambda^{-1}(X_\ell) = \{X_1, X_2\}$, $\lambda^{-1}(Y_1) = \{Z\}$.

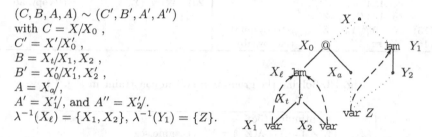

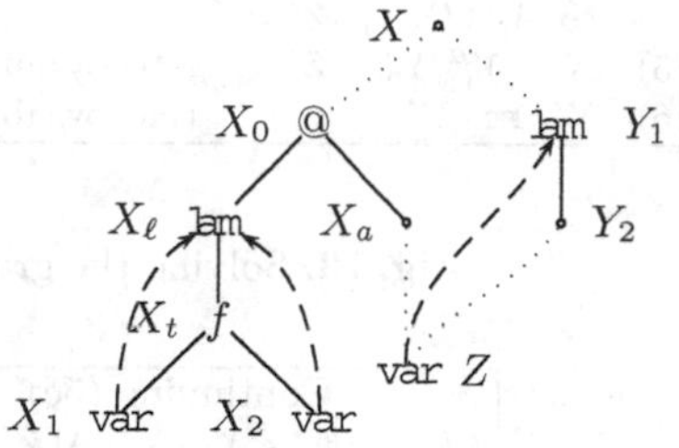

Fig. 12. A group parallelism constraint encoding a non-linear beta reduction step

Theorem 2. *There exists a saturation procedure GP which encompasses all instances of the rule schemata in Fig. 10 and 11, such that each rule of GP is correct, and each GP-solved form of a constraint φ is satisfiable* **(soundness)**, *and for every solution (τ,σ) of φ, GP computes a GP-solved form of φ of which (τ,σ) is a solution* **(completeness)**.

Proving that *GP*-solved forms are satisfiable can be done by constructing a model and variable assignment explicitly. One then has to check that all literals are indeed satisfied, which requires a tedious case distinction. Proving completeness is nontrivial as well, but can be done along the lines of [9]. It is largely independent of the particularities of the rule system we employ.

7 The Procedure in Action

We illustrate the procedure of the previous section by solving the constraint in Fig. 12. It contains a non-linear lambda redex at (C,B,A) (similarly to Fig. 5) and a lambda binder at Y_1 which can either belong to the context C or argument A. The group parallelism constraint $(C,B,A,A) \sim (C',B',A',A'')$ describes a beta-reduction step for the redex (C,B,A).

A record of the solving steps is given in Fig. 13 and 14. We only comment on the main steps. In step (4), we have $Y_1 \triangleleft^* Z \wedge X_0 \triangleleft^* Z$, and as trees do not branch

(1)	$Y_1 \neq X_0$	(D.lab.ineq)
(2)	$Y_1 \lhd^+ Y_2, X_0 \lhd^+ X_a$	(D.lab.dom)
(3)	$Y_1 \lhd^* Z, X_0 \lhd^* Z$	(D.dom.trans)
(4)	$X_0 \lhd^* Y_1 \;\vee\; Y_1 \lhd^* X_0$	(D.distr.notDisj)

(4a) $X_0 \lhd^* Y_1$: (5) $X_0{=}Y_1 \vee X_\ell \lhd^* Y_1 \vee X_a \lhd^* Y_1$ (D.distr.child)	(4b) $Y_1 \lhd^* X_0$: (7) $Y_1{=}X_0 \;\vee\; Y_2 \lhd^* X_0$ (D.distr.child)
(5a) $X_0{=}Y_1$: (5b) $X_\ell \lhd^* Y_1$: ... both lead to **false**	(7a) $Y_1{=}X_0$: (8) **false** (D.clash.ineq)
(5c) $X_a \lhd^* Y_1$: (9) $\mathrm{co}(A,A')(X_a){=}X_1'$ (P.init) (10) $\mathrm{co}(A,A')(Y_1){=}Y_1'$, $\mathrm{co}(A,A')(Y_2){=}Y_2'$, $\mathrm{co}(A,A')(Z){=}Z'$ (P.new) (11) $X_1' \lhd^* Y_1', Y_2' \lhd^* Z'$ (P.copy.dom) (12) Y_1':$\mathrm{lam}\,(Y_2')$ (P.copy.lab) (13) $\mathrm{co}(A,A'')(X_a){=}X_2'$ (P.init) (14) $\mathrm{co}(A,A'')(Y_1){=}Y_1''$, $\mathrm{co}(A,A'')(Y_2){=}Y_2''$, $\mathrm{co}(A,A'')(Z){=}Z''$ (P.new) (15) $X_2' \lhd^* Y_1'', Y_2'' \lhd^* Z''$ (P.copy.dom) (16) Y_1'':$\mathrm{lam}\,(Y_2'')$ (P.copy.lab)	(7b) $Y_2 \lhd^* X_0$: (17) $\mathrm{co}(C,C')(X){=}X'$, $\mathrm{co}(C,C')(X_0){=}X_0'$ (P.init) (18) $\mathrm{co}(C,C')(Y_1){=}Y_1'$, $\mathrm{co}(C,C')(Y_2){=}Y_2'$ (P.new) (19) $X' \lhd^* Y_1', Y_2' \lhd^* X_0'$ (P.copy.dom) (20) Y_1':$\mathrm{lam}\,(Y_2')$ (P.copy.lab)

Fig. 13. Solving the group parallelism constraint in Fig. 12

	Continuing (5c)	
(21)	$\lambda(Z') = Y_1'$, $\lambda(Z'') = Y_1''$ (L.same.seg)	
(22)	$Z{\notin}\mathrm{b}^-(B) \vee Z \in \mathrm{b}^-(B)$ (L.distr.2)	
(22a)	$Z{\notin}\mathrm{b}^-(B)$	(22b) $Z{\notin}\mathrm{b}^-(B)$
(23)	$Z{\notin}\mathrm{b}^-(C) \vee Z \in \mathrm{b}^-(C)$ (L.distr.2)	... **false**
(23a)	$Z{\notin}\mathrm{b}^-(C)$	(23b) $Z \in \mathrm{b}^-(C)$
(24)	$Y_1' {\neq} Y_1'' \vee Y_1' {=} Y_1''$ (P.distr.eq)	... **false**
(24a)	$Y_1' {\neq} Y_1''$	
(25)	$\lambda^{-1}(Z'') \neq Y_1'$	(24b) $Y_1' {=} Y_1''$
(26)	$\lambda^{-1}(Y_1'') = \{Z'\}$ (L.inverse)	... **false**
(27)	$\lambda^{-1}(Y_1'') = \{Z''\}$ (L.inverse)	

Fig. 14. Inverse Binding in case (5c)

upwards, one of Y_1, X_0 must dominate the other. This step effectively guesses whether Y_1, Y_2 are in C or in A. With choice (5c), we make two copies of Y_1 and Y_2 each. This is because A is parallel both to A' and A'': because X_ℓ binds two variables, the argument is copied twice. On the other hand, with (7b) Y_1 and Y_2 are only copied once: they belong to the context C, which is parallel only to C'.

In Fig. 14, we continue case (5c) of Fig. 13, applying the lambda binding rules. All steps from (22) on prepare the determination of $\lambda^{-1}(Y_1')$ and $\lambda^{-1}(Y_1'')$ in (25) and (26). We know $\lambda^{-1}(Y_1){=}\{Z\}$. Steps (22) and (23) determine $S_2 \cup S_3$ to be $\{Y_1', Y_1''\}$ for both (25) and (26). After (24) we know $\lambda(Z'){\neq}Y_1''$ and $\lambda(Z''){\neq}Y_1'$, so we have all we need to infer the correct λ^{-1} information in the last steps.

8 Conclusion and Future Work

We have introduced *beta reduction constraints* and have presented a semi-decision procedure for processing them, in three steps: First, we have extended CLLS by group parallelism constraints. Second, we have expressed β-reduction constraints in this extension of CLLS. Third, we have lifted a known semi-decision procedure for CLLS to also deal with group parallelism constraints. It is an open question to what extent beta reduction constraints can conversely express parallelism.

This gives us a framework for investigating the more general problem of *underspecified beta reduction* [4]: How can we string together several reduction steps, as described by beta reduction constraints, until we arrive at descriptions of normal forms? In this broader setting, we can investigate properties such as confluence and termination on the underspecified level. Another problem, motivated by the application, is to modify the saturation procedure to perform as few case distinctions as possible during underspecified beta reduction. Finally, it will be interesting to find an efficient implementation of this operation, possibly employing concepts such as sharing and constraint programming.

References

1. M. Abadi, L. Cardelli, P.-L. Curien, and J.-J. Lévy. Explicit substitutions. *Journal of Functional Programming*, 1(4):375–416, 1991.
2. R. Backofen, J. Rogers, and K. Vijay-Shanker. A first-order axiomatization of the theory of finite trees. *J. Logic, Language, and Information*, 4(1):5–39, 1995.
3. M. Bodirsky, K. Erk, A. Koller, and J. Niehren. Beta reduction constraints, 2001. Full version. `http://www.ps.uni-sb.de/Papers/abstracts/beta.html`.
4. M. Bodirsky, K. Erk, A. Koller, and J. Niehren. Underspecified beta reduction, 2001. Submitted. `http://www.ps.uni-sb.de/Papers/abstracts/usp-beta.html`.
5. D. Duchier and J. Niehren. Dominance constraints with set operators. In *First International Conference on Computational Logic (CL2000)*, LNCS, July 2000.
6. M. Egg, A. Koller, and J. Niehren. The Constraint Language for Lambda Structures. *Journal of Logic, Language, and Information*, 2001. To appear.
7. M. Egg, J. Niehren, P. Ruhrberg, and F. Xu. Constraints over lambda-structures in semantic underspecification. In *In Proceedings COLING/ACL'98*, 1998.
8. K. Erk, A. Koller, and J. Niehren. Processing underspecified semantic descriptions in CLLS. *Journal of Language & Computation*, 2001. To appear.
9. K. Erk and J. Niehren. Parallelism constraints. In *Int. Conference on Rewriting Techniques and Applications*, volume 1833 of *LNCS*, pages 110–126, 2000.
10. G. P. Huet. A unification algorithm for typed λ-calculus. *Theoretical Computer Science*, 1:27–57, 1975.
11. J. Lamping. An algorithm for optimal lambda calculus reduction. In *ACM Symposium on Principles of Programming Languages*, pages 16–30, 1990.
12. J. Lévy. Linear second order unification. In *7th Int. Conference on Rewriting Techniques and Applications*, volume 1103 of *LNCS*, pages 332–346, 1996.
13. M. P. Marcus, D. Hindle, and M. M. Fleck. D-theory: Talking about talking about trees. In *Proceedings of the 21st ACL*, pages 129–136, 1983.
14. M. Pinkal. Radical underspecification. In *Proc. 10th Amsterdam Colloquium*, 1996.

15. U. Reyle. Dealing with ambiguities by underspecification: construction, representation, and deduction. *Journal of Semantics*, 10:123–179, 1993.
16. M. Schmidt-Schauß and K. Schulz. Solvability of context equations with two context variables is decidable. In *Int. Conf. on Automated Deduction*, LNCS, 1999.
17. K. van Deemter and S. Peters. *Semantic Ambiguity and Underspecification.* CSLI Press, Stanford, 1996.

From Higher-Order to First-Order Rewriting (Extended Abstract)

Eduardo Bonelli[1,2], Delia Kesner[2], and Alejandro Ríos[1]

[1] Departamento de Computación - Facultad de Ciencias Exactas y Naturales, Universidad de Buenos Aires, Argentina. {ebonelli@dc.uba.ar,rios@dc.uba.ar}
[2] LRI (CNRS URA 410) - Bât 490, Université de Paris-Sud, 91405 Orsay Cedex, France. kesner@lri.fr

Abstract. We show how higher-order rewriting may be encoded into first-order rewriting modulo an equational theory $\mathcal{E}$. We obtain a characterization of the class of higher-order rewriting systems which can be encoded by first-order rewriting modulo an empty theory (that is, $\mathcal{E} = \emptyset$). This class includes of course the λ-calculus. Our technique does not rely on a particular substitution calculus but on a set of abstract properties to be verified by the substitution calculus used in the translation.

1 Introduction

Higher-order substitution is a complex operation that consists in the replacement of variables by terms in the context of languages having variable bindings. These bound variables can be annotated by de Bruijn indices so that the renaming operation (α-conversion) which is necessary to carry out higher-order substitution can be avoided. However, substitution is still a complicated notion, which cannot be expressed by simple replacement (a.k.a. grafting) of variables as is done in first-order theories. To solve this problem, many researchers became interested in the formalization of higher-order substitution by *explicit substitutions*, so that higher-order systems/formalisms could be expressible in first-order systems/formalisms: the notion of variable binding is dropped because substitution becomes replacement. A well-known example of the combination of de Bruijn indices and explicit substitutions is the formulation of different first-order calculi for the λ-calculus [1,4,17,15,6,22], which is the paradigmatic example of a higher-order (term) rewriting system. Other examples are the translations of higher-order unification to first-order unification modulo [11], higher-order logic to first-order logic modulo [12], higher-order theorem proving to first-order theorem proving modulo [9], etc.

All these translations have a theoretical interest because the expressive power of higher and first-order formalisms is put in evidence, but another practical issue arises, that of the possibility of transferring results developed in the first-order framework to the higher-order one.

The goal of this paper is to give a translation of higher-order rewrite systems (*HORS*) to a first-order formalism. As a consequence, properties and techniques

A. Middeldorp (Ed.): RTA 2001, LNCS 2051, pp. 47–62, 2001.

developed for first-order rewriting could be exported to the higher-order formalism. For example, techniques concerning confluence, termination, completion, evaluation strategies, etc. This is very interesting since, on one hand it is still not clear how to transfer techniques such as dependency pairs [2], semantic labelling [26] or completion [3] to the higher-order framework, on the other hand, termination techniques such as RPO for higher-order systems [14] turn out to be much more complicated than their respective first-order versions [7,16].

The main difficulty encountered in our translation can be intuitively explained by the fact that in higher-order rewriting a metavariable occurring on the right-hand side in a higher-order rewrite rule may occur in a different *binding context* on the left-hand side: for example, in the usual presentation of the extensional rule for functional types in λ-calculus (η):$\lambda\alpha.app(X,\alpha)\longrightarrow X$, the occurrence of X on the right-hand side of the rule, which does not appear in any binding context, is related to the occurrence of X on the left-hand side, which appears inside a binding context. The immediate consequence of this fact is that the first-order translation of the rule (η) cannot be defined as the naive translation taking both sides of the rule independently. This would give the first-order rule $\lambda(app(X,\underline{1}))\longrightarrow X$, which does not reflect the intended semantics and hence the translation would be incorrect.

As mentioned before, the need for (α)-conversion immediately disappears when de Bruijn notation is considered. Following the example recently introduced, one can express the (η)-rule in a higher-order de Bruijn setting, such as for example the $SERS_{DB}$ (Simplified Expression Reduction Systems) formalism [5], by the rule (η_{dB}): $\lambda(app(X_\alpha,\underline{1}))\longrightarrow X_\epsilon$. The notation used to translate the metavariable X into the de Bruijn formalism enforces the fact that the occurrence of X on the right-hand side of the rule (η) does not appear in a binding context, so it is translated as X_ϵ where ϵ represents an empty binding path, while the occurrence of X on the left-hand side of the rule appears inside a binding context, so it is translated as X_α, where α represents a binding path of length 1 surrounding the metavariable X.

Now, the term $\lambda(app(\underline{3},\underline{1}))$ reduces to $\underline{2}$ via the (η_{dB}) rule. In an explicit substitution setting, we have the alternative formulation:

$$(\eta_{fo}) : \lambda(app(X[\uparrow],\underline{1}))\longrightarrow X$$

However, in order for the metaterm $X[\uparrow]$ to match the subterm $\underline{3}$ first-order matching no longer suffices: we need $\mathcal{E}$-matching, that is, matching modulo an equational theory $\mathcal{E}$. For an appropriate substitution calculus $\mathcal{E}$ we would need to solve the equation $\underline{3} \stackrel{?}{=}_{\mathcal{E}} X[\uparrow]$. Equivalently, we could make use of the theory of conditional rewriting: $\lambda(app(Y,\underline{1}))\longrightarrow X$ where $Y =_{\mathcal{E}} X[\uparrow]$. Another less evident example is given by a commutation rule as for example

$$(C) : Implies(\exists\alpha.\forall\beta.X, \forall\beta.\exists\alpha.X)\longrightarrow true$$

which expresses that the formula appearing as the first argument of the *Implies* function symbol implies the one in the second argument. The naive translation to first-order $Implies(\exists(\forall(X)),\forall(\exists(X)))\longrightarrow true$ is evidently not correct, so that we take its translation in the de Bruijn higher-order formalism $SERS_{DB}$ and then translate it to first-order via the conversion presented in this work obtaining:

$$(C_{fo}) : Implies(\exists(\forall(X)), \forall(\exists(X[\underline{2} \cdot \underline{1} \cdot id]))) \longrightarrow true$$

Now, the rule (C_{fo}) has exactly the same semantics as the original higher-order rule (C). This difficulty does not seem to appear in other problems dealing with translations from higher-order to first-order recently mentioned.

In this work we shall see how higher-order rewriting may be systematically reduced to first-order rewriting modulo an equational theory $\mathcal{E}$. To do this, we choose to work with Expression Reduction Systems [18], and in particular, with de Bruijn based $SERS_{DB}$ as defined in [5] which facilitates the translation of higher-order systems to first-order ones. However, we claim that the same translation could be applied to other higher-order rewriting formalisms existing in the literature. We obtain a characterization of the class of $SERS_{DB}$ (including λ-calculus) for which a translation to a full ($\mathcal{E} = \emptyset$) first-order rewrite system exists. Thus the result mentioned above on the λ-calculus becomes a particular case of our work.

To the best of our knowledge there is just one formalism, called *XRS* [24], which studies higher-order rewrite formalisms based on de Bruijn index notation and explicit substitutions. The formalism *XRS*, which is a first-order formalism, is presented as a generalization of the first-order $\sigma_{\Uparrow}$-calculus [13] to higher-order rewriting and not as a first-order formulation of higher-order rewriting. As a consequence, many well-known higher-order rewriting systems cannot be expressed in such a formalism [5]. Not only do we provide a first-order presentation of higher-order rewriting, but we do not attach to the translation any particular substitution calculus. Instead, we have chosen to work with an abstract formulation of substitution calculi, as done for example in [17] to deal with confluence proofs of λ-calculi with explicit substitutions. As a consequence, the method we propose can be put to work in the presence of different calculi of explicit substitution such as σ [1], $\sigma_{\Uparrow}$ [13], υ [4], f [17], d [17], s [15], χ [20].

The paper is organized as follows. Section 2 recalls the formalism of higher-order rewriting with de Bruijn indices defined in [5], and Section 3 defines a first-order syntax which will be used as target calculus in the conversion procedure given in Section 4. Properties of the conversion procedure are studied in Section 4.1: the conversion is a translation from higher-order rewriting to first-order rewriting modulo, the translation is conservative and finally we give the syntactical criterion to be used in order to decide if a given higher-order system can be translated into a full first-order one (a first-order system modulo an empty theory). We conclude in Section 5.

Due to lack of space proofs are omitted and only the main results are stated. For further details the reader is referred to the full version accessible by ftp at `ftp://ftp.lri.fr/LRI/articles/kesner/ho-to-fo.ps.gz`.

2 The Higher-Order Framework

We briefly recall here the de Bruijn indices based higher-order rewrite formalism called Simplified Expression Reduction Systems with de Bruijn Indices ($SERS_{DB}$) which was introduced in [5]. In full precision we shall work with $SERS_{DB}$ without i(ndex)-metavariables (c.f. full version for details).

Definition 1 (Labels). *A* label *is a finite sequence of symbols of an alphabet. We shall use $k, l, l_i, \ldots$ to denote arbitrary labels and ϵ for the empty label. If α is a symbol and l is a label then $\alpha \in l$ means that the symbol α* appears in *the label l. Other notations are $|l|$ for the* length *of l and $\mathtt{at}(l, n)$ for the n-th element of l assuming $n \leq |l|$. Also, if α occurs (at least once) in l then $\mathtt{pos}(\alpha, l)$ denotes the* position of the first occurrence *of α in l. A* simple label *is a label without repeated symbols.*

Definition 2 (de Bruijn signature). *Consider the denumerable and disjoint infinite sets:*

- $\{\alpha_1, \alpha_2, \alpha_3, \ldots\}$ *a set of symbols called* binder indicators, *denoted $\alpha, \beta, \ldots$,*
- $\{X_l^1, X_l^2, X_l^3, \ldots\}$ *a set of* t-metavariables *(t for term), where l ranges over the set of labels built over binder indicators, denoted $X_l, Y_l, Z_l, \ldots$,*
- $\{f_1, f_2, f_3, \ldots\}$ *a set of* function symbols *equipped with a fixed (possibly zero) arity, denoted $f, g, h, \ldots$,*
- $\{\lambda_1, \lambda_2, \lambda_3, \ldots\}$ *a set of* binder symbols *equipped with a fixed (non-zero) arity, denoted $\lambda, \mu, \nu, \xi, \ldots$.*

Definition 3 (de Bruijn pre-metaterms). *The set of* de Bruijn pre-metaterms, *denoted $\mathcal{PMT}_{db}$, is defined by the following two-sorted grammar:*

$$\begin{array}{ll} \textit{metaindices} & I ::= 1 \mid \mathtt{S}(I) \\ \textit{pre-metaterms} & A ::= I \mid X_l \mid f(A, \ldots, A) \mid \xi(A, \ldots, A) \mid A[\![A]\!] \end{array}$$

We shall use $A, B, A_i, \ldots$ to denote de Bruijn pre-metaterms. The symbol $.[\![.]\!]$ is called *de Bruijn metasubstitution operator.* We assume the convention that $\mathtt{S}^0(1) = 1$ and $\mathtt{S}^{j+1}(n) = \mathtt{S}(\mathtt{S}^j(n))$. As usually done for indices, we shall abbreviate $\mathtt{S}^{j-1}(1)$ as $\underline{j}$.

We use $MVar(A)$ to denote the set of metavariables of the de Bruijn pre-metaterm A. An *X-based* t-metavariable is a t-metavariable of the form X_l for some label l, we say in that case that X is the *name of* X_l. We shall use $NMVar(A)$ to denote the set of names of metavariables in A. In order to say that a t-metavariable X_l occurs in a pre-metaterm A we write $X_l \in A$.

We shall need the notion of *metaterm* (well-formed pre-metaterm). The first motivation is to guarantee that labels of t-metavariables are correct w.r.t the context in which they appear, the second one is to ensure that indices like $\mathtt{S}^i(1)$ correspond to bound variables. Indeed, pre-metaterms like $\xi(X_{\alpha\beta})$ and $\xi(\xi(\underline{4}))$ shall not make sense for us, and hence shall not be considered well-formed. Well-formed pre-metaterms shall be used to describe rewrite rules.

Definition 4 (de Bruijn metaterms). *A pre-metaterm $A \in \mathcal{PMT}_{db}$ is said to be a* metaterm *iff the predicate $\mathcal{WF}(A)$ holds where $\mathcal{WF}(A) =_{def} \mathcal{WF}_{\epsilon}(A)$ and $\mathcal{WF}_l(A)$ is defined as follows:*

- $\mathcal{WF}_l(\mathtt{S}^j(1))$ *iff* $j+1 \leq |l|$
- $\mathcal{WF}_l(X_k)$ *iff* $l = k$ *and* l *is a simple label*
- $\mathcal{WF}_l(f(A_1, \ldots, A_n))$ *iff for all* $1 \leq i \leq n$ *we have* $\mathcal{WF}_l(A_i)$
- $\mathcal{WF}_l(\xi(A_1, \ldots, A_n))$ *iff there exists* $\alpha \notin l$ *s.t. for all* $1 \leq i \leq n$, $\mathcal{WF}_{\alpha l}(A_i)$
- $\mathcal{WF}_l(A_1[\![A_2]\!])$ *iff* $\mathcal{WF}_l(A_2)$ *and there exists* $\alpha \notin l$ *such that* $\mathcal{WF}_{\alpha l}(A_1)$

Example 1. Pre-metaterms $\xi(X_\alpha, \lambda(Y_{\beta\alpha}, \underline{2}))$ and $g(\lambda(\xi(h)))$ are metaterms, while $f(\underline{1}, \xi(X_\beta))$, $\lambda(\xi(X_{\alpha\alpha}))$ and $\xi(X_\alpha, X_\beta)$ (with $\alpha \neq \beta$) are not.

Definition 5 (de Bruijn terms and de Bruijn contexts). *The set of* de Bruijn terms, *denoted $\mathcal{T}_{db}$, and the set of de Bruijn contexts are defined by:*

de Bruijn indices	n	$::= 1 \mid \mathtt{S}(n)$
de Bruijn terms	a	$::= n \mid f(a, \ldots, a) \mid \xi(a, \ldots, a)$
de Bruijn contexts	E	$::= \Box \mid f(a, \ldots, E, \ldots, a) \mid \xi(a, \ldots, E, \ldots, a)$

We use $a, b, a_i, b_i, \ldots$ for de Bruijn terms and $E, F, \ldots$ for de Bruijn contexts. The *binder path number* of a context is the number of binders between the $\Box$ and the root. For example the binder path of $E = f(\underline{3}, \xi(\underline{1}, \lambda(\underline{2}, \Box, \underline{3})), \underline{2})$ is 2.

Remark that de Bruijn terms are also de Bruijn pre-metaterms, that is, $\mathcal{T}_{db} \subset \mathcal{PMT}_{db}$, although note that some de Bruijn terms may not be de Bruijn metaterms, i.e. may not be well-formed de Bruijn pre-metaterms, e.g. $\xi(\xi(\underline{4}))$. The result of substituting a term b for the index $n \geq 1$ in a term a is denoted $a\{\!\{n \leftarrow b\}\!\}$. This is defined as usual [15]. We now recall the definition of rewrite rules, valuations, their validity, and reduction in $SERS_{DB}$.

Definition 6 (de Bruijn rewrite rule). *A* de Bruijn rewrite rule *is a pair of de Bruijn metaterms (L, R) (also written $L \longrightarrow R$) such that the first symbol in L is a function symbol or a binder symbol, $NMVar(R) \subseteq NMVar(L)$, and the metasubstitution operator $.[\![.]\!]$ does not occur in L.*

Definition 7 (de Bruijn valuation). *A* de Bruijn valuation $\overline{\kappa}$ *is a (partial) function from t-metavariables to de Bruijn terms. A valuation $\overline{\kappa}$ determines in a unique way a function κ (also called valuation) on pre-metaterms as follows:*

$$\kappa \underline{n} =_{def} \underline{n} \qquad \kappa X_l =_{def} \overline{\kappa} X_l$$

$$\begin{aligned} \kappa f(A_1, \ldots, A_n) &=_{def} f(\kappa A_1, \ldots, \kappa A_n) \\ \kappa \xi(A_1, \ldots, A_n) &=_{def} \xi(\kappa A_1, \ldots, \kappa A_n) \\ \kappa(A_1[\![A_2]\!]) &=_{def} \kappa(A_1)\{\!\{1 \leftarrow \kappa A_2\}\!\} \end{aligned}$$

We write $Dom(\kappa)$ for the set $\{X_l \mid \kappa X_l \text{ is defined}\}$, called the *domain* of κ.

We now introduce the notion of *value function* which is used to give semantics to metavariables with labels in the $SERS_{DB}$ formalism. The goal pursued by the labels of metavariables is that of incorporating "context" information as a defining part of a metavariable. A typical example is given by a rule like $\mathcal{C} : \xi(\xi(X_{\beta\alpha}))\longrightarrow \xi(\xi(X_{\alpha\beta}))$ where the X-occurrence on the *RHS* of the rule denotes a permutation of the binding context of the X-occurrence on the *LHS*.

As a consequence, we must verify that the terms substituted for every occurrence of a fixed metavariable coincide "modulo" their corresponding context. Dealing with such notion of "coherence" of substitutions in a de Bruijn formalism is also present in other formalisms but in a more restricted form. Thus for example a pre-cooking function[1] is used in [9] to avoid variable capture in the higher-order unification procedure. In *XRS* [24] the notions of binding arity and pseudo-binding arity are introduced to take into account the binder path number of rewrite rules. Our notion of "coherence" is implemented with *valid valuations* (cf. Definition 9) and it turns out to be more general than the solutions proposed in [9] and [24].

Definition 8 (Value function). *Let a be a de Bruijn term and l be a label of binder indicators. We define the value function $Value(l,a)$ as $Value^0(l,a)$ where*

$$Value^i(l,\underline{n}) =_{def} \begin{cases} \underline{n} & \text{if } n \le i \\ \mathtt{at}(l, n-i) & \text{if } 0 < n-i \le |l| \\ x_{n-i-|l|} & \text{if } n-i > |l| \end{cases}$$

$$Value^i(l, f(a_1,\dots,a_n)) =_{def} f(Value^i(l,a_1),\dots Value^i(l,a_n))$$
$$Value^i(l, \xi(a_1,\dots,a_n)) =_{def} \xi(Value^{i+1}(l,a_1),\dots,Value^{i+1}(l,a_n))$$

It is worth noting that $Value^i(l,\underline{n})$ may give three different kinds of results. This is just a technical resource. Indeed, $Value(\alpha\beta,\xi(f(\underline{3},\underline{1}))) = \xi(f(\beta,\underline{1})) = Value(\beta\alpha,\xi(f(\underline{2},\underline{1})))$ and $Value(\epsilon, f(\xi(\underline{1}),\lambda(\underline{2}))) = f(\xi(\underline{1}),\lambda(x_1)) \neq f(\xi(\underline{1}),\lambda(\alpha)) = Value(\alpha, f(\xi(\underline{1}),\lambda(\underline{2})))$. Thus the function $Value(l,a)$ interprets the de Bruijn term a in an l-context: bound indices are left untouched, free indices referring to the l-context are replaced by the corresponding binder indicator and the remaining free indices are replaced by adequate variable names.

Definition 9 (Valid de Bruijn valuation). *A de Bruijn valuation κ is said to be* valid *if for every pair of t-metavariables X_l and $X_{l'}$ in $Dom(\kappa)$ we have $Value(l,\kappa X_l) = Value(l',\kappa X_{l'})$. Likewise, we say that a de Bruijn valuation κ is* valid for a rewrite rule (L,R) *if for every pair of t-metavariables X_l and $X_{l'}$ in (L,R) we have $Value(l,\kappa X_l) = Value(l',\kappa X_{l'})$.*

Example 2. Let us consider the de Bruijn rule $\mathcal{C} : \xi(\xi(X_{\beta\alpha}))\longrightarrow \xi(\xi(X_{\alpha\beta}))$. We have that $\kappa = \{X_{\beta\alpha}/\underline{2}, X_{\alpha\beta}/\underline{1}\}$ is valid since $Value(\beta\alpha,\underline{2}) = \alpha = Value(\alpha\beta,\underline{1})$.

[1] The pre-cooking function translates a de Bruijn λ-term with metavariables into a $\lambda\sigma$-term by suffixing each metavariable X with as many explicit shift operators as the binder path number of the context obtained by replacing X by $\Box$. This avoids variable capture when the higher-order unification procedure finds solutions for the t-metavariables.

As already mentioned the η-contraction rule $\lambda x.app(X,x) \longrightarrow X$ can be expressed in the $SERS_{DB}$ formalism as the rule (η_{dB}) $\lambda(app(X_\alpha,\underline{1})) \longrightarrow X_\epsilon$. Our formalism, like other *HORS* in the literature, allows us to use rules like η_{dB} because valid valuations will test for coherence of values.

Definition 10 ($SERS_{DB}$-rewriting). *Let $\mathcal{R}$ be a set of de Bruijn rewrite rules and a, b be de Bruijn terms. We say that a* $\mathcal{R}$-reduces or rewrites *to b, written $a \longrightarrow_{\mathcal{R}} b$, iff there is a de Bruijn rule $(L,R) \in \mathcal{R}$, a de Bruijn valuation κ valid for (L,R), and a de Bruijn context E such that $a = E[\kappa L]$ and $b = E[\kappa R]$.*

Thus, the term $\lambda(app(\lambda(app(\underline{1},\underline{3})),\underline{1}))$ rewrites by the η_{dB} rule to $\lambda(app(\underline{1},\underline{2}))$, using the (valid) valuation
$\kappa = \{X_\alpha/\lambda(app(\underline{1},\underline{3})), X_\epsilon/\lambda(app(\underline{1},\underline{2}))\}$.

3 The First-Order Framework

In this section we introduce the first-order formalism called Explicit Expression Reduction Systems (*ExERS*) used to translate higher-order rewriting systems based on de Bruijn indices into first-order ones.

Definition 11. *A* substitution declaration *is a (possibly empty) word over the alphabet $\{\mathtt{T},\mathtt{S}\}$. The symbol $\mathtt{T}$ is used to denote terms and $\mathtt{S}$ to denote substitutions. A* substitution signature *is a set Γ_s of* substitution symbols *equipped with an arity n and a substitution declaration of length n. We use $\sigma : (w)$ where $w \in \{\mathtt{T},\mathtt{S}\}^n$ if the substitution symbol σ has arity n and substitution declaration w. We use ϵ to denote the empty word.*

Definition 12 (*ExERS* term algebra). *An ExERS signature is a set $\Gamma = \Gamma_f \cup \Gamma_b \cup \Gamma_s$ where $\Gamma_f = \{f_1,\ldots,f_n\}$ is a set of function symbols, $\Gamma_b = \{\lambda_1,\ldots,\lambda_n\}$ is a set of binder symbols, Γ_s a substitution signature and Γ_f, Γ_b and Γ_s are pairwise disjoint. Both binder and function symbols come equipped with an arity. Given a set of (term) variables $\mathcal{V} = \{X_1, X_2, \ldots\}$, the term algebra of an ExERS of signature Γ generated by $\mathcal{V}$, is denoted by $\mathcal{T}$ and contains all the objects (denoted by letters o and p) generated by the following grammar:*

`indices`	$n ::= 1 \mid \mathtt{S}(n)$
`terms (T)`	$a ::= X \mid n \mid a[s] \mid f(a_1,\ldots,a_n) \mid \xi(a_1,\ldots,a_n)$
`substitutions (S)`	$s ::= \sigma(o_1,\ldots,o_n)$

where X ranges over $\mathcal{V}$, f over Γ_f, ξ over Γ_b, and σ over Γ_s. The arguments of σ are assumed to respect the sorts prescribed in its substitution declaration and function and binder symbols are assumed to respect their arities.

Letters $a, b, c, \ldots$ and $s, s_i, \ldots$ are used for terms and substitutions, respectively. The $.[.]$ operator is called the *substitution operator*. Binder symbols and substitution operators are considered as having binding power. We shall use $a[s]^n$

to abbreviate $a[s]\ldots[s]$ (n-times). Terms without occurrences of the substitution operator (resp. objects in $\mathcal{V}$) are called *pure* (resp. *ground*) terms. A *context* is a ground term with one (and only one) occurrence of a distinguished term variable called a "hole" (and denoted $\Box$). Letters $E, E_i, \ldots$ are used for contexts. The notion of *binder path number* is defined for pure contexts exactly as in the case of de Bruijn contexts.

The formalism of *ExERS* that we are going to use in order to encode higher-order rewriting consists of two sets of rewrite rules:

1. A set of *proper rewrite rules* governing the behaviour of the function and binder symbols in the signature.
2. A set of substitution rules, called the *substitution calculus* governing the behaviour of the substitution symbols in the signature, and used for propagating and performing/eliminating term substitutions.

Let us define these two concepts formally.

Definition 13 (Substitution macros). *Let Γ_s be a substitution signature. The following symbols not included in Γ_s are called* substitution macros*: cons : (TS), lift : (S), id : (ϵ) and shift^j : (ϵ) for $j \geq 1$. We shall abbreviate shift^1 by shift. Also, if $j \geq 0$ then lift$^j(s)$ stands for s if $j = 0$ and for lift(lift$^{j-1}(s)$) otherwise. Furthermore, cons$(a_1, \ldots, a_i, s)$ stands for cons$(a_1, \ldots$ cons$(a_i, s))$.*

Definition 14 (Term rewrite and equational systems). *Let Γ be an ExERS signature. An equation is a pair of terms $L \doteq R$ over Γ such that L and R have the same sort and a* term rewrite rule *is a pair of terms (L, R) over Γ such that (1) L and R have the same sort, (2) the head symbol of the LHS of the rule is a function or a binder symbol, and (3) the set of variables of the LHS includes those of the RHS. An equational (resp. term rewrite) system is a set of equations (resp. term rewrite rules).*

Definition 15 (Substitution calculus). *A* substitution calculus *over an ExERS* signature *Γ consists of a set $\mathcal{W}$ of rewrite rules, and an interpretation of each substitution macro as some combination of substitution symbols from Γ_s of corresponding signature. Definition 16 shall require certain properties for these interpretations to be considered meaningful.*

An example of a substitution calculus is σ [1] with *cons*$(t, s) =_{def} t \cdot s$, *lift*$(s) =_{def} \underline{1} \cdot (s \circ \uparrow)$, *id* $=_{def}$ *id* and *shift*$^j =_{def} \uparrow \circ \ldots (\uparrow \circ \uparrow)$, where $\uparrow$ appears j times.

Definition 16 (Basic substitution calculus). *A substitution calculus $\mathcal{W}$ over Γ is said to be* basic *if the following conditions are satisfied:*

1. *$\mathcal{W}$ is complete (strongly normalizing and confluent) over the ground terms in $\mathcal{T}$. We use $\mathcal{W}(a)$ to indicate the unique $\mathcal{W}$-normal form of a.*
2. *$\mathcal{W}$-normal forms of ground terms are pure terms.*

3. *For each* $f \in \Gamma_f$ *and* $\xi \in \Gamma_b$: $\mathcal{W}(f(a_1,\ldots,a_n)) = f(\mathcal{W}(a_1),\ldots,\mathcal{W}(a_n))$ *and* $\mathcal{W}(\xi(a_1,\ldots,a_n)) = \xi(\mathcal{W}(a_1),\ldots,\mathcal{W}(a_n))$.
4. *Rules for propagating substitutions over function symbols and binders are contained in* $\mathcal{W}$, *for each* $f \in \Gamma_f$ *and* $\xi \in \Gamma_b$:

$$\begin{array}{lll} (func_f) & f(X^1,\ldots,X^n)[s] & \longrightarrow f(X^1[s],\ldots,X^n[s]) \\ (bind_\xi) & \xi(X^1,\ldots,X^n)[s] & \longrightarrow \xi(X^1[lift(s)],\ldots,X^n[lift(s)]) \end{array}$$

5. *For every substitution* s, $\underline{1}[lift(s)] =_{\mathcal{W}} \underline{1}$.
6. *For every substitution* s *and every* $m \geq 0$, $\underline{m+1}[lift(s)] =_{\mathcal{W}} \underline{m}[s][shift]$.
7. *For every term* a *and substitution* s *we have* $\underline{1}[cons(a,s)] =_{\mathcal{W}} a$.
8. *For every term* a, *substitution* s, $m \geq 0$ *we have* $\underline{m+1}[cons(a,s)] =_{\mathcal{W}} \underline{m}[s]$.
9. *For every* $m, j \geq 1$ *we have* $\underline{m}[shift^j] =_{\mathcal{W}} \underline{m+j}$.
10. *For every term* a *we have* $a[id] =_{\mathcal{W}} a$.

Example 3. The σ [1], $\sigma_{\Uparrow}$ [13] and ϕ [22] calculi are basic substitution calculi where the set of function and binder symbols are $\{app\}$ and $\{\lambda\}$, respectively.

The reader may have noted that the macro-based presentation of substitution calculi makes use of parallel substitutions (since $cons(.,.)$ has substitution declaration TS). Nevertheless, the results presented in this work may be achieved via a macro-based presentation using a simpler set of substitutions (such as for example the one used in [17]), where $scons(.)$ has substitution declaration T and the macro $shift^i$ is only defined for $i = 1$. Indeed, the expression $b[cons(a_1,\ldots,a_n,shift^j)]$ could be denoted by the expression

$$b[lift^n(shift)]^j[scons(a_1[shift]^{n-1})]\ldots[scons(a_n)]$$

Definition 17 (*ExERS* **and** *FExERS*). *Let* Γ *be an ExERS signature,* $\mathcal{W}$ *a basic substitution calculus over* Γ *and* $\mathcal{R}$ *a set of term rewrite rules. If each rule of* $\mathcal{R}$ *has sort* T *then* $\mathcal{R}_{\mathcal{W}} =_{def} (\Gamma,\mathcal{R},\mathcal{W})$ *is called an Explicit Expression Reduction System (ExERS). If, in addition, the LHS of each rule in* $\mathcal{R}$ *contains no occurrences of the substitution operator* $.[.]$ *then* $\mathcal{R}_{\mathcal{W}}$ *is called a Fully Explicit Expression Reduction System (FExERS).*

Since reduction in $SERS_{DB}$ only takes place on terms, and first-order term rewrite systems will be used to simulate higher-order reduction, all the rules of a term rewrite system $\mathcal{R}$ are assumed to have sort T. However, rewrite rules of $\mathcal{W}$ may have any sort.

Example 4. Consider the signature Γ formed by $\Gamma_f = \{app\}$, $\Gamma_b = \{\lambda\}$ and Γ_s any substitution signature. Let $\mathcal{W}$ be a basic substitution calculus over Γ. Then for $\mathcal{R} : app(\lambda(X), Y) \longrightarrow_{\beta db} X[cons(Y, id)]$ we have that $\mathcal{R}_{\mathcal{W}}$ is an *FExERS*, and for $\mathcal{R}' : \mathcal{R} \cup \{\lambda(app(X[shift], \underline{1})) \longrightarrow_{\eta db} X\}$, $\mathcal{R}'_{\mathcal{W}}$ is an *ExERS*.

Reduction in an *ExERS* $\mathcal{R}_{\mathcal{W}}$ is first-order reduction in $\mathcal{R}$ modulo $\mathcal{W}$-equality. In contrast, reduction in a *FExERS* $\mathcal{R}_{\mathcal{W}}$ is just first-order reduction in $\mathcal{R} \cup \mathcal{W}$. Before defining these notions more precisely we recall the definition of assignment.

Definition 18 (Assignment (a.k.a. grafting)). *Let $\overline{\rho}$ be a (partial) function mapping variables in $\mathcal{V}$ to ground terms. We define an* assignment *ρ as the unique homeomorphic extension of $\overline{\rho}$ over the set $\mathcal{T}$.*

Definition 19 (Reduction and Equality). *Let o and p be two ground terms of sort* T *or* S. *Given a rewrite system $\mathcal{R}$, we say that o rewrites to p in one step, denoted $o \longrightarrow_{\mathcal{R}} p$, iff $o = E[\rho L]$ and $p = E[\rho R]$ for some assignment ρ, some context E and some rewrite rule (L, R) in $\mathcal{R}$. We shall use $\stackrel{*}{\longrightarrow}_{\mathcal{R}}$ to denote the reflexive transitive closure of the one-step rewrite relation.*

Given an equational system $\mathcal{E}$, we say that o equals p modulo $\mathcal{E}$ in one step, denoted $o =^1_{\mathcal{E}} p$, iff $o = E[\rho L]$ and $p = E[\rho R]$ for some assignment ρ, some context E and some equation $L \doteq R$ in $\mathcal{E}$. We use $=_{\mathcal{E}}$ to denote the reflexive symmetric transitive closure of $=^1_{\mathcal{E}}$, and say that o equals p modulo $\mathcal{E}$ if $o =_{\mathcal{E}} p$.

Definition 20 (*ExERS* and *FExERS*-rewriting). *Let $\mathcal{R}_{\mathcal{W}}$ be an ExERS, $\mathcal{R}'_{\mathcal{W}}$ a FExERS and o, p ground terms of sort* S *or* T. *We say that o $\mathcal{R}_{\mathcal{W}}$*-reduces or rewrites to *p, written $o \longrightarrow_{\mathcal{R}_{\mathcal{W}}} p$, iff $o \longrightarrow_{\mathcal{R}/\mathcal{W}} p$ (i.e. $o =_{\mathcal{W}} o' \longrightarrow_{\mathcal{R}} p' =_{\mathcal{W}} p$); and o $\mathcal{R}'_{\mathcal{W}}$*-reduces to *$p$, iff $o \longrightarrow_{\mathcal{R}' \cup \mathcal{W}} p$.*

3.1 Properties of Basic Substitution Calculi

This subsection takes a look at properties enjoyed by basic substitution calculi and introduces a condition called the *Scheme* [17]. Basic substitution calculi satisfying the scheme ease inductive reasoning when proving properties over them without compromising the genericity achieved by the macro-based presentation.

Lemma 1 (Behavior of Substitutions in Basic Substitution Calculi). *Let $\mathcal{W}$ be a basic substitution calculus and $m \geq 1$.*

1. *For all $n \geq 0$ and s in* S: $\underline{m}[\mathit{lift}^n(s)] =_{\mathcal{W}} \begin{cases} \underline{m-n}[s][\mathit{shift}]^n & \text{if } m > n \\ \underline{m} & \text{if } m \leq n \end{cases}$
2. *For all $n \geq m \geq 1$ and all terms $a_1, \ldots, a_n$: $\underline{m}[\mathit{cons}(a_1, \ldots, a_n, s)] =_{\mathcal{W}} a_m$*
3. *For all pure terms a, b and $m \geq 1$: $a\{\!\{m \leftarrow b\}\!\} =_{\mathcal{W}} a[\mathit{lift}^{m-1}(\mathit{cons}(b, \mathit{id}))]$.*

Definition 21 (The Scheme). *We say that a basic substitution calculus $\mathcal{W}$ obeys* the scheme *iff for every index $\underline{m}$ and every substitution symbol $\sigma \in \Gamma_s$ of arity q one of the following two conditions hold:*

1. *There exists a de Bruijn index $\underline{n}$, positive numbers $i_1, \ldots, i_r$ ($r \geq 0$) and substitutions $u_1, \ldots, u_k$ ($k \geq 0$) such that*
   - *$1 \leq i_1, \ldots, i_r \leq q$ and all the i_j's are distinct*
   - *for all $o_1, \ldots, o_q$ we have: $\underline{m}[\sigma(o_1, \ldots, o_q)] =_{\mathcal{W}} \underline{n}[o_{i_1}] \ldots [o_{i_r}][u_1] \ldots [u_k]$*
2. *There exists an index i ($1 \leq i \leq q$) such that for all $o_1, \ldots, o_q$ we have: $\underline{m}[\sigma(o_1, \ldots, o_q)] =_{\mathcal{W}} o_i$*

We assume these equations to be well-typed: whenever the first case holds, then $o_{i_1}, \ldots, o_{i_r}$ are substitutions, whenever the second case holds, o_i is of sort T.

Example 5. Example of calculi satisfying the scheme are σ, $\sigma_{\Uparrow}$, υ, f and d [17].

4 From Higher-Order to First-Order Rewriting

In this section we give an algorithm, referred to as the Conversion Procedure, to translate any higher-order rewrite system in the formalism $SERS_{DB}$ to a first-order *ExERS*. The Conversion Procedure is somewhat involved since several conditions, mainly related to the labels of t-metavariables, must be met in order for a substitution to be admitted as *valid*. The idea is to replace all occurrences of t-metavariables X_l by a first-order variable X followed by an appropriate *index-adjusting explicit substitution* which computes valid valuations.

We first give the conversion rules of the translation, then we prove its properties in Section 4.1.

Definition 22 (Binding allowance). *Let M be a metaterm and $\{X_{l_1}, \ldots, X_{l_n}\}$ be the set of all the t-metavariables with name X occurring in M. Then, the* binding allowance of X in M, *noted* $\mathtt{Ba}_M(X)$, *is the set* $\bigcap_{i=1}^{n} l_i$. *Likewise, we define the* binding allowance of X in a rule (L, R), *written* $\mathtt{Ba}_{(L,R)}(X)$.

Example 6. Let $M = f(\xi(X_\alpha), \xi(\lambda(X_{\beta\alpha})), \nu(\xi(X_{\alpha\gamma})))$, then $\mathtt{Ba}_M(X) = \{\alpha\}$.

Definition 23 (Shifting index). *Let M be a metaterm, X_l a t-metavariable occurring in M, and i a position in l. The* shifting index determined by X_l in M at position i, *denoted* $\mathtt{Sh}(M, X_l, i)$, *is defined as*

$$\mathtt{Sh}(M, X_l, i) =_{def} |\{j \mid \mathtt{at}(l, j) \notin \mathtt{Ba}_M(X), j \in 1..i-1\}|$$

$\mathtt{Sh}(M, X_l, i)$ is just the total number of binder indicators in l at positions $1..i-1$ that do not belong to $\mathtt{Ba}_M(X)$ (thus $\mathtt{Sh}(M, X_l, 1)$ is always 0). Likewise, we define the shifting index determined by X_l in a rule (L, R) at position i, *written* $\mathtt{Sh}((L, R), X_l, i)$.

Example 7. Let $M = f(\xi(X_\alpha), \xi(\lambda(X_{\beta\alpha})), \nu(\xi(X_{\alpha\gamma})))$. Then $\mathtt{Sh}(M, X_{\beta\alpha}, 2) = 1$ and $\mathtt{Sh}(M, X_\alpha, 1) = \mathtt{Sh}(M, X_{\alpha\gamma}, 2) = 0$.

Definition 24 (Pivot). *Let (L, R) be a $SERS_{DB}$-rewrite rule and let us suppose that $\{X_{l_1}, \ldots, X_{l_n}\}$ is the set of all X-based t-metavariables in (L, R). If $\mathtt{Ba}_{(L,R)}(X) \neq \emptyset$, then X_{l_j} for some $j \in 1..n$ is called an* (X-based) pivot *if $|l_j| \le |l_i|$ for all $i \in 1..n$, and $X_{l_j} \in L$ whenever possible. A* pivot set *for a rewrite rule (L, R) is a set of pivot t-metavariables, one for each name X in L such that $\mathtt{Ba}_{(L,R)}(X) \neq \emptyset$. This notion extends to a set of rewrite rules as expected.*

Note that Definition 24 admits the existence of more than one X-based pivot t-metavariable. A pivot set for (L, R) fixes a t-metavariable for each t-metavariable name having a non-empty binding allowance.

Example 8. Both t-metavariables $X_{\alpha\beta}$ and $X_{\beta\alpha}$ can be chosen as X-based pivot in the rewrite rule $\mathcal{R}$: *Implies*$(\exists(\forall(X_{\alpha\beta})), \forall(\exists(X_{\beta\alpha}))) \longrightarrow$ *true*. In the rewrite rule $\mathcal{R}'$: $f(Y_\epsilon, \lambda(\xi(X_{\alpha\beta})), \nu(\lambda(X_{\beta\alpha}))) \longrightarrow \mu(X_\alpha, Y_\alpha)$ the t-metavariable X_α is the only possible X-based pivot.

Definition 25 (Conversion of t-metavariables). *Consider a $SERS_{DB}$-rewrite rule (L, R) and a pivot set for (L, R). We consider the following cases for every t-metavariable name X occurring in L:*

1. $\mathtt{Ba}_{(L,R)}(X) = \emptyset$. *Then replace each t-metavariable X_l in (L, R) by the metaterm $X[shift^{|l|}]$, and those t-metavariables X_l with $l = \epsilon$ simply by X. This shall allow for example the rule $f(\lambda(app(X_\alpha, \underline{1}), X_\epsilon)) \longrightarrow X_\epsilon$ to be converted to $f(\lambda(app(X[shift], \underline{1}), X)) \longrightarrow X$.*
2. $\mathtt{Ba}_{(L,R)}(X) = \{\beta_1, \ldots, \beta_m\}$ *with $m > 0$. Let X_l be the pivot t-metavariable for X given by the hypothesis. We replace all occurrences of a t-metavariable X_k in (L, R) by the term $X[cons(b_1, \ldots, b_{|l|}, shift^j)]$ where $j =_{def} |k| + |l \setminus \mathtt{Ba}_{(L,R)}(X)|$ and the b_i's depend on whether X_k is a pivot t-metavariable or not. As an optimization and in the particular case that the resulting term $X[cons(b_1, \ldots, b_{|l|}, shift^j)]$ is of the form $cons(\underline{1}, \ldots, \underline{|l|}, shift^{|l|})$, then we simply replace X_k by X. The substitution $cons(b_1, \ldots, b_{|l|}, shift^j)$ is called the* index-adjusting substitution *corresponding to X_k and it is defined as follows:*
   a) *if X_k is the pivot (hence $l = k$), then $b_i = \underline{i}$ if $\mathtt{at}(l, i) \in \mathtt{Ba}_{(L,R)}(X)$ and $b_i = \underline{|l| + 1 + \mathtt{Sh}((L, R), X_l, i)}$ if $\mathtt{at}(l, i) \notin \mathtt{Ba}_{(L,R)}(X)$.*
   b) *if X_k is not the pivot then $b_i = \underline{\mathtt{pos}(\beta_h, k)}$ if $i = \mathtt{pos}(\beta_h, l)$ for some $\beta_h \in \mathtt{Ba}_{(L,R)}(X)$ and $b_i = \underline{|k| + 1 + \mathtt{Sh}((L, R), X_l, i)}$ otherwise.*

Note that for an index-adjusting substitution $X[cons(b_1, \ldots, b_{|l|}, shift^j)]$ each b_i is a distinct de Bruijn index and less than or equal to j. Substitutions of this form have been called pattern substitutions in [10], where unification of higher-order patterns via explicit substitutions is studied.

Definition 26 (The Conversion Procedure). *Given a $SERS_{DB}$ $\mathcal{R}$ the following actions are taken:*

1. **Convert rules.** *The transformation of Definition 25 is applied to all rules in $\mathcal{R}$ prior selection of some set of pivot sets for $\mathcal{R}$.*
2. **Replace metasubstitution operator.** *All submetaterms of the form $M[\![N]\!]$ in $\mathcal{R}$ are replaced by the term substitution operator $M[cons(N, id)]$.*

Example 9. Below we present some examples of conversion of rules. We have fixed $\mathcal{W}$ to be the σ-calculus.

$SERS_{DB}$ rule	*Pivot selected*	*Transformed rule*
$\lambda(app(X_\alpha, \underline{1})) \longrightarrow X_\epsilon$	-	$\lambda(app(X[\uparrow], \underline{1})) \longrightarrow X$
$\lambda(\lambda(X_{\alpha\beta})) \longrightarrow \lambda(\lambda(X_{\beta\alpha}))$	$X_{\alpha\beta}$	$\lambda(\lambda(X)) \longrightarrow \lambda(\lambda(X[\underline{2} \cdot \underline{1} \cdot (\uparrow \circ \uparrow)]))$
$f(\lambda(\lambda(X_{\alpha\beta})), \lambda(\lambda(X_{\beta\alpha}))) \longrightarrow \lambda(X_\gamma)$	-	$f(\lambda(\lambda(X[\uparrow \circ \uparrow])), \lambda(\lambda(X[\uparrow \circ \uparrow]))) \longrightarrow \lambda(X[\uparrow])$
$app(\lambda(X_\alpha), Z_\epsilon) \longrightarrow_{\beta_{db}} X_\alpha[\![Z_\epsilon]\!]$	X_α, Z_ϵ on *LHS*	$app(\lambda(X), Z) \longrightarrow X[cons(Z, id)]$

Let (L, R) be a $SERS_{DB}$-rule and P a pivot set for (L, R). We write $\mathcal{C}_P(L, R)$ for the result of applying the conversion of Definition 26 to (L, R) with pivot set P. We refer to $\mathcal{C}_P(L, R)$ as the *converted version of* (L, R) *via* P.

Note that if the $SERS_{DB}$-rewrite rule (L, R) which is input to the Conversion Procedure is such that for every name X in (L, R) there is a label l with *all* metavariables in (L, R) of the form X_l, then all X_l are replaced simply by X. This is the case of the rule β_{db} of Example 9.

Also, observe that if we replace our *cons*(.,.) macro by a *scons*(.) of substitution declaration T as defined in [17] then the "Replace metasubstitution operator" step in Definition 26 converts a metaterm of the form $M[\![N]\!]$ into $M[scons(N)]$, yielding first-order systems based on substitution calculi, such as $\lambda\upsilon$, which do not implement parallel substitution.

The resulting system of the Conversion Procedure is coded as an *ExERS*, a framework for defining first-order rewrite systems where $\mathcal{E}$-matching is used, $\mathcal{E}$ being an equational theory governing the behaviour of the index adjustment substitutions. Moreover, if it is possible, an *ExERS* may further be coded as a *FExERS* where reduction is defined on first-order terms and matching is just syntactic first-order matching, obtaining a *full first-order system*.

Definition 27 (First-order version of $\mathcal{R}$). *Let Γ be an ExERS signature and let $\mathcal{R}$ be a $SERS_{DB}$. Consider the system $fo(\mathcal{R})$ obtained by applying the Conversion Procedure to $\mathcal{R}$ and let $\mathcal{W}$ be a substitution calculus over Γ. Then the ExERS $fo(\mathcal{R})_{\mathcal{W}}$ is called a* first order-version *of $\mathcal{R}$.*

In what follows we shall assume given some fixed *basic* substitution calculus $\mathcal{W}$. Thus, given a $SERS_{DB}$ $\mathcal{R}$ we shall speak of *the* first-order version of $\mathcal{R}$. This requires considering pivot selection, an issue we take up next.

Assume given some rewrite rule (L, R) and different pivot sets P and Q for this rule. It is clear that $\mathcal{C}_P(L, R)$ and $\mathcal{C}_Q(L, R)$ shall not be identical. Nevertheless, we may show that the reduction relation generated by both of these converted rewrite rules is identical.

Proposition 1. *Let (L, R) be a $SERS_{DB}$-rewrite rule and let P and Q be different pivot sets for this rule. Then the rewrite relation generated by both $\mathcal{C}_P(L, R)$ and $\mathcal{C}_Q(L, R)$ are identical.*

4.1 Properties of the Conversion

The Conversion Procedure satisfies two important properties: each higher-order rewrite step may be simulated by first-order rewriting (*simulation*) and rewrite steps in the first-order version of a higher-order system $\mathcal{R}$ can be projected in $\mathcal{R}$ (*conservation*).

Proposition 2 (Simulation). *Let $\mathcal{R}$ be an $SERS_{DB}$ and let $fo(\mathcal{R})_{\mathcal{W}}$ be its first-order version. Suppose $a \longrightarrow_{\mathcal{R}} b$. If $fo(\mathcal{R})_{\mathcal{W}}$ is an ExERS then $a \longrightarrow_{fo(\mathcal{R})/\mathcal{W}} b$. If $fo(\mathcal{R})_{\mathcal{W}}$ is a FExERS then $a \longrightarrow_{fo(\mathcal{R})} \circ \stackrel{*}{\longrightarrow}_{\mathcal{W}} b$ where $\circ$ denotes relation composition.*

Proposition 3 (Conservation). *Let $\mathcal{R}$ be a $SERS_{DB}$ and $fo(\mathcal{R})_{\mathcal{W}}$ its first-order version with $\mathcal{W}$ satisfying the scheme. If $a \longrightarrow_{fo(\mathcal{R})_{\mathcal{W}}} b$ then $\mathcal{W}(a) \stackrel{*}{\longrightarrow}_{\mathcal{R}} \mathcal{W}(b)$.*

4.2 Essentially First-Order *HORS*

This last subsection gives a very simple syntactical criterion that can be used to decide if a given higher-order rewrite system can be converted into a full first-order rewrite system (modulo an empty theory). In particular, we can check that many higher-order calculi in the literature, e.g. λ-calculus, verify this property.

Definition 28 (Essentially first-order *HORS*). *A $SERS_{DB}$ $\mathcal{R}$ is called* essentially first-order *if $fo(\mathcal{R})_{\mathcal{W}}$ is a FExERS for $\mathcal{W}$ a basic substitution calculus.*

Definition 29 (fo-condition). *A $SERS_{DB}$ $\mathcal{R}$ satisfies the* fo-condition *if every rewrite rule $(L, R) \in \mathcal{R}$ satisfies: for every name X in L let $X_{l_1}, \ldots, X_{l_n}$ be all the X-based t-metavariables in L, then $l_1 = l_2 \ldots = l_n$ and (the underlying set of) l_1 is $\mathtt{Ba}_{(L,R)}(X)$, and for all $X_k \in R$ we have $|k| \geq |l_1|$.*

In the above definition note that $l_1 = l_2 \ldots = l_n$ means that labels $l_1, \ldots, l_n$ must be *identical* (for example $\alpha\beta \neq \beta\alpha$). Also, by Definition 6, l_1 is simple.

Example 10. Consider the β_{db}-calculus: $app(\lambda(X_\alpha), Z_\epsilon) \longrightarrow_{\beta_{db}} X_\alpha[\![Z_\epsilon]\!]$. The β_{db}-calculus satisfies the fo-condition.

Proposition 4 puts forward the importance of the fo-condition. Its proof relies on a close inspection of the Conversion Procedure.

Proposition 4. *Let $\mathcal{R}$ be a $SERS_{DB}$ satisfying the fo-condition. Then $\mathcal{R}$ is essentially first-order.*

Note that many results on higher-order systems (e.g. perpetuality [19], standardization [21]) require *left-linearity* (a metavariable may occur at most once on the *LHS* of a rewrite rule), and *fully-extendedness or locality* (if a metavariable $X(t_1, \ldots, t_n)$ occurs on the *LHS* of a rewrite rule then $t_1, \ldots, t_n$ is the list of variables bound above it). The reader may find it interesting to observe that these conditions together seem to imply the fo-condition. A proof of this fact would require either developing the results of this work in the above mentioned *HORS* or via some suitable translation to the $SERS_{DB}$ formalism.

5 Conclusions and Future Work

This work presents an encoding of higher-order term rewriting systems into first-order rewriting systems modulo an equational theory. This equational theory takes care of the substitution process. The encoding has furthermore allowed us to identify in a simple syntactical manner, via the so-called fo-condition, a class of *HORS* that are fully first-order in that they may be encoded as first-order rewrite systems modulo an empty equational theory. This amounts to incorporating, into the first-order notion of reduction, not only the computation of substitutions but also the higher-order (pattern) matching process. It is fair to say that a higher-order rewrite system satisfying this condition requires a simple matching process, in contrast to those that do not satisfy this condition (such as the $\lambda\beta\eta$-calculus). Other syntactical restrictions, such as linearity and locality, imposed on higher-order rewrite systems in [19,21] in order to reason about their properties can be related to the fo-condition in a very simple way. This justifies that the fo-condition, even if obtained very technically in this paper, may be seen as an interpretation of what a well-behaved higher-order rewrite system is.

Moreover, this encoding has been achieved by working with a general presentation of substitution calculi rather than dealing with some particular substitution calculus. Any calculus of explicit substitutions satisfying this general presentation based on macros will do.

Some further research directions are summarized below:

- As already mentioned, the encoding opens up the possibility of transferring results concerning confluence, termination, completion, evaluation strategies, implementation techniques, etc. from the first-order framework to the higher-order framework.
- Given a $SERS_{DB}$ $\mathcal{R}$ note that the *LHS*s of rules in $fo(\mathcal{R})$ may contain occurrences of the substitution operator (pattern substitutions). It would be interesting to deal with pattern substitutions and "regular" term substitutions (those arising from the conversion of the de Bruijn metasubstitution operator $.[\![.]\!]$) as different substitution operators at the object-level. This would neatly separate the explicit matching computation from that of the usual substitution replacing terms for variables.
- This work has been developed in a type-free framework. The notion of type is central to Computer Science. This calls for a detailed study of the encoding process dealing with typed higher-order rewrite systems such as *HRS* [23].
- The ideas presented in this paper could be used to relax conditions in [9] where only rewrite systems with atomic propositions on the *LHS*s of rules are considered.

Acknowledgements. We thank Bruno Guillaume for helpful remarks.

References

1. M. Abadi, L. Cardelli, P.-L. Curien, and J.-J. Lévy. Explicit substitutions. *Journal of Functional Programming*, 4(1):375–416, 1991.
2. T. Arts and J. Giesl. Termination of term rewriting using dependency pairs. *Theoretical Computer Science*, 236:133–178, 2000.
3. L. Bachmair and N. Dershowitz. Critical pair criteria for completion. *Journal of Symbolic Computation*, 6(1):1–18, 1988.
4. Z. Benaissa, D. Briaud, P. Lescanne, and J. Rouyer-Degli. λv, a calculus of explicit substitutions which preserves strong normalisation. *Journal of Functional Programming*, 6(5):699–722, 1996.
5. E. Bonelli, D. Kesner, and A. Ríos. A de Bruijn notation for higher-order rewriting. In *Proc. of the Eleventh Int. Conference on Rewriting Techniques and Applications*, Norwich, UK, July 2000.
6. R. David and B. Guillaume. A λ-calculus with explicit weakening and explicit substitutions. *Journal of Mathematical Structures in Computer Science*, 2000. To appear.

7. N. Dershowitz. Orderings for term rewriting systems. *Theoretical Computer Science*, 17(3):279–301, 1982.
8. N. Dershowitz and J-P. Jouannaud. Rewrite systems. In J. van Leeuwen, editor, *Handbook of Theoretical Computer Science*, volume B, pages 243–309. North-Holland, 1990.
9. G. Dowek, T. Hardin and C. Kirchner. Theorem proving modulo. Technical Report RR 3400, INRIA, 1998.
10. G. Dowek, T. Hardin, C. Kirchner and F. Pfenning. Unification via Explicit Substitutions: The Case of Higher-Order Patterns. Technical Report RR3591. INRIA. 1998.
11. G. Dowek, T. Hardin and C. Kirchner. Higher-order unification via explicit substitutions. *Information and Computation*, 157:183–235, 2000.
12. G. Dowek, T. Hardin and C. Kirchner. Hol-lambda-sigma: an intentional first-order expression of higher-order logic. *Mathematical Structures in Computer Science*, 11:1–25, 2001.
13. T. Hardin and J-J. Lévy. A confluent calculus of substitutions. In *France-Japan Artificial Intelligence and Computer Science Symposium*, 1989.
14. J-P. Jouannaud and A. Rubio. The higher-order recursive path ordering. In *Fourteenth Annual IEEE Symposium on Logic in Computer Science*, Trento, Italy, 1999.
15. F. Kamareddine and A. Ríos. A λ-calculus à la de Bruijn with explicit substitutions. In *Proc. of the Int. Symposium on Programming Language Implementation and Logic Programming (PLILP)*, LNCS 982. 1995.
16. S. Kamin and J. J. Lévy. Attempts for generalizing the recursive path orderings. University of Illinois, 1980.
17. D. Kesner. Confluence of extensional and non-extensional lambda-calculi with explicit substitutions. *Theoretical Computer Science*, 238(1-2):183–220, 2000.
18. Z. Khasidashvili. Expression Reduction Systems. In *Proc. of I. Vekua Institute of Applied Mathematics*, volume 36, Tbilisi, 1990.
19. Z. Khasidashvili and M. Ogawa and V. van Oostrom. Uniform Normalization beyond Orthogonality. Submitted for publication, 2000.
20. P. Lescanne and J. Rouyer-Degli. Explicit substitutions with de Bruijn levels. In *Proc. of the Sixth Int. Conference on Rewriting Techniques and Applications*, LNCS 914, 1995.
21. P-A. Melliès. Description Abstraite des Systèmes de Réécriture. Thèse de l'Université Paris VII, December 1996.
22. C. Muñoz. A left-linear variant of $\lambda\sigma$. In Michael Hanus, J. Heering, and K. (Karl) Meinke, editors, *Proc. of the 6th Int. Conference on Algebraic and Logic Programming (ALP'97)*, LNCS 1298. 1997.
23. T. Nipkow. Higher-order critical pairs. in *Proc. of the Sixth Annual IEEE Symposium on Logic in Computer Science*, 1991.
24. B. Pagano. *Des Calculs de Susbtitution Explicite et leur application à la compilation des langages fonctionnels.* PhD thesis, Université Paris VI, 1998.
25. R. Pollack. Closure under alpha-conversion. In Henk Barendregt and Tobias Nipkow, editors, *Types for Proofs and Programs (TYPES)*, LNCS 806. 1993.
26. H. Zantema. Termination of Term Rewriting by Semantic Labelling. Fundamenta Informaticae, volume 24, pages 89-105, 1995.

Combining Pattern *E*-Unification Algorithms*

Alexandre Boudet and Evelyne Contejean

LRI, CNRS UMR 8623, Bât. 490, Université Paris-Sud, Centre d'Orsay, 91405 Orsay Cedex, France

Abstract. We present an algorithm for unification of higher-order patterns modulo combinations of disjoint first-order equational theories. This algorithm is highly non-deterministic, in the spirit of those by Schmidt-Schauß [20] and Baader-Schulz [1] in the first-order case. We redefine the properties required for elementary pattern unification algorithms of pure problems in this context, then we show that some theories of interest have elementary unification algorithms fitting our requirements. This provides a unification algorithm for patterns modulo the combination of theories such as the free theory, commutativity, one-sided distributivity, associativity-commutativity and some of its extensions, including Abelian groups.

Introduction

Patterns have been defined by Miller [18] in order to provide a compromise between simply-typed lambda-terms for which unification is known to be undecidable [12,9] and mere first-order terms which are deprived of any abstraction mechanism. A *pattern* is a term of the simply-typed lambda-calculus in which the arguments of a free variable are all pairwise distinct bound variables. Patterns are close to first-order terms in that the free variables (with their bound variables as only permitted arguments) are at the leaves. Under this rather drastic restriction, unification becomes decidable and unitary:

Theorem 1 ([18]). *It is decidable whether two patterns are unifiable, and there exists an algorithm which computes a most general unifier of any two unifiable patterns.*

Yet, patterns are useful in practice for defining higher-order pattern rewrite systems [17,6], or for defining functions by cases in functional programming languages. Some efforts have been devoted to the study of languages combining functional programming (lambda-calculus) and algebraic programming (term rewriting systems) [15,13,7].

In this paper we provide a nondeterministic algorithm for combining elementary equational patterns unification algorithms. This is the object of section 2. In sections 3 and 4, we show that such elementary unification algorithms

* This research was supported in part by the EWG CCL, and the RNRT project Calife.

A. Middeldorp (Ed.): RTA 2001, LNCS 2051, pp. 63–76, 2001.

exist for theories such as the free theory, commutativity, one-sided distributivity, as well as associativity-commutativity and its common extensions including Abelian groups.

Our method does not consist of using a first-order unification algorithm for the combined equational theories extended to the case of patterns. Such an approach has been used by Qian & Wang [19], but considering a first-order unification algorithm as a black box leads to incompleteness (see example in [5]). What we need here is a *pattern* unification algorithm for each of the theories to be combined *plus* a combination algorithm for the elementary E_i-pattern unification algorithm. Evidence of this need is that contrarily as in the first-order case, the unifier set of the pure equation $\lambda xy.Fxy = \lambda xy.Fyx$ modulo the free theory (no equational axioms) changes if one adds say a commutative axiom $x+y = y+x$ (see [19]). The requirements we have on elementary pattern unification algorithms are very much in the spirit of those needed in the first-order case, yet there are relevant differences due in particular to the possible presence of equations with the same free variable at the head on both sides (like $\lambda xy.Fxy = \lambda xy.Fyx$).

The worst difficulties, for the combination part as well as for the elementary unification algorithms, come from such equations which have no minimal complete sets of E-unifiers, even for theories which are finitary unifying in the first-order case. On top of this, the solutions of such equations introduce terms which are not patterns. For this reason, we will never attempt to solve such equations explicitly, but we will keep them as *constraints*. The output of our algorithm is a (DAG-) solved form constrained by some equations of the above form and *compatible* with them. As the algorithm by Baader and Schulz [1], ours can be used for combining decision procedures.

1 Preliminaries

We assume the reader is familiar with simply-typed lambda-calculus, and equational unification. Some background is available in *e.g.* [10,14] for lambda-calculus and E-unification.

1.1 Patterns and Equational Theories

Miller [18] has defined the *patterns* as those terms of the simply-typed lambda-calculus in which the arguments of a free variables are (η-equivalent to) pairwise distinct bound variables: $\lambda xyz.f(H(x,y), H(x,z))$ and $\lambda x.F(\lambda z.x(z))$[1] are patterns while $\lambda xy.G(x,x,y)$, $\lambda xy.H(x,f(y))$ and $\lambda xy.H(F(x),y)$ are not patterns. We shall use the following notations: the sequence of variables $x_1,\ldots,x_n$ will be written $\overline{x_n}$ or even $\overline{x}$ if n is not relevant. Hence $\lambda x_1 \cdots \lambda x_n.s$ will be written $\lambda\overline{x_n}.s$, or even $\lambda\overline{x}.s$. If in a same expression $\overline{x}$ appears several times

[1] We will always write such a pattern in the (η-equivalent) form $\lambda x.F(x)$, where the argument of the free variable F is *indeed* a bound variable.

it denotes the same sequence of variables. If π is a permutation of $\{1,\ldots,n\}$, $\overline{x}^{\pi}$ denotes the sequence $x_{\pi(1)},\ldots,x_{\pi(n)}$. In the following, we shall use either $\lambda\overline{x}.F(\overline{x}^{\pi})$ or the α-equivalent term $\lambda\overline{y}^{\varphi}.F(\overline{y})$, where $\varphi=\pi^{-1}$, in order to denote $\lambda x_1\cdots\lambda x_n.F(x_{\pi(1)},\ldots,x_{\pi(n)})$. The curly-bracketed expression $\{\overline{x_n}\}$ denotes the (multi) set $\{x_1,\ldots,x_n\}$. In addition, we will use the notation $t(u_1,\ldots,u_n)$ or $t(\overline{u_n})$ for $(\cdots(t\ u_1)\cdots u_n)$. The free variables of a term t are denoted by $\mathcal{FV}(t)$.

$t|_p$ is the subterm of t at position p. The notation $t[u]_p$ stands for a term t with a subterm u at position p, $t[u_1,\ldots,u_n]$ for a term t having subterms $u_1,\ldots,u_n$.

Unless otherwise stated, we assume that the terms are in η-long β-normal form [10], the β and η rules being respectively oriented as follows: $(\lambda x.M)N\rightarrow_\beta M\{x\mapsto N\}$ and $F\rightarrow_{\eta\uparrow}\lambda\overline{x_n}.F(\overline{x_n})$ if the type of F is $\alpha_1\rightarrow\ldots\rightarrow\alpha_n\rightarrow\alpha$, and α is a base type. In this case, F is said to have *arity* n. The η-long β-normal form of a term t is denoted by $t\updownarrow_\beta^\eta$.

A *substitution* σ is a mapping from a finite set of variables to terms of the same type, written $\sigma=\{X_1\mapsto t_1,\ldots,X_n\mapsto t_n\}$. The set $\{X_1,\ldots,X_n\}$ is called the *Domain* of σ and denoted by $\mathcal{D}\mathrm{om}(\sigma)$.

The equational theories we consider here are the usual first-order equational theories: given a set E of (unordered) first-order axioms built over a signature $\mathcal{F}$, $=_E$ is the least congruence[2] containing all the identities $l\sigma=r\sigma$ where $l=r\in E$ and σ is a suitably typed substitution. $=_{\eta\beta E}$ is then the least congruence containing $=_E$, $=_\eta$ and $=_\beta$.

The following is a key theorem by Tannen. It allows us to restrict our attention to $=_E$ for deciding η-β-E-equivalence of terms in η-long, β-normal form:

Theorem 2 ([7]). *Let E be an equational theory and s and t two terms. Then* $s=_{\eta\beta E}t\Longleftrightarrow s\updownarrow_\beta^\eta=_E t\updownarrow_\beta^\eta$.

1.2 Unification Problems

Unification problems are formulas built-up using only the equality predicate $=$ (between *terms*), conjunctions, disjunctions and existential quantifiers. The solutions of $s=t$ are the substitutions σ such that $s\sigma=_{\eta\beta E}t\sigma$. This definition extends the natural way to unification problems. We restrict our attention to problems of the form $(\exists\overline{X})\ s_1=t_1\wedge\cdots\wedge s_n=t_n$, the only disjunctions being implicitly introduced by the non-deterministic rules.

Terminology. In the following, *free variable* denotes an occurrence of a variable which is not λ-bound and *bound variable* an occurrence of a variable which is λ-bound. To specify the status of a free variable with respect to existential quantifications, we will explicitly write *existentially quantified* or *not existentially quantified.* In the sequel, upper-case F, G, X,... will denote free variables, a, b, f, g,... constants, and x, y, z, x_1,... bound variables.

Without loss of generality, we assume that the left-hand sides and right-hand sides of the equations have the same prefix of λ-bindings. This is made possible

[2] compatible *also* with application and λ-abstraction in our context.

(by using α-conversion if necessary) because the two terms have to be in η-long β-normal form and of the same type. In other terms, we will assume that the equations are of the form $\lambda\overline{x}.s = \lambda\overline{x}.t$ where s and t do not have an abstraction at the top.

Definition 1. *A* flexible *pattern is a term of the form $\lambda\overline{x}.F(\overline{y})$ where F is a free variable and $\{\overline{y}\} \subseteq \{\overline{x}\}$. A* flex-flex *equation is an equation between two flexible patterns. An equation is* quasi-solved *if it is of the form $\lambda\overline{x_k}.F(\overline{y_n}) = \lambda\overline{x_k}.s$ and $\mathcal{FV}(s) \cap \{\overline{x_k}\} \subseteq \{\overline{y_n}\}$ and $F \notin \mathcal{FV}(s) \cup \{\overline{x_k}\}$. A variable is* solved *in a unification problem if it occurs only once as the left-hand side of a quasi-solved equation.*

Lemma 1. *If the equation $\lambda\overline{x_k}.F(\overline{y_n}) = \lambda\overline{x_k}.s$ is quasi-solved, then it is has the same solutions as $\lambda\overline{y_n}.F(\overline{y_n}) = \lambda\overline{y_n}.s$ and (by η-equivalence) as $F = \lambda\overline{y_n}.s$. A most general unifier of such an equation is $\{F \mapsto \lambda\overline{y_n}.s\}$.*

For the sake of readability, we will often write a quasi-solved equation in the form $F = \lambda\overline{y_n}.s$ instead of $\lambda\overline{x_k}.F(\overline{y_n}) = \lambda\overline{x_k}.s$.

Definition 2. *A* DAG-solved form *is a problem of the form $(\exists Y_1 \cdots Y_m)\ X_1 = s_1 \wedge \cdots \wedge X_n = s_n$ where for $1 \leq i \leq n$, X_i and s_i have the same type, and $X_i \neq X_j$ for $i \neq j$ and $X_i \notin \mathcal{FV}(s_j)$ for $i \leq j$. A* solved form *is a problem of the form $(\exists Y_1 \cdots Y_m)\ X_1 = s_1 \wedge \cdots \wedge X_n = s_n$ where for $1 \leq i \leq n$, X_i and s_i have the same type, X_i is not existentially quantified, and X_i has exactly one occurrence.*

A solved form is obtained from a DAG-solved form by applying as long as possible the rules:

Quasi-solved $\lambda\overline{x_k}.F(\overline{y_n}) = \lambda\overline{x_k}.s \wedge P \quad \to F = \lambda\overline{y_n}.s \wedge P$

Replacement $F = \lambda\overline{y_n}.s \wedge P \quad \to F = \lambda\overline{y_n}.s \wedge P\{F \mapsto \lambda\overline{y_n}.s\}$
if F has a free occurrence in P.

EQE $(\exists F)\ F = t \wedge P \quad \to P \quad$ if F has no free occurrence in P.

1.3 Combinations of Equational Theories

As in the first-order case [22,16,8,5], we will consider a *combination of equational theories*. We assume that $E_0, \ldots, E_n$ are equational theories over disjoint signatures $\mathcal{F}_0, \ldots, \mathcal{F}_n$, and we will provide a unification algorithm for the theory E presented by the reunion of the presentations, provided an *elementary E_i-unification algorithm* for patterns is known for each E_i. As in the first-order case, we have some further assumptions on elementary E_i-unification that will be made precise later.

Definition 3. *The* theory *of a bound variable, or of a free algebraic symbol* i.e., *a symbol which does not appear in any axiom is the free theory E_0. The theory of an algebraic symbol $f \in \mathcal{F}_i$ is E_i. A variable F is E_i-instantiated in a unification problem P if it occurs in a quasi-solved equation $F = s$ where the head of s has theory E_i. A variable F is E_i-instantiated by a substitution σ if the head of $F\sigma$ has theory E_i.*

We first give two well-known rules in order to obtain equations between *pure terms* only.

Definition 4. *The subterm u of the term $t[u]_p$ is an* alien subterm *in t if it occurs as an argument of a symbol from a different theory from its head symbol.*

VA $\lambda\overline{x}.s[u]_p = \lambda\overline{x}.t \quad \to \exists F\ \lambda\overline{x}.s[F]_p = \lambda\overline{x}.t \wedge \lambda\overline{y}.F(\overline{y}) = \lambda\overline{y}.u$
if u is an alien subterm and $\{\overline{y}\} = \{\overline{x}\} \cap \mathcal{FV}(u)$ and F is a new variable of appropriate type.

Split $\lambda\overline{x}.\gamma(\overline{s}) = \lambda\overline{x}.\delta(\overline{t}) \quad \to \exists F\ \lambda\overline{x}.F(\overline{x}) = \lambda\overline{x}.\gamma(\overline{s}) \wedge \lambda\overline{x}.F(\overline{x}) = \lambda\overline{x}.\delta(\overline{t})$
if γ and δ are not free variables and belong to different theories, where F is a new variable of appropriate type.

The above rules obviously terminate and yield a problem which is equivalent to the original problem and where all the equations are *pure, i.e.,* containing only symbols from a same theory. We can now split a unification problem into several pure subproblems:

Definition 5. *A unification problem P will be written in the form*

$$P \equiv (\exists\overline{X})\ P_F \wedge P_V \wedge P_0 \wedge P_1 \wedge \cdots \wedge P_n$$

where

- *P_F is the set of equations of the form $\lambda\overline{x_n}.F(\overline{x_n}) = \lambda\overline{x_n}.F(\overline{x_n}^{\pi})$, where π is a permutation of $\{1, \ldots, n\}$. Such equations will be called* frozen[3].
- *P_V is the set of equations of the form $\lambda\overline{x}.F(\overline{y}) = \lambda\overline{x}.G(\overline{z})$, where F and G are different free variables.*
- *$P_0, \ldots, P_n$ are pure problems in the theories $E_0, \ldots, E_n$ respectively.*

2 A Combination Algorithm

In this section, we will present a non-deterministic algorithm, one step beyond that by Baader and Schulz [1], who extend the method initiated by Schmidt-Schauß [20] for unification in combinations of equational theories. In particular,

[3] These equations are always trivially solvable, for example by $F \mapsto \lambda\overline{x_n}.C$, where C is a variable of the appropriate base type, but we will never solve them explicitly because they have no minimal complete sets of unifiers and their solutions introduce terms which are not patterns, see [19,5].

we will guess a projection for each free variable, and then restrict our attention to *constant-preserving substitutions*, like we suggested in [4].

The aim of these rules is to guess in advance the form of a unifier, and to make the algorithm fail when the solutions of the problem do not correspond to the choices that are being considered currently. The drawback of such an approach is a blow-up in the complexity, but it allows to avoid recursion, hence guaranteeing termination.

After the variable abstraction step, we want to guess, for each free variable F of arity k (*i.e.*, whose η-long form is $\lambda\overline{x_k}.F(x_1,\ldots,x_k)$) which of the bound variables $x_1,\ldots,x_k$ will effectively participate in the solution.

Definition 6. *A* constant-preserving substitution *is a substitution σ such that for all $F \in Dom(\sigma)$ if $F\sigma\updownarrow_\beta^\eta = \lambda\overline{x_k}.s$ then every variable of $\overline{x_k}$ has a free occurrence in s. A* projection *is a substitution of the form*

$$\sigma = \{F \mapsto \lambda\overline{x_k}.F'(\overline{y_m}) \mid F \in Dom(\sigma),\ \{\overline{y_m}\} \subseteq \{\overline{x_k}\}\}$$

Lemma 2. *For every substitution σ, there exist a projection π and a constant-preserving substitution θ such that $\sigma\updownarrow_\beta^\eta = (\pi\theta)\updownarrow_\beta^\eta$.*

Lemma 3. *The equation $\lambda\overline{x}.s = \lambda\overline{x}.t$ where $\{\overline{x}\} \cap \mathcal{FV}(s) \neq \{\overline{x}\} \cap \mathcal{FV}(t)$ has no constant-preserving E-solution. In particular, the equation $\lambda\overline{x}.F(\overline{y}) = \lambda\overline{x}.G(\overline{z})$, where $\{\overline{y}\}$ and $\{\overline{z}\}$ are not the same set, has no constant-preserving E-solution.*

This will allow us to choose (and apply) a projection for the free variables and to discard in the sequel the problems whose solutions are not constant-preserving.

2.1 Non-deterministic Choices

At this point, we will guess through "don't know" nondeterministic choices some properties of the solutions. The idea is that once a choice has been made, some failure rules will apply when the solutions of the current problem correspond to other choices. This method initiated by Schmidt-Schauß [20] is obviously correct because the solutions of the problems that are discarded this way correspond to another choice and will be computed in the corresponding branch.

The following transformations have to be successively applied to the problem:

C.1 Choose a projection for the free variables

We first guess for a given solution σ, the projection of which σ will be an instance by a constant-preserving substitution, in the conditions of Lemma 2. This is achieved by applying nondeterministically the following rule to some of the free variables F of the problem:

Project $P \rightarrow (\exists F')\ F = \lambda\overline{x_n}.F'(\overline{y_k}) \wedge P\{F \mapsto \lambda\overline{x_n}.F'(\overline{y_k})\}$
where F has arity n and F' is a new variable and $\{\overline{y_k}\} \subset \{\overline{x_n}\}$

After this step, we can restrict our attention to constant-preserving solutions.

C.2 Choose some flex-flex equations
We now guess the equations of the form $\lambda\overline{x}.F(\overline{y}) = \lambda\overline{x}.G(\overline{z})$ that will be satisfied by a solution σ. This is done by applying the following rule to some pairs $\{F, G\}$ of the free variables of the problem:

FF$_{\neq}$ $P \rightarrow F = \lambda\overline{x}.G(\overline{x}^{\pi}) \wedge P\{F \mapsto \lambda\overline{x}.G(\overline{x}^{\pi})\}$
where π is a permutation of $\{1, \ldots, n\}$, F has type $\tau_1 \rightarrow \cdots \rightarrow \tau_n \rightarrow \tau$, G has type $\tau_{\pi(1)} \rightarrow \cdots \rightarrow \tau_{\pi(n)} \rightarrow \tau$, $F \neq G$ and F and G occur in P.

We restrict the application of this rule to pairs of variables of the same arity and of the same type (up to a permutation of the types of the arguments), because after applying **Project**, the only flex-flex equations admitting constant-preserving solutions are of this form.

C.3 Choose the permutations on the arguments of the variables
For each free variable F of type $\tau_1 \rightarrow \cdots \rightarrow \tau_n \rightarrow \tau$, we choose the group of permutations Perm(F) such that a solution σ satisfies $\lambda\overline{x_n}.F\sigma(\overline{x_n}) =_{\eta\beta E} \lambda\overline{x_n}.F\sigma(\overline{x_n}^{\pi})$ for each $\pi \in$ Perm(F). For this, we apply the following rule to some of the free variables F of the problem:

FF$_=$ $P \rightarrow \lambda\overline{x_n}.F(\overline{x_n}) = \lambda\overline{x_n}.F(\overline{x_n}^{\pi}) \wedge P$
where F is a free variable of P of type $\tau_1 \rightarrow \cdots \rightarrow \tau_n \rightarrow \tau$ and π is a permutation such that $\tau_{\pi(i)} = \tau_i$ for $1 \leq i \leq n$.

C.4 Apply as long as possible the following transformation:

Coalesce
$\lambda\overline{x_k}.F(\overline{y_n}) = \lambda\overline{x_k}.G(\overline{z_n}) \wedge P \rightarrow F = \lambda\overline{y_n}.G(\overline{z_n}) \wedge P\{F \mapsto \lambda\overline{y_n}.G(\overline{z_n})\}$
if $F \neq G$ and $F, G \in \mathcal{FV}(P)$, where $\overline{y}_n$ is a permutation of $\overline{z_n}$.

After **Project** has been applied, the arity of the values of the variables is fixed, hence two variables may be identified only if they have the same arity. Note that $F = \lambda\overline{y_n}.G(\overline{z_n})$ is solved after the application of **Coalesce**. After this step, we have an equivalence relation on the variables, and a notion of representative:

Definition 7. *Two variables F and G are* identified *in P if they appear in an equation $\lambda\overline{x}.F(\overline{y}) = \lambda\overline{x}.G(\overline{z})$ of P. The relation $=_V$ is the least equivalence containing any pair of identified variables.*
Assume that **Coalesce** *has been applied as long as possible to P. In an equivalence class of $=_V$, only a single variable may occur more than once in P. When such a variable exists, it is chosen as a* representative *for all variables in that class. Otherwise, the representative is chosen arbitrarily in the equivalence class.*

C.5 Choose a theory for the representatives
We now guess for each representative F, the theory E_i such that F is allowed to have the head symbol of its value by a solution in E_i. Again, this was already done by Schmidt-Schauß [20] and by Baader and Schulz.

C.6 Choose an ordering on representatives
Finally, we guess a total strict ordering compatible with the occur-check relation defined by $F{<}G$ if $G\sigma$ is a proper subterm of $F\sigma$. Choose a total ordering $<_{oc}$ on the representatives of the variables of the problem. This is exactly what Baader and Schulz do in the first-order case [1], reflecting the fact that if σ is a finite solution, then the occur-check relation must be acyclic.

2.2 Solving Pure Subproblems

We make now precise our assumptions on the elementary E_i unification algorithms. First, we take note of the fact that there is not much to do with the frozen equations of P_F:

Frozen Equations
Although they are always trivially solvable in the free theory, we will never try to solve the equations of the form $\lambda\overline{x}.F(\overline{x})=\lambda\overline{x}.F(\overline{x}^{\pi})$ of P_F. These equations will be kept as *constraints* because they do not have finite complete sets of unifiers even for theories which have finitary first-order unification, and their solutions introduce terms which are not patterns. Here is an example by Qian and Wang:

Example 1 ([19]). Consider the equation $\lambda xy.F(x,y)=\lambda xy.F(y,x)$ in the AC-theory of $+$. For $m \geq 0$, the substitution $\sigma_m = \{F \mapsto \lambda xy.G_m(H_1(x,y) + H_1(y,x), \ldots, H_m(x,y) + H_m(y,x))\}$ is an AC-unifier of the above equation. On the other hand, every solution of e is an instance of some σ_i. In addition σ_{n+1} is strictly more general than σ_n.

Hence, AC-unification of patterns is not only infinitary, but *nullary*, in the sense that some problems do not have *minimal* complete sets of AC-unifiers [21]. All we can do is to make sure that the frozen equations in P_F are *compatible* with a (DAG-) solved form of the problem:

Definition 8. *Given a conjunction P_F of flex-flex equations of the form $\lambda\overline{x}.F(\overline{x}) = \lambda\overline{x}.F(\overline{x}^{\pi})$, we will write $P_F \models s =_{\eta\beta E} t$ if $s\updownarrow^{\eta}_{\beta} = t\updownarrow^{\eta}_{\beta}$ can be proved using the axioms of E and the equations $\lambda\overline{x}.F(\overline{x})=\lambda\overline{x}.F(\overline{x}^{\pi})$ of P_F, where F is treated like a free algebraic symbol. A substitution σ is* compatible *with P_F if for all equation $\lambda\overline{x}.F(\overline{x})=\lambda\overline{x}.F(\overline{x}^{\pi})$ of P_F, $P_F \models \lambda\overline{x}.F\sigma(\overline{x}) =_{\eta\beta E} \lambda\overline{x}.F\sigma(\overline{x}^{\pi})$.*

Lemma 4. *If a substitution σ (seen as a conjunction of equations) is compatible with P_F as defined above, then the E-solutions of $\sigma \wedge P_F$ are the substitutions $\sigma\theta$, where θ is an E-solution of P_F.*

Definition 9. *Given a unification problem P, with no flex-flex equations and a conjunction P_F of (frozen) equations of the form $\lambda\overline{x}.F(\overline{x}) = \lambda\overline{x}.F(\overline{x}^\pi)$, a* constrained E-solved form *of P is $P' \wedge P'_F$ where*

- *P' is a solved form with mgu σ, containing no equations of the form $\lambda\overline{x}.F(\overline{x}) = \lambda\overline{x}.F(\overline{x}^\pi)$.*
- *P'_F contains P_F plus some equations of the form $\lambda\overline{x}.G(\overline{x}) = \lambda\overline{x}.G(\overline{x}^\pi)$, where G is a new variable not E-instantiated in P'.*
- *$P'_F \models s\sigma =_{\eta\beta E} t\sigma$ for every equation $s = t$ of P.*
- *σ is compatible with P'_F.*

In this case, σ constrained by P'_F, denoted by $\sigma\ |P'_F$ is called a constrained E-unifier *of $P \wedge P_F$. The* solutions *of $\sigma\ |P'_F$ are the substitutions of $\sigma\theta$ where θ is a solution of P'_F.*

Definition 10 (Solve rule for elementary theories).
A **Solve** *rule for the theory E_i is an algorithm that takes as input a problem P_i, pure in E_i and a conjunction P_F of frozen equations (as in definition 5) and that returns P'_i and P'_{iF} such that*

1. *P'_i is a solved form with mgu a constant-preserving substitution σ*[4].
2. *P'_i has no flex-flex equations.*
3. *P'_{iF} contains the equations of P_F plus some only flexible-flexible equations of the form $\lambda\overline{x}.F(\overline{x}) = \lambda\overline{x}.F(\overline{x}^\pi)$, where $F \notin \mathcal{FV}(P_i) \cup Dom(\sigma)$.*
4. *F can be E_i-instantiated by σ only if E_i has been chosen as the theory of F at the step* **C.5**
5. *$F\sigma$ can be of the form $\lambda\overline{x}.c[G(\cdots)]$, where $F, G \in \mathcal{FV}(P_i)$ and another theory than E_i has been chosen for G at the step* **C.5**, *only if $F <_{oc} G$, for the ordering chosen at the step* **C.6**.
6. *σ is compatible with P'_F.*
7. *$P'_{iF} \models \lambda\overline{x}.s =_{E_i} \lambda\overline{x}.t$ for all the equations $\lambda\overline{x}.s =_{E_i} \lambda\overline{x}.t$ of P_i.*

Proposition 1. *Let $s = t$ be an equation, pure in the theory E_i, and let σ be an E-solution of $s = t$. Then there exists a set of equations P_{perm} of the form $\lambda\overline{x}^\pi.F(\overline{x}) = \lambda\overline{x}^\varphi.F(\overline{x})$, and two substitutions σ_{E_i} and θ such that*

- *$\sigma =_E \sigma_{E_i}\theta$.*
- *σ_{E_i} is pure in the theory E_i,*
- *θ is an E-solution of P_{perm}.*
- *$P_{perm} \models s\sigma_{E_i} =_{\eta\beta E_i} t\sigma_{E_i}$,*
- *if $F \in \mathcal{D}om(\sigma)$ and there exists a permutation π such that $\lambda\overline{x}.F(\overline{x})\sigma =_{\eta\beta E} \lambda\overline{x}.F(\overline{x}^\pi)\sigma$, then $P_{perm} \models \lambda\overline{x}.F(\overline{x})\sigma_{E_i} =_{\eta\beta E_i} \lambda\overline{x}.F(\overline{x}^\pi)\sigma_{E_i}$.*

The result is obtained by using Theorem 2 and adapting the proof of Theorem 5.1 of [3], which is the corresponding theorem in the first-order case.

[4] The correctness will be preserved if one allows non-constant-preserving substitutions, but the redundancy of the complete sets of unifiers will be increased in this case.

The Algorithm

ALGORITHM FOR PATTERN UNIFICATION MODULO $E_0 \cup \cdots \cup E_n$

1. Apply as long as possible the rules **VA** and **Split** of section 1.
2. Perform successively the steps **C.1** to **C.6**.
3. Apply a **Solve** rule for theory E_i to each P_i accordingly to definition 10.
4. Return $P'_0 \wedge P'_1 \wedge \cdots \wedge P'_n \wedge P_F \wedge \bigwedge_{1\leq i\leq n} P'_{iF}$.

Theorem 3. *Given an equational theory $E = E_0 \cup \cdots \cup E_n$, where the E_is are defined over disjoint signatures $\mathcal{F}_0, \ldots, \mathcal{F}_n$ and a unification problem P, containing only algebraic symbols of $\mathcal{F}_0 \cup \cdots \cup \mathcal{F}_n$,*

- *The above algorithm returns a constrained DAG-E-solved form of P.*
- *Every E-unifier of P is a solution of a constrained DAG-solved form computed by the above algorithm.*

Corollary 1. *Unifiability of higher-order patterns is decidable in combinations of theories having a* **Solve** *rule.*

3 A Solve Rule for Some Syntactic Theories

In [4], we show how to do pattern unification for a narrow class of theories: a subset of the simple syntactic theories. For lack of space, we just give here some hints on how to design a **Solve** rule for the free theory, the theory of left-distributivity LD and the commutativity C. These three theories are *simple* theories, *i.e.*, they have no equality between a term and one of its proper subterms. As it is well-known from the works on first-order unification, compound cycles or theory conflicts cannot be solved in such theories. It is easy to show that the following two rules are correct for simple theories:

Clash $F = s \quad \to \bot$
if F is E_i-instantiated and E_j, $j \neq i$ has been chosen for F at **C.5**.

Cycle $F = c[G] \quad \to \bot$
if c is a non-empty context and $F \not<_{oc} G$ for the ordering chosen at **C.6**.

Now, the free theory and LD have their symbols *decomposable* (*i.e.*, $f(s_1, \ldots, s_n) =_E f(t_1, \ldots, t_n)$ iff $s_i =_E t_i$) and C enjoys a similar property: $s_1 + s_2 =_C t_1 + t_2$ iff $s_1 =_C t_1 \wedge s_2 =_C t_2$ or $s_1 =_C t_2 \wedge s_2 =_C t_1$. Hence, the rules for testing the compatibility of a solved form with a frozen equation are:

Fail $\lambda\overline{x}.F(\overline{x}) = \lambda\overline{x}.F(\overline{x^\pi}) \quad \rightarrow \quad \bot$
if F is not a new variable and $\pi \notin \text{Perm}(F)$ as chosen at the step **C.3**.

Dec-Propagate
$F = \lambda\overline{x}.f(\overline{s_n}) \wedge \lambda\overline{x}.F(\overline{x}) = \lambda\overline{x}.F(\overline{x^\pi}) \quad \rightarrow$
$\quad F = \lambda\overline{x}.f(\overline{s_n}) \bigwedge_{1 \le i \le n} \lambda\overline{x}.s_i = \lambda\overline{x^\pi}.s_i$
if f is a decomposable constant or a bound variable.

C-Propagate
$F = \lambda\overline{x}.s_1 + s_2 \wedge \lambda\overline{x}.F(\overline{x}) = \lambda\overline{x}.F(\overline{x^\pi}) \quad \rightarrow$
$\quad F = \lambda\overline{x}.s_1 + s_2 \;\wedge\; ((\lambda\overline{x}.s_1 = \lambda\overline{x^\pi}.s_1 \;\wedge\; \lambda\overline{x}.s_2 = \lambda\overline{x^\pi}.s_2)$
$\quad\quad \vee\; (\lambda\overline{x}.s_1 = \lambda\overline{x^\pi}.s_2 \;\wedge\; \lambda\overline{x}.s_2 = \lambda\overline{x^\pi}.s_1))$
if $+$ is a commutative algebraic symbol.

The **Mutate** rule of [4], together with the rule **Coalesce** and a failure rule when two (non-new) variables are identified allow us to compute a solved form satisfying the conditions 1 to 3 and 7 of definition 10. The first of the two above sets of rules allows us to fulfill conditions 4 and 5, and the second, condition 6.

4 From AC to Abelian Groups

In this section, we consider the associativity-commutativity, AC and some of its usual extensions ACU (AC with unit), AG (the Abelian groups) and ACUN (ACU with nilpotence). For lack of space, we only give the flavor of a **Solve** rule for these theories. Some more details can be found for AC in our previous paper [5].

In the first order case, the unification algorithm consists of counting the number of times an immediate subterm from another theory occurs in each side of an equation: both sides must have the same number of occurrences. We associate with each algebraic variable x an integer variable x_t representing the number of times the value of x contains the term t as an immediate subterm, and we translate each equation between two terms into a linear equation over the integers. These linear equations have to be solved over different integer domains depending on the considered theory. Then the solutions for the unification problem are built from the integer solutions, modulo some restrictions, in order to get some "well-formed" terms, and a complete set of unifiers. Thanks to Theorem 2, the same approach can also be used in the pattern case, as shown in [5] for the AC case. The main difference comes from the bound variables: if $\lambda\overline{x}.F(\overline{x})$ introduces a term $t(\overline{x})$, then $\lambda\overline{x}.F(\overline{x^\pi})$ introduces $t(\overline{x^\pi})$, and we do not know *a priori* whether $t(\overline{x})$ and $t(\overline{x^\pi})$ are equal or not. This is exemplified below:

4.1 An Example of AC(+)-Unification Problem

Consider the equation $\mathcal{E} \equiv \lambda\overline{x_3}.2F(\overline{x_3}) + F(\overline{x_3}^\pi) + 9G(\overline{x_3}) = \lambda\overline{x_3}.2H(\overline{x_3})$ where $\pi = \{1 \mapsto 2; 2 \mapsto 3; 3 \mapsto 1\}$, to be solved modulo AC(+). If F introduces α times the term $t(\overline{x_3})$, then F^π introduces α times the term $t(\overline{x_3}^\pi)$. If $t(\overline{x_3})$ and $t(\overline{x_3}^\pi)$

are distinct, we have to count also the number of times F introduces $t(\overline{x_3}^{\pi})$, and finally, we have to count the number of $t(\overline{x_3}^{\pi^2})$. Then we can stop since π^3 is the identity. Let us denote by $\mathcal{G}(\pi)$ the group of permutations generated by π. The above unification problem is translated into 2 subsystems:

$$S_{\mathcal{G}(\pi)} = 2\alpha \ + \alpha \ + 9\beta \ = 2\gamma \qquad\qquad S_{\{id\}} = \begin{cases} 2\alpha'_{id} + \alpha'_{\pi^2} + 9\beta'_{id} = 2\gamma'_{id} \\ 2\alpha'_{\pi} + \alpha'_{id} + 9\beta'_{\pi} = 2\gamma'_{\pi} \\ 2\alpha'_{\pi^2} + \alpha'_{\pi} + 9\beta'_{\pi^2} = 2\gamma'_{\pi^2} \end{cases}$$

where F (resp. G, H) introduces α (resp. β, γ) times a term $t(\overline{x_3})$ such that $\lambda\overline{x_3}.t(\overline{x_3}) =_E \lambda\overline{x_3}.t(\overline{x_3}^{\pi})$, and α'_{φ} (resp. β'_{φ}, γ'_{φ}) times $s(\overline{x_3}^{\varphi})$, $\varphi = id, \pi, \pi^2$ where $\lambda\overline{x_3}.s(\overline{x_3}) \neq_E \lambda\overline{x_3}.s(\overline{x_3}^{\pi})$. These two systems are solved over *non-negative* integers, and as in the first order case, the unifiers are built from the Diophantine solutions, with the main difference that in the pattern case the introduced variables L_is corresponding to a solution of $S_{\mathcal{G}(\pi)}$ are constrained by $\mathrm{Perm}(L_i) = \mathcal{G}(\pi)$.

In the same spirit, this can be done in the extensions of AC, such as ACU, AG and ACUN, where the equations are solved by counting how many times a variable introduces a given term.

4.2 Handling the Additional Constraints for a Solve Rule

Let us assume now that the problem to be solved is a part of a combination problem and have to be solved modulo the conditions of definition 10. Each variable F comes with some additional assumptions, such as the theory in which it can be instantiated, and its group of permutations $\mathrm{Perm}(F)$ as defined in section 3. These constraints are also translated into linear constraints over integers. Indeed the constraint $\mathrm{Perm}(H) = \mathcal{H}$ corresponds to the equations $\lambda\overline{x_3}.H(\overline{x_3}) = \lambda\overline{x_3}.H(\overline{x_3}^{\psi})$, where $\psi \in \mathcal{H}$, and these equations are translated into a system of linear equations over variables exactly in the same way as before, except that we have to consider all subgroups of the permutation group generated by π and $\mathcal{H}$ as possible invariants for an introduced term.

A constraint like "the variable G cannot be instantiated in the considered theory" means here that the number of terms from another theory introduced by its value is exactly one. Only one among all of the integer variables corresponding to G has to be equal to 1, the others being null.

A constraint like $H \not<_{oc} G$ will be treated in a second step, after one has built the solutions to the unification problem. If we get a solution where G occurs in the value of H, this is due to the fact that G is (equal to a new variable which is) associated with a solution which has some non-zero values for some integer variables corresponding to H. In the AC and ACU cases, such a integer solution has to be discarded, while in the AG and ACUN cases, this problem can be fixed by computing a particular solution such that these integer variables are null.

In all the 4 cases of AC, ACU, AG and ACUN, we are able to get a solved form which satisfies the additional hypotheses of the **Solve** rule.

5 Conclusion

We believe that with the emergence of higher-order rewriting and higher-order logic programming, there will be a use for pattern unification modulo equational theories. The algorithm that we proposed here is meant to provide a decidability result: it will not behave satisfactorily in practice, due to the heavy nondeterminism. It will be necessary to investigate how to reduce the nondeterminism as we did in the first-order case in [2,3]. Another issue of interest will be to develop matching algorithms which should be dramatically more efficient in practice.

Although our method for elementary unification works well for the AC-like theories, we do not have a general method for ensuring the compatibility of a unifier with an equation of the form $\lambda xy.F(x,y) = \lambda xy.F(y,x)$. For instance, the known methods for unification in Boolean rings do not use equations over the integers, and such equations do not translate naturally as shown in the previous section. Actually, we conjecture that there exists a theory with decidable unification of problems with linear constant restriction (the equivalent in the first-order case of our **Solve** rule [1]) and undecidable pattern unification.

References

1. Franz Baader and Klaus Schulz. Unification in the Union of Disjoint Equational Theories: Combining Decision Procedures. *Journal of Symbolic Computation*, 21(2), February 1996.
2. Alexandre Boudet. *Unification dans les Mélanges de Théories équationnelles*. Thèse de doctorat, Université Paris-Sud, Orsay, France, February 1990.
3. Alexandre Boudet. Combining unification algorithms. *Journal of Symbolic Computation*, 16:597–626, 1993.
4. Alexandre Boudet. Unification of higher-order patterns modulo simple syntactic equational theories. *Discrete Mathematics and Theoretical Computer Science*, 4(1):11–30, 2000.
5. Alexandre Boudet and Evelyne Contejean. AC-unification of higher-order patterns. In Gert Smolka, editor, *Principles and Practice of Constraint Programming*, volume 1330 of *Lecture Notes in Computer Science*, pages 267–281, Linz, Austria, October 1997. Springer-Verlag.
6. Alexandre Boudet and Evelyne Contejean. About the Confluence of Equational Pattern Rewrite Systems. In C. and H. Kirchner, editors, *15th International Conference on Automated Deduction*, volume 1421 of *Lecture Notes in Artificial Intelligence*, pages 88–102. Springer-Verlag, 1998.
7. Val Breazu-Tannen. Combining algebra and higher-order types. In *Proc. 3rd IEEE Symp. Logic in Computer Science, Edinburgh*, July 1988.
8. François Fages. Associative-commutative unification. *Journal of Symbolic Computation*, 3(3), June 1987.
9. Warren D. Goldfarb. Note on the undecidability of the second-order unification problem. *Theoretical Computer Science*, 13:225–230, 1981.
10. R. Hindley and J. Seldin. *Introduction to Combinators and λ-calculus*. Cambridge University Press, 1986.

11. Jieh Hsiang and Michaël Rusinowitch. On word problems in equational theories. In Thomas Ottmann, editor, *14th International Colloquium on Automata, Languages and Programming*, volume 267 of *Lecture Notes in Computer Science*, pages 54–71, Karlsruhe, Germany, July 1987. Springer-Verlag.
12. Gérard Huet. *Résolution d'équations dans les langages d'ordre* $1, 2, \ldots \omega$. Thèse d'Etat, Univ. Paris 7, 1976.
13. Jean-Pierre Jouannaud. Executable higher-order algebraic specifications. In C. Choffrut and M. Jantzen, editors, *Proc. 8th Symp. on Theoretical Aspects of Computer Science, Hamburg, LNCS 480*, pages 16–25, February 1991.
14. Jean-Pierre Jouannaud and Claude Kirchner. Solving equations in abstract algebras: A rule-based survey of unification. In Jean-Louis Lassez and Gordon Plotkin, editors, *Computational Logic: Essays in Honor of Alan Robinson*. MIT-Press, 1991.
15. Jean-Pierre Jouannaud and Mitsuhiro Okada. Executable higher-order algebraic specification languages. In *Proc. 6th IEEE Symp. Logic in Computer Science, Amsterdam*, pages 350–361, 1991.
16. M. Livesey and Jörg H. Siekmann. Unification of bags and sets. Research report, Institut fur Informatik I, Universität Karlsruhe, West Germany, 1976.
17. Richard Mayr and Tobias Nipkow. Higher-order rewrite systems and their confluence. *Theoretical Computer Science*, 192(1):3–29, February 1998.
18. D. Miller. A logic programming language with lambda-abstraction, function variables, and simple unification. In P. Schroeder-Heister, editor, *Extensions of Logic Programming*. LNCS 475, Springer Verlag, 1991.
19. Zhenyu Qian and Kang Wang. Modular AC-Unification of Higher-Order Patterns. In Jean-Pierre Jouannaud, editor, *First International Conference on Constraints in Computational Logics*, volume 845 of *Lecture Notes in Computer Science*, pages 105–120, München, Germany, September 1994. Springer-Verlag.
20. M. Schmidt-Schauß. Unification in a combination of arbitrary disjoint equational theories. *Journal of Symbolic Computation*, 1990. Special issue on Unification.
21. Jörg H. Siekmann. Unification theory. *Journal of Symbolic Computation*, 7(3 & 4), 1989. Special issue on unification, part one.
22. M. E. Stickel. A complete unification algorithm for associative-commutative functions. In *Proceedings 4th International Joint Conference on Artificial Intelligence, Tbilissi (USSR)*, pages 71–76, 1975.

Matching Power

Horatiu Cirstea, Claude Kirchner, and Luigi Liquori

LORIA INRIA INPL ENSMN
54506 Vandoeuvre-lès-Nancy BP 239 Cedex France
{Horatiu.Cirstea,Claude.Kirchner,Luigi.Liquori}@loria.fr
www.loria.fr/{~cirstea,~ckirchne,~lliquori}

Abstract. In this paper we give a simple and uniform presentation of the rewriting calculus, also called *Rho Calculus*. In addition to its simplicity, this formulation explicitly allows us to encode complex structures such as lists, sets, and objects. We provide extensive examples of the calculus, and we focus on its ability to represent some object oriented calculi, namely the *Lambda Calculus of Objects* of Fisher, Honsell, and Mitchell, and the *Object Calculus* of Abadi and Cardelli. Furthermore, the calculus allows us to get object oriented constructions unreachable in other calculi. *In summa*, we intend to show that because of its matching ability, the Rho Calculus represents a *lingua franca* to naturally encode many paradigms of computations. This enlightens the capabilities of the rewriting calculus based language ELAN to be used as a logical as well as powerful semantical framework.

1 Introduction

Matching is a feature provided implicitly in many, and explicitly in few, programming languages. In this paper, by making matching a "first-class" concept, we present, experiment with, and show the expressive power of a new version of the rewriting calculus, also called Rho Calculus (ρCal).

The ability to discriminate patterns is one of the main basic mechanisms the human reasoning is based on; as one commonly says "one picture is better than a thousand explanations". Indeed, the ability to recognize patterns, *i.e.* pattern matching, is present since the beginning of information processing modeling. Instances of it can be traced back to pattern recognition and it has been extensively studied when dealing with strings [26], trees [19] or feature objects [2].

Matching occurs implicitly in many languages through the parameter passing mechanism but often as a very simple instance, and explicitly in languages like PROLOG and ML, where it can be quite sophisticated [28,27]. It is somewhat astonishing that one of the most commonly used model of computation, the lambda calculus, uses only trivial pattern matching. This has been extended, initially for programming concerns, either by the introduction of patterns in lambda calculi [33,36,11], or by the introduction of matching and rewrite rules in functional programming languages. And indeed, many works address the integration of term rewriting with lambda calculus, either by enriching first-order rewriting with higher-order capabilities, or by adding to lambda calculus

A. Middeldorp (Ed.): RTA 2001, LNCS 2051, pp. 77–92, 2001.

algebraic features allowing one, in particular, to deal with equality in an efficient way. In the first case, we find the works on CRS [25] and other higher-order rewriting systems [39,29], in the second case the works on combination of lambda calculus with term rewriting [30,4,16,22], to mention only a few.

Embedding more information in the matching process makes it appropriate to deal with complex tasks like program transformations [20] or theorem proving [32]. In that direction, matching in elaborated theories has been also studied extensively, either in equational theories [21,5] or in higher-order logic [12,31], where it is still an open problem at order five.

Matching allows one to discriminate between alternatives. Once the patterns are recognized, the action to be taken on the appropriate pattern should be described, and this is what rewriting is designed for. The corresponding pattern is thus rewritten in an appropriate instance of a new one. The mechanism that describes this process is the *rewriting calculus*. Its main design concept is to make all the basic ingredients of rewriting explicit objects, in particular the notions of *rule application* and *result*. By making the application explicit, the calculus emphasizes on one hand the fundamental role of matching, and on the other hand the intrinsic higher-order nature of rewriting. By making the results explicit, the Rho Calculus has the ability to handle non-determinism in the sense of a collection of results: an empty collection of results represents an application failure, a singleton represents a deterministic result, and a collection with more than one element represents a non-deterministic choice between the elements of the collection.

For example, assuming a, b, and c to be different constants, in the rewriting calculus the application of the rewrite rule $a \to b$ on the term a is expressed by the term $(a \to b)\bullet a$. This term is evaluated into the term b. The application of same rule on the term c is expressed by the term $(a \to b)\bullet c$ and it evaluated to $null$, therefore memorizing the fact that the rule is not applicable since the term a does not match the term c. Of course one can use variables and the simplest way to do it is just with rewrite rules whose left-hand side is a single variable term, like in $(X \to b)\bullet c$. Such a rule always applies and the previous term evaluates to c. This trivial rule application corresponds indeed exactly to the lambda calculus application: the previous term could be written as $(\lambda X.b)\, c$. This enlightens one of the nice feature of the rewriting calculus to abstract not only on variables like in the previous example, but also on arbitrary terms, including non-linear ones.

This matching power of the calculus provides important expressivity capabilities. As suggested by the previous example, it embeds lambda calculus, but also permits the representation of term rewrite derivations, even in the conditional case. More generally, it allows us to describe traversal, evaluation and search strategies like leftmost innermost, or breath first [7]. In [7], we have shown that *conditional* rewrite rules of the form "$l \to r$ if c" can be faithfully represented in the rewriting calculus as $l \to (\mathsf{True} \to r)\bullet(strat\bullet c)$, where $strat$ is a suitable term representing the normalization strategy of the condition.

Rewriting is central in several programming languages developed since the seventies. Amongst the main ones let us mention OBJ [18], ASF+SDF [35],

Maude [10], CafeOBJ [15], Stratego [38], and ELAN [24,3,34] which has been at the origin of some of the main concepts of the rewriting calculus. In turn, the Rho Calculus provides a natural semantics to such languages, and in particular to ELAN, covering the notion of rule application strategy, a fundamental concept of the language.

In this paper, we give the newest description of the Rho Calculus, as introduced in [9]. It provides a simplified version of the evaluation rules of the calculus as well as a generic and explicit handling of result structures, a point left open in the previous works [6,7].

The contributions of this paper are therefore the following:

- we provide a broad set of examples showing the expressiveness of the Rho Calculus obtained mainly thanks to its "matching power" and how this makes it suitable to uniformly model various paradigms of computation;
- we show how the matching power of the Rho Calculus allows us to encode two major object-calculi which have strongly influenced the type-theoretical research of the last five years: the *Object Calculus* ($\varsigma\mathcal{O}bj$) of Abadi and Cardelli [1] and the *Lambda Calculus of Objects* of Fisher, Honsell, and Mitchell [14] ($\lambda\mathcal{O}bj$). Moreover, we show two examples in Rho Calculus that cannot be encoded in the above calculi.

Road Map of the Paper. The paper is structured as follows: in Section 2, we present the syntax and the small-step semantics of the Rho Calculus; Section 3 presents a *plethora* of examples describing the power of matching; Section 4 presents the encoding of the Lambda Calculus of Objects and of the Object Calculus in the Rho Calculus. Conclusions and further works are finally discussed in Section 5. An extended version of the paper can be found in [8].

2 Syntax and Semantics

Notational Conventions. In this paper, the symbol t ranges over the set $\mathcal{T}$ of terms, the symbols $S, X, Y, Z, \ldots$ range over the infinite set $\mathcal{V}$ of variables, the symbols $null, \oplus, \circ, a, b, \ldots, z, 0, 1, 2, \ldots$ range over the infinite set $\mathcal{C}$ of constants of fixed arity. All symbols can be indexed. The symbol $\equiv$ denotes syntactic identity of objects like terms or substitutions. We work modulo α-conversion, and we follow the Barendregt convention that free and bound variables have different names.

2.1 Syntax

The syntax of the ρCal is defined as follows:

$$t ::= a \mid X \mid t \to t \mid t \bullet t \mid \qquad \text{plain terms}$$

$$null \mid t, t \qquad \text{structured terms}$$

The main intuition behind this syntax is that a rewrite rule $t \to t$ is an *abstraction*, the left-hand-side of which determines the bound variables and some pattern structure. The application of a ρCal-term on another ρCal-term is represented by "$\bullet$". The terms can be grouped together into a structure built using the "," operator and, according to the theory behind this operator, different structures can be obtained. The term *null* denotes an empty structure.

We assume that the application operator "$\bullet$" associates to the left, while the "$\rightarrow$" and the "," operators associate to the right. The priority of the application "$\bullet$" is higher than that of the "$\rightarrow$" operator which is in turn of higher priority than the "," operator.

Definition 1 (Some Type Signatures and Abbreviations).

$\rightarrow : \mathcal{T} \times \mathcal{T} \Rightarrow \mathcal{T}$		$t_1.t_2$	$\triangleq t_1 \bullet t_2 \bullet t_1$	*self-application*
$\bullet\ : \mathcal{T} \times \mathcal{T} \Rightarrow \mathcal{T}$	*and*	$t(t_1 \ldots t_n)$	$\triangleq t \bullet t_1 \ldots \bullet t_n$	*function-application* ($n \in \mathbb{N}$)
$,\ : \mathcal{T} \times \mathcal{T} \Rightarrow \mathcal{T}$		$(t_i)^{i=1\ldots n}$	$\triangleq t_1, \ldots, t_n$	*structure* ($n \in \mathbb{N}$)

We draw the attention of the reader on the main difference between "$\bullet$" denoting the *application*, and "." denoting the object-oriented *self-application* operator.

2.2 Matching Theories

An important parameter of the ρCal is the matching theory $\mathbb{T}$. We give below examples of theories $\mathbb{T}$ defined equationally.

Definition 2 (Matching theories).

- *the Empty theory* $\mathbb{T}_\emptyset$ *of equality (up to* α*-conversion) is defined as the following inference rules:*

$$\frac{t_1 = t_2 \quad t_2 = t_3}{t_1 = t_3}\ (Tra) \qquad \frac{t_1 = t_2}{t_2 = t_1}\ (Sym) \qquad \frac{t_1 = t_2}{t_3[t_1]_p = t_3[t_2]_p}\ (Ctx) \qquad \frac{}{t = t}\ (Ref)$$

where $t_1[t_2]_p$ *denotes the term* t_1 *with the term* t_2 *at position* p*. The* α*-conversion definition follows the standard intuition and is made precise for* ρCal *in [7].*
- *the theory of Commutativity* $\mathbb{T}_{C(f)}$ *(resp. Associativity* $\mathbb{T}_{A(f)}$*) is defined as* $\mathbb{T}_\emptyset$ *plus the following inference rules:*

$$\frac{}{f(t_1\ t_2) = f(t_2\ t_1)}\ (Com) \qquad \frac{}{f(f(t_1\ t_2)\ t_3) = f(t_1\ f(t_2\ t_3))}\ (Ass)$$

- *the theory of Idempotency* $\mathbb{T}_{I(f)}$ *is defined as* $\mathbb{T}_\emptyset$ *plus the axiom* $f(t\ t) = t$.
- *the theory of Neutral Element* $\mathbb{T}_{N(f^0)}$ *is defined as* $\mathbb{T}_\emptyset$ *plus the following inference rules:*

$$\frac{}{f(0\ t) = t}\ (0_{Left}) \qquad \frac{}{f(t\ 0) = t}\ (0_{Right})$$

- *the theory of the Lambda Calculus of Objects,* $\mathbb{T}_{\lambda Obj}$*, is obtained by considering the symbol "," as associative and null as its neutral element, i.e.:*

$$\mathbb{T}_{\lambda Obj} = \mathbb{T}_{A(,)} \cup \mathbb{T}_{N(,^{null})}$$

– *the theory of the Object Calculus,* $\mathbb{T}_{\varsigma Obj}$*, is obtained by considering the symbol "," as associative and commutative and null as its neutral element, i.e.:*

$$\mathbb{T}_{\varsigma Obj} = \mathbb{T}_{A(,)} \cup \mathbb{T}_{C(,)} \cup \mathbb{T}_{N(,null)} = \mathbb{T}_{\lambda Obj} \cup \mathbb{T}_{C(,)}$$

Other interesting theories can be built from the above ones, such as *e.g.* $\mathbb{T}_{MSet(f,nil)}$, and $\mathbb{T}_{Set(f,nil)}$ [7]. For the sake of completeness, we include in the paper the definition of syntactic matching, which can also be found in [7], together with more explanatory examples.

Definition 3 (Syntactic Matching). *For a given theory* $\mathbb{T}$ *over* ρCal*-terms:*

1. *a* $\mathbb{T}$-match equation *is a formula of the form* $t_1 \ll_{\mathbb{T}} t_2$*;*
2. *a* substitution σ *is a solution of the* $\mathbb{T}$*-match equation* $t_1 \ll_{\mathbb{T}} t_2$ *if* $\sigma t_1 =_{\mathbb{T}} t_2$*;*
3. *a* $\mathbb{T}$-matching system *is a conjunction of* $\mathbb{T}$*-match equations;*
4. *a substitution* σ *is a solution of a* $\mathbb{T}$*-matching system if it is a solution of all the* $\mathbb{T}$*-match equations in it;*
5. *a* $\mathbb{T}$*-matching system is* trivial *when all substitutions are solution of it and we denote by* $\mathbb{F}$ *a* $\mathbb{T}$*-matching system without solution;*
6. *we define the function Sol on a* $\mathbb{T}$*-matching system* T *as returning the* $\prec$*-ordered*[1] *list of all* $\mathbb{T}$*-matches of* T *when* T *is not trivial and the list containing only* $\sigma_{\mathbf{id}}$*, where* $\sigma_{\mathbf{id}}$ *is the identity substitution, when* T *is trivial.*

Notice that when the matching algorithm fails (*i.e.* returns $\mathbb{F}$), the function $\mathcal{S}ol$ returns the empty list. A more detailed discussion on decidability of matching can be found in [7].

For example, in $\mathbb{T}_\emptyset$, the matching substitution from a ρCal-term t_1 to a ρCal-term t_2 can be computed by the rewrite system presented in Figure 1, where the symbol $\wedge$ is assumed to be associative and commutative, and $\diamond_1$, $\diamond_2$ are either constant symbols or the prefix notations of "," or "•" or "→".

Starting from a matching system T, the application of this rule set terminates and returns either $\mathbb{F}$ when there are no substitutions solving the system, or a system T' in "normal form" from which the solution can be trivially inferred [23]. This set of rules could be extended to deal with more elaborated theories like commutativity.

2.3 Operational Semantics

For a given total ordering $\prec$ on substitutions (which is left implicit in the notation) and a theory $\mathbb{T}$, the operational semantics is defined by the computational rules given in Figure 2. The central idea of the main rule of the calculus (ρ) is that the application of a rewrite rule $t_1 \to t_2$ at the root position of a term t_3, consists in computing all the solutions of the matching equation ($t_1 \ll_{\mathbb{T}} t_3$) in the theory $\mathbb{T}$ and applying all the substitutions from the $\prec$-ordered list returned by the function $Sol(t_1 \ll_{\mathbb{T}} t_3)$ to the term t_2. When there

[1] We consider a total order $\prec$ on the set of substitutions [7].

$$
\begin{array}{lll}
\diamond_1(t_1 \ldots t_n) \ll_{\mathbb{T}_\emptyset} \diamond_2(t'_1 \ldots t'_m) & \leadsto & \begin{cases} \bigwedge_{i=1\ldots n} t_i \ll_{\mathbb{T}_\emptyset} t'_i & \text{if } \diamond_1 \equiv \diamond_2 \text{ and } n = m \\ \mathbb{F} & \text{otherwise} \end{cases} \\
(X \ll_{\mathbb{T}_\emptyset} t) \wedge (X \ll_{\mathbb{T}_\emptyset} t') & \leadsto & \begin{cases} X \ll_{\mathbb{T}_\emptyset} t & \text{if } t =_{\mathbb{T}_\emptyset} t' \\ \mathbb{F} & \text{otherwise} \end{cases} \\
t \ll_{\mathbb{T}_\emptyset} X & \leadsto & \mathbb{F} \qquad\qquad\qquad \text{if } t \notin \mathcal{V} \\
\mathbb{F} \wedge (t \ll_{\mathbb{T}_\emptyset} t') & \leadsto & \mathbb{F}
\end{array}
$$

Fig. 1. Rules for Syntactic Matching

$$
\begin{array}{llll}
(\rho) & (t_1 \to t_2)^\bullet t_3 & \mapsto_{\mathbb{T}} & \begin{cases} null & \text{if } t_1 \ll_{\mathbb{T}} t_3 \text{ has no solution} \\ \sigma_1 t_2, \ldots, \sigma_n t_2 & \text{if } \sigma_i \in \mathcal{S}ol(t_1 \ll_{\mathbb{T}} t_3), \sigma_1 \prec \sigma_{i+1},\ n \leq \infty \end{cases} \\
(\epsilon) & (t_1, t_2)^\bullet t_3 & \mapsto_{\mathbb{T}} & t_1{}^\bullet t_3, t_2{}^\bullet t_3 \\
(\nu) & null^\bullet t & \mapsto_{\mathbb{T}} & null
\end{array}
$$

Fig. 2. Evaluation rules of the ρCal

is no solution for the matching equation $t_1 \ll_{\mathbb{T}} t_3$, the special constant *null* is obtained as result of the application. Notice that in some theories, there could be an infinite set of solutions to the matching problem $(t_1 \ll_{\mathbb{T}} t_3)$; possible ways to deal with the infinitary case are described in [6].

The other rules (ϵ) and (ν) deal with the distributivity of the application on the structures whose constructors are "," and *null*. When the theory $\mathbb{T}$ is clear from the context, its denotation will be omitted. Notice that if t_1 is a variable, then the (ρ)-rule corresponds exactly to the (β)-rule of the lambda calculus.

With respect to the previous presentation of the Rho Calculus [7], we have modified the notation of the application operator which was denoted $[\_](\_)$, but more importantly, the evaluation rules have been simplified on one hand, and generalized to deal with generic result structures on the other hand.

As usual, given a theory $\mathbb{T}$, we denote by $=_\rho$ the smallest reflexive, symmetric, and transitive relation containing $\mapsto_{\mathbb{T}}$, stable by context and substitution. When working modulo reasonably powerful theories $\mathbb{T}$, the evaluation rules of the ρCal are confluent:

Theorem 1 (Confluence in $\mathbb{T}_\emptyset$). *Given a term t_1 such that all its abstractions contain no arrow in the first argument, if $t_1 \mapsto\!\!\!\!\to_{\mathbb{T}_\emptyset} t_2$ and $t_1 \mapsto\!\!\!\!\to_{\mathbb{T}_\emptyset} t_3$ then there exists a term t_4 such that $t_2 \mapsto\!\!\!\!\to_{\mathbb{T}_\emptyset} t_4$ and $t_3 \mapsto\!\!\!\!\to_{\mathbb{T}_\emptyset} t_4$.*

Proof. *The proof follows the same lines defined in [7].*

3 Examples in Rho Calculus

In the following section we present some simple examples intended to help the reader in the understanding of the behavior of the Rho Calculus.

Example 1 (In $\mathbb{T}_\emptyset$).

1. The application of the simple rewrite rule $a \to b$ to a, *i.e.* $(a \to b)\bullet a$, is evaluated to b since $Sol(a \ll_{\mathbb{T}_\emptyset} a) = \sigma_{\mathbf{id}}$ and $\sigma_{\mathbf{id}} b \equiv b$;
2. The matching between the left-hand side of the rule and the argument can also fail and in this case the result of the application is the constant *null*, *i.e.*: $(a \to b)\bullet c \overset{\rho}{\mapsto} null$;
3. When the left-hand side of a rewrite rule is not a ground term, the matching can yield a substitution different from $\sigma_{\mathbf{id}}$, e.g. $(X \to X)\bullet a \overset{\rho}{\mapsto} [X/a]X \equiv a$;
4. The non-deterministic application of two rewrite rules is represented by the application of the structure containing the respective rules: $(X \to X(a), Y \to Y(b))\bullet c \overset{\epsilon}{\mapsto} (X \to X(a))\bullet c, (Y \to Y(b))\bullet c \overset{\rho}{\mapsto\!\!\!\to} [X/c]X(a), [Y/c]Y(b) \equiv c(a), c(b)$;
5. The selection of the field cx inside the record structure $(cx \to 0, cy \to 0)$ evaluates to the term $(0, null)$, *i.e.*: $(cx \to 0, cy \to 0)\bullet cx \overset{\epsilon}{\mapsto} (cx \to 0)\bullet cx, (cy \to 0)\bullet cx \overset{\rho}{\mapsto\!\!\!\to} (0, null)$;
6. Functions are first-class entities in the ρCal: $(X \to (X\bullet a))\bullet\ (Y \to Y) \overset{\rho}{\mapsto} (Y \to Y)\bullet a \overset{\rho}{\mapsto} a$;
7. The lambda calculus with *patterns* [33] can be easily represented in the ρCal. For instance, the lambda-term $\lambda Pair(X\,Y).X$, can be represented and reduced as follows: $(Pair(X\ Y) \to X)\bullet Pair(a\ b) \overset{\rho}{\mapsto} [X/a, Y/b]X \equiv a$;
8. Starting from the fixed-point combinators of the lambda calculus, we can define a ρCal-term that applies recursively a given ρCal-term. We use the classical fixed-point $Y_\lambda \triangleq (A_\lambda\ A_\lambda)$ with $A_\lambda \triangleq \lambda X.\lambda Y.Y(XXY)$, which can be translated as $Y_\rho \triangleq A_\rho\bullet A_\rho$ with $A_\rho \triangleq X \to Y \to Y\bullet(X\bullet X\bullet Y)$. Then: $Y_\rho\bullet t \triangleq A_\rho\bullet A_\rho\bullet t \triangleq (X \to Y \to Y\bullet(X\bullet X\bullet Y))\bullet A_\rho\bullet t \overset{\rho}{\mapsto} (Y \to Y\bullet(A_\rho\bullet A_\rho\bullet Y))\bullet t \overset{\rho}{\mapsto} t\bullet(A_\rho\bullet A_\rho\bullet t) \triangleq t\bullet(Y_\rho\bullet t)$. Starting from the Y_ρ term, we can define more elaborated terms describing, for example, the repeated application of a given term or normalization strategies according to a given rewrite rule [7];
9. Let $car \triangleq X, Y \to X$, $cdr \triangleq X, Y \to Y$, $cons \triangleq X \to Y \to (X, Y)$. It is easy to check that $car(a, b, c, null) \mapsto\!\!\!\to a$, and that $cdr(a, b, c, null) \mapsto\!\!\!\to b, c, null$, and that $cons(d\ a, b, c, null) \mapsto\!\!\!\to d, a, b, c, null$.

Example 2 (In $\mathbb{T}_A$, $\mathbb{T}_C$, $\mathbb{T}_{AC}$, and $\mathbb{T}_{N(f^\circ)}$).

1. ($\mathbb{T}_{A(\circ)}$) The application of the rewrite rule $\circ(X\ Y) \to X$ to $\circ(a\ \circ(b\ \circ(c\ d)))$ reduces, thanks to the associativity of $\circ$, to $(a, \circ(a\ b), \circ(a\ \circ(b\ c)))$;
2. ($\mathbb{T}_{C(\oplus)}$) The application of the rewrite rule $\oplus(X\ Y) \to X$ to $\oplus(a\ b)$ reduces, thanks to the associativity-commutativity of $\oplus$, to (a, b), a structure representing all possible results;
3. ($\mathbb{T}_{AC(\oplus)}$) The application of the rewrite rule $\oplus(X\ \oplus(X\ Y)) \to \oplus(X\ Y)$ to $\oplus(a\ \oplus(b\ \oplus(c\ \oplus(a\ d))))$ reduces to $\oplus(a\ \oplus(b\ \oplus(c\ d)))$. The search for the two equal elements is done by matching thanks to the associativity-commutativity of the $\oplus$ operator, while the elimination of doubles is performed by the rewrite rule;

4. ($\mathbb{T}_{N(f^o)}$) Using a theory with a neutral element allows us to "ignore" variables from the rewrite rules. For example, the rewrite rule $X \oplus a \oplus Y \to X \oplus b \oplus Y$ replaces an a with a b in a structure built using the "$\oplus$" operator and containing one or more elements. The application of the previous rewrite rule to $b \oplus a \oplus b$ reduces to $b \oplus b \oplus b$ and the same rule applied to a leads to b, since $a =_{\mathbb{T}_{N(\oplus^o)}} 0 \oplus a \oplus 0$.

The next example shows how the object oriented paradigm can be easily captured in the $\mathbb{T}_{\lambda Obj}$. In particular we focus our example on the usage of the pseudo-variable `this` which is crucial for sending messages inside method bodies. In the ρCal, a method is seen as a term of the shape $m \to S \to t_m$, where m is the name of the method, S is a variable playing the role of `this` and t_m is the body of the method that can contain free occurrences of S. Sending a message m to a structure (*i.e.* an object) t is represented via the alias $t.m$, *i.e.* $t \bullet m \bullet t$. Intuitively, if the method m exists in the structure t, then its body t_m can be executed with the binding of S to the object itself. This type of application is also called, in the object-oriented jargon, *self-application*, and it is fundamental for modeling mutual recursion between methods inside an object.

Example 3 (In $\mathbb{T}_{\lambda Obj}$).

1. This example presents a simple object t with only one method a that do not effectively use the variable S. Let $t \triangleq a \to S \to b$. Then, $t.a \triangleq t \bullet a \bullet t \overset{\rho}{\mapsto} (\sigma_{\mathrm{id}}(S \to b)) \bullet t \equiv (S \to b) \bullet t \overset{\rho}{\mapsto} b$;
2. This example presents an object t with a non-terminating method ω. Let $t \triangleq \omega \to S \to S.\omega$. Then, $t.\omega \overset{\rho}{\mapsto} (S \to S.\omega) \bullet t \overset{\rho}{\mapsto} t.\omega \mapsto\!\!\!\to \ldots$;
3. We consider another object with a non-terminating behavior consisting of two methods *ping* and *pong*, one calling the other via the variable S. Let $t \triangleq (ping \to S \to S.pong, pong \to S \to S.ping)$. Then, $t.ping \triangleq t \bullet ping \bullet t \overset{\epsilon}{\mapsto} ((ping \to S \to S.pong) \bullet ping, (pong \to S \to S.ping) \bullet ping) \bullet t \overset{\rho}{\mapsto\!\!\!\to} (S \to S.pong) \bullet t, null =_{\mathbb{T}_{\lambda Obj}} (S \to S.pong) \bullet t \overset{\rho}{\mapsto} t.pong \mapsto\!\!\!\to t.ping \mapsto\!\!\!\to \ldots$

In the above example, we can notice how natural the use of matching is for directly selecting the method name. Starting from these simple examples, we can now imagine how matching can be use in its full generality (*i.e.* allowing variables as well as appropriate equational theories) in order to deal with more general objects and methods. The purpose of the rest of this paper is to make these aspects precise.

4 Object-Based in Rho Calculus

In this section we focus on two major object-calculi which have influenced the type-theoretical research of the last five years:

- The *Lambda Calculus of Objects* of Fisher, Honsell, and Mitchell [14];
- The *Object Calculus* of Abadi and Cardelli [1].

Both calculi are *prototype-based i.e.* they are based on the notion of "objects" and not of "classes". Nevertheless, classes can be easily encoded as suitable objects. Those calculi have been extensively studied in a "typed" setting where the main objective was to conceive sound type systems capturing the unfortunate run-time error `message-not-understood` which happen when we send a message m to an object which do not have the method m in its interface.

As previously shown in Example 3, structured-terms are well suited to represent objects and to model the special pseudovariable `this`. In order to support the intuition, we start by showing the way some classical examples of objects can be easily expressed in the ρCal.

Example 4 (A Point Object Encoding in $\mathbb{T}_{\lambda\mathcal{O}bj}$). Given the symbols val, get, set and v (used to denote pairs), an object $Point$ is encoded in ρCal by

$$val \to S \to v(1\ 1), get \to S \to S.val, set \to S \to v(X\ Y) \to (S, val \to S' \to v(X\ Y))$$

The term $Point$ represents an object with an attribute val and two methods get and set. The method get gives access to the attribute, while method set is used for modifying the attribute by adding the new value at the end of the object. In this context, it is easy to check that $Point.get \mapsto\!\!\!\!\twoheadrightarrow v(1\ 1)$, and $Point.set(v(2\ 2)) \mapsto\!\!\!\!\twoheadrightarrow Point, (val \to S' \to v(2\ 2))$, and $Point.set(v(2\ 2)).get \mapsto\!\!\!\!\twoheadrightarrow v(1\ 1), v(2\ 2)$. Worthy of notice is that:

1. The call $Point.set(v(2\ 2))$ produces a result which consists of the old $Point$ and the new (modified) value for the attribute val, *i.e.* $val \to S' \to v(2\ 2)$;
2. The call $Point.set(v(2\ 2)).get$ produces a structure composed of two elements, the former representing the value of val before the execution of set (*i.e.* before a side effect), and the latter one after the execution of set;
3. A trivial strategy to recover determinism is to consider only the last value from the list of results, *i.e.* $v(2\ 2)$. From this point of view, the ρCal can be also understood as a formalism to study side effects in imperative calculi;
4. A way to fix imperative features is to modify the encoding of the method set by considering the term $kill_n \triangleq (X, n \to Z, Y) \to X, Y$, and by defining the new object $Point_{imp}$ as
$$val \to \ldots, get \to \ldots, set \to S \to v(X\ Y) \to (kill_{val}(S), val \to S' \to v(X\ Y))$$
such that $Point_{imp}.get \mapsto\!\!\!\!\twoheadrightarrow v(1\ 1)$, and $Point_{imp}.set(v(2\ 2)) \mapsto\!\!\!\!\twoheadrightarrow val \to S' \to v(2\ 2), get \to \ldots, set \to \ldots$, and $Point_{imp}.set(v(2\ 2)).get \mapsto\!\!\!\!\twoheadrightarrow v(2\ 2)$;
5. The moral of this example is that the encoding of objects into the ρCal can strongly modify the behavior of a computation.

In the next example we present the encoding of the Fisher, Honsell, and Mitchell fixed-point operator [14] and its generalization in the ρCal.

Example 5 (A Fixed Point Object). Assume symbols rec and f. The fixed-point object Fix_f for f can be represented in the ρCal as $Fix_f \triangleq rec \to S \to f(S.rec)$. It is not hard to verify that $Fix_f.rec \mapsto\!\!\!\!\twoheadrightarrow f(Fix_f.rec)$. This fixed point can be generalized as $Fix \triangleq rec \to S \to X \to X(S.rec(X))$ and its behavior will be $Fix.rec(f) \mapsto\!\!\!\!\twoheadrightarrow f(Fix.rec(f))$.

4.1 The Lambda Calculus of Objects

We now present a translation of the Lambda Calculus of Objects of Fisher, Honsell, and Mitchell [14] into the ρCal. This calculus is an untyped lambda calculus with constants enriched with object primitives. A new object can be created by modifying and/or extending an existing prototype object; the result is a new object which inherits all the methods and fields of the prototype. This calculus is trivially computationally complete, since the lambda calculus is built in the calculus itself. The syntax and the small-step semantics of $\lambda\mathcal{O}bj$ we present in this paper are inspired by the work of [17].

Syntax and Operational Semantics. The syntax of the calculus is defined as follows:

$$\begin{aligned} M,N ::= {} & \lambda X.M \mid MN \mid X \mid c \mid \\ & \langle\rangle \mid \langle M \leftarrow n = N\rangle \mid \langle M \leftarrow\!\!+ n = N\rangle \mid M \Leftarrow n \mid \mathit{Sel}(M,m,N) \end{aligned}$$

Let $\leftarrow\!\!*$ be either $\leftarrow$ or $\leftarrow\!\!+$; the small-step semantics is defined by

$$\begin{array}{llll} (\lambda X.M)\,N & \mapsto_{\lambda\mathcal{O}bj} [X/N]M & \mathit{Sel}(\langle M\leftarrow\!\!*\, n = N\rangle, n, P) & \mapsto_{\lambda\mathcal{O}bj} NP \\ M \Leftarrow m & \mapsto_{\lambda\mathcal{O}bj} \mathit{Sel}(M,m,M) & \mathit{Sel}(\langle M\leftarrow\!\!*\, n = N\rangle, m, P) & \mapsto_{\lambda\mathcal{O}bj} \mathit{Sel}(M,m,P) \end{array}$$

The main operation on objects is method invocation, whose reduction is defined by the second rule. Sending a message m to an object M, containing a method m, reduces to $\mathit{Sel}(M,m,M)$. More generally, in the expression $\mathit{Sel}(M,m,N)$, the term N represents the receiver (or recipient) of the message, the constant m is the message we want to send to the receiver of the message, and the term M is (or reduces to) a proper sub-object of N.

By looking at the last two rewrite rules, one may note that the *Sel* function "scans" the recipient of the message until it finds the definition of the method we want to use; when it finds the body of the method, it applies this body to the recipient of the message. The operational semantics in [14] was based on a more elaborate *bookkeeping* relation which transforms the receiver (*i.e.* an ordered list of methods) into another equivalent object where the method we are calling is always the last overridden one.

As a simple example of the calculus, we show an object which has the capability to extend itself simply by receiving a message which encodes the method to be added.

Example 6 (An object with "self-extension"). Consider the object *Self_ext* [17] $\langle\langle\rangle \leftarrow\!\!+ \mathit{add\_n} = \lambda S.\langle S \leftarrow\!\!+ n = \lambda S'.1\rangle\rangle$. If we send the message *add_n* to *Self_ext*, then we get $\mathit{Self\_ext} \Leftarrow \mathit{add\_n} \mapsto_{\lambda\mathcal{O}bj} \mathit{Sel}(\mathit{Self\_ext}, \mathit{add\_n}, \mathit{Self\_ext}) \mapsto_{\lambda\mathcal{O}bj} (\lambda S.\langle S \leftarrow\!\!+ n = \lambda S'.1\rangle)\mathit{Self\_ext} \mapsto_{\lambda\mathcal{O}bj} \langle \mathit{Self\_ext} \leftarrow\!\!+ n = \lambda S'.1\rangle$, resulting in the method n being added to *Self_ext*.

The Translation of λObj into ρCal. The translation of a λObj-term into a corresponding ρCal-term is quite trivial and can be done in the theory $\mathbb{T}_{\lambda Obj}$ where the symbol "," is associative and *null* is its neutral element. Intuitively, an object in λObj is translated into a simple structure in ρCal. The choice we made for object override is an imperative one, *i.e.* we *delete* the method we are overriding using the *kill* function defined in Example 4. The translation is defined as follows:

$$\begin{array}{llll}
[\![c]\!] & \triangleq c & [\![\langle\rangle]\!] & \triangleq null \\
[\![X]\!] & \triangleq X & [\![\langle M \leftarrow n = N\rangle]\!] & \triangleq kill_n([\![M]\!]), n \to [\![N]\!] \\
[\![\lambda X.M]\!] & \triangleq X \to [\![M]\!] & [\![\langle M \leftarrow\!\!\!+ n = N\rangle]\!] & \triangleq [\![M]\!], n \to [\![N]\!] \\
[\![MN]\!] & \triangleq [\![M]\!]\bullet[\![N]\!] & [\![M \Leftarrow m]\!] & \triangleq [\![M]\!].m \triangleq [\![M]\!]\bullet m\bullet[\![M]\!] \\
 & & [\![Sel(M,m,N)]\!] & \triangleq [\![M]\!]\bullet m\bullet[\![N]\!]
\end{array}$$

For instance, Example 7 shows an example of a simple computation in λObj and the corresponding translation into the ρCal, and Example 8 presents the translation of the *Self_ext* object into the ρCal.

Example 7 (A Simple Computation). Let *Point* be the simple diagonal point $\langle\langle\langle\rangle \leftarrow\!\!\!+ x = \lambda S.S \Leftarrow y\rangle \leftarrow\!\!\!+ y = \lambda S.1\rangle$. Then $Point \Leftarrow x \mapsto_{\lambda Obj} Sel(Point, x, Point) \mapsto_{\lambda Obj} Sel(\langle\langle\rangle \leftarrow\!\!\!+ x = \lambda S.S \Leftarrow y\rangle, x, Point) \mapsto_{\lambda Obj} (\lambda S.S \Leftarrow y)Point \mapsto_{\lambda Obj} Point \Leftarrow y \mapsto_{\lambda Obj} Sel(Point, y, Point) \mapsto_{\lambda Obj} (\lambda S.1)Point \mapsto_{\lambda Obj} 1$.

The above computation in λObj can be easily translated into a corresponding computation in ρCal using $t \triangleq [\![Point]\!] \triangleq x \to S \to S.y, y \to S \to 1$ as follows: $[\![Point \Leftarrow x]\!] \triangleq t.x \triangleq (x \to S \to S.y, y \to S \to 1)\bullet x\bullet t \mapsto\!\!\!\!\to (S \to S.y, null)\bullet t =_{\mathbb{T}_{\lambda Obj}} (S \to S.y)\bullet t \mapsto t.y \triangleq (x \to S \to S.y,\ y \to S \to 1)\bullet y\bullet t \mapsto\!\!\!\!\to (null, S \to 1)\bullet t =_{\mathbb{T}_{\lambda Obj}} (S \to 1)\bullet t \mapsto 1$.

Example 8 (Translation of Self_ext). The object *Self_ext* can be easily translated in the ρCal as $t_1 \triangleq [\![Self\_ext]\!] \triangleq add\_n \to S \to (S, n \to S' \to 1)$. Then: $(t_1.add\_n).n \mapsto ((S \to (S, n \to S' \to 1))\bullet t_1).n \mapsto (t_1, n \to S' \to 1).n \triangleq \underbrace{(add\_n \to \ldots, n \to S' \to 1)}_{t_2}.n \mapsto\!\!\!\!\to null, (S' \to 1)\bullet t_2 =_{\mathbb{T}_{\lambda Obj}} (S' \to 1)\bullet t_2 \mapsto 1$.

The translation into the ρCal can be proved correct when the theory $\mathbb{T}_{\lambda Obj}$ is considered:

Theorem 2 (Translation of λObj into ρCal). *If $M \mapsto\!\!\!\!\to_{\lambda Obj} N$, then $[\![M]\!] \mapsto\!\!\!\!\to_{\mathbb{T}_{\lambda Obj}} [\![N]\!]$.*

4.2 The Object Calculus

The Object Calculus [1] is a calculus where the only existing entities are the objects; it is computationally complete since λ-calculus, fixed points and complex structures can be easily encoded within it. A large collection of variants (functional and imperative, typed and untyped) for this calculus are presented in the book and in the literature.

Syntax and Operational Semantics. The syntax of the object calculus is defined as follows:

$$a, b ::= X \mid [m_i = \varsigma(X)b_i]^{i=1\ldots n} \mid a.m \mid a.m := \varsigma(X)b$$

Let $a \triangleq [m_i = \varsigma(X)b_i]^{i=1\ldots n}$; the small-step semantics is:

$$\begin{array}{lll} a.m_j & \mapsto_{\varsigma Obj} [X/a]b_j & j = 1\ldots n \\ a.m_j := \varsigma(X)b & \mapsto_{\varsigma Obj} [m_i = \varsigma(X)b_i, m_j = \varsigma(X)b]^{i=1\ldots n\setminus\{j\}} & j = 1\ldots n \end{array}$$

The Translation into ρCal. The translation of an ςObj-term into a corresponding ρCal-term is quite similar to the one of λObj, and can be done in the theory $\mathbb{T}_{\varsigma Obj}$ where the symbol "," is associative and commutative, and *null* is its neutral element. Given the function $kill_m \triangleq (X, m \to Y) \to X$, and the alias $(t_1.m := t_2) \triangleq (kill_m(t_1), m \to t_2)$, the translation is defined as follows:

$$\begin{array}{llll} [\![X]\!] & \triangleq X & [\![\, [m_i = \varsigma(X)b_i]^{i=1\ldots n} \,]\!] & \triangleq (m_i \to X \to [\![b_i]\!])^{i=1\ldots n} \\ [\![a.m_j]\!] & \triangleq [\![a]\!].m_j & [\![a.m := \varsigma(X)b]\!] & \triangleq [\![a]\!].m := X \to [\![b]\!] \end{array}$$

As a simple example, we present the usual Abadi and Cardelli's encoding of the Point class [1].

Example 9 (A Point Class). The object *PClass* is defined in ςObj as follows:

$$[new = \varsigma(S)o, val = \lambda S'.\ v(1\ 1), get = \lambda S'.\ (S'.val), set = \lambda S'.\lambda N.\ S'.val := N]$$

with $o \triangleq [val = \varsigma(S')(S.val)(S'), get = \varsigma(S')(S.get)(S'), set = \varsigma(S')(S.set)(S')]$, and it is translated into the ρCal as follows:

$$\begin{array}{l} new \to S \to t, val \to S \to S' \to v(1\ 1), \\ get \to S \to S' \to S'.val, set \to S \to S' \to v(X\ Y) \to (S'.val := S'' \to v(X\ Y)) \\ \text{with } t \triangleq (val \to S' \to (S.val)\bullet S', get \to S' \to (S.get)\bullet S', set \to S' \to (S.set)\bullet S') \end{array}$$

It is not hard to verify that $[\![PClass]\!].new \mapsto\!\!\!\rightarrow_{\mathbb{T}_{\varsigma Obj}} Point_{imp}$.

As another example, we present the Abadi and Cardelli's fixed point object operator. To do this we recall the usual encoding of lambda calculus in ςObj:

$$\begin{array}{ll} [\![S]\!] & \triangleq S \\ [\![M\,N]\!] & \triangleq [\![M]\!] \circ [\![N]\!] \end{array} \qquad [\![\lambda S.M]\!] \triangleq [arg = \varsigma(S)S.arg, val = \varsigma(S)[S.arg/S][\![M]\!]]$$

and the alias $p \circ q \triangleq (p.arg := \varsigma(S)q).val$, which represents the encoding of the function application.

Example 10 (Another Fixed-Point Object). In ςObj, the generic fixed-point object $Fix \triangleq [arg = \varsigma(S)S.arg, val = \varsigma(S)((S.arg).arg := \varsigma(S')S.val).val]$, can be translated into ρCal as:

$$Fix \triangleq arg \to S \to S.arg, val \to S \to (kill_{arg}(S.arg), arg \to S' \to S.val).val$$

Using the aliases $t_1 \circ t_2 \triangleq (t_1.arg := S \to t_2).val$, and $Fix_f \triangleq Fix.arg := S \to f$, we can prove that $Fix \circ f \equiv Fix_f.val \mapsto\!\!\!\rightarrow ((Fix_f.arg).arg := S' \to Fix_f.val).val \mapsto\!\!\!\rightarrow (f.arg := S' \to Fix \circ f).val \equiv f \circ (Fix \circ f)$.

The translation into the ρCal can be proved correct when the theory $\mathbb{T}_{\varsigma Obj}$ is considered:

Theorem 3 (Translation of ςObj into ρCal). *If* $M \mapsto_{\varsigma Obj} N$, *then* $[\![M]\!] \mapsto\!\!\!\!\rightarrow_{\mathbb{T}_{\varsigma Obj}} [\![N]\!]$.

The following example shows that the expressivity of ρCal is strictly stronger than the two previous calculi of objects as they cannot be translated neither in λObj nor in ςObj. In fact, we can easily consider "labels" and "bodies" as *first-class entities* that can be passed as function arguments.

Example 11 (The Daemon and the Para object).

1. Assume $Para$ be $a \to S \to b, par(X) \to S \to S.X$. This object has a method $par(X)$ which seeks for a method name that is assigned to the variable X and then sends this method to the object itself. Then: $Para.(par(a)) \triangleq Para\bullet(par(a))\bullet Para \mapsto\!\!\!\!\rightarrow (S \to S.a)\bullet Para \mapsto Para.a \mapsto\!\!\!\!\rightarrow b$.
2. Assume $Daemon$ be $set \to S \to X \to (X, set \to S' \to Y \to (Y, S'))$. The set method of $Daemon$ is used to create an object completely from scratch by receiving from outside all the components of a method, namely, the labels and the bodies. Once the object is installed, it has the capability to extend itself upon the reception of the same message set. In some sense the "power" of $Daemon$ has been inherited by the created object. Then: $Daemon.set(x \to S \to 3) \triangleq Daemon\bullet set\bullet Daemon\bullet(x \to S \to 3) \mapsto\!\!\!\!\rightarrow (X \to (X, set \to S' \to Y \to (Y, S')))\bullet(x \to S \to 3) \mapsto x \to S \to 3, set \to S' \to Y \to (Y, S') \triangleq t$, and $t.set(y \to S \to 4) \mapsto\!\!\!\!\rightarrow y \to S \to 4, t$.

One may wonder if some reductions in the ρCal can be translated into a suitable computation either in the λObj calculus or in the ςObj calculus: we can distinguish the following cases:

- reductions *à la* lambda calculus, like $(X \to t_1)\bullet t_2 \mapsto [X/t_2]t_1$, *i.e.* with trivial matching, are directly translated in either λObj and ςObj;
- reductions which use non-trivial matching, like $(a \to b)\bullet c \mapsto null$, can be translated in either λObj and ςObj, modulo an encoding of the underlined matching theory (taking into account also possible failures), as in $Sol(a \ll_{\mathbb{T}} c) = null$;
- reductions which use structures, like $(X \to a, X \to b)\bullet c \mapsto (a, b)$, can be translated in either λObj and ςObj, modulo a non trivial encoding of the nondeterministic features intrinsic to the ρCal.

Therefore, when using more elaborate theories and structures, the encoding becomes at least as difficult as the matching problem underlining the theory itself.

5 Conclusions and Further Work

We have presented a new version of the Rho Calculus and shown that its embedded matching power permits us to uniformly and naturally encode various calculi including the Lambda Calculus of Objects and the Object Calculus.

This presentation of the Rho Calculus inherits from the ideas and concepts of the first proposed one [6,7], it simplifies the rules of the calculus and improves the way the results are handled. This allows us first to encode object oriented calculi in a very natural and simple way but further to design new powerful object oriented features, like parameterized methods or self creating objects. Based on this new generic approach, an implementation of objects is under way in the Rho-based language ELAN [13]. More generally, rewrite based languages like ASF+SDF, CafeOBJ, Maude, Stratego, or ELAN, could benefit from a Rho-based semantics that gives a first-class status to rewrite rules and to their application.

We are now planning to work on several directions. First, on giving a big-step semantics in order to define a deterministic evaluation strategy, when needed. Then, the calculus could be further generalized by the explicit use of constraints. For the moment, the (ρ) rule calls for the solutions set of the relevant matching constraint; this could be replaced by an appropriate constrained term, in the spirit of constraint programming. We are also exploring an elaborated type system allowing in particular to type self-applications. As we have seen, the are many possible applications of the framework; a track that we have not yet mention in this paper concerns encoding concurrency in the spirit of the early work of Viry [37].

Independently of these ongoing works, we believe that the matching power of the Rho Calculus could be widely used, thanks to its expressiveness and simplicity, as a new model of computation.

Acknowledgement. We thank the referees for their constructive remarks, Hubert Dubois and all the members of the ELAN group for their comments and interactions on the topics of the Rho Calculus.

References

1. M. Abadi and L. Cardelli. *A Theory of Objects.* Springer Verlag, 1996.
2. H. Ait-Kaci, A. Podelski, and G. Smolka. A feature constraint system for logic programming with entailment. *Theoretical Computer Science*, 122(1-2):263–283, 1994.
3. P. Borovanský, C. Kirchner, H. Kirchner, P.-E. Moreau, and C. Ringeissen. An overview of ELAN. In *Proc. of WRLA*, volume 15. Electronic Notes in Theoretical Computer Science, 1998.
4. V. Breazu-Tannen. Combining algebra and higher-order types. In *Proc. of LICS*, pages 82–90, 1988.
5. H.-J. Bürckert. Matching — A special case of unification? *Journal of Symbolic Computation*, 8(5):523–536, 1989.
6. H. Cirstea. *Calcul de réécriture : fondements et applications.* Thèse de Doctorat d'Université, Université Henri Poincaré - Nancy I, 2000.
7. H. Cirstea and C. Kirchner. The rewriting calculus — Part I *and* II. *Logic Journal of the Interest Group in Pure and Applied Logics*, 9(3):427–498, 2001.
8. H. Cirstea, C. Kirchner, and L. Liquori. Matching Power. Technical Report A00-R-363, LORIA, Nancy, 2000.

9. H. Cirstea, C. Kirchner, and L. Liquori. The Rho Cube. In *Proc. of FOSSACS*, volume 2030 of *LNCS*, pages 166–180. Springer-Verlag, 2001.
10. M. Clavel, S. Eker, P. Lincoln, and J. Meseguer. Principles of Maude. In *Proc. of WRLA*, volume 4. Electronic Notes in Theoretical Computer Science, 1996.
11. L. Colson. Une structure de données pour le λ-calcul typé. Private Communication, 1988.
12. G. Dowek. Third order matching is decidable. *Annals of Pure and Applied Logic*, 69:135–155, 1994.
13. Hubert Dubois and Hélène Kirchner. Objects, rules and strategies in ELAN. , 2000. Submitted.
14. K. Fisher, F. Honsell, and J. C. Mitchell. A Lambda Calculus of Objects and Method Specialization. *Nordic Journal of Computing*, 1(1):3–37, 1994.
15. K. Futatsugi and A. Nakagawa. An overview of CAFE specification environment – an algebraic approach for creating, verifying, and maintaining formal specifications over networks. In *Proc. of FEM*, 1997.
16. J. Gallier and V. Breazu-Tannen. Polymorphic rewriting conserves algebraic strong normalization and confluence. In *Proc. of ICALP*, volume 372 of *LNCS*, pages 137–150. Springer-Verlag, 1989.
17. P. Di Gianantonio, F. Honsell, and L. Liquori. A Lambda Calculus of Objects with Self-inflicted Extension. In *Proc. of OOPSLA*, pages 166–178. The ACM Press, 1998.
18. J. A. Goguen, C. Kirchner, H. Kirchner, A. Mégrelis, J. Meseguer, and T. Winkler. An introduction to OBJ-3. In *Proc. of CTRS*, volume 308 of *LNCS*, pages 258–263. Springer-Verlag, 1987.
19. C. M. Hoffmann and M. J. O'Donnell. Pattern matching in trees. *Journal of the ACM*, 29(1):68–95, 1982.
20. G. Huet and B. Lang. Proving and applying program transformations expressed with second-order patterns. *Acta Informatica*, 11:31–55, 1978.
21. J.-M. Hullot. Associative-commutative pattern matching. In *Proc. of IJCAI*, 1979.
22. J.P. Jouannaud and M. Okada. Abstract data type systems. *Theoretical Computer Science*, 173(2):349–391, 1997.
23. C. Kirchner and H. Kirchner. Rewriting, solving, proving. A preliminary version of a book available at `www.loria.fr/~ckirchne/rsp.ps.gz`, 1999.
24. C. Kirchner, H. Kirchner, and M. Vittek. Designing constraint logic programming languages using computational systems. In *Principles and Practice of Constraint Programming. The Newport Papers.*, chapter 8, pages 131–158. The MIT press, 1995.
25. J.W. Klop, V. van Oostrom, and F. van Raamsdonk. Combinatory reduction systems: introduction and survey. *Theoretical Computer Science*, 121:279–308, 1993.
26. Donald E. Knuth, J. Morris, and V. Pratt. Fast pattern matching in strings. *SIAM Journal of Computing*, 6(2):323–350, 1977.
27. A. Laville. Lazy pattern matching in the ML language. In *Proc. FCT & TCS*, volume 287 of *LNCS*, pages 400–419. Springer-Verlag, 1987.
28. D Miller. A logic programming language with lambda-abstraction, function variables, and simple unification. In *Proc. of ELP*, volume 475 of *LNCS*, pages 253–281. Springer-Verlag, 1991.
29. Tobias Nipkow and Christian Prehofer. Higher-order rewriting and equational reasoning. In W. Bibel and P. Schmitt, editors, *Automated Deduction — A Basis for Applications. Volume I: Foundations.* Kluwer, 1998.

30. M. Okada. Strong normalizability for the combined system of the typed λ calculus and an arbitrary convergent term rewrite system. In *Proc. of ISSAC*, pages 357–363. ACM Press, 1989.
31. V. Padovani. Decidability of fourth-order matching. *Mathematical Structures in Computer Science*, 3(10):361–372, 2000.
32. G. Peterson and M. E. Stickel. Complete sets of reductions for some equational theories. *Journal of the ACM*, 28:233–264, 1981.
33. S. Peyton-Jones. *The implementation of functional programming languages.* Prentice Hall, Inc., 1987.
34. Équipe Protheo. The Elan Home Page, 2001. `http://elan.loria.fr`.
35. A. van Deursen, J. Heering, and P. Klint. *Language Prototyping.* World Scientific, 1996.
36. V. van Oostrom. Lambda calculus with patterns. Technical Report IR-228, Vrije Universiteit, November 1990.
37. P. Viry. Input/Output for ELAN. In *Proc. of WRLA*, volume 4. Electronic Notes in Theoretical Computer Science, 1996.
38. E. Visser and Z.e.A. Benaissa. A core language for rewriting. In *Proc. of WRLA*, volume 15. Electronic Notes in Theoretical Computer Science, 1998.
39. D. A. Wolfram. *The Clausal Theory of Types*, volume 21 of *Cambridge Tracts in Theoretical Computer Science.* Cambridge University Press, 1993.

Dependency Pairs for Equational Rewriting*

Jürgen Giesl[1] and Deepak Kapur[2]

[1] LuFG Informatik II, RWTH Aachen, Ahornstr. 55, 52074 Aachen, Germany
giesl@informatik.rwth-aachen.de
[2] Computer Science Dept., University of New Mexico, Albuquerque, NM 87131, USA
kapur@cs.unm.edu

Abstract. The dependency pair technique of Arts and Giesl [1,2,3] for termination proofs of term rewrite systems (TRSs) is extended to rewriting modulo equations. Up to now, such an extension was only known in the special case of *AC*-rewriting [15,17]. In contrast to that, the proposed technique works for arbitrary non-collapsing equations (satisfying a certain linearity condition). With the proposed approach, it is now possible to perform automated termination proofs for many systems where this was not possible before. In other words, the power of dependency pairs can now also be used for rewriting modulo equations.

1 Introduction

Termination of term rewriting (e.g., [1,2,3,9,22]) and termination of rewriting modulo associativity and commutativity equations (e.g., [8,13,14,20,21]) have been extensively studied. For equations other than *AC*-axioms, however, there are only a few techniques available to prove termination (e.g., [6,10,16,18]).

This paper presents an extension of the dependency pair approach [1,2,3] to rewriting modulo equations. In the special case of *AC*-axioms, our technique corresponds to the methods of [15,17], but in contrast to these methods, our technique can also be used if the equations are not *AC*-axioms. This allows much more automated termination proofs for equational rewrite systems than those possible with directly applying simplification orderings for equational rewriting (like equational polynomial orderings or *AC*-versions of path orderings).

We first review dependency pairs for ordinary term rewriting in Sect. 2. In Sect. 3, we show why a straightforward extension of dependency pairs to rewriting modulo equations is not possible. Therefore, we follow an idea similar to the one of [17] for *AC*-axioms: We consider a restricted form of equational rewriting, which is more suitable for termination proofs with dependency pairs.

In Sect. 4, we show how to ensure that termination of this restricted equational rewrite relation is equivalent to termination of full rewriting modulo equations. Under certain conditions on the equations $\mathcal{E}$, we show how to compute an extended rewrite system $Ext_{\mathcal{E}}(\mathcal{R})$ from the given TRS $\mathcal{R}$ such that the restricted rewrite relation of $Ext_{\mathcal{E}}(\mathcal{R})$ modulo $\mathcal{E}$ is terminating iff $\mathcal{R}$ is terminating modulo

* Supported by the Deutsche Forschungsgemeinschaft Grant GI 274/4-1 and the National Science Foundation Grants nos. CCR-9996150, CDA-9503064, CCR-9712396.

A. Middeldorp (Ed.): RTA 2001, LNCS 2051, pp. 93–107, 2001.

$\mathcal{E}$. This is proved for (almost) arbitrary $\mathcal{E}$-rewriting, thus generalizing a related result for AC-rewriting. This general result may be of independent interest, and may also be useful in investigating other properties of $\mathcal{E}$-rewriting. Finally, in Sect. 5, we extend the dependency pair approach to rewriting modulo equations.

2 Dependency Pairs for Ordinary Rewriting

The dependency pair approach allows the use of standard methods like simplification orderings [9,22] for automated termination proofs where they were not applicable before. In this section we briefly summarize the basic concepts of this approach. All results in this section are due to Arts and Giesl and we refer to [1,2,3] for further details, refinements, and explanations.

In contrast to the standard techniques for termination proofs, which compare left and right-hand sides of rules, in this approach one concentrates on the subterms in the right-hand sides that have a defined[1] root symbol, because these are the only terms responsible for starting new reductions.

More precisely, for every rule $f(s_1, \ldots, s_n) \to C[g(t_1, \ldots, t_m)]$ (where f and g are defined symbols), we compare the argument tuples $s_1, \ldots, s_n$ and $t_1, \ldots, t_m$. To avoid the handling of tuples, for every defined symbol f, we introduce a fresh *tuple* symbol F. To ease readability, we assume that the original signature consists of lower case function symbols only, whereas the tuple symbols are denoted by the corresponding upper case symbols. Now instead of the tuples $s_1, \ldots, s_n$ and $t_1, \ldots, t_m$ we compare the *terms* $F(s_1, \ldots, s_n)$ and $G(t_1, \ldots, t_m)$.

Definition 1 (Dependency Pair [1,2,3]). *If $f(s_1,\ldots,s_n) \to C[g(t_1,\ldots,t_m)]$ is a rule of a TRS $\mathcal{R}$ and g is a defined symbol, then $\langle F(s_1, \ldots, s_n), G(t_1, \ldots, t_m)\rangle$ is a* dependency pair *of $\mathcal{R}$.*

Example 2. As an example, consider the TRS $\{\mathsf{a} + \mathsf{b} \to \mathsf{a} + (\mathsf{b} + \mathsf{c})\}$, cf. [17]. Termination of this system cannot be shown by simplification orderings, since the left-hand side of the rule is embedded in the right-hand side. In this system, the defined symbol is $+$ and thus, we obtain the dependency pairs $\langle \mathsf{P}(\mathsf{a}, \mathsf{b}), \mathsf{P}(\mathsf{a}, \mathsf{b}+\mathsf{c})\rangle$ and $\langle \mathsf{P}(\mathsf{a}, \mathsf{b}), \mathsf{P}(\mathsf{b}, \mathsf{c})\rangle$ (where P is the tuple symbol for the plus-function "+").

Arts and Giesl developed the following new termination criterion. As usual, a quasi-ordering $\succsim$ is a reflexive and transitive relation, and we say that an ordering $>$ is *compatible* with $\succsim$ if we have $> \circ \succsim \subseteq >$ or $\succsim \circ > \subseteq >$.

Theorem 3 (Termination with Dependency Pairs [1,2,3]). *A TRS $\mathcal{R}$ is terminating iff there exists a weakly monotonic quasi-ordering $\succsim$ and a well-founded ordering $>$ compatible with $\succsim$, where both $\succsim$ and $>$ are closed under substitution, such that*

(1) $s > t$ for all dependency pairs $\langle s, t\rangle$ of $\mathcal{R}$ and
(2) $l \succsim r$ for all rules $l \to r$ of $\mathcal{R}$.

---

[1] Root symbols of left-hand sides are *defined* and all other functions are *constructors*.

Consider the TRS from Ex. 2 again. In order to prove its termination according to Thm. 3, we have to find a suitable quasi-ordering $\succsim$ and ordering $>$ such that $\mathsf{P}(\mathsf{a},\mathsf{b}) > \mathsf{P}(\mathsf{a},\mathsf{b}+\mathsf{c})$, $\mathsf{P}(\mathsf{a},\mathsf{b}) > \mathsf{P}(\mathsf{b},\mathsf{c})$, and $\mathsf{a}+\mathsf{b} \succsim \mathsf{a}+(\mathsf{b}+\mathsf{c})$.

Most standard orderings amenable to automation are *strongly* monotonic (cf. e.g. [9,22]), whereas here we only need *weak* monotonicity. Hence, before synthesizing a suitable ordering, some of the arguments of function symbols may be eliminated, cf. [3]. For example, in our inequalities, one may eliminate the first argument of $+$. Then every term $s+t$ in the inequalities is replaced by $+'(t)$ (where $+'$ is a new unary function symbol). By comparing the terms resulting from this replacement instead of the original terms, we can take advantage of the fact that $+$ does not have to be strongly monotonic in its first argument. Note that there are only finitely many possibilities to eliminate arguments of function symbols. Therefore all these possibilities can be checked automatically.

In this way, we obtain the inequalities $\mathsf{P}(\mathsf{a},\mathsf{b}) > \mathsf{P}(\mathsf{a},+'(\mathsf{c}))$, $\mathsf{P}(\mathsf{a},\mathsf{b}) > \mathsf{P}(\mathsf{b},\mathsf{c})$, and $+'(\mathsf{b}) \succsim +'(+'(\mathsf{c}))$. These inequalities are satisfied by the recursive path ordering (rpo) [9] with the precedence $\mathsf{a} \sqsupset \mathsf{b} \sqsupset \mathsf{c} \sqsupset +'$ (i.e., we choose $\succsim$ to be $\succsim_{rpo}$ and $>$ to be $\succ_{rpo}$). So termination of this TRS can now be proved automatically. For implementations of the dependency pair approach see [4,7].

3 Rewriting Modulo Equations

For a set $\mathcal{E}$ of equations between terms, we write $s \to_{\mathcal{E}} t$ if there exist an equation $l \approx r$ in $\mathcal{E}$, a substitution σ, and a context C such that $s = C[l\sigma]$ and $t = C[r\sigma]$. The symmetric closure of $\to_{\mathcal{E}}$ is denoted by $\vdash\dashv_{\mathcal{E}}$ and the transitive reflexive closure of $\vdash\dashv_{\mathcal{E}}$ is denoted by $\sim_{\mathcal{E}}$. In the following, we restrict ourselves to equations $\mathcal{E}$ where $\sim_{\mathcal{E}}$ is decidable.

Definition 4 (Rewriting Modulo Equations). *Let $\mathcal{R}$ be a TRS and let $\mathcal{E}$ be a set of equations. A term s rewrites to a term t modulo $\mathcal{E}$, denoted $s \to_{\mathcal{R}/\mathcal{E}} t$, iff there exist terms s' and t' such that $s \sim_{\mathcal{E}} s' \to_{\mathcal{R}} t' \sim_{\mathcal{E}} t$. The TRS $\mathcal{R}$ is called terminating modulo $\mathcal{E}$ iff there does not exist an infinite $\to_{\mathcal{R}/\mathcal{E}}$ reduction.*

Example 5. An interesting special case are equations $\mathcal{E}$ which state that certain function symbols are associative and commutative (*AC*). As an example, consider the TRS $\mathcal{R} = \{\mathsf{a}+\mathsf{b} \to \mathsf{a}+(\mathsf{b}+\mathsf{c})\}$ again and let $\mathcal{E}$ consist of the associativity and commutativity axioms for $+$, i.e., $\mathcal{E} = \{x_1 + x_2 \approx x_2 + x_1, x_1 + (x_2 + x_3) \approx (x_1 + x_2) + x_3\}$, cf. [17]. $\mathcal{R}$ is not terminating modulo $\mathcal{E}$, since we have

$$\mathsf{a}+\mathsf{b} \to_{\mathcal{R}} \mathsf{a}+(\mathsf{b}+\mathsf{c}) \sim_{\mathcal{E}} (\mathsf{a}+\mathsf{b})+\mathsf{c} \to_{\mathcal{R}} (\mathsf{a}+(\mathsf{b}+\mathsf{c}))+\mathsf{c} \sim_{\mathcal{E}} ((\mathsf{a}+\mathsf{b})+\mathsf{c})+\mathsf{c} \to_{\mathcal{R}} \ldots$$

There are, however, many other sets of equations $\mathcal{E}$ apart from associativity and commutativity, which are also important in practice, cf. [11]. Hence, our aim is to extend dependency pairs to rewriting modulo (almost) arbitrary equations.

The soundness of dependency pairs for ordinary rewriting relies on the fact that whenever a term starts an infinite reduction, then one can also construct an infinite reduction where only *terminating or minimal non-terminating subterms* are reduced (i.e., one only applies rules to redexes without proper non-terminating subterms). The contexts of minimal non-terminating redexes can be completely disregarded. If a rule is applied at the root position of a minimal non-terminating subterm s (i.e., $s \to_{\mathcal{R}}^{\epsilon} t$ where ϵ denotes the root position), then s and each minimal non-terminating subterm t' of t correspond to a dependency pair. Hence, Thm. 3 (1) implies $s > t'$. If a rule is applied at a non-root position of a minimal non-terminating subterm s (i.e., $s \to_{\mathcal{R}}^{>\epsilon} t$), then we have $s \gtrsim t$ by Thm. 3 (2). However, due to the minimality of s, after finitely many such non-root rewrite steps, a rule must be applied at the root position of the minimal non-terminating term. Thus, every infinite reduction of minimal non-terminating subterms corresponds to an infinite $>$-sequence. This contradicts the well-foundedness of $>$.

So for ordinary rewriting, any infinite reduction from a minimal non-terminating subterm involves an $\mathcal{R}$-reduction at the root position. But as observed in [15], when extending the dependency pair approach to rewriting modulo equations, this is no longer true. For an illustration, consider Ex. 5 again, where $\mathsf{a} + (\mathsf{b} + \mathsf{c})$ is a minimal non-terminating term. However, in its infinite $\mathcal{R}/\mathcal{E}$-reduction no $\mathcal{R}$-step is ever applicable at the root position. (Instead one applies an $\mathcal{E}$-step at the root position and further $\mathcal{R}$- and $\mathcal{E}$-steps below the root.)

In the rest of the paper, from a rewrite system $\mathcal{R}$, we generate a new rewrite system $\mathcal{R}'$ with the following three properties: (i) the termination of a weaker form of rewriting by $\mathcal{R}'$ modulo $\mathcal{E}$ is equivalent to the termination of $\mathcal{R}$ modulo $\mathcal{E}$, (ii) every infinite reduction of a minimal non-terminating term in this weaker form of rewriting by $\mathcal{R}'$ modulo $\mathcal{E}$ involves a reduction step at the root level, and (iii) every such minimal non-terminating term has an infinite reduction where the variables of the $\mathcal{R}'$-rules are instantiated with terminating terms only.

4 $\mathcal{E}$-Extended Rewriting

We showed why the dependency pair approach cannot be extended to rewriting modulo equations directly. As a solution for this problem, we propose to consider a restricted form of rewriting modulo equations, i.e., the so-called *$\mathcal{E}$-extended $\mathcal{R}$-rewrite relation* $\to_{\mathcal{E}\backslash\mathcal{R}}$. (This approach was already taken in [17] for rewriting modulo AC.) The relation $\to_{\mathcal{E}\backslash\mathcal{R}}$ was originally introduced in [19] in order to circumvent the problems with infinite or impractically large $\mathcal{E}$-equivalence classes.[2]

Definition 6 ($\mathcal{E}$-extended $\mathcal{R}$-rewriting [19]). *Let $\mathcal{R}$ be a TRS and let $\mathcal{E}$ be a set of equations. The $\mathcal{E}$-extended $\mathcal{R}$-rewrite relation is defined as $s \to_{\mathcal{E}\backslash\mathcal{R}}^{\pi} t$ iff $s|_\pi \sim_{\mathcal{E}} l\sigma$ and $t = s[r\sigma]_\pi$ for some rule $l \to r$ in $\mathcal{R}$, some position π of s, and some substitution σ. We also write $\to_{\mathcal{E}\backslash\mathcal{R}}$ instead of $\to_{\mathcal{E}\backslash\mathcal{R}}^{\pi}$.*

To demonstrate the difference between $\to_{\mathcal{R}/\mathcal{E}}$ and $\to_{\mathcal{E}\backslash\mathcal{R}}$, consider Ex. 5 again. We have already seen that $\to_{\mathcal{R}/\mathcal{E}}$ is not terminating, since $\mathsf{a} + \mathsf{b} \to_{\mathcal{R}/\mathcal{E}} (\mathsf{a} + \mathsf{b}) + \mathsf{c} \to_{\mathcal{R}/\mathcal{E}} ((\mathsf{a} + \mathsf{b}) + \mathsf{c}) + \mathsf{c} \to_{\mathcal{R}/\mathcal{E}} \ldots$ But $\to_{\mathcal{E}\backslash\mathcal{R}}$ is terminating, because $\mathsf{a} + \mathsf{b} \to_{\mathcal{E}\backslash\mathcal{R}} \mathsf{a} + (\mathsf{b} + \mathsf{c})$, which is a normal form w.r.t. $\to_{\mathcal{E}\backslash\mathcal{R}}$.

[2] In [12], the relation $\to_{\mathcal{E}\backslash\mathcal{R}}$ is denoted "$\to_{\mathcal{R},\mathcal{E}}$".

The above example also demonstrates that in general, termination of $\rightarrow_{\mathcal{E}\backslash\mathcal{R}}$ is not sufficient for termination of $\rightarrow_{\mathcal{R}/\mathcal{E}}$. In this section we will show how termination of $\rightarrow_{\mathcal{R}/\mathcal{E}}$ can nevertheless be ensured by only regarding an $\mathcal{E}$-extended rewrite relation induced by a larger $\mathcal{R}' \supseteq \mathcal{R}$.

For the special case of AC-rewriting, this problem can be solved by extending $\mathcal{R}$ as follows: Let $\mathcal{G}$ be the set of all AC-symbols and

$$Ext_{AC(\mathcal{G})} = \mathcal{R} \cup \{f(l,y) \rightarrow f(r,y) \mid l \rightarrow r \in \mathcal{R},\ root(l) = f \in \mathcal{G}\},$$

where y is a new variable not occurring in the respective rule $l \rightarrow r$. A similar extension has also been used in previous work on extending dependency pairs to AC-rewriting [17]. The reason is that for AC-equations $\mathcal{E}$, the termination of $\rightarrow_{\mathcal{R}/\mathcal{E}}$ is in fact equivalent to the termination of $\rightarrow_{\mathcal{E}\backslash Ext_{AC(\mathcal{G})}(\mathcal{R})}$.

For Ex. 5, we obtain $Ext_{AC(\mathcal{G})}(\mathcal{R}) = \{\mathsf{a}+\mathsf{b} \rightarrow \mathsf{a}+(\mathsf{b}+\mathsf{c}), (\mathsf{a}+\mathsf{b})+y \rightarrow (\mathsf{a}+(\mathsf{b}+\mathsf{c}))+y\}$. Thus, in order to prove termination of $\rightarrow_{\mathcal{R}/\mathcal{E}}$, it is now sufficient to verify termination of $\rightarrow_{\mathcal{E}\backslash Ext_{AC(\mathcal{G})}(\mathcal{R})}$.

The above extension of [19] only works for AC-axioms $\mathcal{E}$. A later paper [12] treats arbitrary equations, but it does not contain any definition for extensions $Ext_{\mathcal{E}}(\mathcal{R})$, and termination of $\rightarrow_{\mathcal{R}/\mathcal{E}}$ is always a prerequisite in [12]. The reason is that [12] and also subsequent work on symmetrization and coherence were devoted to the development of completion algorithms (i.e., here the goal was to generate a convergent rewrite system and not to investigate the termination behavior of possibly non-terminating TRSs). Thus, these papers did not compare the termination behavior of full rewriting modulo equations with the termination of restricted versions of rewriting modulo equations. In fact, [12] focuses on the notion of coherence, which is not suitable for our purpose since coherence of $\mathcal{E}\backslash\mathcal{R}$ modulo $\mathcal{E}$ does not imply that termination of $\rightarrow_{\mathcal{R}/\mathcal{E}}$ is equivalent to termination of $\rightarrow_{\mathcal{E}\backslash\mathcal{R}}$.[3]

To extend dependency pairs to rewriting modulo non-AC-equations $\mathcal{E}$, we have to compute extensions $Ext_{\mathcal{E}}(\mathcal{R})$ such that termination of $\rightarrow_{\mathcal{R}/\mathcal{E}}$ is equivalent to termination of $\rightarrow_{\mathcal{E}\backslash Ext_{\mathcal{E}}(\mathcal{R})}$. The only restriction we will impose on the equations in $\mathcal{E}$ is that they must have *identical unique variables.* This requirement is satisfied by most practical examples where $\mathcal{R}/\mathcal{E}$ is terminating. As usual, a term t is called *linear* if no variable occurs more than once in t.

Definition 7 (Equations with Identical Unique Variables [19]). *An equation $u \approx v$ is said to have* identical unique variables *if u and v are both linear and the variables in u are the same as the variables in v.*

Let $uni_{\mathcal{E}}(s,t)$ denote a complete set of $\mathcal{E}$-unifiers of two terms s and t. As usual, δ is an *$\mathcal{E}$-unifier* of s and t iff $s\delta \sim_{\mathcal{E}} t\delta$ and a set $uni_{\mathcal{E}}(s,t)$ of $\mathcal{E}$-unifiers is *complete* iff for every $\mathcal{E}$-unifier δ there exists a $\sigma \in uni_{\mathcal{E}}(s,t)$ and a substitution

[3] In [12], $\mathcal{E}\backslash\mathcal{R}$ is coherent modulo $\mathcal{E}$ iff for all terms s, t, u, we have that $s \sim_{\mathcal{E}} t \rightarrow^{+}_{\mathcal{E}\backslash\mathcal{R}} u$ implies $s \rightarrow^{+}_{\mathcal{E}\backslash\mathcal{R}} v \sim_{\mathcal{E}} w \leftarrow^{*}_{\mathcal{E}\backslash\mathcal{R}} u$ for some v, w. Consider $\mathcal{R} = \{\mathsf{a}+\mathsf{b} \rightarrow \mathsf{a}+(\mathsf{b}+\mathsf{c}),\ x+y \rightarrow \mathsf{d}\}$ with $\mathcal{E}$ being the AC-axioms for $+$. The above system is coherent, since $s \sim_{\mathcal{E}} t \rightarrow^{+}_{\mathcal{E}\backslash\mathcal{R}} u$ implies $s \rightarrow^{+}_{\mathcal{R}} \mathsf{d} \leftarrow^{*}_{\mathcal{R}} u$. However, $\rightarrow_{\mathcal{E}\backslash\mathcal{R}}$ is terminating but $\rightarrow_{\mathcal{R}/\mathcal{E}}$ is not terminating.

ρ such that $\delta \sim_{\mathcal{E}} \sigma\rho$, cf. [5]. ("$\sigma\rho$" is the composition of σ and ρ where σ is applied first and "$\delta \sim_{\mathcal{E}} \sigma\rho$" means that for all variables x we have $x\delta \sim_{\mathcal{E}} x\sigma\rho$.)

To construct $Ext_{\mathcal{E}}(\mathcal{R})$, we consider all overlaps between equations $u \approx v$ or $v \approx u$ from $\mathcal{E}$ and rules $l \to r$ from $\mathcal{R}$. More precisely, we check whether a non-variable subterm $v|_\pi$ of v $\mathcal{E}$-unifies with l (where we always assume that rules in $\mathcal{R}$ are variable disjoint from equations in $\mathcal{E}$). In this case one adds the rules $(v[l]_\pi)\sigma \to (v[r]_\pi)\sigma$ for all $\sigma \in uni_{\mathcal{E}}(v|_\pi, l)$.[4] In Ex. 5, the subterm $x_1 + x_2$ of the right-hand side of $x_1 + (x_2 + x_3) \approx (x_1 + x_2) + x_3$ unifies with the left-hand side of the only rule $\mathsf{a} + \mathsf{b} \to \mathsf{a} + (\mathsf{b} + \mathsf{c})$. Thus, in the extension of $\mathcal{R}$, we obtain the rule $(\mathsf{a} + \mathsf{b}) + y \to (\mathsf{a} + (\mathsf{b} + \mathsf{c})) + y$.

$Ext_{\mathcal{E}}(\mathcal{R})$ is built via a kind of fixpoint construction, i.e., we also have to consider overlaps between equations of $\mathcal{E}$ and the newly constructed rules of $Ext_{\mathcal{E}}(\mathcal{R})$. For example, the subterm $x_1 + x_2$ also unifies with the left-hand side of the new rule $(\mathsf{a} + \mathsf{b}) + y \to (\mathsf{a} + (\mathsf{b} + \mathsf{c})) + y$. Thus, one would now construct a new rule $((\mathsf{a} + \mathsf{b}) + y) + z \to ((\mathsf{a} + (\mathsf{b} + \mathsf{c})) + y) + z$.

Obviously, in this way one obtains an infinite number of rules by subsequently overlapping equations with the newly constructed rules. However, in order to use $Ext_{\mathcal{E}}(\mathcal{R})$ for automated termination proofs, our aim is to restrict ourselves to finitely many rules. It turns out that we do not have to include new rules $(v[l]_\pi)\sigma \to (v[r]_\pi)\sigma$ in $Ext_{\mathcal{E}}(\mathcal{R})$ if $u\sigma \to^{\pi'}_{\mathcal{E}\backslash Ext_{\mathcal{E}}(\mathcal{R})} q \sim_{\mathcal{E}} (v[r]_\pi)\sigma$ already holds for some position π' of u and some term q (using just the old rules of $Ext_{\mathcal{E}}(\mathcal{R})$).

When constructing the rule $((\mathsf{a} + \mathsf{b}) + y) + z \to ((\mathsf{a} + (\mathsf{b} + \mathsf{c})) + y) + z$ above, the equation $u \approx v$ used was $x_1 + (x_2 + x_3) \approx (x_1 + x_2) + x_3$ and the unifier σ replaced x_1 by $(\mathsf{a} + \mathsf{b})$ and x_2 by y. Hence, here $u\sigma$ is the term $(\mathsf{a} + \mathsf{b}) + (y + x_3)$. But this term reduces with $\to^{1}_{\mathcal{E}\backslash Ext_{\mathcal{E}}(\mathcal{R})}$ to $(\mathsf{a} + (\mathsf{b} + \mathsf{c})) + (y + x_3)$ which is indeed $\sim_{\mathcal{E}}$-equivalent to $(v[r]_\pi)\sigma$, i.e., to $((\mathsf{a} + (\mathsf{b} + \mathsf{c})) + y) + x_3$. Thus, we do not have to include the rule $((\mathsf{a} + \mathsf{b}) + y) + z \to ((\mathsf{a} + (\mathsf{b} + \mathsf{c})) + y) + z$ in $Ext_{\mathcal{E}}(\mathcal{R})$.

The following definition shows how suitable extensions can be computed for arbitrary equations with identical unique variables. It will turn out that with these extensions one can indeed simulate $\to_{\mathcal{R}/\mathcal{E}}$ by $\to_{\mathcal{E}\backslash Ext_{\mathcal{E}}(\mathcal{R})}$, i.e., $s \to_{\mathcal{R}/\mathcal{E}} t$ implies $s \to_{\mathcal{E}\backslash Ext_{\mathcal{E}}(\mathcal{R})} t'$ for some $t' \sim_{\mathcal{E}} t$. This constitutes a crucial contribution of the paper, since it is the main requirement needed in order to extend dependency pairs to rewriting modulo equations.

Definition 8 (Extending $\mathcal{R}$ for Arbitrary Equations). *Let $\mathcal{R}$ be a TRS and let $\mathcal{E}$ be a set of equations. Let $\mathcal{R}'$ be a set containing only rules of the form $C[l\sigma] \to C[r\sigma]$ (where C is a context, σ is a substitution, and $l \to r \in \mathcal{R}$). $\mathcal{R}'$ is an* extension *of $\mathcal{R}$ for the equations $\mathcal{E}$ iff*

(a) $\mathcal{R} \subseteq \mathcal{R}'$ and

[4] Obviously, $uni_{\mathcal{E}}(v|_\pi, l)$ always exists, but it can be infinite in general. So when automating our approach for equational termination proofs, we have to restrict ourselves to equations $\mathcal{E}$ where $uni_{\mathcal{E}}(v|_\pi, l)$ can be chosen to be finite for all subterms $v|_\pi$ of equations and left-hand sides of rules l. This includes all sets $\mathcal{E}$ of finitary unification type, but our restriction is weaker, since we only need finiteness for certain terms $v|_\pi$ and l.

(b) *for all* $l \to r \in \mathcal{R}'$, $u \approx v \in \mathcal{E}$ *and* $v \approx u \in \mathcal{E}$, *all positions* π *of* v *and* $\sigma \in uni_{\mathcal{E}}(v|_\pi, l)$, *there is a position* π' *in* u *and a* $q \sim_{\mathcal{E}} (v[r]_\pi)\sigma$ *with* $u\sigma \to^{\pi'}_{\mathcal{E}\setminus\mathcal{R}'} q$.

In the following, let $Ext_{\mathcal{E}}(\mathcal{R})$ always denote an arbitrary extension of $\mathcal{R}$ for $\mathcal{E}$.

In order to satisfy Condition (b) of Def. 8, it is always sufficient to add the rule $(v[l]_\pi)\sigma \to (v[r]_\pi)\sigma$ to $\mathcal{R}'$. The reason is that then we have $u\sigma \to^{\epsilon}_{\mathcal{E}\setminus\mathcal{R}'} (v[r]_\pi)\sigma$. But if $u\sigma \to^{\pi'}_{\mathcal{E}\setminus\mathcal{R}'} q \sim_{\mathcal{E}} (v[r]_\pi)\sigma$ already holds with the other rules of $\mathcal{R}'$, then the rule $(v[l]_\pi)\sigma \to (v[r]_\pi)\sigma$ does not have to be added to $\mathcal{R}'$.

Condition (b) of Def. 8 also makes sure that as long as the equations have identical unique variables, we do not have to consider overlaps at variable positions.[5] The reason is that if $v|_\pi$ is a variable $x \in \mathcal{V}$, then we have $u\sigma = u[x\sigma]_{\pi'} \sim_{\mathcal{E}} u[l\sigma]_{\pi'} \to_{\mathcal{R}} u[r\sigma]_{\pi'} \sim_{\mathcal{E}} v[r\sigma]_\pi = (v[r]_\pi)\sigma$, where π' is the position of x in u. Hence, such rules $(v[l]_\pi)\sigma \to (v[r]_\pi)\sigma$ do not have to be included in $\mathcal{R}'$.

Overlaps at root positions do not have to be considered either. To see this, assume that π is the top position ϵ of v, i.e., that $v\sigma \sim_{\mathcal{E}} l\sigma$. In this case we have $u\sigma \sim_{\mathcal{E}} v\sigma \sim_{\mathcal{E}} l\sigma \to_{\mathcal{R}} r\sigma$ and thus, $u\sigma \to^{\epsilon}_{\mathcal{E}\setminus\mathcal{R}} r\sigma = (v[r]_\pi)\sigma$. So again, such rules $(v[l]_\pi) \to (v[r]_\pi)\sigma$ do not have to be included in $\mathcal{R}'$.

The following procedure is used to compute extensions. Here, we assume both $\mathcal{R}$ and $\mathcal{E}$ to be finite, where the equations $\mathcal{E}$ must have identical unique variables.

1. $\mathcal{R}' := \mathcal{R}$
2. For all $l \to r \in \mathcal{R}'$,
   all $u \approx v$ or $v \approx u$ from $\mathcal{E}$,
   and all positions π of v where $\pi \neq \epsilon$ and $v|_\pi \notin \mathcal{V}$ do:
   2.1. Let $\Sigma := uni_{\mathcal{E}}(v|_\pi, l)$.
   2.2. For all $\sigma \in \Sigma$ do:
      2.2.1. Let $T := \{q \mid u\sigma \to^{\pi'}_{\mathcal{E}\setminus\mathcal{R}'} q$ for a position π' of $u\}$.
      2.2.2. If there exists a $q \in T$ with $(v[r]_\pi)\sigma \sim_{\mathcal{E}} q$, then $\Sigma := \Sigma \setminus \{\sigma\}$.
   2.3. $\mathcal{R}' := \mathcal{R}' \cup \{(v[l]_\pi)\sigma \to (v[r]_\pi)\sigma \mid \sigma \in \Sigma\}$.

This algorithm has the following properties:

(a) If in Step 2.1, $uni_{\mathcal{E}}(v|_\pi, l)$ is finite and computable, then every step in the algorithm is computable.
(b) If the algorithm terminates, then the final value of $\mathcal{R}'$ is an extension of $\mathcal{R}$ for the equations $\mathcal{E}$.

With the TRS of Ex. 5, $Ext_{\mathcal{E}}(\mathcal{R}) = \{\mathsf{a} + \mathsf{b} \to \mathsf{a} + (\mathsf{b} + \mathsf{c}), (\mathsf{a} + \mathsf{b}) + y \to (\mathsf{a} + (\mathsf{b} + \mathsf{c})) + y\}$. In general, if $\mathcal{E}$ only consists of AC-axioms for some function symbols $\mathcal{G}$, then Def. 8 "coincides" with the well-known extension for AC-axioms, i.e., $\mathcal{R}' = \mathcal{R} \cup \{f(l, y) \to f(r, y) \mid l \to r \in \mathcal{R}, root(l) = f \in \mathcal{G}\}$ satisfies the

[5] Note that considering overlaps at variable positions as well would still not allow us to treat equations with non-linear terms. As an example regard $\mathcal{E} = \{\mathsf{f}(x) \approx \mathsf{g}(x, x)\}$ and $\mathcal{R} = \{\mathsf{g}(\mathsf{a}, \mathsf{b}) \to \mathsf{f}(\mathsf{a}), \mathsf{a} \to \mathsf{b}\}$. Here, $\to_{\mathcal{E}\setminus Ext_{\mathcal{E}}(\mathcal{R})}$ is well founded although $\mathcal{R}$ is not terminating modulo $\mathcal{E}$.

conditions (a) and (b) of Def. 8. So in case of AC-equations, our approach indeed corresponds to the approaches of [15,17]. However, Def. 8 can also be used for other forms of equations.

Example 9. As an example, consider the following system from [18].

$$\mathcal{R} = \{ \quad x - \mathsf{0} \to x, \qquad \mathsf{s}(x) - \mathsf{s}(y) \to x - y, \qquad \mathsf{0} \div \mathsf{s}(y) \to \mathsf{0}, \qquad \mathsf{s}(x) \div \mathsf{s}(y) \to \mathsf{s}((x-y) \div \mathsf{s}(y))\} \qquad \mathcal{E} = \{(u \div v) \div w \approx (u \div w) \div v\}$$

By overlapping the subterm $u \div w$ in the right-hand side of the equation with the left-hand sides of the last two rules we obtain

$$Ext_{\mathcal{E}}(\mathcal{R}) = \mathcal{R} \cup \{ \quad (\mathsf{0} \div \mathsf{s}(y)) \div z \to \mathsf{0} \div z, \quad (\mathsf{s}(x) \div \mathsf{s}(y)) \div z \to \mathsf{s}((x-y) \div \mathsf{s}(y)) \div z \}.$$

Note that these are indeed all the rules of $Ext_{\mathcal{E}}(\mathcal{R})$. Overlapping the subterm $u \div v$ of the equation's left-hand side with the third rule would result in $(\mathsf{0} \div \mathsf{s}(y)) \div z' \to \mathsf{0} \div z'$. But this new rule does not have to be included in $Ext_{\mathcal{E}}(\mathcal{R})$, since the corresponding other term of the equation, $(\mathsf{0} \div z') \div \mathsf{s}(y)$, would $\to^{\epsilon}_{\mathcal{E}\backslash Ext_{\mathcal{E}}(\mathcal{R})}$-reduce with the rule $(\mathsf{0} \div \mathsf{s}(y)) \div z \to \mathsf{0} \div z$ to $\mathsf{0} \div z'$. Overlapping $u \div v$ with the left-hand side of the fourth rule is also superfluous.

Similarly, overlaps with the new rules $(\mathsf{0} \div \mathsf{s}(y)) \div z \to \mathsf{0} \div z$ or $(\mathsf{s}(x) \div \mathsf{s}(y)) \div z \to \mathsf{s}((x-y) \div \mathsf{s}(y)) \div z$ also do not give rise to additional rules in $Ext_{\mathcal{E}}(\mathcal{R})$. To see this, overlap the subterm $u \div w$ in the right-hand side of the equation with the left-hand side of $(\mathsf{0} \div \mathsf{s}(y)) \div z \to \mathsf{0} \div z$. This gives the rule $((\mathsf{0} \div \mathsf{s}(y)) \div z) \div z' \to (\mathsf{0} \div z) \div z'$. However, the corresponding other term of the equation is $((\mathsf{0} \div \mathsf{s}(y)) \div z') \div z$. This reduces at position 1 (or position 11) to $(\mathsf{0} \div z') \div z$, which is $\mathcal{E}$-equivalent to $(\mathsf{0} \div z) \div z'$. Overlaps with the other new rule $(\mathsf{s}(x) \div \mathsf{s}(y)) \div z \to \mathsf{s}((x-y) \div \mathsf{s}(y)) \div z$ are not needed either.

Nevertheless, the above algorithm for computing extensions does not always terminate. For example, for $\mathcal{R} = \{\mathsf{a}(x) \to \mathsf{c}(x)\}$, $\mathcal{E} = \{\mathsf{a}(\mathsf{b}(\mathsf{a}(x))) \approx \mathsf{b}(\mathsf{a}(\mathsf{b}(x)))\}$, it can be shown that all extensions $Ext_{\mathcal{E}}(\mathcal{R})$ are infinite.

We prove below that $Ext_{\mathcal{E}}(\mathcal{R})$ (according to Def. 8) has the desired property needed to reduce rewriting modulo equations to $\mathcal{E}$-extended rewriting. The following important lemma states that whenever s rewrites to t with $\to_{\mathcal{R}/\mathcal{E}}$ modulo $\mathcal{E}$, then s also rewrites with $\to_{\mathcal{E}\backslash Ext_{\mathcal{E}}(\mathcal{R})}$ to a term which is $\mathcal{E}$-equivalent to t.[6]

Lemma 10 (Connection between $\to_{\mathcal{R}/\mathcal{E}}$ and $\to_{\mathcal{E}\backslash Ext_{\mathcal{E}}(\mathcal{R})}$). *Let $\mathcal{R}$ be a TRS and let $\mathcal{E}$ be a set of equations with identical unique variables. If $s \to_{\mathcal{R}/\mathcal{E}} t$, then there exists a term $t' \sim_{\mathcal{E}} t$ such that $s \to_{\mathcal{E}\backslash Ext_{\mathcal{E}}(\mathcal{R})} t'$.*

[6] Our extension $Ext_{\mathcal{E}}$ has some similarities to the construction of contexts in [23]. However, in contrast to [23] we also consider the rules of $\mathcal{R}'$ in Condition (b) of Def. 8 in order to reduce the number of rules in $Ext_{\mathcal{E}}$. Moreover, in [23] equations may also be non-linear (and thus, Lemma 10 does not hold there).

Proof. Let $s \to_{\mathcal{R}/\mathcal{E}} t$, i.e., there exist terms $s_0, \ldots, s_n, p$ with $n \geq 0$ such that $s = s_n \vdash\dashv_{\mathcal{E}} s_{n-1} \vdash\dashv_{\mathcal{E}} \ldots \vdash\dashv_{\mathcal{E}} s_0 \to_{\mathcal{R}} p \sim_{\mathcal{E}} t$. For the lemma, it suffices to show that there is a $t' \sim_{\mathcal{E}} p$ such that $s \to_{\mathcal{E}\backslash Ext_{\mathcal{E}}(\mathcal{R})} t'$, since $t' \sim_{\mathcal{E}} p$ implies $t' \sim_{\mathcal{E}} t$.

We perform induction on n. If $n = 0$, we have $s = s_n = s_0 \to_{\mathcal{R}} p$. This implies $s \to_{\mathcal{E}\backslash Ext_{\mathcal{E}}(\mathcal{R})} p$ since $\mathcal{R} \subseteq Ext_{\mathcal{E}}(\mathcal{R})$. So with $t' = p$ the claim is proved.

If $n > 0$, the induction hypothesis implies $s = s_n \vdash\dashv_{\mathcal{E}} s_{n-1} \to_{\mathcal{E}\backslash Ext_{\mathcal{E}}(\mathcal{R})} t'$ such that $t' \sim_{\mathcal{E}} p$. So there exists an equation $u \approx v$ or $v \approx u$ from $\mathcal{E}$ and a rule $l \to r$ from $Ext_{\mathcal{E}}(\mathcal{R})$ such that $s|_\tau = u\delta$, $s_{n-1} = s[v\delta]_\tau$, $s_{n-1}|_\xi \sim_{\mathcal{E}} l\delta$, and $t' = s_{n-1}[r\delta]_\xi$ for positions τ and ξ and a substitution δ. We can use the same substitution δ for instantiating the equation $u \approx v$ (or $v \approx u$) and the rule $l \to r$, since equations and rules are assumed variable disjoint. We now perform a case analysis depending on the relationship of the positions τ and ξ.

Case 1: $\tau = \xi\pi$ for some π. In this case, we have $s|_\xi = s|_\xi[u\delta]_\pi \vdash\dashv_{\mathcal{E}} s|_\xi[v\delta]_\pi = s_{n-1}|_\xi \sim_{\mathcal{E}} l\delta$. This implies $s \to_{\mathcal{E}\backslash Ext_{\mathcal{E}}(\mathcal{R})} s[r\delta]_\xi = s_{n-1}[r\delta]_\xi = t'$, as desired.

Case 2: $\tau \bot \xi$. Now we have $s|_\xi = s_{n-1}|_\xi \sim_{\mathcal{E}} l\delta$ and thus, $s \to_{\mathcal{E}\backslash Ext_{\mathcal{E}}(\mathcal{R})} s[r\delta]_\xi = s[r\delta]_\xi[u\delta]_\tau \vdash\dashv_{\mathcal{E}} s[r\delta]_\xi[v\delta]_\tau = s[v\delta]_\tau[r\delta]_\xi = s_{n-1}[r\delta]_\xi = t'$.

Case 3: $\xi = \tau\pi$ for some π. Thus, $(v\delta)|_\pi \sim_{\mathcal{E}} l\delta$. We distinguish two sub-cases.

Case 3.1: $u\delta \to_{\mathcal{E}\backslash Ext_{\mathcal{E}}(\mathcal{R})} q \sim_{\mathcal{E}} (v[r]_\pi)\delta$ for some term q. This implies $s = s[u\delta]_\tau \to_{\mathcal{E}\backslash Ext_{\mathcal{E}}(\mathcal{R})} s[q]_\tau \sim_{\mathcal{E}} s[v[r]_\pi\delta]_\tau = (s[v\delta]_\tau)[r\delta]_\xi = s_{n-1}[r\delta]_\xi = t'$.

Case 3.2: Otherwise. First assume that $\pi = \pi_1\pi_2$ where $v|_{\pi_1}$ is a variable x. Hence, $(v\delta)|_\pi = \delta(x)|_{\pi_2}$. Let $\delta'(y) = \delta(y)$ for $y \neq x$ and let $\delta'(x) = \delta(x)[r\delta]_{\pi_2}$. Since $u \approx v$ (or $v \approx u$) is an equation with identical unique variables, x also occurs in u at some position π'. This implies $u\delta|_{\pi'\pi_2} = \delta(x)|_{\pi_2} \sim_{\mathcal{E}} l\delta \to_{Ext_{\mathcal{E}}(\mathcal{R})} r\delta$. Hence, we obtain $u\delta \to^{\pi'\pi_2}_{\mathcal{E}\backslash Ext_{\mathcal{E}}(\mathcal{R})} u\delta[r\delta]_{\pi'\pi_2} = u\delta' \sim_{\mathcal{E}} v\delta' = (v[r]_\pi)\delta$ in contradiction to the condition of Case 3.2.

Hence, π is a position of v and $v|_\pi$ is not a variable. Thus, $(v\delta)|_\pi = v|_\pi\delta \sim_{\mathcal{E}} l\delta$. Since rules and equations are assumed variable disjoint, the subterm $v|_\pi$ $\mathcal{E}$-unifies with l. Thus, there exists a $\sigma \in uni_{\mathcal{E}}(v|_\pi, l)$ such that $\delta \sim_{\mathcal{E}} \sigma\rho$.

Due to the Condition (b) of Def. 8, there is a term q' such that $u\sigma \to^{\pi'}_{\mathcal{E}\backslash Ext_{\mathcal{E}}(\mathcal{R})} q' \sim_{\mathcal{E}} (v[r]_\pi)\sigma$. Since π' is a position in u, we have $u|_{\pi'}\sigma \sim_{\mathcal{E}} \circ \to_{Ext_{\mathcal{E}}(\mathcal{R})} q''$, where $q' = u\sigma[q'']_{\pi'}$. This also implies $u|_{\pi'}\delta \sim_{\mathcal{E}} u|_{\pi'}\sigma\rho \sim_{\mathcal{E}} \circ \to_{Ext_{\mathcal{E}}(\mathcal{R})} q''\rho$, and thus $u\delta \to^{\pi'}_{\mathcal{E}\backslash Ext_{\mathcal{E}}(\mathcal{R})} u\delta[q''\rho]_{\pi'} \sim_{\mathcal{E}} u\sigma[q'']_{\pi'}\rho = q'\rho \sim_{\mathcal{E}} (v[r]_\pi)\sigma\rho \sim_{\mathcal{E}} (v[r]_\pi)\delta$. This is a contradiction to the condition of Case 3.2. □

The following theorem shows that $Ext_{\mathcal{E}}$ indeed has the desired property.

Theorem 11 (Termination of $\mathcal{R}/\mathcal{E}$ by $\mathcal{E}$-Extended Rewriting). *Let $\mathcal{R}$ be a TRS, let $\mathcal{E}$ be a set of equations with identical unique variables, and let t be a term. Then t does not start an infinite $\to_{\mathcal{R}/\mathcal{E}}$-reduction iff t does not start an infinite $\to_{\mathcal{E}\backslash Ext_{\mathcal{E}}(\mathcal{R})}$-reduction. So in particular, $\mathcal{R}$ is terminating modulo $\mathcal{E}$ (i.e., $\to_{\mathcal{R}/\mathcal{E}}$ is well founded) iff $\to_{\mathcal{E}\backslash Ext_{\mathcal{E}}(\mathcal{R})}$ is well founded.*

Proof. The "only if" direction is straightforward because $\to_{Ext_{\mathcal{E}}(\mathcal{R})} = \to_{\mathcal{R}}$ and therefore, $\to_{\mathcal{E}\backslash Ext_{\mathcal{E}}(\mathcal{R})} \subseteq \to_{Ext_{\mathcal{E}}(\mathcal{R})/\mathcal{E}} = \to_{\mathcal{R}/\mathcal{E}}$.

For the "if" direction, assume that t starts an infinite $\to_{\mathcal{R}/\mathcal{E}}$-reduction

$$t = t_0 \to_{\mathcal{R}/\mathcal{E}} t_1 \to_{\mathcal{R}/\mathcal{E}} t_2 \to_{\mathcal{R}/\mathcal{E}} \cdots$$

For every $i \in \mathbb{N}$, let f_{i+1} be a function from terms to terms such that for every $t'_i \sim_{\mathcal{E}} t_i$, $f_{i+1}(t'_i)$ is a term $\mathcal{E}$-equivalent to t_{i+1} such that $t'_i \to_{\mathcal{E}\backslash Ext_{\mathcal{E}}(\mathcal{R})} f_{i+1}(t'_i)$. These functions f_{i+1} must exist due to Lemma 10, since $t'_i \sim_{\mathcal{E}} t_i$ and $t_i \to_{\mathcal{R}/\mathcal{E}} t_{i+1}$ implies $t'_i \to_{\mathcal{R}/\mathcal{E}} t_{i+1}$. Hence, t starts an infinite $\to_{\mathcal{E}\backslash Ext_{\mathcal{E}}(\mathcal{R})}$-reduction:

$$t \to_{\mathcal{E}\backslash Ext_{\mathcal{E}}(\mathcal{R})} f_1(t) \to_{\mathcal{E}\backslash Ext_{\mathcal{E}}(\mathcal{R})} f_2(f_1(t)) \to_{\mathcal{E}\backslash Ext_{\mathcal{E}}(\mathcal{R})} f_3(f_2(f_1(t))) \to_{\mathcal{E}\backslash Ext_{\mathcal{E}}(\mathcal{R})} \cdots \quad \square$$

5 Dependency Pairs for Rewriting Modulo Equations

In this section we finally extend the dependency pair approach to rewriting modulo equations: To show that $\mathcal{R}$ modulo $\mathcal{E}$ terminates, one first constructs the extension $Ext_{\mathcal{E}}(\mathcal{R})$ of $\mathcal{R}$. Subsequently, dependency pairs can be used to prove well-foundedness of $\to_{\mathcal{E}\backslash Ext_{\mathcal{E}}(\mathcal{R})}$ (which is equivalent to termination of $\mathcal{R}$ modulo $\mathcal{E}$). The idea for the extension of the dependency pair approach is simply to modify Thm. 3 as follows.

1. The equations should be satisfied by the equivalence $\sim$ corresponding to the quasi-ordering $\succsim$, i.e., we demand $u \sim v$ for all equations $u \approx v$ in $\mathcal{E}$.
2. A similar requirement is needed for equations $u \approx v$ when the root symbols of u and v are replaced by the corresponding tuple symbols. We denote tuples of terms $s_1, \ldots, s_n$ by $\boldsymbol{s}$ and for any term $t = f(\boldsymbol{s})$ with a defined root symbol f, let $t^\sharp$ be the term $F(\boldsymbol{s})$. Hence, we also have to demand $u^\sharp \sim v^\sharp$.
3. The notion of "defined symbols" must be changed accordingly. As before, all root symbols of left-hand sides of rules are regarded as being defined, but if there is an equation $f(\boldsymbol{u}) = g(\boldsymbol{v})$ in $\mathcal{E}$ and f is defined, then g must be considered defined as well, as otherwise we would not be able to trace the redex in a reduction by only regarding subterms with defined root symbols.

Definition 12 (Defined Symbols for Rewriting Modulo Equations). *Let $\mathcal{R}$ be a TRS and let $\mathcal{E}$ be a set of equations. Then the set of* defined *symbols $\mathcal{D}$ of $\mathcal{R}/\mathcal{E}$ is the smallest set such that $\mathcal{D} = \{root(l) \mid l \to r \in \mathcal{R}\} \cup \{root(v) \mid u \approx v \in \mathcal{E}$ or $v \approx u \in \mathcal{E}, root(u) \in \mathcal{D}\}$.*

The constraints of the dependency pair approach as sketched above are not yet sufficient for termination of $\to_{\mathcal{E}\backslash\mathcal{R}}$ as the following example illustrates.

Example 13. Consider $\mathcal{R} = \{\mathsf{f}(x) \to x\}$ and $\mathcal{E} = \{\mathsf{f}(\mathsf{a}) \approx \mathsf{a}\}$. There is no dependency pair in this example and thus, the only constraints would be $\mathsf{f}(x) \succsim x$, $\mathsf{f}(\mathsf{a}) \sim \mathsf{a}$, and $\mathsf{F}(\mathsf{a}) \sim \mathsf{A}$. Obviously, these constraints are satisfiable (by using an equivalence relation $\sim$ where all terms are equal). However, $\to_{\mathcal{E}\backslash\mathcal{R}}$ is not terminating since we have $\mathsf{a} \vdash\!\dashv_{\mathcal{E}} \mathsf{f}(\mathsf{a}) \to_{\mathcal{R}} \mathsf{a} \vdash\!\dashv_{\mathcal{E}} \mathsf{f}(\mathsf{a}) \to_{\mathcal{R}} \mathsf{a} \vdash\!\dashv_{\mathcal{E}} \ldots$

The soundness of the dependency pair approach for ordinary rewriting (Thm. 3) relies on the fact that an infinite reduction from a minimal non-terminating term can be achieved by applying only normalized instantiations of $\mathcal{R}$-rules. But for $\mathcal{E}$-extended rewriting (or full rewriting modulo equations), this is not true any more. For instance, the minimal non-terminating subterm a in Ex. 13 is first modified by applying an $\mathcal{E}$-equation (resulting in f(a)) and then an $\mathcal{R}$-rule is applied whose variable is instantiated with the non-terminating term a. Hence, the problem is that the new minimal non-terminating subterm a which results from application of the $\mathcal{R}$-rule does not correspond to the right-hand side of a dependency pair, because this minimal non-terminating subterm is completely inside the instantiation of a *variable* of the $\mathcal{R}$-rule. With ordinary rewriting, this situation can never occur.

In Ex. 13, the problem can be avoided by adding a suitable *instance* of the rule $\mathsf{f}(x) \to x$ (viz. $\mathsf{f}(\mathsf{a}) \to \mathsf{a}$) to $\mathcal{R}$, since this instance is used in the infinite reduction. Now there would be a dependency pair $\langle \mathsf{F}(\mathsf{a}), \mathsf{A} \rangle$ and with the additional constraint $\mathsf{F}(\mathsf{a}) > \mathsf{A}$ the resulting inequalities are no longer satisfiable.

The following definition shows how to add the right instantiations of the rules in $\mathcal{R}$ in order to allow a sound application of dependency pairs. As usual, a substitution ν is called a *variable renaming* iff the range of ν only contains variables and if $\nu(x) \neq \nu(y)$ for $x \neq y$.

Definition 14 (Adding Instantiations). *Given a TRS $\mathcal{R}$, a set $\mathcal{E}$ of equations, let $\mathcal{R}'$ be a set containing only rules of the form $l\sigma \to r\sigma$ (where σ is a substitution and $l \to r \in \mathcal{R}$). $\mathcal{R}'$ is an* instantiation of $\mathcal{R}$ for the equations $\mathcal{E}$ *iff*

(a) $\mathcal{R} \subseteq \mathcal{R}'$,
(b) for all $l \to r \in \mathcal{R}$, all $u \approx v \in \mathcal{E}$ and $v \approx u \in \mathcal{E}$, and all $\sigma \in uni_{\mathcal{E}}(v, l)$, there exists a rule $l' \to r' \in \mathcal{R}'$ and a variable renaming ν such that $l\sigma \sim_{\mathcal{E}} l'\nu$ and $r\sigma \sim_{\mathcal{E}} r'\nu$.

In the following, let $Ins_{\mathcal{E}}(\mathcal{R})$ always denote an instantiation of $\mathcal{R}$ for $\mathcal{E}$.

Unlike extensions $Ext_{\mathcal{E}}(\mathcal{R})$, instantiations $Ins_{\mathcal{E}}(\mathcal{R})$ are never infinite if $\mathcal{R}$ and $\mathcal{E}$ are finite and if $uni_{\mathcal{E}}(v, l)$ is always finite (i.e., they are not defined via a fixpoint construction). In fact, one might even demand that for all $l \to r \in \mathcal{R}$, all equations, and all σ from the corresponding complete set of $\mathcal{E}$-unifiers, $Ins_{\mathcal{E}}(\mathcal{R})$ should contain $l\sigma \to r\sigma$. The condition that it is enough if some $\mathcal{E}$-equivalent variable-renamed rule is already contained in $Ins_{\mathcal{E}}(\mathcal{R})$ is only added for efficiency considerations in order to reduce the number of rules in $Ins_{\mathcal{E}}(\mathcal{R})$. Even without this condition, $Ins_{\mathcal{E}}(\mathcal{R})$ would still be finite and all the following theorems would hold as well.

However, the above instantiation technique only serves its purpose if there are no collapsing equations (i.e., no equations $u \approx v$ or $v \approx u$ with $v \in \mathcal{V}$).

Example 15. Consider $\mathcal{R} = \{\mathsf{f}(x) \to x\}$ and $\mathcal{E} = \{\mathsf{f}(x) \approx x\}$. Note that $Ins_{\mathcal{E}}(\mathcal{R}) = \mathcal{R}$. Although $\to_{\mathcal{E}\backslash\mathcal{R}}$ is clearly not terminating, the dependency pair approach would falsely prove termination of $\to_{\mathcal{E}\backslash\mathcal{R}}$, since there is no dependency pair.

Now we can present the main result of the paper.

Theorem 16 (Termination of Equational Rewriting using Dependency Pairs). *Let $\mathcal{R}$ be a TRS and let $\mathcal{E}$ be a set of non-collapsing equations with identical unique variables. $\mathcal{R}$ is terminating modulo $\mathcal{E}$ (i.e., $\to_{\mathcal{R}/\mathcal{E}}$ is well founded) if there exists a weakly monotonic quasi-ordering $\succsim$ and a well-founded ordering $>$ compatible with $\succsim$ where both $\succsim$ and $>$ are closed under substitution, such that*

(1) $s > t$ *for all dependency pairs* $\langle s, t\rangle$ *of* $Ins_{\mathcal{E}}(Ext_{\mathcal{E}}(\mathcal{R}))$,
(2) $l \succsim r$ *for all rules* $l \to r$ *of* $\mathcal{R}$,
(3) $u \sim v$ *for all equations* $u \approx v$ *of* $\mathcal{E}$, *and*
(4) $u^\sharp \sim v^\sharp$ *for all equations* $u \approx v$ *of* $\mathcal{E}$ *where* $root(u)$ *and* $root(v)$ *are defined.*

Proof. Suppose that there is a term t with an infinite $\to_{\mathcal{R}/\mathcal{E}}$-reduction. Thm. 11 implies that t also has an infinite $\to_{\mathcal{E}\backslash Ext_{\mathcal{E}}(\mathcal{R})}$-reduction. By a minimality argument, $t = C[t']$, where t' is an *minimal non-terminating* term (i.e., t' is non-terminating, but all its subterms only have finite $\to_{\mathcal{E}\backslash Ext_{\mathcal{E}}(\mathcal{R})}$-reductions). We will show that there exists a term t_1 with $t \to^+_{\mathcal{E}\backslash Ext_{\mathcal{E}}(\mathcal{R})} t_1$, t_1 contains a minimal non-terminating subterm t'_1, and $t'^\sharp \succsim \circ > t_1'^\sharp$. By repeated application of this construction we obtain an infinite sequence $t \to^+_{\mathcal{E}\backslash Ext_{\mathcal{E}}(\mathcal{R})} t_1 \to^+_{\mathcal{E}\backslash Ext_{\mathcal{E}}(\mathcal{R})} t_2 \to^+_{\mathcal{E}\backslash Ext_{\mathcal{E}}(\mathcal{R})} \ldots$ such that $t'^\sharp \succsim \circ > t_1'^\sharp \succsim \circ > t_2'^\sharp \succsim \circ > \ldots$. This, however, is a contradiction to the well-foundedness of $>$.

Let t' have the form $f(\boldsymbol{u})$. In the infinite $\to_{\mathcal{E}\backslash Ext_{\mathcal{E}}(\mathcal{R})}$-reduction of $f(\boldsymbol{u})$, first some $\to_{\mathcal{E}\backslash Ext_{\mathcal{E}}(\mathcal{R})}$-steps may be applied to $\boldsymbol{u}$ which yields new terms $\boldsymbol{v}$. Note that due to the definition of $\mathcal{E}$-extended rewriting, in these reductions, no $\mathcal{E}$-steps can be applied outside of $\boldsymbol{u}$. Due to the termination of $\boldsymbol{u}$, after a finite number of those steps, an $\to_{\mathcal{E}\backslash Ext_{\mathcal{E}}(\mathcal{R})}$-step must be applied on the root position of $f(\boldsymbol{v})$.

Thus, there exists a rule $l \to r \in Ext_{\mathcal{E}}(\mathcal{R})$ such that $f(\boldsymbol{v}) \sim_{\mathcal{E}} l\alpha$ and hence, the reduction yields $r\alpha$. Now the infinite $\to_{\mathcal{E}\backslash Ext_{\mathcal{E}}(\mathcal{R})}$-reduction continues with $r\alpha$, i.e., the term $r\alpha$ starts an infinite $\to_{\mathcal{E}\backslash Ext_{\mathcal{E}}(\mathcal{R})}$-reduction, too. So up to now the reduction has the following form (where $\to_{Ext_{\mathcal{E}}(\mathcal{R})}$ equals $\to_{\mathcal{R}}$):

$$t = C[f(\boldsymbol{u})] \to^*_{\mathcal{E}\backslash Ext_{\mathcal{E}}(\mathcal{R})} C[f(\boldsymbol{v})] \sim_{\mathcal{E}} C[l\alpha] \to_{Ext_{\mathcal{E}}(\mathcal{R})} C[r\alpha].$$

We perform a case analysis depending on the positions of $\mathcal{E}$-steps in $f(\boldsymbol{v}) \sim_{\mathcal{E}} l\alpha$.

First consider the case where all $\mathcal{E}$-steps in $f(\boldsymbol{v}) \sim_{\mathcal{E}} l\alpha$ take place below the root. Then we have $l = f(\boldsymbol{w})$ and $\boldsymbol{v} \sim_{\mathcal{E}} \boldsymbol{w}\alpha$. Let $t_1 := C[r\alpha]$. Note that $\boldsymbol{v}$ do not start infinite $\to_{\mathcal{E}\backslash Ext_{\mathcal{E}}(\mathcal{R})}$-reductions and by Thm. 11, they do not start infinite $\to_{\mathcal{R}/\mathcal{E}}$-reductions either. But then $\boldsymbol{w}\alpha$ also cannot start infinite $\to_{\mathcal{R}/\mathcal{E}}$-reductions and therefore they also do not start infinite $\to_{\mathcal{E}\backslash Ext_{\mathcal{E}}(\mathcal{R})}$-reductions. This implies that for all variables x occurring in $f(\boldsymbol{w})$ the terms $\alpha(x)$ are terminating. Thus, since $r\alpha$ starts an infinite reduction, there occurs a non-variable subterm s in r, such that $t'_1 := s\alpha$ is a minimal non-terminating term. Since $\langle l^\sharp, s^\sharp\rangle$ is a dependency pair, we obtain $t'^\sharp = F(\boldsymbol{u}) \succsim F(\boldsymbol{v}) \sim l^\sharp\alpha > s^\sharp\alpha = t_1'^\sharp$. Here, $F(\boldsymbol{u}) \succsim F(\boldsymbol{v})$ holds since $\boldsymbol{u} \to^*_{\mathcal{E}\backslash Ext_{\mathcal{E}}(\mathcal{R})} \boldsymbol{v}$ and since $l \succsim r$ for every rule $l \to r \in Ext_{\mathcal{E}}(\mathcal{R})$.

Now we consider the case where there are $\mathcal{E}$-steps in $f(\boldsymbol{v}) \sim_{\mathcal{E}} l\alpha$ at the root position. Thus we have $f(\boldsymbol{v}) \sim_{\mathcal{E}} f(\boldsymbol{q}) \vdash\!\dashv_{\mathcal{E}} p \sim_{\mathcal{E}} l\alpha$, where $f(\boldsymbol{q}) \vdash\!\dashv_{\mathcal{E}} p$ is the first $\mathcal{E}$-step at the root position. In other words, there is an equation $u \approx v$ or $v \approx u$ in $\mathcal{E}$ such that $f(\boldsymbol{q})$ is an instantiation of v.

Note that since $\boldsymbol{v} \sim_{\mathcal{E}} \boldsymbol{q}$, the terms $\boldsymbol{q}$ only have finite $\to_{\mathcal{E}\backslash Ext_{\mathcal{E}}(\mathcal{R})}$-reductions (the argumentation is similar as in the first case). Let δ be the substitution which operates like α on the variables of l and which yields $v\delta = f(\boldsymbol{q})$. Thus, δ is an $\mathcal{E}$-unifier of l and v. Since l is $\mathcal{E}$-unifiable with v, there also exists a corresponding complete $\mathcal{E}$-unifier σ from $uni_{\mathcal{E}}(l, v)$. Thus, there is also a substitution ρ such that $\delta \sim_{\mathcal{E}} \sigma\rho$. As l is a left-hand side of a rule from $Ext_{\mathcal{E}}(\mathcal{R})$, there is a rule

$l' \to r'$ in $Ins_{\mathcal{E}}(Ext_{\mathcal{E}}(\mathcal{R}))$ and a variable renaming ν such that $l\sigma \sim_{\mathcal{E}} l'\nu$ and $r\sigma \sim_{\mathcal{E}} r'\nu$.

Hence, $v\sigma\rho \sim_{\mathcal{E}} v\delta = f(\boldsymbol{q})$, $l'\nu\rho \sim_{\mathcal{E}} l\sigma\rho \sim_{\mathcal{E}} l\delta = l\alpha$, and $r'\nu\rho \sim_{\mathcal{E}} r\sigma\rho \sim_{\mathcal{E}} r\delta = r\alpha$. So instead we now consider the following reduction (where $\to_{Ins_{\mathcal{E}}(Ext_{\mathcal{E}}(\mathcal{R}))}$ equals $\to_{\mathcal{R}}$):

$$t = C[f(\boldsymbol{u})] \to^{*}_{\mathcal{E}\backslash Ext_{\mathcal{E}}(\mathcal{R})} C[f(\boldsymbol{v})] \sim_{\mathcal{E}} C[l'\nu\rho] \to_{Ins_{\mathcal{E}}(Ext_{\mathcal{E}}(\mathcal{R}))} C[r'\nu\rho] = t_1.$$

Since all proper subterms of $v\delta$ only have finite $\to_{\mathcal{R}/\mathcal{E}}$-reductions, for all variables x of $l'\nu$, the term $x\rho$ only has finite $\to_{\mathcal{R}/\mathcal{E}}$-reductions and hence, also only finite $\to_{\mathcal{E}\backslash Ext_{\mathcal{E}}(\mathcal{R})}$-reductions. To see this, note that since all equations have identical unique variables, $v\sigma \sim_{\mathcal{E}} l\sigma \sim_{\mathcal{E}} l'\nu$ implies that all variables of $l'\nu$ also occur in $v\sigma$. Thus, if x is a variable from $l'\nu$, then there exists a variable y in v such that x occurs in $y\sigma$. Since $\mathcal{E}$ does not contain collapsing equations, y is a proper subterm of v and thus, $y\delta$ is a proper subterm of $v\delta$. As all proper subterms of $v\delta$ only have finite $\to_{\mathcal{R}/\mathcal{E}}$-reductions, this implies that $y\delta$ only has finite $\to_{\mathcal{R}/\mathcal{E}}$-reductions, too. But then, since $y\delta \sim_{\mathcal{E}} y\sigma\rho$, the term $y\sigma\rho$ only has finite $\to_{\mathcal{R}/\mathcal{E}}$-reductions, too. Then this also holds for all subterms of $y\sigma\rho$, i.e., all $\to_{\mathcal{R}/\mathcal{E}}$-reductions of $x\rho$ are also finite.

So for all variables x of l', $x\nu\rho$ only has finite $\to_{\mathcal{E}\backslash Ext_{\mathcal{E}}(\mathcal{R})}$-reductions. (Note that this only holds because ν is just a variable renaming.) Since $r\alpha$ starts an infinite $\to_{\mathcal{E}\backslash Ext_{\mathcal{E}}(\mathcal{R})}$-reduction, $r'\nu\rho \sim_{\mathcal{E}} r\alpha$ must start an infinite $\to_{\mathcal{R}/\mathcal{E}}$-reduction (and hence, an infinite $\to_{\mathcal{E}\backslash Ext_{\mathcal{E}}(\mathcal{R})}$-reduction) as well. As for all variables x of r', $x\nu\rho$ is $\to_{\mathcal{E}\backslash Ext_{\mathcal{E}}(\mathcal{R})}$-terminating, there must be a non-variable subterm s of r', such that $t_1' := s\nu\rho$ is a minimal non-terminating term. As $\langle l'^{\sharp}, s^{\sharp}\rangle$ is a dependency pair, we obtain $t'^{\sharp} = F(\boldsymbol{u}) \succsim F(\boldsymbol{v}) \sim_{\mathcal{E}} l'^{\sharp}\nu\rho > s^{\sharp}\nu\rho = t_1'^{\sharp}$. Here, $F(\boldsymbol{v}) \sim_{\mathcal{E}} l'^{\sharp}\nu\rho$ is a consequence of Condition (4). □

Now termination of the division-system (Ex. 9) can be proved by dependency pairs. Here we have $Ins_{\mathcal{E}}(Ext_{\mathcal{E}}(\mathcal{R})) = Ext_{\mathcal{E}}(\mathcal{R})$ and thus, the resulting constraints are

$$\begin{array}{ll} \mathsf{M}(\mathsf{s}(x),\mathsf{s}(y)) > \mathsf{M}(x,y) & \mathsf{Q}(0 \div \mathsf{s}(y), z) > \mathsf{Q}(0,z) \\ \mathsf{Q}(\mathsf{s}(x),\mathsf{s}(y)) > \mathsf{M}(x,y) & \mathsf{Q}(\mathsf{s}(x) \div \mathsf{s}(y), z) > \mathsf{M}(x,y) \\ \mathsf{Q}(\mathsf{s}(x),\mathsf{s}(y)) > \mathsf{Q}(x-y,\mathsf{s}(y)) & \mathsf{Q}(\mathsf{s}(x) \div \mathsf{s}(y), z) > \mathsf{Q}(x-y,\mathsf{s}(y)) \\ & \mathsf{Q}(\mathsf{s}(x) \div \mathsf{s}(y), z) > \mathsf{Q}(\mathsf{s}((x-y) \div \mathsf{s}(y)), z) \end{array}$$

as well as $l \succsim r$ for all rules $l \to r$, $(u \div v) \div w \sim (u \div w) \div v$, and $\mathsf{Q}(u \div v, w) \sim \mathsf{Q}(u \div w, v)$. (Here, M and Q are the tuple symbols for the minus-symbol "$-$" and the quot-symbol "$\div$".) As explained in Sect. 2 one may again eliminate arguments of function symbols before searching for suitable orderings. In this example we will eliminate the second arguments of $-$, $\div$, M, and Q (i.e., every term $s - t$ is replaced by $-'(s)$, etc.). Then the resulting inequalities are satisfied by the rpo with the precedence $\div' \sqsupset \mathsf{s} \sqsupset -'$, $\mathsf{Q}' \sqsupset \mathsf{M}'$. Thus, with the method of the present paper, one can now verify termination of this example *automatically* for the first time. This example also demonstrates that by using dependency pairs, termination of equational rewriting can sometimes even be shown by *ordinary* base orderings (e.g., the ordinary rpo which on its own cannot be used for rewriting modulo equations).

6 Conclusion

We have extended the dependency pair approach to equational rewriting. In the special case of AC-axioms, our method is similar to the ones previously presented in [15,17]. In fact, as long as the equations only consist of AC-axioms, one can show that using the instances $Ins_{\mathcal{E}}$ in Thm. 16 is not necessary.[7] (Hence, such a concept cannot be found in [17]). However, even then the only additional inequalities resulting from $Ins_{\mathcal{E}}$ are instantiations of other inequalities already present and inequalities which are special cases of an AC-deletion property (which is satisfied by all known AC-orderings and similar to the one required in [15]). This indicates that in practical examples with AC-axioms, our technique is at least as powerful as the ones of [15,17] (actually, we conjecture that for AC-examples, these three techniques are virtually equally powerful). But compared to the approaches of [15,17], our technique has a more elegant treatment of tuple symbols. (For example, if the TRS contains a rule $\mathsf{f}(t_1, t_2) \to \mathsf{g}(\mathsf{f}(s_1, s_2), s_3)$ were f and g are defined AC-symbols, then we do not have to extend the TRS by rules with tuple symbols like $\mathsf{f}(t_1, t_2) \to \mathsf{G}(\mathsf{f}(s_1, s_2), s_2)$ in [17]. Moreover, we do not need dependency pairs where tuple symbols occur outside the root position such as $\langle \mathsf{F}(\mathsf{F}(t_1, t_2), y), \ldots \rangle$ in [17] and [15] and $\langle \mathsf{F}(t_1, t_2), \mathsf{G}(\mathsf{F}(s_1, s_2), s_3) \rangle$ in [15]. Finally, we also do not need the "AC-marked condition" $\mathsf{F}(\mathsf{f}(x, y), z) \sim \mathsf{F}(\mathsf{F}(x, y), z)$ of [15].) But most significantly, unlike [15,17] our technique works for *arbitrary* non-collapsing equations $\mathcal{E}$ with identical unique variables where $\mathcal{E}$-unification is finitary (for subterms of equations and left-hand sides of rules). Obviously, an implementation of our technique also requires $\mathcal{E}$-unification algorithms [5] for the concrete sets of equations $\mathcal{E}$ under consideration.

In [1,2,3], Arts and Giesl presented the *dependency graph* refinement which is based on the observation that it is possible to treat subsets of the dependency pairs separately. This refinement carries over to the equational case in a straightforward way (by using $\mathcal{E}$-unification to compute an estimation of this graph). For details on this refinement and for further examples to demonstrate the power and the usefulness of our technique, the reader is referred to [11].

Acknowledgments. We thank A. Middeldorp, T. Arts, and the referees for comments.

References

1. T. Arts and J. Giesl, Automatically Proving Termination where Simplification Orderings Fail, in *Proc. TAPSOFT '97*, LNCS 1214, 261-272, 1997.
2. T. Arts and J. Giesl, Modularity of Termination Using Dependency Pairs, in *Proc. RTA '98*, LNCS 1379, 226-240, 1998.
3. T. Arts and J. Giesl, Termination of Term Rewriting Using Dependency Pairs, *Theoretical Computer Science*, 236:133-178, 2000.

[7] Then in the proof of Thm. 16, instead of a minimal non-terminating term t' one regards a term t' which is non-terminating and minimal up to some extra f-occurrences on the top (where f is an AC-symbol).

4. T. Arts, System Description: The Dependency Pair Method, in *Proc. RTA '00*, LNCS 1833, 261-264, 2000.
5. F. Baader and W. Snyder, Unification Theory, in *Handbook of Automated Reasoning*, J. A. Robinson and A. Voronkov (eds.), Elsevier. To appear.
6. A. Ben Cherifa and P. Lescanne, Termination of Rewriting Systems by Polynomial Interpretations and its Implementation, *Sc. Comp. Prog.*, 9(2):137-159, 1987.
7. CiME 2. Pre-release available at http://www.lri.fr/~demons/cime-2.0.html.
8. C. Delor and L. Puel, Extension of the Associative Path Ordering to a Chain of Associative Commutative Symbols, in *Proc. RTA '93*, LNCS 690, 389-404, 1993.
9. N. Dershowitz, Termination of Rewriting, *J. Symbolic Computation*, 3:69-116, 1987.
10. M. C. F. Ferreira, Dummy Elimination in Equational Rewriting, in *Proc. RTA '96*, LNCS 1103, 78-92, 1996.
11. J. Giesl and D. Kapur, Dependency Pairs for Equational Rewriting, Technical Report TR-CS-2000-53, University of New Mexico, USA, 2000. Available from `http://www.cs.unm.edu/soe/cs/tech_reports`
12. J.-P. Jouannaud and H. Kirchner, Completion of a Set of Rules Modulo a Set of Equations, *SIAM Journal on Computing*, 15(4):1155-1194, 1986.
13. D. Kapur and G. Sivakumar, A Total Ground Path Ordering for Proving Termination of *AC*-Rewrite Systems, in *Proc. RTA '97*, LNCS 1231, 142-156, 1997.
14. D. Kapur and G. Sivakumar, Proving Associative-Commutative Termination Using RPO-Compatible Orderings, in *Proc. Automated Deduction in Classical and Non-Classical Logics*, LNAI 1761, 40-62, 2000.
15. K. Kusakari and Y. Toyama, On Proving *AC*-Termination by *AC*-Dependency Pairs, Research Report IS-RR-98-0026F, School of Information Science, JAIST, Japan, 1998. Revised version in K. Kusakari, Termination, *AC*-Termination and Dependency Pairs of Term Rewriting Systems, PhD Thesis, JAIST, Japan, 2000.
16. J.-P. Jouannaud and M. Muñoz, Termination of a Set of Rules Modulo a Set of Equations, in *Proc. 7th CADE*, LNCS 170, 175-193, 1984.
17. C. Marché and X. Urbain, Termination of Associative-Commutative Rewriting by Dependency Pairs, in *Proc. RTA '98*, LNCS 1379, 241-255, 1998.
18. H. Ohsaki, A. Middeldorp, and J. Giesl, Equational Termination by Semantic Labelling, in *Proc. CSL '00*, LNCS 1862, 457-471, 2000.
19. G. E. Peterson and M. E. Stickel, Complete Sets of Reductions for Some Equational Theories, *Journal of the ACM*, 28(2):233-264, 1981.
20. A. Rubio and R. Nieuwenhuis, A Total *AC*-Compatible Ordering based on RPO, *Theoretical Computer Science*, 142:209-227, 1995.
21. A. Rubio, A Fully Syntactic *AC*-RPO, *Proc. RTA-99*, LNCS 1631, 133-147, 1999.
22. J. Steinbach, Simplification Orderings: History of Results, *Fundamenta Informaticae*, 24:47-87, 1995.
23. L. Vigneron, Positive Deduction modulo Regular Theories, in *Proc. CSL '95*, LNCS 1092, 468-485, 1995.

Termination Proofs by Context-Dependent Interpretations

Dieter Hofbauer

Universität Gh Kassel, Fachbereich 17 Mathematik/Informatik,
D–34109 Kassel, Germany
dieter@theory.informatik.uni-kassel.de

Abstract. Proving termination of a rewrite system by an interpretation over the natural numbers directly implies an upper bound on the derivational complexity of the system. In this way, however, the derivation height of terms is often heavily overestimated.
Here we present a generalization of termination proofs by interpretations that can avoid this drawback of the traditional approach. A number of simple examples illustrate how to achieve tight or even optimal bounds on the derivation height. The method is general enough to capture cases where simplification orderings fail.

1 Introduction

Proving termination of a rewrite system by an interpretation into some well-founded domain is perhaps the most natural approach among the numerous methods developed in the last few decades. Early references are [21,17], but the idea of termination proofs by interpretations can at least be traced back to Turing [30,24].

We are interested in extracting upper bounds on the derivational complexity of rewrite systems from termination proofs. Proving termination by an interpretation over the natural numbers always directly yields such a bound for the system [12,13,23]. In particular, this is true for polynomial interpretations [18,9,16] (cf [3]), but also for interpretations using other subrecursive functions classes as in [20], for instance. Upper bound results have been also obtained for most of the standard reduction orders like multiset path orders [13], lexicographic path orders [32], Knuth-Bendix orders [16,15,29,19], or for all terminating ground systems [14]. All these results are proven either by combinatorial considerations or, more interestingly in the context of the present paper, by transforming termination proofs via 'syntactic' orders into proofs via interpretations into the natural numbers.

The major disadvantage of using monotone interpretations for upper bound results is that in this way the derivational complexity is often heavily overestimated. A natural question therefore is how to prove polynomial upper bounds (if they exist), or even asymptotically optimal bounds for polynomials of small degree. As one remedy it was suggested to consider syntactic restrictions of some

A. Middeldorp (Ed.): RTA 2001, LNCS 2051, pp. 108–121, 2001.

standard termination orders like multiset or lexicographic path orders [22,4]. Another approach uses incremental termination proofs as described in [7]. In this paper, in contrast, we present a generalization of the notion of interpretations in order to obtain tight or even optimal upper bounds. For introductions to term rewriting, especially to termination of rewriting, we refer to [6,1,5,33,34,35].

An *interpretation* for a rewrite system R is an order-preserving mapping τ from $(\mathcal{T}_\Sigma, \xrightarrow{+}_R)$ into some (partially) ordered set $(A, >)$, that is, a mapping $\tau : \mathcal{T}_\Sigma \to A$ satisfying $\to_R \subseteq >_\tau$, where $>_\tau$ is the order on $\mathcal{T}_\Sigma$ induced by τ (i.e., $s >_\tau t$ iff $\tau(s) > \tau(t)$). Clearly, if $(A, >)$ is well-founded then $(\mathcal{T}_\Sigma, \xrightarrow{+}_R)$ is well-founded too, and we have a termination proof for R via τ.

Often, A has the structure of a Σ-Algebra, that is, for each n-ary symbol $f \in \Sigma$ $(n \geq 0)$ there is an associated mapping $f_\tau : A^n \to A$. In this case the mapping $\tau : \mathcal{T}_\Sigma \to A$ can be chosen as the unique Σ-homomorphism from the term algebra into A, where

$$\tau(f(t_1, \ldots, t_n)) = f_\tau(\tau(t_1), \ldots, \tau(t_n)). \tag{1}$$

Such a homomorphic interpretation τ into a Σ-algebra A is particularly convenient if A is a strictly monotone ordered algebra (cf. [31], p. 265), that is, if $(A, >)$ is a poset, and all its operations are strictly monotone in each argument: $a_j > a'_j$ implies $f_\tau(a_1, \ldots, a_j, \ldots, a_n) > f_\tau(a_1, \ldots, a'_j, \ldots, a_n)$ for $a_i, a'_j \in A$, $f \in \Sigma_n$, $1 \leq j \leq n$. Then $>_\tau$ is compatible with Σ-contexts: if $s >_\tau t$ then $c[s] >_\tau c[t]$ for ground terms s, t and ground contexts c. Thus proving termination of a rewrite system R via τ amounts to check $\ell\gamma >_\tau r\gamma$ for each rule $\ell \to r$ in R and each ground substitution γ only. In this situation, τ is called a *strictly monotone interpretation* for R.

We are interested in extracting bounds on the length of derivation sequences from termination proofs by interpretations. Let R be a terminating rewrite system where $\to_R$ is finitely branching (which is always true for finite systems). Then the *derivation height* function with respect to R on $\mathcal{T}_\Sigma$ is defined by

$$\mathrm{dh}_R(t) = \max\{n \in \mathbb{N} \mid \exists s : t \to_R^n s\},$$

that is, $\mathrm{dh}_R(t)$ is the height of t modulo $\xrightarrow{+}_R$, see [8]. The *derivational complexity* of R is the function $\mathrm{dc}_R : \mathbb{N} \to \mathbb{N}$ with

$$\mathrm{dc}_R(n) = \max\{\mathrm{dh}_R(t) \mid \mathsf{size}(t) \leq n\}.$$

In case we found a mapping $\tau : \mathcal{T}_\Sigma \to \mathbb{N}$ with $\to_R \subseteq >_\tau$, that is, if

$$s \to_R t \quad \text{implies} \quad \tau(s) - \tau(t) \geq 1 \tag{2}$$

for $s, t \in \mathcal{T}_\Sigma$, then clearly

$$\mathrm{dh}_R(t) \leq \tau(t). \tag{3}$$

Typically, however, the derivation height is heavily overestimated by this inequality. As a simple introductory example consider the term rewrite system R with the one rule

$$a(b(x)) \to b(a(x))$$

over a signature containing the two unary function symbols a, b and the constant symbol c (in order to guarantee the existence of ground terms). Here we obtain a termination proof by a strictly monotone interpretation over the natural numbers $\mathbb{N} = \{0, 1, \dots\}$ by choosing

$$a_\tau(n) = 2n, \qquad b_\tau(n) = 1 + n, \qquad c_\tau = 0. \tag{4}$$

Strict monotonicity of a_τ and b_τ means $a_\tau(n+1) > a_\tau(n)$ and $b_\tau(n+1) > b_\tau(n)$ for $n \in \mathbb{N}$; equivalently,

$$a_\tau(n+1) - a_\tau(n) \geq 1, \qquad b_\tau(n+1) - b_\tau(n) \geq 1. \tag{5}$$

It remains to show $\ell\gamma >_\tau r\gamma$ for the above rewrite rule $\ell \to r$ and any ground substitution γ. Indeed, $\tau(a(b(t))) - \tau(b(a(t))) = 2(1+\tau(t)) - (1+2\tau(t)) = 1$ for $t \in \mathcal{T}_\Sigma$. Here, for example terms of the form a^*b^*c we get

$$\tau(a^n b^m c) = 2^n \cdot m,$$

whereas, as is easily seen,

$$\mathrm{dh}_R(a^n b^m c) = n \cdot m.$$

Thus, τ can be exponential in the size of terms while dc_R is polynomially bounded. The explanation is that, although rewrite steps at the root of terms cause a decrease of only 1 under the interpretation, rewrite steps further down in the tree exhibit an arbitrarily large decrease. For instance, the single rewrite step $a^n abc \to_R a^n bac$ is reflected by $\tau(a^n abc) - \tau(a^n bac) = 2^{n+1} - 2^n = 2^n$.

Before formally defining context-dependent interpretations in the next section we will already now illustrate their use in obtaining tight bounds on the derivational complexity for the introductory example. Crucial for our approach is that we consider a family of interpretations $\tau[\Delta]$ rather than a single one, where the parameter[1] Δ is a positive real number, that is, $\Delta \in \mathbb{R}^+$. Furthermore, the domain $\mathbb{N}$ is replaced by the domain $\mathbb{R}_0^+$ of non-negative real numbers, and the standard well-ordering on $\mathbb{N}$ is replaced by a family of well-founded (partial) orderings $>_\Delta$ on $\mathbb{R}_0^+$ for $\Delta > 0$ with

$$z' >_\Delta z \quad \text{iff} \quad z' - z \geq \Delta.$$

[1] It is a matter of taste whether we work with $\Delta > 0$ or with δ, $0 < \delta < 1$, by the bijection $\Delta \mapsto \delta/(1-\delta)$ and $\delta \mapsto \Delta/(1+\Delta)$.

Similar to the traditional approach, if each mapping $\tau[\Delta] : \mathcal{T}_\Sigma \to \mathbb{R}_0^+$ into the order $(\mathbb{R}_0^+, >_\Delta)$ satisfies $\to_R \subseteq >_{\tau[\Delta]}$, that is, if

$$s \to_R t \quad \text{implies} \quad \tau[\Delta](s) - \tau[\Delta](t) \geq \Delta \tag{2'}$$

for $s, t \in \mathcal{T}_\Sigma$, then R is terminating and $\text{dh}_R(t) \leq \tau[\Delta](t)/\Delta$, hence

$$\text{dh}_R(t) \leq \inf_{\Delta>0} \frac{\tau[\Delta](t)}{\Delta}. \tag{3'}$$

for $t \in \mathcal{T}_\Sigma$. In order to guarantee (2') we again proceed in analogy to the traditional approach. We generalize (1) in defining the function $\tau : \mathbb{R}^+ \times \mathcal{T}_\Sigma \to \mathbb{R}_0^+$ by

$$\tau[\Delta](f(t_1, \ldots, t_n)) = f_\tau[\Delta]\big(\tau[f_\tau^1(\Delta)](t_1), \ldots, \tau[f_\tau^n(\Delta)](t_n)\big), \tag{1'}$$

using functions $f_\tau : \mathbb{R}^+ \times (\mathbb{R}_0^+)^n \to \mathbb{R}_0^+$ and $f_\tau^i : \mathbb{R}^+ \to \mathbb{R}^+$ for each n-ary symbol $f \in \Sigma$ and $1 \leq i \leq n$. (We write $\tau[\Delta](t)$ instead of $\tau(\Delta, t)$ and $f_\tau[\Delta](z_1, \ldots, z_n)$ instead of $f_\tau(\Delta, z_1, \ldots, z_n)$ to emphasize the special role of the first argument.) Returning to our example, we modify (4) and obtain

$$a_\tau[\Delta](z) = (1+\Delta)z, \qquad b_\tau[\Delta](z) = 1 + z, \qquad c_\tau[\Delta] = 0. \tag{4'}$$

This will be the only 'creative' step in finding a context-dependent interpretation for R. But how then to choose the functions a_τ^1 and b_τ^1? As we will explain in more detail in the next section, *Δ-monotonicity* of f_τ is required in our approach. Informally, this says that $f_\tau[\Delta]$ propagates a difference of at least Δ provided a difference of at least $f_\tau^i(\Delta)$ in argument position i is given:

$$\begin{aligned} a_\tau[\Delta](z + a_\tau^1(\Delta)) - a_\tau[\Delta](z) &\geq \Delta, \\ b_\tau[\Delta](z + b_\tau^1(\Delta)) - b_\tau[\Delta](z) &\geq \Delta. \end{aligned} \tag{5'}$$

Solving (5') gives $(1+\Delta)(z + a_\tau^1(\Delta)) - (1+\Delta)z = (1+\Delta)a_\tau^1(\Delta) \geq \Delta$, that is, $a_\tau^1(\Delta) \geq \Delta/(1+\Delta)$, and $(1 + z + b_\tau^1(\Delta)) - (1+z) \geq \Delta$, that is, $b_\tau^1(\Delta) \geq \Delta$. Therefore, the most natural choice is

$$a_\tau^1(\Delta) = \frac{\Delta}{1+\Delta}, \qquad b_\tau^1(\Delta) = \Delta.$$

To sum up, we found in a rather systematic way the following interpretation:

$$\begin{aligned} \tau[\Delta](a(t)) &= (1+\Delta) \cdot \tau[\tfrac{\Delta}{1+\Delta}](t), \\ \tau[\Delta](b(t)) &= 1 + \tau[\Delta](t), \\ \tau[\Delta](c) &= 0. \end{aligned}$$

All that remains to show is $\tau[\Delta](a(b(t))) - \tau[\Delta](b(a(t))) \geq \Delta$ for $t \in \mathcal{T}_\Sigma$. Indeed,

$$\tau[\Delta](a(b(t))) - \tau[\Delta](b(a(t))) =$$
$$(1+\Delta)(1 + \tau[\tfrac{\Delta}{1+\Delta}](t)) - (1 + (1+\Delta)\tau[\tfrac{\Delta}{1+\Delta}](t)) = \Delta.$$

For the terms $a^n b^m c$ discussed above we obtain[2]

$$\tau[\Delta](a^n b^m c) = (1 + \Delta n)m,$$

thus

$$\inf_{\Delta>0} \frac{\tau[\Delta](a^n b^m c)}{\Delta} = \inf_{\Delta>0} (\frac{1}{\Delta} + n)m = n \cdot m = \mathrm{dh}_R(a^n b^m c).$$

So for these terms, the above context-dependent interpretation does yield the precise derivation height. Even better yet, this is true for all ground terms; a proof of $\inf_{\Delta>0} \tau[\Delta](t)/\Delta = \mathrm{dh}_R(t)$ for $t \in \mathcal{T}_\Sigma$ can be found in appendix 5.1 (the present system R corresponds to R_1 there, cf Section 3.2).

2 Context-Dependent Interpretations

In this section, context-dependent interpretations are introduced more formally, and it is shown how the traditional approach appears as a special case.

A context-dependent interpretation consists of functions $f_\tau : \mathbb{R}^+ \times (\mathbb{R}_0^+)^n \to \mathbb{R}_0^+$ and $f_\tau^i : \mathbb{R}^+ \to \mathbb{R}^+$ for each n-ary symbol $f \in \Sigma$ and $1 \leq i \leq n$. They induce a function $\tau : \mathbb{R}^+ \times \mathcal{T}_\Sigma \to \mathbb{R}_0^+$ by

$$\tau[\Delta](f(t_1, \ldots, t_n)) = f_\tau[\Delta](\tau[f_\tau^1(\Delta)](t_1), \ldots, \tau[f_\tau^n(\Delta)](t_n)). \qquad (1')$$

We always assume *Δ-monotonicity* of f_τ, that is,

$$\text{if } z_i' - z_i \geq f_\tau^i(\Delta) \text{ then } f_\tau[\Delta](z_1, \ldots, z_i', \ldots, z_n) - f_\tau[\Delta](z_1, \ldots, z_i, \ldots, z_n) \geq \Delta \qquad (6)$$

for $z_i', z_i \in \mathbb{R}_0^+$. Note that this does not imply weak monotonicity of $f_\tau[\Delta]$ (seen as n-ary function). However, if weak monotonicity is given, that is, if $z_i' \geq z_i$ implies $f_\tau[\Delta](\ldots, z_i', \ldots) \geq f_\tau[\Delta](\ldots, z_i, \ldots)$, then (6) specializes to

$$f_\tau[\Delta](z_1, \ldots, z_i + f_\tau^i(\Delta), \ldots, z_n) - f_\tau[\Delta](z_1, \ldots, z_i, \ldots, z_n) \geq \Delta.$$

Further assume that τ is *compatible* with a rewrite system R, that is, for each rule $\ell \to r$ in R and each ground substitution γ,

$$\tau[\Delta](\ell\gamma) - \tau[\Delta](r\gamma) \geq \Delta. \qquad (7)$$

Under these two conditions the following property holds.

Lemma 1. *For $s, t \in \mathcal{T}_\Sigma$ and $\Delta \in \mathbb{R}^+$,*

$$s \to_R t \quad \textit{implies} \quad \tau[\Delta](s) - \tau[\Delta](t) \geq \Delta.$$

---

[2] By induction: Clearly $\tau[\Delta](b^m c) = m$ and $\tau[\Delta](a^{n+1} b^m c) = (1+\Delta)\tau[\frac{\Delta}{1+\Delta}](a^n b^m c) = (1 + \Delta)(1 + \frac{\Delta}{1+\Delta} n)m$ (by the induction hypothesis) $= (1 + \Delta(n + 1))m$.

Proof. By induction on s. If the rewrite step takes place at the root position in s, the claim follows from (7). Otherwise, assume a rewrite step at position $i.p$ in s. Then $s = f(s_1, \ldots, s_i, \ldots, s_n)$, $t = f(s_1, \ldots, t_i, \ldots, s_n)$, and $s_i \to_R t_i$. From the induction hypothesis,

$$\tau[f_\tau^i(\Delta)](s_i) - \tau[f_\tau^i(\Delta)](t_i) \geq f_\tau^i(\Delta).$$

Therefore, by Δ-monotonicity,

$$\begin{aligned}
\tau[\Delta](s) - \tau[\Delta](t) &= \tau[\Delta](f(\ldots, s_i, \ldots)) - \tau[\Delta](f(\ldots, t_i, \ldots)) \\
&= f_\tau[\Delta](\ldots, \tau[f_\tau^i(\Delta)](s_i), \ldots) - f_\tau[\Delta](\ldots, \tau[f_\tau^i(\Delta)](t_i), \ldots) \\
&\geq \Delta.
\end{aligned}$$

□

Proposition. *Let τ be a context-dependent interpretation compatible with a rewrite system R. Then R is terminating and (3') holds.*

REMARK. A more general version of this result is obtained by choosing an appropriate domain $D \subseteq \mathbb{R}^+$ such that Δ-monotonicity (6) holds for $\Delta \in D$ and $z_i', z_i \in \bigcup_{\Delta \in D} \tau[\Delta](\mathcal{T}_\Sigma)$, such that R-compatibility (7) holds for $\Delta \in D$, and such that $f_\tau^i(D) \subseteq D$ for each function f_τ^i. Then by the same reasoning we get

$$\mathrm{dh}_R(t) \leq \inf_{\Delta \in D} \frac{\tau[\Delta](t)}{\Delta}.$$

2.1 A Special Case

The traditional (non context-dependent) approach appears as a special case. Let us assume that termination of a rewrite system R is provable by a strictly monotone interpretation over the natural numbers, that is, $\tau(\ell\gamma) > \tau(r\gamma)$ for rules $\ell \to r$ in R and ground substitutions γ, where $\tau : \mathbb{N} \to \mathbb{N}$ is induced by a family of strictly monotone functions f_τ ($f \in \Sigma$). Extend each n-ary function f_τ from $\mathbb{N}$ to $\mathbb{R}_0^+$, preserving not only strict monotonicity but also the property (which over the natural numbers coincides with strict monotonicity) that $f_\tau(\ldots, z_i + 1, \ldots) - f_\tau(\ldots, z_i, \ldots) \geq 1$; this is always possible by choosing[3]

$$f_\tau(k_1 + x_1, \ldots, k_n + x_n) = \sum_{b_i \in \{0,1\}} \left(f_\tau(k_1 + b_1, \ldots, k_n + b_n) \cdot \prod_{i=1}^{n} \left((1 - b_i)(1 - x_i) + b_i x_i \right) \right)$$

for $k_i \in \mathbb{N}$ and $0 \leq x_i < 1$. Now define functions $f_\tau[\Delta]$ and f_τ^i by

$$f_\tau[\Delta](z_1, \ldots, z_n) = \Delta f_\tau(z_1/\Delta, \ldots, z_n/\Delta),$$
$$f_\tau^i(\Delta) = \Delta.$$

Then we get the following properties.

[3] For instance, for $n = 2$ we get $f_\tau(k_1 + x_1, k_2 + x_2) = f_\tau(k_1, k_2)(1 - x_1)(1 - x_2) + f_\tau(k_1, k_2 + 1)(1 - x_1)x_2 + f_\tau(k_1 + 1, k_2)x_1(1 - x_2) + f_\tau(k_1 + 1, k_2 + 1)x_1x_2$.

Lemma 2. *(i)* $\tau[\Delta](t) = \Delta\tau(t)$*, hence* $\inf_{\Delta>0} \tau[\Delta](t)/\Delta = \tau(t)$ *for* $t \in \mathcal{T}_\Sigma$*. (ii)* τ *is compatible with* R *(7). (iii)* Δ*-monotonicity (6) holds true.*

Proof. (i) By induction on t:

$$\begin{aligned}\tau[\Delta](f(\dots,t_i,\dots)) &= f_\tau[\Delta](\dots,\tau[f_\tau^i(\Delta)](t_i),\dots)\\ &= f_\tau[\Delta](\dots,\tau[\Delta](t_i),\dots)\\ &= f_\tau[\Delta](\dots,\Delta\tau(t_i),\dots) \quad \text{(by ind. hypothesis)}\\ &= \Delta f_\tau(\dots,\tau(t_i),\dots)\\ &= \Delta\tau(f(\dots,t_i,\dots)).\end{aligned}$$

(ii) We know $\tau(\ell\gamma)-\tau(r\gamma) \geq 1$, thus $\tau[\Delta](\ell\gamma)-\tau[\Delta](r\gamma) = \Delta\tau(\ell\gamma)-\Delta\tau(r\gamma) \geq \Delta$ by (i). (iii) Strict monotonicity of $f_\tau[\Delta]$ is implied by strict monotonicity of f_τ, so it suffices to show $f_\tau[\Delta](\dots,z_i+f_\tau^i(\Delta),\dots) - f_\tau[\Delta](\dots,z_i,\dots) \geq \Delta$. This is equivalent to $\Delta f_\tau(\dots,(z_i+\Delta)/\Delta,\dots) - \Delta f_\tau(\dots,z_i/\Delta,\dots) \geq \Delta$, thus follows from $f_\tau(\dots,z_i/\Delta+1,\dots) - f_\tau(\dots,z_i/\Delta,\dots) \geq 1$. □

3 More Examples

3.1 Associativity

As another example we study termination and derivational complexity of the associativity rule

$$(x \circ y) \circ z \to x \circ (y \circ z)$$

over a signature containing the binary function symbol $\circ$ and the constant symbol c (again in order to guarantee the existence of ground terms); let us call the system R, as always. Termination by a strictly monotone interpretation $\tau : \mathcal{T}_\Sigma \to \mathbb{N}$ can be easily checked for the choice

$$\circ_\tau(n_1,n_2) = 2n_1 + n_2 + 1, \qquad c_\tau = 0.$$

We use the same heuristics we found useful before: Replace in the interpretation functions (some occurrence of) $k+1$ by $k+\Delta$. This first step in finding a context-dependent interpretation yields

$$\circ_\tau[\Delta](z_1,z_2) = (1+\Delta)z_1 + z_2 + 1, \qquad c_\tau[\Delta] = 0.$$

And again, the functions $\circ_\tau^i$ can be found by solving the Δ-monotonicity requirements

$$\begin{aligned}\circ_\tau[\Delta](z_1+\circ_\tau^1(\Delta),z_2) - \circ_\tau[\Delta](z_1,z_2) \geq \Delta,\\ \circ_\tau[\Delta](z_1,z_2+\circ_\tau^2(\Delta)) - \circ_\tau[\Delta](z_1,z_2) \geq \Delta.\end{aligned}$$

We get $((1+\Delta)(z_1+\mathrm{o}_\tau^1(\Delta))+z_2+1)-((1+\Delta)z_1+z_2+1)=(1+\Delta)\mathrm{o}_\tau^1(\Delta)\geq\Delta$, that is, $\mathrm{o}_\tau^1(\Delta)\geq\Delta/(1+\Delta)$, and $((1+\Delta)z_1+z_2+\mathrm{o}_\tau^2(\Delta)+1)-((1+\Delta)z_1+z_2+1)=\mathrm{o}_\tau^2(\Delta)\geq\Delta$, therefore we choose

$$\mathrm{o}_\tau^1(\Delta)=\frac{\Delta}{1+\Delta},\qquad \mathrm{o}_\tau^2(\Delta)=\Delta,$$

resulting in the following context-dependent interpretation:

$$\tau[\Delta](s\circ t)=(1+\Delta)\cdot\tau[\frac{\Delta}{1+\Delta}](s)+\tau[\Delta](t)+1,\qquad \tau[\Delta](c)=0.$$

Lemma 3. *For* $r,s,t\in\mathcal{T}_\Sigma$, $\tau[\Delta]((r\circ s)\circ t)-\tau[\Delta](r\circ(s\circ t))\geq\Delta$.

Proof. By definition of τ,

$$\begin{aligned}
&\tau[\Delta]((r\circ s)\circ t)-\tau[\Delta](r\circ(s\circ t))=\\
&\Big((1+\Delta)\cdot\tau[\frac{\Delta}{1+\Delta}](r\circ s)+\tau[\Delta](t)+1\Big)-\Big((1+\Delta)\cdot\tau[\frac{\Delta}{1+\Delta}](r)+\tau[\Delta](s\circ t)+1\Big)=\\
&\Big((1+\Delta)(\frac{1+2\Delta}{1+\Delta}\tau[\frac{\Delta}{1+2\Delta}](r)+\tau[\frac{\Delta}{1+\Delta}](s)+1)+\tau[\Delta](t)+1\Big)-\\
&\Big((1+\Delta)\cdot\tau[\frac{\Delta}{1+\Delta}](r)+(1+\Delta)\cdot\tau[\frac{\Delta}{1+\Delta}](s)+\tau[\Delta](t)+2\Big)=\\
&(1+2\Delta)\tau[\frac{\Delta}{1+2\Delta}](r)+\Delta-(1+\Delta)\tau[\frac{\Delta}{1+\Delta}](r).
\end{aligned}$$

Thus, we have to prove

$$(1+2\Delta)\tau[\frac{\Delta}{1+2\Delta}](r)-(1+\Delta)\tau[\frac{\Delta}{1+\Delta}](r)\geq 0. \tag{8}$$

By induction on r: For $r=c$ we get $0-0\geq 0$. Note that this is the only place where $\tau[\Delta](c)$ comes into play. For $r=s\circ t$ we get

$$\begin{aligned}
&(1+2\Delta)\tau[\frac{\Delta}{1+2\Delta}](s\circ t)-(1+\Delta)\tau[\frac{\Delta}{1+\Delta}](s\circ t)=\\
&\Big((1+3\Delta)\tau[\frac{\Delta}{1+3\Delta}](s)-(1+2\Delta)\tau[\frac{\Delta}{1+2\Delta}](s)\Big)+\\
&\Big((1+2\Delta)\tau[\frac{\Delta}{1+2\Delta}](t)-(1+\Delta)\tau[\frac{\Delta}{1+\Delta}](t)\Big)+\Big((1+2\Delta)-(1+\Delta)\Big).
\end{aligned}$$

The first and the second difference are non-negative by the induction hypothesis (for the first one substituting $\Delta/(1+\Delta)$ for Δ in (8)), and the third difference is non-negative as $\Delta>0$. □

As example terms consider R_n, L_n (right and left combs respectively), and B_n (full binary trees), defined by $R_0=L_0=B_0=c$ and

$$R_{n+1}=c\circ R_n,\qquad L_{n+1}=L_n\circ c,\qquad B_{n+1}=B_n\circ B_n.$$

It is not difficult to verify that $\tau[\Delta](R_n) = n$,

$$\tau[\Delta](L_n) = n + \Delta\frac{n(n-1)}{2}, \quad \tau[\Delta](B_n) = 2^n - 1 + \Delta(2^{n-1}(n-2)+1),$$

thus by our key observation (3') we obtain $\mathrm{dh}_R(R_n) \leq 0$,

$$\mathrm{dh}_R(L_n) \leq \frac{n(n-1)}{2}, \qquad \mathrm{dh}_R(B_n) \leq 2^{n-1}(n-2)+1.$$

This upper bound is obviously optimal for R_n. That this is also true for L_n and B_n could easily be seen by considering innermost reductions. In this example, however, we can again more generally prove $\inf_{\Delta>0} \tau[\Delta](t)/\Delta = \mathrm{dh}_R(t)$ for all ground terms t, see appendix 5.2.

3.2 Introductory Example, Cont'd

Up to now we have seen examples with polynomially bounded derivational complexity only. A generalization of the first example will lead to (tight) exponential bounds. Consider a family of one-rule systems R_k for $k > 0$ with the rule

$$a(b(x)) \to b^k(a(x)).$$

A straightforward stictly monotone interpretation for R_k is $a_\tau(n) = (k+1)n$, $b_\tau(n) = 1 + n$, $c_\tau = 0$. This suggests the context-dependent interpretation

$$a_\tau(n) = (k+\Delta)n, \qquad b_\tau(n) = 1 + n, \qquad c_\tau = 0.$$

Again, solving (5') gives $(k+\Delta)(z + a_\tau^1(\Delta)) - (k+\Delta)z = (k+\Delta)a_\tau^1(\Delta) \geq \Delta$ and $(1 + z + b_\tau^1(\Delta)) - (1+z) \geq \Delta$, thus we choose

$$a_\tau^1(\Delta) = \frac{\Delta}{k+\Delta}, \qquad b_\tau^1(\Delta) = \Delta$$

and obtain the context-dependent interpretation $\tau[\Delta](c) = 0$,

$$\tau[\Delta](a(t)) = (k+\Delta)\cdot\tau[\frac{\Delta}{k+\Delta}](t), \qquad \tau[\Delta](b(t)) = 1 + \tau[\Delta](t).$$

Verification of $\tau[\Delta](a(b(t))) - \tau[\Delta](b^k(a(t))) = \Delta$ is simple. For terms $a^n b^m c$ we obtain, for $k > 1$,

$$\tau[\Delta](a^n b^m c) = (k^n + \frac{k^n - 1}{k-1}\cdot\Delta)m,$$

thus

$$\inf_{\Delta>0} \frac{\tau[\Delta](a^n b^m c)}{\Delta} = \frac{k^n-1}{k-1}\cdot m = \mathrm{dh}_{R_k}(a^n b^m c).$$

For a proof of $\inf_{\Delta>0} \tau[\Delta](t)/\Delta = \mathrm{dh}_{R_k}(t)$ for all terms t see appendix 5.1.

3.3 Where Simplification Orderings Fail

Here we show that our approach is general enough to handle rewrite systems that are not simply terminating. The example below illustrates that this is possible even without involving context dependency, due to weak monotonicity and Δ-monotonicity for fixed Δ. Consider

$$a(a(x)) \to a(b(a(x)))$$

over signature $\{a, b, c\}$ from the introductory example. First, define (non context-dependent) interpretation functions $a_\tau, b_\tau : \mathbb{R}_0^+ \to \mathbb{R}_0^+$ by

$$\begin{aligned} a_\tau(z) &= n + 1/2 && \text{if } n - 1 < z \leq n, \\ b_\tau(z) &= n && \text{if } n - 1/2 < z \leq n + 1/2, \end{aligned}$$

where $z \in \mathbb{R}_0^+$, $n \in \mathbb{N}$, and let $c_\tau = 0$. Both functions are weakly monotone, and '1-monotonicity' holds:

$$a_\tau(z+1) - a_\tau(z) \geq 1, \qquad b_\tau(z+1) - b_\tau(z) \geq 1.$$

Note that over natural numbers, the requirement $b_\tau(n+1) - b_\tau(n) \geq 1$ implies $b_\tau(n) \geq n$. This is not the case for the domain of real numbers. For instance, $b_\tau(1/2) = 0$, thus $\tau(b(a(c))) < \tau(a(c))$, violating any kind of subterm property.

Lemma 4. *$\tau(a(a(t))) - \tau(a(b(a(t)))) \geq 1$ for $t \in \mathcal{T}_\Sigma$.*

Proof. We consider two cases as $\tau(\mathcal{T}_\Sigma) = \{n, n{+}1/2 \mid n \in \mathbb{N}\}$. For $z = n{+}1/2$ we have $a_\tau(a_\tau(z)) = a_\tau(n+3/2) = n + 5/2$ and $a_\tau(b_\tau(a_\tau(z))) = a_\tau(b_\tau(n+3/2)) = a_\tau(n+1) = n + 3/2$, for $z = n$ we have $a_\tau(a_\tau(z)) = a_\tau(n+1/2) = n + 3/2$ and $a_\tau(b_\tau(a_\tau(z))) = a_\tau(b_\tau(n+1/2)) = a_\tau(n) = n + 1/2$. □

Now, similar to the considerations in section 2.1 this can easily be turned into a context-dependent interpretation in the proper sense (without really depending on the context, though). Thus the rewrite system is terminating, and $\mathrm{dh}_R(t) \leq \tau(t)$. This bound is optimal as we have

$$\mathrm{dh}_R(t) = \lfloor \tau(t) \rfloor.$$

4 Conclusion

By the approach presented in this paper one can sometimes obtain tight upper bounds on the derivational complexity of rewrite systems. We think that there is a certain potential to further extend this idea. In particular, we believe that semi-automatic assistance in finding complexity bounds might be possible by a two step procedure as follows. First, find the interpretation, for instance using variants of methods presented for traditional interpretations. We have seen that in a number of simple examples, variants of automatically generated polynomial interpretations are appropriate; heuristics for finding termination proofs by interpretations or their implementation is addressed in [2,28,25,26,27,10,11], among others. Second, extract complexity bounds from the interpretation. For recursively defined families of terms, or for all terms, a computer algebra system might be helpful here.

We finally want to point out that a-priori-knowledge of the derivation lengths function is only used for proving that the bounds obtained for the few simple examples are optimal. Our approach allows one to obtain better upper bounds than previously just without that knowledge, therefore these proofs are relegated to the appendix.

Acknowledgements. I am grateful to the referees for their helpful remarks.

References

1. F. Baader and T. Nipkow. *Term Rewriting and All That.* Cambridge University Press, 1998.
2. A. Ben Cherifa and P. Lescanne. Termination of rewriting systems by polynomial interpretations and its implementation. *Science of Computer Programming*, 9:137–159, 1987.
3. A. Cichon and P. Lescanne. Polynomial interpretations and the complexity of algorithms. In D. Kapur, editor, *Proc. 11th Int. Conf. on Automated Deduction CADE-11*, volume 607 of *Lecture Notes in Artificial Intelligence*, pages 139–147. Springer-Verlag, 1992.
4. A. Cichon and J.-Y. Marion. Light LPO. Technical Report 99-R-138, LORIA, Nancy, France, 1999.
5. N. Dershowitz. Termination of rewriting. *Journal of Symbolic Computation*, 3(1–2):69–115, 1987.
6. N. Dershowitz and J.-P. Jouannaud. Rewrite systems. In J. van Leeuwen, editor, *Handbook of Theoretical Computer Science*, volume B, pages 243–320. Elsevier Science Publishers, 1990.
7. F. Drewes and C. Lautemann. Incremental termination proofs and the length of derivations. In R.V. Book, editor, *Proc. 4th Int. Conf. on Rewriting Techniques and Applications RTA-91*, volume 488 of *Lecture Notes in Computer Science*, pages 49–61. Springer-Verlag, 1991.
8. R. Fraïssé. *Theory of Relations*, volume 118 of *Studies in Logic and the Foundations of Mathematics.* North-Holland, Amsterdam, 1986.
9. O. Geupel. *Terminationsbeweise bei Termersetzungssystemen.* Diplomarbeit, Technische Universität Dresden, Sektion Mathematik, 1988.
10. J. Giesl. Generating polynomial orderings for termination proofs. In J. Hsiang, editor, *Proc. 6th Int. Conf. on Rewriting Techniques and Applications RTA-95*, volume 914 of *Lecture Notes in Computer Science*, pages 426–431. Springer-Verlag, 1995.
11. J. Giesl. POLO – a system for termination proofs using polynomial orderings. Technical Report 95/24, Technische Hochschule Darmstadt, 1995.
12. D. Hofbauer. *Termination proofs and derivation lengths in term rewriting systems.* Dissertation, Technische Universität Berlin, Germany, 1991. Available as Technical Report 92-46, TU Berlin, 1992.

13. D. Hofbauer. Termination proofs by multiset path orderings imply primitive recursive derivation lengths. *Theoretical Computer Science*, 105(1):129–140, 1992.
14. D. Hofbauer. Termination proofs for ground rewrite systems – interpretations and derivational complexity. Mathematische Schriften 24/00, Universität Gh Kassel, Germany, 2000. To appear in *Applicable Algebra in Engineering, Communication and Computing.*
15. D. Hofbauer. An upper bound on the derivational complexity of Knuth-Bendix orderings. Mathematische Schriften 28/00, Universität Gh Kassel, Germany, 2000. To appear in *Information and Computation.*
16. D. Hofbauer and C. Lautemann. Termination proofs and the length of derivations. In N. Dershowitz, editor, *Proc. 3rd Int. Conf. on Rewriting Techniques and Applications RTA-89*, volume 355 of *Lecture Notes in Computer Science*, pages 167–177. Springer-Verlag, 1989.
17. D. S. Lankford. On proving term rewriting systems are Noetherian. Memo MTP-3, Mathematics Department, Louisiana Tech. University, Ruston, LA, May 1979.
18. C. Lautemann. A note on polynomial interpretation. *Bulletin of the European Association for Theoretical Computer Science*, 36:129–131, Oct. 1988.
19. I. Lepper. Derivation lengths and order types of Knuth-Bendix orders. Technical report, Westfälische Wilhelms-Universität Münster, Institut für Mathematische Logik und Grundlagenforschung, Germany, May 2000. Available as `http://wwwmath.uni-muenster.de/logik/publ/pre/5.html`, to appear in *Theoretical Computer Science.*
20. P. Lescanne. Termination of rewrite systems by elementary interpretations. *Formal Aspects of Computing*, 7(1):77–90, 1995.
21. Z. Manna and S. Ness. On the termination of Markov algorithms. In *Proc. 3rd Hawaii Int. Conf. on System Science*, pages 789–792, Honolulu, 1970.
22. J.-Y. Marion. Analysing the implicit complexity of programs. Technical Report 99-R-106, LORIA, Nancy, France, 1999.
23. V. C. S. Meeussen and H. Zantema. Derivation lengths in term rewriting from interpretations in the naturals. Technical Report RUU-CS-92-43, Utrecht University, The Netherlands, 1992.
24. F. L. Morris and C. B. Jones. An early program proof by Alan Turing. *Annals of the History of Computing*, 6(2):139–143, 1984.
25. J. Steinbach. Termination proofs of rewriting systems – Heuristics for generating polynomial orderings. Technical Report SR-91-14, Fachbereich Informatik, Universität Kaiserslautern, Germany, 1991.
26. J. Steinbach. Proving polynomials positive. In R. Shyamasundar, editor, *Proc. Foundations of Software Technology and Theoretical Computer Science*, volume 652 of *Lecture Notes in Computer Science*, pages 191–202. Springer-Verlag, 1992.
27. J. Steinbach. Generating polynomial orderings. *Information Processing Letters*, 49(2):85–93, 1994.
28. J. Steinbach and M. Zehnter. Vademecum of polynomial orderings. Technical Report SR-90-03, Fachbereich Informatik, Universität Kaiserslautern, Germany, 1990.
29. H. Touzet. *Propriétés combinatoires pour la terminaison de systèmes de réécriture.* PhD thesis, Université Henri Poincaré – Nancy 1, France, 1997.
30. A. M. Turing. Checking a large routine. In *Report of a Conference on High Speed Automatic Calculating Machines*, pages 67–69. University Mathematics Lab, Cambridge University, 1949.
31. W. Wechler. *Universal Algebra for Computer Scientists*, volume 25 of *EATCS Monographs on Theoretical Computer Science.* Springer-Verlag, Berlin, 1992.

32. A. Weiermann. Termination proofs for term rewriting systems by lexicographic path orderings imply multiply recursive derivation lengths. *Theoretical Computer Science*, 139(1–2):355–362, 1995.
33. H. Zantema. Termination of term rewriting: interpretation and type elimination. *Journal of Symbolic Computation*, 17(1):23–50, 1994.
34. H. Zantema. Termination of term rewriting. Technical Report RUU-CS-2000-04, Utrecht University, The Netherlands, 2000.
35. H. Zantema. The termination hierarchy for term rewriting. To appear in *Applicable Algebra in Engineering, Communication and Computing.*

5 Appendix

5.1 Introductory Example

In this section let R abbreviate R_k, and let dh and $\downarrow$ denote dh_{R_k} and $\downarrow_{R_k}$ respectively (in case term t has a unique normal form with respect to R, as always in this example, it is denoted by $t\downarrow_R$). We prove the following claim for $\Delta \in \mathbb{R}^+$ and $t \in \mathcal{T}_\Sigma$ by induction on t, where $|t|_f$ denotes the number of occurrences of symbol f in term t:

$$\tau[\Delta](t) = |t\downarrow|_b + \Delta \mathrm{dh}(t).$$

Proof. For $t = c$ we get $\tau[\Delta](c) = 0 = |c\downarrow|_b + \Delta\mathrm{dh}(c)$ as $|c\downarrow|_b = \mathrm{dh}(c) = 0$. For $t = b(s)$ we know $|b(s)\downarrow|_b = 1 + |s\downarrow|_b$ and $\mathrm{dh}(b(s)) = \mathrm{dh}(s)$. Thus, using the induction hypothesis,

$$\begin{aligned}\tau[\Delta](b(s)) &= 1 + \tau[\Delta](s) \\ &= 1 + |s\downarrow|_b + \Delta\mathrm{dh}(s) \\ &= |b(s)\downarrow|_b + \Delta\mathrm{dh}(b(s)).\end{aligned}$$

For $t = a(s)$ we know $|a(s)\downarrow|_b = k|s\downarrow|_b$ and $\mathrm{dh}(a(s)) = \mathrm{dh}(s) + |s\downarrow|_b$ from $s\downarrow \in b^*a^*c$. Hence, using the induction hypothesis,

$$\begin{aligned}\tau[\Delta](a(s)) &= (k+\Delta)\tau[\frac{\Delta}{k+\Delta}](s) \\ &= (k+\Delta)(|s\downarrow|_b + \frac{\Delta}{k+\Delta}\mathrm{dh}(s)) \\ &= k|s\downarrow|_b + \Delta\mathrm{dh}(s) + \Delta|s\downarrow|_b \\ &= |a(s)\downarrow|_b + \Delta\mathrm{dh}(a(s)).\end{aligned}$$

□

As a direct consequence,

$$\inf_{\Delta>0} \frac{\tau[\Delta](t)}{\Delta} = \inf_{\Delta>0} \left(\frac{|t\downarrow|_b}{\Delta} + \mathrm{dh}(t)\right) = \mathrm{dh}(t).$$

5.2 Associativity

Similarly, we prove for $\Delta \in \mathbb{R}^+$ and $t \in \mathcal{T}_\Sigma$, where dh abbreviates dh_R,

$$\tau[\Delta](t) \leq |t|_c - 1 + \Delta \mathrm{dh}(t).$$

Proof. For $t = c$ we have $\tau[\Delta](c) = 0 = |c|_c - 1 + \Delta \mathrm{dh}(c)$ as $|c|_c = 1$ and $\mathrm{dh}(c) = 0$. For $t = r \circ s$ we know that

$$\mathrm{dh}(r \circ s) \geq \mathrm{dh}(r) + \mathrm{dh}(s) + |r|_c - 1$$

since $\mathrm{dh}(r \circ s) \geq \mathrm{dh}(r) + \mathrm{dh}(s) + \mathrm{dh}(R_{|r|_c - 1} \circ R_{|s|_c - 1})$ and $\mathrm{dh}(R_n \circ t) \geq n + \mathrm{dh}(t)$ and $\mathrm{dh}(R_n) = 0$. (Note that $R_{|t|_c - 1}$ is the unique normal form of a term t.) Thus, applying the induction hypothesis twice,

$$\begin{aligned}
\tau[\Delta](r \circ s) &= (1 + \Delta)\tau[\frac{\Delta}{1 + \Delta}](r) + \tau[\Delta](s) + 1 \\
&\leq (1 + \Delta)\Big(|r|_c - 1 + \frac{\Delta}{1 + \Delta}\mathrm{dh}(r)\Big) + |s|_c - 1 + \Delta \mathrm{dh}(s) + 1 \\
&= |r|_c + |s|_c - 1 + \Delta(\mathrm{dh}(r) + \mathrm{dh}(s) + |r|_c - 1) \\
&\leq |r \circ s|_c - 1 + \Delta \mathrm{dh}(r \circ s).
\end{aligned}$$

□

Again, as a consequence we obtain

$$\inf_{\Delta > 0} \frac{\tau[\Delta](t)}{\Delta} \leq \inf_{\Delta > 0} \left(\frac{|t|_c - 1}{\Delta} + \mathrm{dh}(t)\right) = \mathrm{dh}(t).$$

Uniform Normalisation beyond Orthogonality

Zurab Khasidashvili[1], Mizuhito Ogawa[2], and Vincent van Oostrom[3]

[1] Department of Mathematics and Computer Science, Bar-Ilan University
Ramat-Gan 52900, Israel
khasidz@cs.bu.ac.il

[2] Japan Science and Technology Corporation, PRESTO and
NTT Communication Science Laboratories
3-1 Morinosato-Wakamiya, Atsugi, Kanagawa 243-0198, Japan
mizuhito@theory.brl.ntt.co.jp

[3] Department of Philosophy, Utrecht University
P.O. Box 80089, 3508 TB Utrecht, The Netherlands
oostrom@phil.uu.nl

Abstract. A rewrite system is called uniformly normalising if all its steps are perpetual, i.e. are such that if $s \to t$ and s has an infinite reduction, then t has one too. For such systems termination ($\mathtt{SN}$) is equivalent to normalisation ($\mathtt{W\,N}$). A well-known fact is uniform normalisation of *orthogonal non-erasing* term rewrite systems, e.g. the λI-calculus. In the present paper both restrictions are analysed. Orthogonality is seen to pertain to the linear part and non-erasingness to the non-linear part of rewrite steps. Based on this analysis, a modular proof method for uniform normalisation is presented which allows to go beyond orthogonality. The method is shown applicable to biclosed first- and second-order term rewrite systems as well as to a λ-calculus with explicit substitutions.

1 Introduction

Two classical results in the study of uniform normalisation are:

- the λI-calculus is uniformly normalising [7, p. 20, 7 XXV], and
- non-erasing steps are perpetual in orthogonal TRSs [14, Thm. II.5.9.6].

In previous work we have put these results and many variations on them in a unifying framework [13]. At the heart of that paper is the result (Thm. 3.16) that a term s not in normal form contains a redex which is external for *any* reduction from s.[1] Since external redexes need not exist in rewrite systems having critical pairs, the result does not apply to these. The method presented here, is based instead on the existence of redexes which are external for all reductions which are permutation equivalent to a *given* reduction. Since this so-called standardisation theorem holds for all left-linear rewrite systems, with or without critical pairs, the resulting framework is more general. It is applied to obtain

[1] According to [11, p. 404], a redex at position p is external to a reduction if in the reduction no redex is contracted above p to which the redex did not contribute.

A. Middeldorp (Ed.): RTA 2001, LNCS 2051, pp. 122–136, 2001.

uniform normalisation results for abstract rewrite systems (ARSs), first-order term rewrite systems (TRSs) and second-order term rewrite systems (P$_2$RS) in Sect. 2, 3 and 4, respectively. In each section, the proof method is presented for the orthogonal case first, deriving traditional results. We then vary on it, relaxing the orthogonality restriction. This leads to new uniform normalisation results for biclosed rewrite systems (e.g. Cor. 2, 5, 6, and 8). In Sect. 5 uniform normalisation for λx^-, a prototypical λ-calculus with explicit substitutions, is shown to hold, extending earlier work of [6] who only shows it for the explicit substitution part x of the calculus. The proof boils down to an analysis of the (only) critical pair of λx^- and uses a particularly simple proof of preservation of strong normalisation for λx^-, also based on the standardisation theorem.

2 Abstract Rewriting

Although trivial, the results in this section and their proofs form the heart of the following sections. Moreover, they are applicable to various concrete (linear) rewrite systems, for instance to interaction nets [16]. The reader is assumed to be familiar with *abstract rewrite systems* (ARSs, [15, Chap. 1] or [1, Chap. 2]).

Definition 1. *Let a be an object of an abstract rewrite system. a is* terminating *(strongly normalising,* SN*) if no infinite reductions are possible from it. We use* ∞ *to denote the complement of* SN*. a is* normalising *(weakly normalising,* WN*) if some reduction to normal form is possible from it.*

Definition 2. *A rewrite step* $s \to t$ *is* critical *if* $s \in \infty$ *and* $t \in \mathsf{SN}$*, and* perpetual *otherwise. A rewrite system is* uniformly normalising *if there are no critical steps.*

First, note that a rewrite system is uniformly normalising iff $\mathsf{WN} \subseteq \mathsf{SN}$ holds. Moreover, uniform normalisation holds for deterministic rewrite systems.

Definition 3. *A* fork *in a rewrite system is pair of steps* $t_1 \leftarrow s \to t_2$*. It is called* trivial *if* $t_1 = t_2$*. A rewrite system is* deterministic *if all forks are trivial, and* non-deterministic *otherwise.*

To analyse uniform normalisation for non-deterministic rewrite systems it thus seems worthwhile to study their non-trivial forks.

Definition 4. *A rewrite system is* linear orthogonal *if every fork* $t_1 \leftarrow s \to t_2$ *is either trivial or* square*, that is,* $t_1 \to s' \leftarrow t_2$ *for some* s' *[1, Exc. 2.33].*

We will show the *fundamental theorem of perpetuality*:

Theorem 1 (FTP). *Steps are perpetual in linear orthogonal rewrite systems.*

Corollary 1. *Linear orthogonal rewrite systems are uniformly normalising.*

In the next section we will show (Lem. 1) that the abstract rewrite system associated to a term rewrite system which is linear and orthogonal, is linear orthogonal. Linear orthogonality is a weakening of the diamond property [1, Def. 2.7.8], and a strengthening of subcommutativity [15, Def. 1.1.(v)] and of the balanced weak Church-Rosser property [25, Def. 3.1], whence:

Proof. (of Thm. 1) Suppose $s \in \infty$ and $s \to t$. We need to show $t \in \infty$. By the first assumption, there exists an infinite reduction $S : s_0 \to s_1 \to s_2 \to \ldots$, with $s_0 = s$. One can build an infinite reduction T from t as follows: let $t_0 = t$ be the first object of T. By orthogonality we can find for every non-trivial fork $s_{i+1} \leftarrow s_i \to t_i$ a next object t_{i+1} of T such that $s_{i+1} \to t_{i+1} \leftarrow t_i$. Consider a maximal reduction T thus constructed. If T is infinite we are done. If T is finite, it has a final object, say t_n, and a fork $s_{n+1} \leftarrow s_n \to t_n$ exists which is trivial, i.e. $s_{n+1} = t_n$. Hence, T and the infinite reduction S from s_{n+1} on can be concatenated. □

FTP can be brought beyond linear orthogonality. Let $\to^=$ and $\twoheadrightarrow$ denote the reflexive and reflexive-transitive closure of $\to$, respectively.

Definition 5. *A fork $t_1 \leftarrow s \to t_2$ is* closed *if $t_1 \twoheadrightarrow t_2$. A rewrite system is* linear biclosed *if all forks are either closed or square.*[2]

By replacing the appeal to triviality by an appeal to closedness in the proof of FTP, i.e. by replacing $s_{n+1} = t_n$ by $s_{n+1} \twoheadleftarrow t_n$, we get:

Corollary 2. *Linear biclosed rewrite systems are uniformly normalising.*

3 First-Order Term Rewriting

In this section first the uniform normalisation results of Section 2 are instantiated to linear term rewriting. Next, the *fundamental theorem of perpetuality for first-order term rewrite systems* is established:

Theorem 2 (F$_1$TP). *Non-erasing steps are perpetual in orthogonal TRSs.*

Corollary 3. *Non-erasing orthogonal TRSs are uniformly normalising.*

The chief purpose of this section is to illustrate our proof method based on standardisation. Except for the results on biclosed systems, the results obtained are not novel (cf. [15, Lem. 8.11.3.2] and [9, Sect. 3.3]). The reader is assumed to be familiar with *first-order term rewrite systems* (TRSs) as can be found in e.g. [15] or [1]. We summarise some aberrations and additional concepts:

Definition 6.
- *A term is* linear *if any variable occurs at most once in it. Let $\varrho : l \to r$ be a TRS rule. It is* left-linear *(*right-linear*) if l (r) is linear. It is* linear *if $Var(l) = Var(r)$ and both sides are linear. A TRS is (left-,right) linear if all its rules are.*
- *Let $\varrho : l \to r$ be a rule. A variable $x \in Var(l)$ is* erased *by ϱ if it does not occur in r. The rule ϱ is erasing if it erases some variable. A rewrite step is erasing if the applied rule is. A TRS is erasing if some step is.*
- *Let $\varrho : l \to r$ and $\vartheta : g \to d$ be rules which have been renamed apart. Let p be a non-variable position in l. ϱ is said to* overlap ϑ at p *if a unifier σ of $l_{|p}$ and g does exist. If σ is a most general such unifier, then both $\langle l[d]_p^\sigma, r^\sigma\rangle$ and $\langle r^\sigma, l[d]_p^\sigma\rangle$ are* critical pairs at p between ϱ and ϑ. [3]

[2] Beware of the symmetry: if the fork is not square, then both $t_1 \twoheadrightarrow t_2$ and $t_2 \twoheadrightarrow t_1$.

[3] Beware of the symmetry (see the next item and cf. Footnote 2).

– *If for all such critical pairs* $\langle t_1, t_2 \rangle$ *of a left-linear TRS* $\mathcal{R}$ *it holds that:*
$\exists s'\, t_1 \twoheadrightarrow s' \leftarrow^{=} t_2$, *then* $\mathcal{R}$ *is* strongly closed *[10, p. 812]*
$t_1 \twoheadrightarrow t_2$, *then* $\mathcal{R}$ *is* biclosed *[22, p. 70]*
$t_1 = t_2$, *then* $\mathcal{R}$ *is* weakly orthogonal
$t_1 = t_2$ *and* $p = \epsilon$, *then* $\mathcal{R}$ *is* almost orthogonal
$t_1 = t_2$, $p = \epsilon$ *and* $\varrho = \vartheta$, *then* $\mathcal{R}$ *is* orthogonal

Some remarks are in order. First, our critical pairs for a TRS are the critical pairs $\langle s, t\rangle$ of [1, Def. 6.2.1] extended with their opposites ($\langle t, s\rangle$) and the trivial critical pairs between a rule with itself at the head ($\langle r, r\rangle$ for every rule $l \to r$). Next, linearity in our sense implies linearity in the sense of [1, Def. 6.3.1], but not vice versa. Linearity of a step $s = C[l^\sigma] \to C[r^\sigma] = t$ as defined here captures the idea that every symbol in the context-part C or the substitution-part σ in s has a *unique* descendant in t, whereas linearity in the sense of [1, Def. 6.3.1] only guarantees that there is *at most one* descendant in t. Remark:

orth. $\Longrightarrow$ almost orth. $\Longrightarrow$ weakly orth. $\Longrightarrow$ biclosed $\Longrightarrow$ strongly closed

3.1 Linear Term Rewriting

In this subsection the results of Section 2 for abstract rewriting are instantiated to linear term rewriting. First, remark that linear strongly closed TRSs are confluent (combine Lem. 6.3.2, 6.3.3 and 2.7.4 of [1]). Therefore, a linear TRS satisfying any of the above mentioned critical pair criteria is confluent.

Lemma 1. *If* $\mathcal{R}$ *is a linear orthogonal TRS,* $\to_{\mathcal{R}}$ *is a linear orthogonal ARS.*

Proof. The proof is based on the standard critical pair analysis of a fork $t_1 \leftarrow_{\mathcal{R}} s \to_{\mathcal{R}} t_2$ as in [1, Sect. 6.2]. Actually, it is directly obtained from the proof of [1, Lem. 6.3.3], by noting that:

Case 1 (parallel) establishes that the fork is square (joinable into a diamond),
Case 2.1 (nested) also yields that the fork is square,[4] and
Case 2.2 (overlap) can occur only if the steps in the fork arise by applying the same rule at the same position, by orthogonality, so the fork is trivial. □

From Lem. 1 and Cor. 1 we obtain a special case of Corollary 3.

Corollary 4. *Linear orthogonal TRSs are uniformly normalising.*

Lemma 2. *If* $\mathcal{R}$ *is a linear biclosed TRS,* $\to_{\mathcal{R}}$ *is a linear biclosed ARS.*

Proof. The analysis in the proof of Lem. 1 needs to be adapted as follows:

Case 2.2 , the instance of a critical pair, is closed by biclosedness of critical pairs and the fact that rewriting is closed under substitution. □

Corollary 5. *Linear biclosed TRSs are uniformly normalising.*

[4] Note that the case $x \notin \mathit{Var}(r_1)$ cannot happen, due to our notion of linearity.

3.2 Non-linear Term Rewriting

In this subsection the results of the previous subsection are adapted to non-linear TRSs, leading to a proof of $\mathrm{F_1TP}$ (Thm. 2). The adaptation is non-trivial, since uniform normalisation may fail for orthogonal non-linear TRSs.

Example 1. The term $e(a)$ in the TRS $\{a \to a, e(x) \to b\}$ witnesses that orthogonal TRSs need not be uniformly normalising.

Non-linearity of a TRS may be caused by non-left-linearity. Although non-left-linearity in itself is not fatal for uniform normalisation of TRSs (see [9, Chap. 3], e.g. Cor. 3.2.9), it will be in case of second-order rewriting (cf. Ex. 2) and our method cannot deal with it. Hence: We assume TRSs to be left-linear.
Under this assumption, non-linearity may only be caused by some symbol having zero or multiple descendants after a step. The problem in Ex. 1 is seen to arise from the fork $e(a) \leftarrow e(a) \to b$ which is not balancedly joinable: it is neither trivial ($e(a) \neq b$) nor square ($\nexists s'\ e(a) \to s' \leftarrow b$). Erasingness is the only problem. To prove $\mathrm{F_1TP}$, we will make use of the apparent asymmetry in the non-linearity

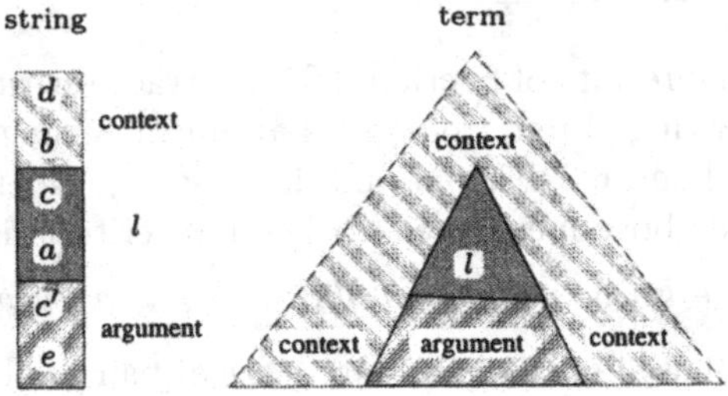

Fig. 1. Split

of term rewrite steps: an occurrence of a left-hand side of a rule $l \to r$ *splits* the surrounding into two parts (see Fig. 1):

- the *context*-part above or parallel to [1, Def. 3.1.3] l, and
- the *argument*-part, below l.

Observe that term rewrite steps in the context-part might replicate the occurrence of the left-hand side l, whereas steps in the argument-part cannot do so. To deal with such replicating steps in the context-part, we will actually prove a strengthening of $\mathrm{F_1TP}$ for parallel steps instead of ordinary steps.

Definition 7. *Let $\varrho : l \to r$ be a TRS rule. s* parallel rewrites *to t using ϱ, $s \mathrel{+\!\!\!\!\to}_{\varrho} t$ [10, p. 814],*[5] *if it holds that $s = C[l^{\sigma_1}, \ldots, l^{\sigma_k}]$ and $t = C[r^{\sigma_1}, \ldots, r^{\sigma_k}]$, for some $k \geq 0$. The step is* erasing *if the rule is. The* context*(*argument*)-part of the step is the part above or parallel to all (below some) occurrences of l.*

[5] Actually our notion is a restriction of his, since we allow only one rule.

To reduce F_1TP to FTP it suffices to reduce to the case where the infinite reduction does not take place (entirely) in the context-part, since then the steps either have overlap or are in the, linear, argument-part. To that end, we want to transform the infinite reduction into an infinite reduction where the steps in the context-part precede the steps in the argument-part.

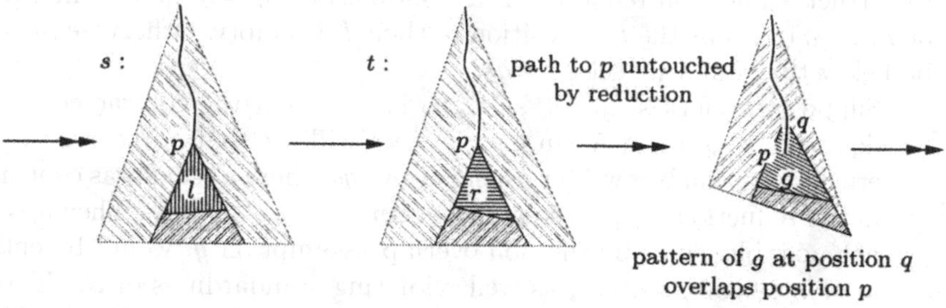

Fig. 2. Standard

Definition 8. *A reduction is* standard *(see Fig. 2) if for any step* $C[l^\sigma]_p \to C[r^\sigma]_p$ *in the reduction,* p *is in the pattern of the first step after that step which is above* p. *That is, if* $D[g^\tau]_q$ *displays the occurrence of the first redex with* $p = qo$, *we have that* o *is a non-variable position in* g.

Theorem 3 (STD). *Any reduction in a TRS can be transformed into a standard one. The transformation preserves infiniteness.*

Proof. The first part of the theorem was shown to hold for orthogonal TRSs in [11, Thm. 3.19] and extended to left-linear TRSs possibly having critical pairs in [8]. That standardisation preserves infiniteness follows from the fact that at some moment along an infinite reduction $S : s_0 \to s_1 \to \ldots$ a redex at minimal position p w.r.t. the prefix order $\leq$ [1, Def. 3.1.3] must be contracted. Say this happens the first time in step $s_i \to_p s_{i+1}$. Permute all steps parallel to p in S after this step resulting in $S_0; S_1$, where S_0 contains only steps below p and ends with a step at position p, and S_1 is infinite. Standardise S_0 into T_0, note that it is non-empty and that concatenating T_0 with *any* standardisation of S_1 will yield a standard reduction by the choice of p. Repeat the process on S_1. □

Proof. (of Thm. 2) Suppose $s \in \infty$ and $s \mathrel{+\!\!\!\!+\!\!\!\to}_{\varrho}^{k} t$ is non-erasing, contracting k redexes w.r.t. rule $\varrho : l \to r$ in parallel. We need to show $t \in \infty$. If $k = 0$, then $t = s \in \infty$. Otherwise, there exists by the first assumption an infinite reduction $S : s_0 \to_{q_0} s_1 \to_{q_1} s_2 \to \ldots$, with $s_0 = s$ and $s_i \to_{q_i} s_{i+1}$ contracting a redex at position q_i w.r.t. rule $\vartheta_i : g_i \to d_i$. By STD S may be assumed standard. Consider the relative positions of the redexes in the fork $s_1 \leftarrow_{q_0} s \mathrel{+\!\!\!\!+\!\!\!\to}_{\varrho} t$.

(context) If g_0 occurs entirely in the context-part of the parallel step, then by the Parallel Moves lemma [1, Lem. 6.4.4] the fork is joinable into $s_1 \mathrel{+\!\!\!\!+\!\!\!\to}_{\varrho}^{k'}$

$t_1 \leftarrow_{q_0} t_0$. Since $t_0 \to t_1$, $s_1 \in \infty$, and $s_1 \twoheadrightarrow\!\!\!\!\!\!+\,_{\varrho} t_1$ is non-erasing, repeating the process will yield an infinite reduction from $t_0 = t$ as desired.

(non-context) Otherwise g_0 must be below one or overlap at least one contracted left-hand side l, say the one at position p. Hence, $s \twoheadrightarrow\!\!\!\!\!\!+^{k} t$ can be decomposed as $s \to_p s' \twoheadrightarrow\!\!\!\!\!\!+^{k-1} t$. We claim $s' \in \infty$. The proof is as for FTP, employing standardness to exclude replication of the pivotal l-redex. Construct a maximal reduction T as follows. Let $t_0 = s'$ be the first object of T. If g_0 overlaps the l at position p, then T is empty. Otherwise, g_0 must be below that l and we set $o_0 = q_0$.

- Suppose the fork $s_{i+1} \leftarrow_{q_i} s_i \to_p t_i$ is such that the contracted redexes do not have overlap. As an invariant we will use that o_i records the outermost position below l (at p) and above q_0 where a redex was contracted in the reduction S up to step i, hence $p \leq o_{i+1} \leq o_i \leq q_0$. Then $q_i < p$ is not possible, since by the non-overlap assumption g_i would be entirely above p, hence above o_i as well, violating standardness of S. Hence, q_i is parallel to or below l (at p). By another appeal to the Parallel Moves lemma the fork can be joined via $s_{i+1} \to_p t_{i+1} \twoheadleftarrow\!\!\!\!\!\!+^{k} t_i$, where $k > 0$ by non-erasingness of $s_i \to t_i$ (†). The invariant is maintained by setting o_{i+1} to q_i if $q_i < o_i$, and to o_i otherwise.

If T is infinite we are done. If T is finite, it has a final object, say t_n, and a fork $s_{n+1} \leftarrow_{q_i,\vartheta_i} s_n \to_p t_n$ such that the redexes have overlap (‡). By the orthogonality assumption we must have $q_n = p$ and $\vartheta_n = \varrho$, hence $s_{n+1} = t_n$. By concatenating T and the infinite reduction S from s_{n+1}, the claim ($s' \in \infty$) is then proven. From the claim, we may repeat the process with an infinite standard reduction from s' and $s' \twoheadrightarrow\!\!\!\!\!\!+^{k-1} t$.

Observe that the (context)-case is the only case producing a rewrite step from t, but it must eventually always apply since the other case decreases k by 1. □

By replacing the appeal to orthogonality by an appeal to biclosedness in the proof of F$_1$TP, i.e. by replacing $s_{n+1} = t_n$ by $s_{n+1} \twoheadleftarrow t_n$, we get:

Theorem 4. *Non-erasing steps are perpetual in biclosed TRSs.*

Corollary 6. *Non-erasing biclosed TRSs are uniformly normalising.*

Note that we are beyond orthogonality since biclosed TRSs need not be confluent. The example is as for strongly closed TRSs [10, p. 814], but note that the latter need not be uniformly normalising! Next, we show [15, Lem. 8.11.3.2].

Definition 9. *A step $C[l^\sigma] \to C[r^\sigma]$ is* ∞-erasing, *if it erases all ∞-variables, that is, if $x \in Var(r)$ then $x^\sigma \in \mathsf{SN}$.*

Theorem 5. *Non-∞-erasing rewrite steps are perpetual in biclosed TRSs.*

Proof. Replace in the proof of Thm. 4 everywhere non-erasingness by non-∞-erasingness. The only thing which fails is the statement resulting from (†):

- By another appeal to the Parallel Moves Lemma the fork can be joined via $s_{i+1} \to_p t_{i+1} \twoheadleftarrow\!\!\!\!\!\!+^{k} t_i$, where $k > 0$ by non-∞-erasingness of $s_i \to t_i$.

We split this case into two new ones depending on whether some argument (instance of variable) to l is ∞ or not.

- In the former case, $t_i \in \infty$ follows directly from non-∞-erasingness.
- In the latter case, $s_i \to s_{i+1}$ may take place in an erased argument, and $s_{i+1} \to_p t_{i+1} = t_i$. But since all arguments to l are SN, this can happen only finitely often and eventually the first case applies. □

In [9] a uniform normalisation result not requiring left-linearity, but having a critical pair condition incomparable to biclosedness was presented.

4 Second-Order Term Rewriting

In this section, the *fundamental theorem of perpetuality for second-order term rewrite systems* is established, by generalising the method of Section 3.

Theorem 6 ($\mathrm{F_2TP}$). *Non-erasing steps are perpetual in orthogonal P_2RSs.*

Corollary 7. *Non-erasing orthogonal P_2RSs are uniformly normalising.*

For ERSs and CRSs these results can be found as [12, Thm. 60] and [14, Cor. II.5.9.4], respectively. The reader is assumed to be familiar with *second-order term rewrite systems* be it in the form of combinatory reduction systems (CRSs [14]), expression reduction systems (ERSs [13]), or higher-order pattern rewrite systems (PRSs [17]). We employ PRSs as defined in [17], but will write $x.s$ instead of $\lambda x.s$, thereby freeing the λ for usage as a function symbol.

Definition 10. – *The* order *of a rewrite rule is the maximal order of the free variables in it. The order of a PRS is the maximal order of the rules in it. P_nRS abbreviates n^{th}-order PRS.*
- *A rule $l \to r$ is* fully-extended *(FE) if for every occurrence $Z(t_1, \ldots, t_n)$ in l of a free variable Z, $t_1, \ldots, t_n$ is the list of variables bound above it.*
- *A rewrite step $s = C[l^\sigma] \to C[r^\sigma] = t$ is* non-erasing *if every symbol from C and σ in s descends [20, Sect. 3.1.1] to some symbol in t.*[6]

The adaptation is non-trivial since uniform normalisation may fail for orthogonal, but third-order or non-left-linear or non-fully-extended systems.

Table 1. Three counterexamples against uniform normalisation of PRSs

third-order	non-fully-extended	non-left-linear
$(\lambda z.M(z))N \to M(N)$	$M(z)\langle z := N\rangle \to M(N)$	$M(x)\langle x := N\rangle \to M(N)$
$fxy.Z(u.x(u), y) \to Z(u.c, \Omega)$	$gxy.Z(y) \to Z(a)$	$g(x.Z(x), x.Z(x)) \to Z(a)$
	$e(x, y) \to c$	$e(x) \to c$
	$f(a) \to f(a)$	$f(a) \to f(a)$

[6] A TRS step is non-erasing in this sense iff it is non-erasing in the sense of Def. 6.

Example 2. (third-order) [13, Ex. 7.1] Consider the 3rd-order PRS in Tab. 1. It is the standard PRS-presentation of the $\lambda\beta$-calculus [17] extended by a rule. $@ : o{\rightarrow}o{\rightarrow}o$ and $\lambda : (o{\rightarrow}o){\rightarrow}o$ are the function symbols and $M : o{\rightarrow}o$ and $N : o$ are the free-variables of the first (β-)rule. We have made @ an implicit binary infix operation and have written $\lambda x.s$ for $\lambda(x.s)$, for the λ-calculus to take a more familiar form. If Ω abbreviates $(\lambda x.xx)(\lambda x.xx)$, the step $fxy.(\lambda u.x(u))y \rightarrow_\beta fxy.x(y)$ is non-erasing but critical.

(non-fully-extended) [13, Ex. 5.9] Consider the non-FE P_2RS in Tab. 1. The step $gxy.e(z,x)\langle z := f(y)\rangle \rightarrow gxy.e(f(y),x)$ is non-erasing but critical.

(non-left-linear) Consider the non-left-linear P_2RS in Tab. 1. The rewrite step $g(y.e(x)\langle x{:=}f(y)\rangle, y.c\langle x{:=}f(y)\rangle) \rightarrow g(y.e(f(y)), y.c\langle x{:=}f(y)\rangle)$ from s to t is non-erasing but critical; t is terminating, but we have the infinite reduction

$$s \rightarrow g(y.c\langle x := f(y)\rangle, y.c\langle x := f(y)\rangle) \rightarrow c\langle x := f(a)\rangle \rightarrow \ldots$$

In each item, the second rule causes failure of uniform normalisation.

Hence, for uniform normalisation to hold some restrictions need to be imposed: We assume PRSs to be left-linear and fully-extended P_2RSs. For TRSs the fully-extendedness condition is vacuous, hence the assumption reduces to left-linearity as in Sect. 3 The restriction to P_2RSs entails no restriction w.r.t. the other formats, since both CRSs and ERSs can be embedded into P_2RSs, by coding metavariables in rules as free variables of type $o{\rightarrow}\ldots{\rightarrow}o{\rightarrow}o$ [23]. To adapt the proof of F_1TP to P_2RSs, we review its two main ingredients. The first one was a notion of simultaneous reduction, extending one-step reduction such that:

- The residual of a non-erasing step after a context-step is non-erasing.

The second ingredient was STD. It guarantees the following property:

- Any redex pattern l which is entirely above a contracted redex is external to the reduction S; in particular, l cannot be replicated along S, it can only be eliminated by contraction of an overlapping redex in S.

Since the residual of a parallel reduction after a step above it is usually not parallel, we switch from $\twoheadrightarrow\!\!\!\!\!\!\!+$ to $\multimap\!\!\!\rightarrow$, where the latter is the (one-rule restriction of the) simultaneous reduction relation of [21, Def. 3.4]. The context-part of such a $\multimap\!\!\!\rightarrow$-step is the part above or parallel to all occurrences of l.

Definition 11. *Let $\varrho : l \rightarrow r$ be a rewrite rule. Write $s \multimap\!\!\!\rightarrow_\varrho t$ if it holds that $s = C[l^{\sigma_1}, \ldots, l^{\sigma_k}]$ and $t = C[r^{\tau_1}, \ldots, r^{\tau_k}]$, where $\sigma_i \multimap\!\!\!\rightarrow_\varrho \tau_i$ for all $1 \leq i \leq k$.*

Lemma 3 (Finiteness of Developments). *(FD [20, Thm. 3.1.45]) Let $s \multimap\!\!\!\rightarrow t$ by simultaneously contracting redexes at positions in P. Repeated contraction of residuals of redexes in P starting from s terminates and ends in t.*

The second lemma on $\multimap\!\!\!\rightarrow$ is a close relative of [13, Lem. 5.1] and establishes the first ingredient above. It fails for P_3RSs as witnessed by the first item of Ex. 2.

Lemma 4 (Parallel Moves). *Let $\varrho : l \rightarrow r$ and $\vartheta : g \rightarrow d$ be PRS rules, with ϑ second-order. If $s' \leftarrow_\vartheta s \multimap\!\!\!\rightarrow_\varrho t$ is a fork such that g is in the context-part of the non-erasing simultaneous step, then the fork is joinable into $s' \multimap\!\!\!\rightarrow_\varrho t' \leftarrow_\vartheta t$, with the simultaneous step non-erasing.*

Proof. Joinability follows by FD. It remains to show non-erasingness. ϑ being of order 2, each free variable Z occurs in g as $Z(x_1,\ldots,x_n)$ with $x_i : o$ and $Z : o\to\ldots\to o\to o$ and in d as $Z(t_1,\ldots,t_n)$ with $t_i : o$. Hence, the residuals in s' of redexes of $s \multimap t$ are first-order substitution instances of them. Then, to show preservation of non-erasingness it suffices to show that $Var(s) \subseteq Var(s^\sigma)$ for any first-order substitution σ, which follows by induction on s. □

Left-linearity and fully-extendedness are sufficient conditions for STD to hold.

Theorem 7 (STD). *Any reduction in a P_2RS can be transformed into a standard one. The transformation preserves infiniteness.*

Proof. The proof of the second part of the theorem is as for TRSs. For a proof of the first part for left-linear fully-extended (orthogonal) CRSs see [18, Sect. 7.7.3] ([26]). By the correspondence between CRSs and P_2RSs this suffices for our purposes. (STD even holds for PRSs [22, Cor. 1.5].) □

Proof. (of Thm. 6) Replace in the proof of Thm. 2 everywhere $\twoheadrightarrow\!\!\!\!\!\!\!+\!\!+$ by $\multimap$. That the (context)-case eventually applies follows by an appeal to FD. □

The proofs of the results below are obtained by analogous modifications.

Theorem 8. *Non-erasing rewrite steps are perpetual in biclosed P_2RSs.*

F_2TP can be strengthened in various ways. Unlike for TRSs, a critical step in a P_2RS need not erase a term in ∞ as witnessed by $e(f(x))\langle x{:=}a\rangle \to c\langle x{:=}a\rangle$ in the PRS $\{M(x)\langle x := N\rangle \to M(N), e(Z) \to c, f(a) \to f(a)\}$. Note that $f(x) \in \mathsf{SN}$, but by contracting the $\_\langle\_ := \_\rangle$-redex a is substituted for x and $f(a) \in \infty$.

Definition 12. *An occurrence of (the head symbol of) a subterm is* potentially infinite *if some descendant [20] of it along some reduction is in ∞. A step is ∞-erasing if it erases all potentially infinite subterms in its arguments.*

For TRSs this notion of ∞-erasingness coincides with the one of Def. 9.

Corollary 8. *Non-∞-erasing rewrite steps are perpetual in biclosed P_2RSs.*

Many variations of this result are possible. We mention two. First, the motivation for this paper originates with [13, Sect. 6.4], where we failed to obtain:

Theorem 9. *([5]) λ-δ_K-calculus is uniformly normalising.*

Proof. By Cor. 8, since λ-δ_K-calculus is weakly orthogonal. □

Second, we show that non-fully-extended P_2RSs may have uniform normalisation. By the same method, P_2RSs where non-fully-extended steps are terminating and postponable have uniform normalisation.

Theorem 10. *Non-∞-erasing steps are perpetual in $\lambda\beta\eta$-calculus [24, Prop. 27].*

Proof. It suffices to remark that η-steps can be postponed after β-steps in a standard reduction [2, Cor. 15.1.6]. Since η is terminating, an infinite standard reduction must contain infinitely many β-steps, hence may be assumed to consist of β's only and the proof of F_2TP goes through unchanged. □

5 λx^-

In this section familiarity with the nameful λ-calculus with explicit substitutions λx^- of [4] is assumed. We define it as a P_2RS and establish the *fundamental theorem of perpetuality for* λx^-:

Theorem 11 (F_xTP). *Non-erasing steps are perpetual in* λx^-.

Definition 13. *The alphabet of* λx^- *[4] consists of the function symbols* $@ : o{\to}o{\to}o$, $\lambda : (o{\to}o){\to}o$ *and* $\_\langle\_ := \_\rangle : (o{\leftarrow}o){\to}o{\to}o$. *As above, we make* $@$ *an implicit infix operator associating to the left. The rules of* λx^- *are (for* $x \neq y$*):*

$$\begin{array}{rcl}(\lambda x.M(x))N & \to_{\mathrm{Beta}} & M(x)\langle x := N\rangle \\ x\langle x := N\rangle & \to_{=} & N \\ y\langle x := N\rangle & \to_{\neq} & y \\ (\lambda y.M(y,x))\langle x := N\rangle & \to_{\lambda} & \lambda y.M(y,x)\langle x := N\rangle \\ (M(x)L(x))\langle x := N\rangle & \to_{@} & M(x)\langle x := N\rangle L(x)\langle x := N\rangle\end{array}$$

The last four rules are the explicit substitution *rules denoted* x, *generating* $\to_x$. $\to_x$ is a terminating and orthogonal P_2RS, hence the normal form of a term s exists uniquely and is denoted by $s{\downarrow_x}$. Note that $s{\downarrow_x}$ is a *pure* λ-term, i.e. it does not contain *closures* ($\_\langle\_ := \_\rangle$-symbols). λx^- implements (only) substitution [4]:

Lemma 5. *1. If* $s =_x t$, *then* $s{\downarrow_x} = t{\downarrow_x}$.
2. If $s \to_{\mathrm{Beta}} t$, *then* $s{\downarrow_x} \twoheadrightarrow_\beta t{\downarrow_x}$.
3. If s *is pure and* $s \to_\beta t$, *then* $s \to_{\mathrm{Beta}} \cdot \to_x^+ t$.

Remark that in the second item the number of β-steps might be zero, but is always positive when the Beta-step is not inside a closure. We call λx^--reductions without steps inside closures *pretty.* λx^- *preserves strong normalisation* in the sense that any pure term which is β-terminating is λx^--terminating.

Lemma 6 (PSN). *[4, Thm. 4.19] If* s *is pure and* β-SN, *then* s *is* λx^--SN.

Proof. Suppose $s \in \infty$. Since λx^- is a fully-extended left-linear sub-P_2RS[7], we may by STD assume an infinite standard reduction $S : s_0 \to s_1 \to \ldots$ from $s = s_0$. We show that we may choose S to be pretty decent, where a reduction is *decent* [4, Def. 4.16] if for every closure $\langle x := t\rangle$ in any term, $t \in$ SN.

(init) s is decent since it is pure.

(step) Suppose $s_i \in \infty$ and s_i is decent. From the shape of the rules we have that 'brackets are king' [19][8]: if any step takes place in t inside some closure $\langle x := t\rangle$ in a standard reduction, then no step above the closure can be performed later in the reduction. This entails that if t is terminating, S need not perform any step inside t. Hence assume $s_i \to s_{i+1}$ is pretty.

[7] It only is a *sub*-P_2RS since the y in the $\to_{\neq}$-rule ranges over variables not over terms.

[8] Thinking of terms as trees representing hierarchies of people, creating a redex above (overruling) someone (the ruler) from below (the people) is a revolution. For closures/brackets this is not possible, whence these are king.

(Beta) Suppose $s_i \to_{\text{Beta}} s_{i+1}$ contracting $(\lambda x.M(x))N$ to $M(x)\langle x := N\rangle$. We may assume that N is terminating since otherwise we could instead perform an infinite reduction on N itself, hence the reduct is decent.

(x) Otherwise, decency is preserved, since x-steps do not create closures.

Since x is terminating S must contain infinitely many Beta-steps. Since S is pretty $S{\downarrow_{\text{x}}}$ is an infinite β-reduction from s by (the remark after) Lem. 5. □

Our method relates to closure-tracking [3] as preventing to curing. Trying to apply it to prove [4, Conj. 6.45], stating that explicification of *redex preserving* CRSs is PSN, led to the following counterexample.

Example 3. Consider the term $s = (\tilde{\lambda}(x.b))a$ in the P_2RS $\mathcal{R}$ with rewrite rules $\{(\tilde{\lambda}x.M(x))N \to M(g(N,N)), a \to b, g(a,b) \to g(a,b)\}$. On the one hand s is terminating, since $s \to b[x{:=}g(a,a)] = b$. On the other hand, explicifying $\mathcal{R}$ will make s infinite, since $g(a,a) \to g(a,b) \to g(a,b)$. The PRS is redex preserving in the sense of [4, Def. 6.44] since any redex in the argument $g(N,N)$ to M occurs in N already. So s is a term for which PSN does not hold.

We expect the conjecture to hold for orthogonal CRSs. For our purpose, uniform normalisation, we will need the following corollary to Lem. 6, on preservation of infinity. It is useful in situations where terms are only the same up to the Substitution Lemma [2, Lem. 2.1.16]: $M(x,y)\langle x := N(y)\rangle\langle y := L\rangle{\downarrow_{\text{x}}} = M(x,y)\langle y := P\rangle\langle x := N\langle y := L\rangle\rangle{\downarrow_{\text{x}}}$.

Corollary 9. *If s is decent and $s{\downarrow_{\text{x}}} = t{\downarrow_{\text{x}}}$, then $s \in \infty$ implies $t \in \infty$.*

How should non-erasingness be defined for λx^-? The naïve attempt falters.

Example 4. From the term $s = ((\lambda x.z)(y\omega))\langle y := \omega\rangle$, where $\omega = \lambda x.xx$, we have a unique terminating reduction starting with a 'non-erasing' Beta-step:

$$s \to_{\text{Beta}} z\langle x := y\omega\rangle\langle y := \omega\rangle \to_{\text{x}} z\langle y := \omega\rangle \to_{\text{x}} z$$

On the other hand, developing $\langle y := \omega\rangle$ yields the term $\omega\omega \in \infty$.

Translating the example into $\lambda\beta$-calculus shows that the culprit is the 'non-erasing' Beta-step, which translates into an erasing β-step. Therefore:

Definition 14. *A λx^--step contracting redex s to t is* erasing *if $s \to t$ is*

$$\begin{array}{ll}(\lambda x.M(x))N \to_{\text{Beta}} M, & \text{with } x \notin Var(M(x){\downarrow_{\text{x}}}), \text{ or}\\ y\langle x := N\rangle \to_{\neq} y & \end{array}$$

Proof. (of Thm. 11) Since λx^- is a sub-P_2RS, it suffices by the proof of F_2TP to consider perpetuality of a step $s \to_{p,\varrho} t$, for some infinite standard reduction $S : s_0 \to_{q_0,\vartheta_0} s_1 \to_{q_1,\vartheta_1} \dots$ starting from $s = s_0$ such that $s_1 \leftarrow s \to t$ is an overlapping fork (case (‡) on p. 128). λx^- has only one non-trivial critical pair. It arises by @ and Beta from $s' = ((\lambda x.M(x,y))N(y))\langle y := P\rangle$, so let $s = C[s']$.

(Beta,@) In case $s \to_{\text{Beta}} C[M(x,y)\langle x := N(y)\rangle\langle y := P\rangle] = s_1$, we note that

$$\begin{aligned} s &\to_{p,@} C[(\lambda x.M(x,y))\langle y := P\rangle N(y)\langle y := P\rangle] = t \\ &\to_{\lambda} C[(\lambda x.M(x,y)\langle y := P\rangle)N(y)\langle y := P\rangle] \\ &\to_{\text{Beta}} C[M(x,y)\langle y := P\rangle\langle x := N(y)\langle y := P\rangle\rangle] = t_1 \end{aligned}$$

Consider a minimal closure in s_1 (or s_1 itself) which is decent and ∞, say at position o. If o is parallel or properly below p, i.e. inside one of $M(x,y)$, $N(y)$ or P, then obviously $t_1 \in \infty$. Otherwise, o is above p and $t_1 \in \infty$ follows from Corollary 9, since $s_{1|o}\!\downarrow_{\mathrm{x}} = t_{1|o}\!\downarrow_{\mathrm{x}}$.

(@,Beta) The case $s \to_{p,\text{Beta}} C[M(x,y)\langle x := N(y)\rangle\langle y := P\rangle] = t$ is more involved. Construct a maximal reduction T as follows. Let $t_0 = t$ be the first term of T and set $o_0 = p$.

- Suppose $s_i \to_{q_i,\vartheta_i} s_{i+1}$ does not contract a redex below o_i. As an invariant we will use that o_i traces the position of @ (initially at p) along S. If q_i is parallel to o_i, then we set $t_i \to_{q_i,\vartheta_i} t_{i+1}$. Otherwise $q_i < o_i$ and by standardness this is only possible in case of an @-step distributing closures over the @ at o_i. Then we set $t_{i+1} = t_i$ and $o_{i+1} = q_i$.

If this process continues, then T is infinite since in case no steps are generated $o_{i+1} < o_i$, hence eventually a step must be generated. If the process stops, say at n, then by construction $s_n = D[u]_{o_n}$ and $t_n = D[v]_{o_n}$, with $u = (\lambda x.M(x,y))\langle \boldsymbol{y} := \boldsymbol{P}\rangle N(y)\langle \boldsymbol{y} := \boldsymbol{P}\rangle$, $v = M(x,y)\langle x := N(y)\rangle\langle \boldsymbol{y} := \boldsymbol{P}\rangle$ and $\langle \boldsymbol{y} := \boldsymbol{P}\rangle$ abbreviates a sequence of closures the first of which is $\langle y := P\rangle$. Per construction, $o_n \leq q_n$ for the step $s_n \to_{q_n} s_{n+1}$ and we are in the 'non-replicating' case: by standardness the @ cannot be replicated along S and it can only be eliminated as part of a Beta-step. Consider a maximal part of S not contracting o_n. Remark that if any of $M(x,y)$, $N(y)$ and $\boldsymbol{P}$ is infinite, then $t_n \in \infty$, so we assume them terminating.

(context) If infinitely many steps parallel to o_i take place, then $D \in \infty$, hence $t_n = D[v] \in \infty$.

(left) Suppose infinitely many steps are in $(\lambda x.M(x,y))\langle \boldsymbol{y}{:=}\boldsymbol{P}\rangle$. This implies $M(x,y)\langle \boldsymbol{y} := \boldsymbol{P}\rangle \in \infty$, hence $M(x,y)\langle \boldsymbol{y} := \boldsymbol{P}\rangle\langle x := N(y)\langle \boldsymbol{y} := \boldsymbol{P}\rangle\rangle \in \infty$, which by Corollary 9 implies $t_n \in \infty$.

(right) Suppose infinitely many steps are in $N(y)\langle \boldsymbol{y} := \boldsymbol{P}\rangle$. By non-erasingness of $s \to_{\text{Beta}} t$, $x \in M(x,y)\!\downarrow_{\mathrm{x}}$ hence

$$\begin{aligned} v &\twoheadrightarrow_{\mathrm{x}} M(x,y)\!\downarrow_{\mathrm{x}}\langle x := N(y)\rangle\langle \boldsymbol{y} := \boldsymbol{P}\rangle \\ &= E[x,\ldots,x]\langle x := N(y)\rangle\langle \boldsymbol{y} := \boldsymbol{P}\rangle \\ &\twoheadrightarrow_{\mathrm{x}} E^*[x\langle x := N(y)\rangle\langle \boldsymbol{y} := \boldsymbol{P}\rangle,\ldots,x\langle x := N(y)\rangle\langle \boldsymbol{y} := \boldsymbol{P}\rangle] \\ &\twoheadrightarrow_{=} E^*[N(y)\langle \boldsymbol{y} := \boldsymbol{P}\rangle,\ldots,N(y)\langle \boldsymbol{y} := \boldsymbol{P}\rangle] \in \infty \end{aligned}$$

where E^* arises by pushing $\langle x{:=}N(y)\rangle\langle \boldsymbol{y}{:=}\boldsymbol{P}\rangle$ through E, and $E[\,,\ldots,\,]$ is a pure λ-calculus context with at least one hole. Hence $t = D[v] \in \infty$.

(Beta) Suppose o_n is Beta-reduced sometime in S. By standardness steps before Beta can be neither in occurrences of the closures $\langle \boldsymbol{y} := \boldsymbol{P}\rangle$ nor in $M(x,y)$, hence we may assume S proceeds as:

$$\begin{aligned} s_n \quad &\twoheadrightarrow_\lambda \quad D[(\lambda x.M(x,y)\langle \boldsymbol{y} := \boldsymbol{P}\rangle)N(y)\langle \boldsymbol{y} := \boldsymbol{P}\rangle] \\ &\rightarrow_{\text{Beta}} D[M(x,y)\langle \boldsymbol{y} := \boldsymbol{P}\rangle\langle x := N(y)\langle \boldsymbol{y} := \boldsymbol{P}\rangle\rangle] = u' \end{aligned}$$

We proceed as in item (Beta,@), using $u'\downarrow_{\mathrm{x}} = v\downarrow_{\mathrm{x}}$ to conclude $v \in \infty$ by Corollary 9. The only exception to this is an infinite reduction from $N(y)\langle \boldsymbol{y} := \boldsymbol{P}\rangle$, but such a reduction can be simulated from v by non-erasingness of the Beta-step as in item (right). □

The proof is structured as before, only di/polluted by explicit substitutions travelling through the pivotal Beta-redex. Again, one can vary on these results. For example, it should not be difficult to show that non-∞-erasing steps are perpetual, where $y\langle x{:=}N\rangle \rightarrow_{\neq} y$ is ∞-erasing if $N \in \infty$ and $(\lambda x.M(x))N \rightarrow_{\text{Beta}} M$ is ∞-erasing if $x \notin \mathit{Var}(\mathrm{x}(M(x)))$ and N contains a *potentially infinite* subterm.

6 Conclusion

The uniform normalisation proofs in literature are mostly based on particular *perpetual* strategies, that is, strategies performing only perpetual steps. Observing that the non-computable[9] such strategies usually yield standard reductions we have based our proof on standardisation, instead of searching for yet another 'improved' perpetual strategy. This effort was successful and resulted in a flexible proof strategy with a simple invariant easily adaptable to a λ-calculus with explicit substitutions. Nevertheless, our results are still very much orthogonality-bound: the biclosedness results arise by tweaking orthogonality and the $\lambda\mathrm{x}^-$ results by interpretation in the, orthogonal, $\lambda\beta$-calculus. It would be interesting to see what can be done for truly non-orthogonal systems. The fully-extendedness and left-linearity restrictions are serious ones, e.g. in the area of process-calculi (scope extrusion) or even already for λx [4], so should be ameliorated.

Acknowledgments. We would like to thank R. Bloo, E. Bonelli, D. Kesner, P.-A. Melliès, A. Visser and the members of PAM at CWI for feedback.

References

[1] F. Baader and T. Nipkow. *Term Rewriting and All That.* CUP, 1998.
[2] H. Barendregt. *The Lambda Calculus, Its Syntax and Semantics.* NH, 1984.
[3] Z. Benaissa, D. Briaud, P. Lescanne, and J. Rouyer-Degli. λυ, a calculus of explicit substitutions which preserves strong normalisation. *JFP*, 6(5):699–722, 1996.
[4] R. Bloo. *Preservation of Termination for Explicit Substitution.* PhD thesis, Technische Universiteit Eindhoven, 1997.

[9] No computable strategy exists which is both perpetual and standard, since then one could for all terms s, t decide whether $\mathtt{SN}(s) \Rightarrow \mathtt{SN}(t)$ or $\mathtt{SN}(t) \Rightarrow \mathtt{SN}(s)$.

[5] C. Böhm and B. Intrigila. The ant-lion paradigm for strong normalization. *I&C*, 114(1):30–49, 1994.

[6] E. Bonelli. Perpetuality in a named lambda calculus with explicit substitutions. MSCS, to appear.

[7] Alonzo Church. *The Calculi of Lambda-Conversion.* PUP, 1941.

[8] Georges Gonthier, Jean-Jacques Lévy, and Paul-André Melliès. An abstract standardisation theorem. In *LICS'92*, pages 72–81, 1992.

[9] Bernhard Gramlich. *Termination and Confluence Properties of Structured Rewrite Systems.* PhD thesis, Universität Kaiserslautern, 1996.

[10] Gérard Huet. Confluent reductions: Abstract properties and applications to term rewriting systems. *JACM*, 27(4):797–821, 1980.

[11] Gérard Huet and Jean-Jacques Lévy. Computations in orthogonal rewriting systems, I. In *Computational Logic: Essays in Honor of Alan Robinson*, pages 395–414. MIT Press, 1991.

[12] Z. Khasidashvili. On the longest perpetual reductions in orthogonal expression reduction systems. TCS, To appear.

[13] Z. Khasidashvili, M. Ogawa, and V. van Oostrom. Perpetuality and uniform normalization in orthogonal rewrite systems. I&C, To appear. `http://www.phil.uu.nl/~oostrom/publication/ps/pun-icv2.ps`.

[14] Jan Willem Klop. *Combinatory Reduction Systems.* PhD thesis, Rijksuniversiteit Utrecht, 1980. Mathematical Centre Tracts 127.

[15] J.W. Klop. Term rewriting systems. In *Handbook of Logic in Computer Science*, volume 2, pages 1–116. OUP, 1992.

[16] Yves Lafont. From proof-nets to interaction nets. In *Advances in Linear Logic*, pages 225–247. CUP, 1995.

[17] Richard Mayr and Tobias Nipkow. Higher-order rewrite systems and their confluence. *Theoretical Computer Science*, 192:3–29, 1998.

[18] Paul-André Melliès. *Description Abstraite des Systèmes de Réécriture.* Thèse de doctorat, Université Paris VII, 1996.

[19] Paul-André Melliès. Personal communication, 1999.

[20] Vincent van Oostrom. *Confluence for Abstract and Higher-Order Rewriting.* Academisch proefschrift, Vrije Universiteit, Amsterdam, 1994.

[21] Vincent van Oostrom. Development closed critical pairs. In *HOA'95*, volume 1074 of *LNCS*, pages 185–200. Springer, 1996.

[22] Vincent van Oostrom. Normalisation in weakly orthogonal rewriting. In *RTA'99*, volume 1631 of *LNCS*, pages 60–74. Springer, 1999.

[23] F. van Raamsdonk. *Confluence and Normalisation for Higher-Order Rewriting.* Academisch proefschrift, Vrije Universiteit, Amsterdam, 1996.

[24] M.H. Sørensen. Effective longest and infinite reduction paths in untyped lambda-calculi. In *CAAP'96*, volume 1059 of *LNCS*, pages 287–301. Springer, 1996.

[25] Yoshihito Toyama. Strong sequentiality of left-linear overlapping term rewriting systems. In *LICS'92*, pages 274–284, 1992.

[26] J.B. Wells and Robert Muller. Standardization and evaluation in combinatory reduction systems, 2000. Working paper.

Verifying Orientability of Rewrite Rules Using the Knuth-Bendix Order

Konstantin Korovin and Andrei Voronkov

University of Manchester
{korovin,voronkov}@cs.man.ac.uk

Abstract. We consider two decision problems related to the Knuth-Bendix order (KBO). The first problem is *orientability*: given a system of rewrite rules R, does there exist some KBO which orients every ground instance of every rewrite rule in R. The second problem is whether a given KBO orients a rewrite rule. This problem can also be reformulated as the problem of solving a single ordering constraint for the KBO. We prove that both problems can be solved in polynomial time. The algorithm builds upon an algorithm for solving systems of homogeneous linear inequalities over integers. Also we show that if a system is orientable using a real-valued KBO, then it is also orientable using an integer-valued KBO.

1 Introduction

In this section we give an informal overview of the results proved in this paper. The formal definitions will be given in the next section.

Let $\succ$ be any order on ground terms and $l \to r$ be a rewrite rule. We say that $\succ$ *orients* $l \to r$, if for every ground instance $l' \to r'$ of $l \to r$ we have $l' \succ r'$. We write $l \succeq r$ if either $l \succ r$ or $l = r$. There are situations where we want to check if there *exists* a simplification order on ground terms that orients a given system of (possible nonground) rewrite rules. We call this problem *orientability*. Orientability can be useful when a theorem prover is run on a new problem for which no suitable simplification order is known, or when termination of a rewrite system is to be established automatically. For a recent survey, see [4]. We consider the orientability problem for the Knuth-Bendix orders (in the sequel *KBO*) [7] on ground terms. We give a polynomial-time algorithm for checking orientability by the KBO. A similar problem of orientability by the nonground version of the real-valued KBO was studied in [6] and an algorithm for orientability was given. We prove that any rewrite rule system orientable by a real-valued KBO is also orientable by an integer-valued KBO. This result also holds for the nonground version of the KBO considered in [6]. In our proofs we use some techniques of [6]. We also show that some rewrite systems could not be oriented by nonground version of the KBO, but can be oriented by our algorithm.

The second problem we consider is solving ordering constraints consisting of a single inequality, over the Knuth-Bendix order. If $\succ$ is total on ground terms,

A. Middeldorp (Ed.): RTA 2001, LNCS 2051, pp. 137–153, 2001.

then the problem of checking if $\succ$ orients $l \to r$ has relation to the problem of solving ordering constraints over $\succ$. Indeed, $\succ$ *does not* orient $l \to r$ if and only if there exists a ground instance $l' \to r'$ of $l \to r$ such that $r' \succeq l'$, i.e., if and only if the ordering constraint $r \succeq l$ has a solution. This means that any procedure for solving ordering constraint consisting of a single inequality can be used for checking whether a given system of rewrite rules is oriented by $\succ$, and vice versa. Using the same technique as for the orientability problem, we show that the problem of solving an ordering constraint consisting of a single inequality for the KBO can be solved in polynomial time.

Algorithms for, and complexity of, orientability for various version of the recursive path ordering were considered in [12,5,11]. The problems of solving ordering constraints for lexicographic and recursive path orderings and for KBO are NP-complete [2,15,14,9]. However, to check if $\succ$ orients $l \to r$, it is sufficient to check solvability of a *single* ordering constraint $r \succeq l$. This problem is NP-complete for LPO [3], and therefore the problem of checking if an LPO orients a rewrite rule is coNP-complete.

2 Preliminaries

A *signature* is a finite set of function symbols with associated arities. In this paper Σ denotes an arbitrary signature. *Constants* are function symbols of the arity 0. We assume that Σ contains at least one constant. We denote variables by x, y, z, constants by a, b, c, d, e, function symbols by f, g, h, and terms by l, r, s, t. Systems of rewrite rules and rewrite rules are defined as usual, see e.g. [1]. An expression E (e.g. a term or a rewrite rule) is called *ground* if no variable occurs in E. Denote the set of natural numbers by $\mathbb{N}$.

We call a *weight function* on Σ any function $w : \Sigma \to \mathbb{N}$ such that (i) $w(a) > 0$ for every constant $a \in \Sigma$, (ii) there exist at most one unary function symbol $f \in \Sigma$ such that $w(f) = 0$. Given a weight function w, we call $w(g)$ the *weight* of g. The *weight* of any ground term t, denoted $|t|$, is defined as follows: for every constant c we have $|c| = w(c)$ and for every function symbol g of a positive arity $|g(t_1, \ldots, t_n)| = w(g) + |t_1| + \ldots + |t_n|$.

A *precedence relation* on Σ is any total order $\gg$ on Σ. A precedence relation $\gg$ is said to be *compatible* with a weight function w if for every unary function symbol f, if $w(f) = 0$, then f is the greatest element w.r.t. $\gg$.

Let w be a weight function on Σ and $\gg$ a precedence relation on Σ compatible with w. The *Knuth-Bendix order induced by* $(w, \gg)$ is the binary relation $\succ$ on the set of ground terms of Σ defined as follows. For all ground terms $t = g(t_1, \ldots, t_n)$ and $s = h(s_1, \ldots, s_k)$ we have $t \succ s$ if one of the following conditions holds:

1. $|t| > |s|$;
2. $|t| = |s|$ and $g \gg h$;
3. $|t| = |s|$, $g = h$ and for some $1 \le i \le n$ we have $t_1 = s_1, \ldots, t_{i-1} = s_{i-1}$ and $t_i \succ s_i$.

The compatibility condition ensures that the Knuth-Bendix order is a simplification order total on ground terms.

In the sequel we will often refer to the least and the greatest terms among the terms of the minimal weight for a given KBO. It is easy to see that every term of the minimal weight is either a constant of the minimal weight, or a term $f^n(c)$, where c is a constant of the minimal weight, and $w(f) = 0$. Therefore, the least term of the minimal weight is always the constant of the minimal weight which is the least among all such constants w.r.t. $\gg$. This constant is also the least term w.r.t. $\succ$.

The greatest term of the minimal weight exists if and only if there is no unary function symbol of the weight 0. In this case, this term is the constant of the minimal weight which is the greatest among such constants w.r.t. $\gg$.

DEFINITION 1 (substitution) A *substitution* is a mapping from a set of variables to the set of terms. A substitution θ is *grounding* for an expression E (i.e., term, rewrite rule etc.) if for every variable x occurring in E the term $\theta(x)$ is ground. We denote by $E\theta$ the expression obtained from E by replacing in it every variable x by $\theta(x)$. A *ground instance* of an expression E is any expression $E\theta$ which is ground.

The following definition is central to this paper.

DEFINITION 2 (orientability) A KBO $\succ$ *orients* a rewrite rule $l \to r$ if for every ground instance $l' \to r'$ of $l \to r$ we have $l' \succ r'$. A KBO *orients a system* R of rewrite rules if it orients every rewrite rule in R.

Note that we define orientability in terms of ground instances of rewrite rules. One can also define orientability using the nonground version of the KBO as originally defined in [7]. But then we obtain a weaker notion (fewer systems can be oriented) as the following example from [8] shows.

Example 1. Consider the rewrite rule $g(x, a, b) \to g(b, b, a)$. For any choice of the weight function w and order $\gg$, $g(x, a, b) \succ g(b, b, a)$ does not hold for the original Knuth-Bendix order with variables. However, this rewrite rule can be oriented by any KBO such that $w(a) \geq w(b)$ and $a \gg b$.

In fact the order based on all ground instances is the greatest simplification order extending the ground KBO to nonground terms.

3 Systems of Homogeneous Linear Inequalities

In our proofs and in the algorithm we will use several properties of homogeneous linear inequalities. The definitions related to systems of linear inequalities can be found in standard textbooks (e.g., [16]). We will denote column vectors of variables by X, integer or real vectors by V, W, integer or real matrices by A, B. Column vectors consisting of 0's will be denoted by $\mathbf{0}$. The set of real numbers is denoted by $\mathbb{R}$, and the set of nonnegative real numbers by $\mathbb{R}^+$.

DEFINITION 3 (homogeneous linear inequalities) A *homogeneous linear inequality* has the form either $VX \geq 0$ or $VX > 0$. A *system of homogeneous linear inequalities* is a finite set of homogeneous linear inequalities.

Solutions (real or integer) to systems of homogeneous linear inequalities are defined as usual.

We will use the following fundamental property of system of homogeneous linear inequalities:

Lemma 1. *Let $AX \geq \mathbf{0}$ be a system of homogeneous linear inequalities, where A is an integer matrix. Then there exists a finite number of integer vectors $V_1, \ldots, V_n$ such that the set of solutions to $AX \geq \mathbf{0}$ is*

$$\{r_1 V_1 + \ldots + r_n V_n \mid r_1, \ldots, r_n \in \mathbb{R}^+\}. \tag{1}$$

The proof can be found in e.g., [16].

The following lemma was proved in [13] for the systems of linear homogeneous inequalities over the real numbers. We will give a simpler proof of it here.

Lemma 2. *Let $AX \geq \mathbf{0}$ be a system of homogeneous linear inequalities where A is an integer matrix and* Sol *be the set of all real solutions to the system. Then the system can be split into two disjoint subsystems $BX \geq \mathbf{0}$ and $CX \geq \mathbf{0}$ such that*

1. *$BV = \mathbf{0}$ for every $V \in$ Sol.*
2. *If C is nonempty then there exists a solution $V \in$ Sol such that $CV > \mathbf{0}$.*

PROOF. By Lemma 1 we can find integer vectors $V_1, \ldots, V_n$ such that the set Sol is (1). We define $BX \geq \mathbf{0}$ to be the system consisting of all inequalities in $WX \geq 0$ in the system such that $WV_i = 0$ for all $i = 1, \ldots, n$; then property 1 is obvious.

Note that the system $CX \geq \mathbf{0}$ consists of the inequalities $WX \geq 0$ such that for some V_i we have $WV_i > 0$. Take V to be $V_1 + \ldots + V_n$, then it is not hard to argue that $CV > \mathbf{0}$. Let $\mathbb{D}$ be a system of homogeneous linear inequalities with a real matrix. We will call the subsystem $BX \geq \mathbf{0}$ of $\mathbb{D}$ the *degenerate subsystem* if the following holds. Denote by C be the matrix of the complement to $BX \geq \mathbf{0}$ in $\mathbb{D}$ and by Sol be the set of all real solutions to $\mathbb{D}$. Then

1. $BV = \mathbf{0}$ for every $V \in$ Sol.
2. If C is nonempty then there exists a solution $V \in$ Sol such that $CV > \mathbf{0}$.

For every system $\mathbb{D}$ of homogeneous linear inequalities the degenerate subsystem of $\mathbb{D}$ will be denoted by $\mathbb{D}^=$. Note that the degenerate subsystem is defined for arbitrary systems, not only those of the form $AX \geq 0$.

Let us now prove another key property of integer systems of homogeneous linear inequalities: the existence of a real solution implies the existence of an integer solution.

Lemma 3. *Let $\mathbb{D}$ be a system of homogeneous linear inequalities with an integer matrix. Let V be a real solution to this system and for some subsystem of $\mathbb{D}$ with the matrix B we have $BV > \mathbf{0}$. Then there exists an integer solution V' to $\mathbb{D}$ for which we also have $BV' > \mathbf{0}$.*

PROOF. Let $\mathbb{D}'$ be obtained from $\mathbb{D}$ by replacement of all strict equalities $WX > 0$ by their nonstrict versions $WX \geq 0$. Take vectors $V_1, \ldots, V_n$ so that the set of solutions to $\mathbb{D}'$ is (1). Evidently, for every inequality $WX \geq 0$ in $BV > \mathbf{0}$ there exists some V_i such that $WV_i > 0$. Define V' as $V_1 + \ldots + V_n$, then it is not hard to argue that $BV' > 0$. We claim that V' is a solution to $\mathbb{D}$. Assume the converse, then there exists an equation $WX > 0$ in $\mathbb{D}$ such that $WV' = 0$. But $WV' = 0$ implies that $WV_i = 0$ for all i, so $WX \geq 0$ cannot belong to $\mathbb{D}'$.

The following Lemma follows from Lemmas 2 and 3.

Lemma 4. *Let $\mathbb{D}$ be a system of homogeneous linear inequalities with an integer matrix and its degenerate subsystem is different from $\mathbb{D}$. Let B be the matrix of the complement of the degenerate subsystem. Then there exists an integer solution V to $\mathbb{D}$ such that $BV > \mathbf{0}$.*

The following result is well-known, see e.g., [16].

Lemma 5. *The existence of a real solution to a system of linear inequalities can be decided in polynomial time.*

This lemma and Lemma 3 imply the following key result.

Lemma 6. *(i) The existence of an integer solution to an integer system of homogeneous linear inequalities can be decided in polynomial time. (ii) If an integer system $\mathbb{D}$ of homogeneous linear inequalities has a solution, then its degenerate subsystem $\mathbb{D}^=$ can be found in polynomial time.*

4 States

In Section 5 we will present an algorithm for ground orientability by the Knuth-Bendix order. This algorithm will work on *states* which generalize systems of rewrite rules in several ways. A state will use a generalization of rewrite rules to tuples of terms and some information about possible solutions.

Let $\succ$ be any order on ground terms. We extend it lexicographically to an order on tuples of ground terms as follows: we write $\langle l_1, \ldots, l_n \rangle \succ \langle r_1, \ldots, r_n \rangle$ if for some $i \in \{1, \ldots, n\}$ we have $l_1 = r_1, \ldots, l_{i-1} = r_{i-1}$ and $l_i \succ r_i$. We call a *tuple inequality* any expression $\langle l_1, \ldots, l_n \rangle > \langle r_1, \ldots, r_n \rangle$. The *length* of this tuple inequality is n.

In the sequel we assume that Σ is a fixed signature and e is a constant not belonging to Σ. The constant e will play the role of a temporary substitute for a constant of the minimal weight. We will present the algorithm for orienting a system of rewrite rules as a sequence of state changes. We call a *state* a tuple $(\mathbb{R}, \mathbb{M}, \mathbb{D}, \mathbb{U}, \mathbb{G}, \mathbb{L}, \gg)$, where

1. $\mathbb{R}$ is a set of tuple inequalities $\langle l_1, \ldots, l_n \rangle > \langle r_1, \ldots, r_n \rangle$, such that every two different tuple inequalities in this set have disjoint variables.
2. $\mathbb{M}$ is a set of variables. This set denotes the variables ranging over the terms of the minimal weight.
3. $\mathbb{D}$ is a system of homogeneous linear inequalities with variables $\{w_g \mid g \in \Sigma \cup \{e\}\}$. This system denotes constraints on the weight function collected so far, and w_e denotes the minimal weight of terms.
4. $\mathbb{U}$ is one of the following values *one* or *undefined*. The value *one* signals that there exists exactly one term of the minimal weight, while *undefined* means that no constraints on the number of elements of the minimal weight have been imposed.
5. $\mathbb{G}$ and $\mathbb{L}$ are sets of constants, each of them contains at most one element. If $d \in \mathbb{G}$ (respectively $d \in \mathbb{L}$), this signals that d is the greatest (respectively least) term among the terms of the minimal weight.
6. $\gg$ is a binary relation on Σ. This relation denotes the subset of the precedence relation computed so far.

Let w be a weight function on Σ, $\gg'$ a precedence relation on Σ compatible with w, and $\succ$ the Knuth-Bendix order induced by $(w, \gg')$. A substitution σ grounding for a set of variables X is said to be *minimal for* X if for every variable $x \in X$ the term $\sigma(x)$ is of the minimal weight. We extend w to e by defining $w(e)$ to be the minimal weight of a constant of Σ.

We say that the pair $(w, \gg')$ is a *solution* to a state $(\mathbb{R}, \mathbb{M}, \mathbb{D}, \mathbb{U}, \mathbb{G}, \mathbb{L}, \gg)$ if

1. The weight function w solves every inequality in $\mathbb{D}$ in the following sense: replacement of each w_g by $w(g)$ gives a tautology.
2. If $\mathbb{U} = one$, then there exists exactly one term of the minimal weight.
3. If $d \in \mathbb{G}$ (respectively $d \in \mathbb{L}$) for some constant d, then d is the greatest (respectively least) term among the terms of the minimal weight. Note that if d is the greatest term of the minimal weight, then the signature contains no unary function symbol of the weight 0.
4. For every tuple inequality $\langle l_1, \ldots, l_n \rangle > \langle r_1, \ldots, r_n \rangle$ in $\mathbb{R}$ and every substitution σ grounding for this tuple inequality and minimal for $\mathbb{M}$ we have $\langle l_1\sigma, \ldots, l_n\sigma \rangle \succ \langle r_1\sigma, \ldots, r_n\sigma \rangle$.
5. $\gg'$ extends $\gg$.

We will now show how to reduce the orientability problem for the systems of rewrite rules to the solvability problem for states.

Let R be a system of rewrite rules such that every two different rules in R have disjoint variables. Denote by $\mathbb{S}_R$ the state $(\mathbb{R}, \mathbb{M}, \mathbb{D}, \mathbb{U}, \mathbb{G}, \mathbb{L}, \gg)$ defined as follows.

1. $\mathbb{R}$ consists of all tuple inequalities $\langle l \rangle > \langle r \rangle$ such that $l \to r$ belongs to R.
2. $\mathbb{M} = \emptyset$.
3. $\mathbb{D}$ consists of (a) all inequalities $w_g \geq 0$, where $g \in \Sigma$ is a nonconstant; (b) the inequality $w_e > 0$ and all inequalities $w_d - w_e \geq 0$, where d is a constant of Σ.

4. $\mathbb{U} = undefined$.
5. $\mathbb{G} = \mathbb{L} = \emptyset$.
6. $\gg$ is the empty binary relation on Σ.

Lemma 7. *Let w be a weight function, $\gg'$ a precedence relation on Σ compatible with w, and $\succ$ the Knuth-Bendix order induced by $(w, \gg')$. Then $\succ$ orients R if and only if $(w, \gg')$ is a solution to $\mathbb{S}_R$.*

The proof is straightforward.

For technical reasons, we will distinguish two kinds of signatures. Essentially, our algorithm depends on whether the weights of terms are restricted or not. For the so-called *nontrivial* signatures, the weights are not restricted. When we present the orientability algorithm for the nontrivial signatures, we will use the fact that terms of sufficiently large weight always exist. For the trivial signatures the algorithm is presented in the full version of this paper [10].

A signature Σ is called *trivial* if it contains no function symbol of arity ≥ 2 and at most one unary function symbol. Note that Σ is nontrivial if and only if it contains either a function symbol of arity ≥ 2 or at least two function symbols of arity 1. The proof of the following lemma can be found in [10].

Lemma 8. *Let Σ be a nontrivial signature and w be a weight function for Σ. Then for every integer m there exists a ground term of the signature Σ such that $|t| > m$.*

5 An Algorithm for Orientability in the Case of Nontrivial Signatures

In this section we only consider nontrivial signatures. An algorithm for trivial signatures is given in [10]. The algorithm given in this section will be illustrated below in Section 5.5 on rewrite rule of Example 1.

Our algorithm works as follows. Given a system of rewrite rules R, we build the initial state $\mathbb{S}_R = (\mathbb{R}, \mathbb{M}, \mathbb{D}, \mathbb{U}, \mathbb{G}, \mathbb{L}, \gg)$. Then we transform $(\mathbb{R}, \mathbb{M}, \mathbb{D}, \mathbb{U}, \mathbb{G}, \mathbb{L}, \gg)$ repeatedly as described below. We call the *size* of the state the total number of occurrences of function symbols and variables in $\mathbb{R}$. Every transformation step will terminate with either success or failure, or else decrease the size of $\mathbb{R}$.

At each step we assume that $\mathbb{R}$ consists of k tuple inequalities

$$\begin{aligned} \langle l_1, L_1 \rangle &> \langle r_1, R_1 \rangle, \\ &\cdots \\ \langle l_k, L_k \rangle &> \langle r_k, R_k \rangle, \end{aligned} \tag{2}$$

such that all of the L_i, R_i are tuples of terms. We denote by $\mathbb{S}^{-i}$ the state obtained from $\mathbb{S}$ by removal of the ith tuple inequality $\langle l_i, L_i \rangle > \langle r_i, R_i \rangle$ from $\mathbb{R}$.

We will label parts of the algorithm, these labels will be used in the proof of its soundness. The algorithm can make a nondeterministic choice, but at most once, and the number of nondeterministic branches is bounded by the number of constants in Σ.

When the set $\mathbb{D}$ of linear inequalities changes, we assume that we check the new set for satisfiability, and terminate with failure if it is unsatisfiable. Likewise, when we change $\gg$, we check if it can be extended to an order and terminate with failure if it cannot.

5.1 The Algorithm

The algorithm works as follows. Every step consists of a number of state transformations, beginning with PREPROCESS defined below. During the algorithm, we will perform two kinds of *consistency checks*:

- The *consistency check on* $\mathbb{D}$ is the check if $\mathbb{D}$ has a solution. If it does not, we terminate with failure.
- The *consistency check on* $\gg$ is the check if $\gg$ can be extended to an order, i.e., the transitive closure $\gg'$ of $\gg$ is irreflexive, i.e., for no $g \in \Sigma$ we have $g \gg' g$. If $\gg$ cannot be extended to an order, we terminate with failure.

It is not hard to argue that both kinds of consistency checks can be performed in polynomial time. The consistency check on $\mathbb{D}$ is polynomial by Lemma 6. The consistency check on $\gg$ is polynomial since the transitive closure of a binary relation can be computed in polynomial time.

PREPROCESS. Do the following transformations while possible. If any tuple inequality in $\mathbb{R}$ has length 0, remove it from $\mathbb{R}$. Otherwise, if $\mathbb{R}$ contains a tuple inequality $\langle l_1, \ldots, l_n \rangle > \langle l_1, \ldots, l_n \rangle$, terminate with failure. Otherwise, if $\mathbb{R}$ contains a tuple inequality $\langle l, l_1, \ldots, l_n \rangle > \langle l, r_1, \ldots, r_n \rangle$, replace it by $\langle l_1, \ldots, l_n \rangle > \langle r_1, \ldots, r_n \rangle$.

If $\mathbb{R}$ becomes empty, proceed to TERMINATE, otherwise continue with MAIN.

MAIN. Now we can assume that in (2) each l_i is a term different from the corresponding term r_i. For every variable x and term t denote by $n(x,t)$ the number of occurrences of x in t. For example, $n(x, g(x, h(y, x))) = 2$. Likewise, for every function symbol $g \in \Sigma$ and term t denote by $n(g,t)$ the number of occurrences of g in t. For example, $n(h, g(x, h(y, x))) = 1$.

(M1) For all x and i such that $n(x, l_i) > n(x, r_i)$, add x to $\mathbb{M}$.

(M2) If for some i there exists a variable $x \notin \mathbb{M}$ such that $n(x, l_i) < n(x, r_i)$, then terminate with failure.

For every pair of terms l, r, denote by $W(l,r)$ the linear inequality obtained as follows. Let v_l and v_r be the numbers of occurrences of variables in l and r respectively. Then

$$W(l,r) = \sum_{g \in \Sigma} (n(g,l) - n(g,r)) w_g + (v_l - v_r) w_e \geq 0. \tag{3}$$

For example, if $l = h(x, f(y))$ and $r = f(g(x, g(x, y)))$, then

$$W(l, r) = w_h - 2 \cdot w_g - w_e \geq 0.$$

(M3) Add to $\mathbb{D}$ all the linear inequalities $W(l_i, r_i)$ for all i and perform the consistency check on $\mathbb{D}$.

Now compute $\mathbb{D}^=$. If $\mathbb{D}^=$ contains none of the inequalities $W(l_i, r_i)$, proceed to TERMINATE. Otherwise, for all i such that $W(l_i, r_i) \in \mathbb{D}^=$ apply the applicable case below, depending on the form of l_i and r_i.

(M4) If (l_i, r_i) has the form $(g(s_1, \ldots, s_n), h(t_1, \ldots, t_p))$, where g is different from h, then extend $\gg$ by adding $g \gg h$ and remove the tuple inequality $\langle l_i, L_i \rangle > \langle r_i, R_i \rangle$ from $\mathbb{R}$. Perform the consistency check on $\gg$.

(M5) If (l_i, r_i) has the form $(g(s_1, \ldots, s_n), g(t_1, \ldots, t_n))$, then replace $\langle l_i, L_i \rangle > \langle r_i, R_i \rangle$ by $\langle s_1, \ldots, s_n, L_i \rangle > \langle t_1, \ldots, t_n, R_i \rangle$.

(M6) If (l_i, r_i) has the form (x, y), where x and y are different variables, do the following. If L_i is empty, then terminate with failure. Otherwise, set $\mathbb{U}$ to *one*, add x to $\mathbb{M}$ and replace $\langle l_i, L_i \rangle > \langle r_i, R_i \rangle$ by $\langle L_i \rangle > \langle R_i \rangle$.

(M7) If (l_i, r_i) has the form (x, t), where t is not a variable, do the following. If t is not a constant, or L_i is empty, then terminate with failure. So assume that t is a constant c. If $\mathbb{L} = \{d\}$ for some d different from c, then terminate with failure. Otherwise, set $\mathbb{L}$ to $\{c\}$. Replace in L_i and R_i the variable x by c, obtaining L'_i and R'_i respectively, and then replace $\langle l_i, L_i \rangle > \langle r_i, R_i \rangle$ by $\langle L'_i \rangle > \langle R'_i \rangle$.

(M8) If (l_i, r_i) has the form (t, x), where t is not a variable, do the following. If t contains x, remove $\langle l_i, L_i \rangle > \langle r_i, R_i \rangle$ from $\mathbb{R}$. Otherwise, if t is a nonconstant or L_i is empty, terminate with failure. Let now t be a constant c. If $\mathbb{G} = \{d\}$ for some d different from c, then terminate with failure. Otherwise, set $\mathbb{G}$ to $\{c\}$. Replace in L_i and R_i the variable x by c, obtaining L'_i and R'_i respectively, and then replace $\langle l_i, L_i \rangle > \langle r_i, R_i \rangle$ by $\langle L'_i \rangle > \langle R'_i \rangle$.

After this step repeat PREPROCESS.

TERMINATE. Let $(\mathbb{R}, \mathbb{M}, \mathbb{D}, \mathbb{U}, \mathbb{G}, \mathbb{L}, \gg)$ be the current state. Do the following.

(T1) If $d \in \mathbb{G}$, then for all constants c different from d such that $w_c - w_e \geq 0$ belongs to $\mathbb{D}^=$ extend $\gg$ by adding $d \gg c$. Likewise, if $c \in \mathbb{L}$, then for all constants d different from c such that $w_d - w_e \geq 0 \in \mathbb{D}^=$ extend $\gg$ by adding $d \gg c$. Perform the consistency check on $\gg$.

(T2) For all f in Σ do the following. If f is a unary function symbol and $w_f \geq 0$ belongs to $\mathbb{D}^=$, then extend $\gg$ by adding $f \gg h$ for all $h \in \Sigma - \{f\}$. Perform the consistency check on $\gg$. If $\mathbb{U} = $ *one* or $\mathbb{G} \neq \emptyset$, then terminate with failure.

(T3) If there exists no constant c such that $w_c - w_e \geq 0$ is in $\mathbb{D}^=$, nondeterministically choose a constant $c \in \Sigma$, add $w_e - w_c \geq 0$ to $\mathbb{D}$, perform the consistency check on $\mathbb{D}$ and repeat PREPROCESS.

(T4) If $\mathbb{U} = $ *one*, then terminate with failure if there exists more than one constant c such that $w_c - w_e \geq 0$ belongs to $\mathbb{D}^=$.

(T5) Terminate with success.

We will show how to build a solution at step (T5) below in Lemma 21.

5.2 Correctness

In this section we prove correctness of the algorithm. In Section 5.3 we show how to find a solution when the algorithm terminates with success. The correctness will follow from a series of lemmas asserting that the transformation steps performed by the algorithm preserve the set of solutions. We will use notation and terminology of the algorithm. We say that a step of the algorithm is *correct* if the set of solutions to the state before this step coincides with the set of solutions after the step. When we prove correctness of a particular step, we will always denote by $\mathbb{S} = (\mathbb{R}, \mathbb{M}, \mathbb{D}, \mathbb{U}, \mathbb{G}, \mathbb{L}, \gg)$ the state before this step, and by $\mathbb{S}' = (\mathbb{R}', \mathbb{M}', \mathbb{D}', \mathbb{U}', \mathbb{G}', \mathbb{L}', \gg')$ the state after this step. When we use substitutions in the proof, we always assume that the substitutions are grounding for the relevant terms.

The following two lemmas can be proved by a straightforward application of the definition of solution to a state.

Lemma 9 (consistency check). *If consistency check on $\mathbb{D}$ or on $\gg$ terminates with failure, then $\mathbb{S}$ has no solution.*

Lemma 10. *Step* PREPROCESS *is correct.*

Let us now analyze MAIN. For every weight function w and precedence relation $\gg$ compatible with w we call a *counterexample* to $\langle l_i, L_i\rangle > \langle r_i, R_i\rangle$ w.r.t. $(w, \gg)$ any substitution σ minimal for $\mathbb{M}$ such that $\langle r_i\sigma, R_i\sigma\rangle \succeq \langle l_i\sigma, L_i\sigma\rangle$ for the order $\succ$ induced by $(w, \gg)$.

The following lemma follows immediately from the definition of solution.

Lemma 11 (counterexample). *If for every solution $(w, \gg)$ to $\mathbb{S}^{-i}$ there exists a counterexample to $\langle l_i, L_i\rangle > \langle r_i, R_i\rangle$ w.r.t. $(w, \gg)$, then $\mathbb{S}$ has no solution. If for every solution $(w, \gg)$ to $\mathbb{S}^{-i}$ there exists no counterexample to the tuple inequality $\langle l_i, L_i\rangle > \langle r_i, R_i\rangle$, then removing this tuple inequality from $\mathbb{R}$ does not change the set of solutions to $\mathbb{S}$.*

This lemma means that we can change $\langle l_i, L_i\rangle > \langle r_i, R_i\rangle$ into a different tuple inequality or change $\mathbb{M}$, if we can prove that this change does not influence the existence of a counterexample.

Let σ be a substitution, x a variable and t a term. We will denote by σ_x^t the substitution defined by

$$\sigma_x^t(y) = \begin{cases} \sigma(y), & \text{if } y \neq x, \\ t, & \text{if } y = x. \end{cases}$$

Lemma 12. *Let for some x and i we have $n(x, l_i) > n(x, r_i)$ and there exists a counterexample σ to $\langle l_i, L_i\rangle > \langle r_i, R_i\rangle$ w.r.t. $(w, \gg)$. Then there exists a counterexample σ' to $\langle l_i, L_i\rangle > \langle r_i, R_i\rangle$ w.r.t. $(w, \gg)$ minimal for $\{x\}$.*

PROOF. Suppose that σ is not minimal for $\{x\}$. Denote by c a minimal constant w.r.t. w and by t the term $x\sigma$. Since σ is not minimal for x, we have $|t| > |c|$. Consider the substitution σ_x^c. Since σ is a counterexample, we have $|r_i\sigma| \geq |l_i\sigma|$. We have

$$|l_i\sigma_x^c| = |l_i\sigma| - n(x, l_i) \cdot (|t| - |c|);$$
$$|r_i\sigma_x^c| = |r_i\sigma| - n(x, r_i) \cdot (|t| - |c|).$$

Then

$$\begin{aligned}|r_i\sigma_x^c| = |r_i\sigma| - n(x, r_i) \cdot (|t| - |c|) &\geq |l_i\sigma| - n(x, r_i) \cdot (|t| - |c|)\\ > |l_i\sigma| - n(x, l_i) \cdot (|t| - |c|) &= |l_i\sigma_x^c|.\end{aligned}$$

Therefore, $|r_i\sigma_x^c| > |l_i\sigma_x^c|$, and so σ_x^c is a counterexample too.

One can immediately see that this lemma implies correctness of step (M1).

Lemma 13. *Step (M1) is correct.*

PROOF. Evidently, every solution to $\mathbb{S}$ is also a solution to $\mathbb{S}'$. But by Lemma 12, every counterexample to $\mathbb{S}$ can be turned into a counterexample to $\mathbb{S}'$, so every solution to $\mathbb{S}'$ is also a solution to $\mathbb{S}$.

Let us now turn to step (M2).

Lemma 14 (M2). *If for some i and $x \notin \mathbb{M}$ we have $n(x, l_i) < n(x, r_i)$, then $\mathbb{S}$ has no solution. Therefore, step (M2) is correct.*

PROOF. We show that for every $(w, \gg)$ there exists a counterexample to $\langle l_i, L_i\rangle > \langle r_i, R_i\rangle$ w.r.t. $(w, \gg)$. Let σ be any substitution grounding for this tuple inequality. Take any term t and consider the substitution σ_x^t. We have

$$|r_i\sigma_x^t| - |l_i\sigma_x^t| = |r_i\sigma| - |l_i\sigma| + (n(x, r_i) - n(x, l_i)) \cdot (|t| - |x\sigma|).$$

By Lemma 8 there exist terms of an arbitrarily large weight, so for a term t of a large enough weight we have $|r_i\sigma_x^t| > |l_i\sigma_x^t|$, and so σ_x^t is a counterexample to $\langle l_i, L_i\rangle > \langle r_i, R_i\rangle$.

Correctness of (M2) is straightforward.

Note that after step (M2) for all i and $x \notin \mathbb{M}$ we have $n(x, l_i) = n(x, r_i)$.

Denote by Θ_c the substitution such that $\Theta_c(x) = c$ for every variable x.

Lemma 15 (M3). *Let for all i and $x \notin \mathbb{M}$ we have $n(x, l_i) = n(x, r_i)$. Every solution $(w, \gg)$ to $\mathbb{S}$ is also a solution to $W(l_i, r_i)$. Therefore, step (M3) is correct.*

PROOF. Let c be a constant of the minimal weight. Consider the substitution Θ_c. Note that this substitution is minimal for $\mathbb{M}$. It follows from the definition of W that $(w, \gg)$ is a solution to $W(l_i, r_i)$ if and only if $|l_i\Theta_c| \geq |r_i\Theta_c|$. But $|l_i\Theta_c| \geq |r_i\Theta_c|$ is a straightforward consequence of the definition of solutions to tuple inequalities.

Correctness of (M3) is straightforward.

Lemma 16. *Let for all $x \notin \mathbb{M}$ we have $n(x, l_i) = n(x, r_i)$. Let also $W(l_i, r_i) \in \mathbb{D}^=$. Then for every solution to $\mathbb{S}^{-i}$ and every substitution σ minimal for $\mathbb{M}$ we have $|l_i\sigma| = |r_i\sigma|$.*

PROOF. Using the fact that for all $x \notin \mathbb{M}$ we have $n(x, l_i) = n(x, r_i)$, it is not hard to argue that $|l_i\sigma| - |r_i\sigma|$ does not depend on σ, whenever σ is minimal for $\mathbb{M}$.

It follows from the definition of W that if $W(l_i, r_i) \in \mathbb{D}^=$, then for every solution to $\mathbb{D}$ (and so for every solution to $\mathbb{S}^{-i}$) we have $|l_i\Theta_c| = |r_i\Theta_c|$. Therefore, $|l_i\sigma| = |r_i\sigma|$ for all substitutions σ minimal for $\mathbb{M}$.

The proof of correctness of steps (M4)–(M8) will use this lemma in the following way. A pair $(w, \gg)$ is a solution to $\mathbb{S}$ if and only if it is a solution to $\mathbb{S}^{-i}$ and a solution to $\langle l_i, L_i\rangle > \langle r_i, R_i\rangle$. Equivalently, $(w, \gg)$ it is a solution to $\mathbb{S}$ if and only if it is a solution to $\mathbb{S}^{-i}$ and for every substitution σ minimal for $\mathbb{M}$ we have $\langle l_i\sigma, L_i\sigma\rangle \succ \langle r_i\sigma, R_i\sigma\rangle$. But by Lemma 16 we have $|l_i\sigma| = |r_i\sigma|$, so $\langle l_i\sigma, L_i\sigma\rangle \succ \langle r_i\sigma, R_i\sigma\rangle$ must be satisfied by either condition 2 or condition 3 of the definition of the Knuth-Bendix order.

This consideration can be summarized as follows.

Lemma 17. *Let for all $x \notin \mathbb{M}$ we have $n(x, l_i) = n(x, r_i)$. Let also $W(l_i, r_i) \in \mathbb{D}^=$. Then $(w, \gg)$ is a solution to $\mathbb{S}$ if and only if it is a solution to $\mathbb{S}^{-i}$ and for every substitution σ minimal for $\mathbb{M}$ the following holds. Let $l_i\sigma = g(t_1, \ldots, t_n)$ and $r_i\sigma = h(s_1, \ldots, s_n)$. Then at least one of the following conditions holds*

1. *$l_i\sigma = r_i\sigma$ and $L_i\sigma \succ R_i\sigma$; or*
2. *$g \gg h$; or*
3. *$g = h$ and for some $1 \leq i \leq n$ we have $t_1\sigma = s_1\sigma, \ldots, t_{i-1}\sigma = s_{i-1}\sigma$ and $t_i\sigma \succ s_i\sigma$.*

Lemma 18. *Steps (M4)–(M8) are correct.*

PROOF. (M4) We know that $l_i = g(s_1, \ldots, s_n)$ and $r_i = h(t_1, \ldots, t_p)$. Take any substitution σ minimal for $\mathbb{M}$. Obviously, $l_i\sigma = r_i\sigma$ is impossible, so $\langle l_i, L_i\rangle\sigma \succ \langle r_i, R_i\rangle\sigma$ if and only if $l_i\sigma \succ r_i\sigma$. By Lemma 17 this holds if and only if $g \gg h$, so step (M4) is correct.

(M5) We know that $l_i = g(s_1, \ldots, s_n)$ and $r_i = g(t_1, \ldots, t_n)$. Note that due to PREPROCESS, $l_i \neq r_i$, so $n \geq 1$. It follows from Lemma 17 that $\langle l_i, L_i\rangle\sigma \succ \langle r_i, R_i\rangle\sigma$ if and only if $\langle s_1, \ldots, s_n, L_i\rangle\sigma \succ \langle t_1, \ldots, t_n, R_i\rangle\sigma$, so step (M5) is correct.

(M6) We know that $l_i = x$ and $r_i = y$, where x, y are different variables. Note that if L_i is empty, then the substitution Θ_c, where c is of the minimal weight, is a counterexample to $\langle x, L_i\rangle > \langle y, R_i\rangle$. So assume that L_i is nonempty and consider two cases.

1. If there exists at least two terms s, t of the minimal weight, then there exists a counterexample to $\langle x, L_i\rangle > \langle y, R_i\rangle$. Indeed, if $s \succ t$, then $y\sigma \succ x\sigma$ for every σ such that $\sigma(x) = t$ and $\sigma(y) = s$.

2. If there exists exactly one term t of the minimal weight, then $x\sigma = y\sigma$ for every σ minimal for $\mathbb{M}$. Therefore, $\langle x, L_i \rangle > \langle y, R_i \rangle$ is equivalent to $\langle L_i \rangle > \langle R_i \rangle$.

In either case it is not hard to argue that step (M6) is correct.

(M7) We know that $l_i = x$ and $r_i = t$. Let c be the least constant in the signature. If $t \neq c$, then Θ_c is obviously a counterexample to $\langle x, L_i \rangle > \langle t, R_i \rangle$. Otherwise $t = c$, then for every counterexample σ we have $\sigma(x) = c$. In either case it is not hard to argue that step (M7) is correct.

(M8) We know that $l_i = t$ and $r_i = x$. Note that $t \neq x$ due to the PREPROCESS step, so if x occurs in t we have $t\sigma \succ x\sigma$ for all σ. Assume now that x does not occur in t. Then $x \in \mathbb{M}$. Consider two cases.

1. t is a nonconstant. For every substitution σ minimal for $\mathbb{M}$ we have $|t\sigma| = |x\sigma|$, hence $t\sigma$ is a nonconstant term of the minimal weight. This implies that the signature contains a unary function symbol f of the weight 0. Take any substitution σ. It is not hard to argue that $\sigma_x^{f(t)\sigma}$ is a counterexample to $\langle t, L_i \rangle > \langle x, R_i \rangle$.
2. t is a constant c. Let d be the greatest constant in the signature among the constants of the minimal weight. If $d \neq c$, then Θ_d is obviously a counterexample to $\langle c, L_i \rangle > \langle x, R_i \rangle$. Otherwise $d = c$, then for every counterexample σ we have $\sigma(x) = c$.

In either case it is not hard to argue that step (M8) is correct.

Let us now analyze steps TERMINATE. Note that for every constant c the inequality $w_c - w_e \geq 0$ belongs to $\mathbb{D}$ and for every function symbol g the inequality $w_g \geq 0$ belongs to $\mathbb{D}$ too.

Lemma 19. *Steps (T1)–(T2) are correct.*

PROOF. (T1) Suppose $d \in \mathbb{G}$, $c \neq d$, and $w_c - w_e \geq 0$ belongs to $\mathbb{D}^=$. Then for every solution to $\mathbb{S}$ we have $w(c) = w(e)$, and therefore c is a constant of the minimal weight. But since for every solution d is the greatest constant among those having the minimal weight, we must have $d \gg c$. The case $c \in \mathbb{L}$ is similar.

(T2) If f is a unary function symbol and $w_f \geq 0$ belongs to $\mathbb{D}^=$, then for every solution $w(f) = 0$. By the definition of the Knuth-Bendix order we must have $f \gg' g$ for all $g \in \Sigma - \{f\}$. But then (i) there exists an infinite number of terms of the minimal weight and (ii) a constant $d \in \mathbb{G}$ cannot be the greatest term of the minimal weight (since for example $f(d) \succ d$ and $|f(d)| = |d|$).

Step (T3) makes a nondeterministic choice, which can replace one state by several states $\mathbb{S}_1, \ldots, \mathbb{S}_n$. We say that such a step is correct if the set of solutions to $\mathbb{S}$ is the union of the sets of solutions to $\mathbb{S}_1, \ldots, \mathbb{S}_n$.

Lemma 20. *Steps (T3)–(T4) are correct.*

PROOF. (T3) Note that w is a solution to $w_e - w_c \geq 0$ if and only if $w(c)$ is the minimal weight, so addition of $w_e - w_c \geq 0$ to $\mathbb{D}$ amounts to stating that c has the minimal weight. Evidently, for every solution, there must be a constant c of the minimal weight, so the step is correct.

(T4) Suppose $\mathbb{U} = one$, then for every solution there exists a unique term of the minimal weight. If, for a constant c, $w_c - w_e \geq 0$ belongs to $\mathbb{D}^=$, then c must be a term of the minimal weight. Therefore, there cannot be more than one such a constant c.

5.3 Extracting a Solution

In this section we will show how to find a solution when the algorithm terminates with success.

Lemma 21. *Step (T5) is correct.*

PROOF. To prove correctness of (T5) we have to show the existence of solution. In fact, we will show how to build a particular solution.

Note that when we terminate at step (T5), the system $\mathbb{D}$ is solvable, since it was solvable initially and we performed consistency checks on every change of $\mathbb{D}$.

By Lemma 4 there exists an integer solution w to $\mathbb{D}$ which is also a solution to the strict versions of every inequality in $\mathbb{D} - \mathbb{D}^=$. Likewise, there exists a linear order $\gg'$ extending $\gg$, since we performed consistency checks on every change of $\gg$. We claim that $(w, \gg')$ is a solution to $(\mathbb{R}, \mathbb{M}, \mathbb{D}, \mathbb{U}, \mathbb{G}, \mathbb{L}, \gg)$. To this end we have to show that w is weight function, $\gg'$ is compatible with w and all items 1–5 of the definition of solution are satisfied.

Let us first show that w is a weight function. Note that $\mathbb{D}$ contains all inequalities $w_g \geq 0$, where $g \in \Sigma$ is a nonconstant, the inequality $w_e > 0$ and the inequalities $w_d - w_e \geq 0$ for all constant $d \in \Sigma$. So to show that w is a weight function it remains to show that at most one unary function symbol f has weight 0. Indeed, if there were two such function symbols f_1 and f_2, then at step (T2) we would add both $f_1 \gg f_2$ and $f_2 \gg f_1$, but the following consistency check on $\gg$ would fail.

The proof that $\gg$ is compatible with w is similar.

Denote by $\succ$ the Knuth-Bendix order induced by $(w, \gg')$.

1. *The weight function w solves every inequality in $\mathbb{D}$.* This follows immediately from our construction, if we show that $w(e)$ is the minimal weight. Let us show that w_e is the minimal weight. Indeed, since $\mathbb{D}$ initially contains the inequalities $w_c - w_e \geq 0$ for all constants c, we have that $w(e)$ is less than or equal to the minimal weight. By step (T3), there exists a constant c such that $w_c - w_e \geq 0$ is in $\mathbb{D}^=$, hence $w(c) = w(e)$, and so $w(e)$ is greater than or equal to the minimal weight. Evidently, $w(e)$ cannot be greater than the minimal weight, since $\mathbb{D}$ contains the inequalities $w_c - w_e \geq 0$ for all constants c.
2. *If $\mathbb{U} = one$, then there exists exactly one term of the minimal weight.* Assume $\mathbb{U} = one$. We have to show that (i) there exists no unary function symbol f of the minimal weight and (ii) there exists exactly one constant of the minimal weight. Let f be a unary function symbol. By our construction, $w_f \geq 0$ belongs to $\mathbb{D}$. By step (T2) $w_f \geq 0$ does not belong to $\mathbb{D}^=$, so by the

definition of w we have $w(f) > 0$. By our construction, $w_c - w_e \geq 0$ belongs to $\mathbb{D}$ for every constant c. By step (T4), at most one of such inequalities belongs to $\mathbb{D}^=$. Note that if $w_c - w_e \geq 0$ does not belong to $\mathbb{D}^=$, then $w(c) - w(e) > 0$ by the construction of w. Therefore, there exists at most one constant of the minimal weight.

3. *If $d \in \mathbb{G}$ (respectively $d \in \mathbb{L}$) for some constant d, then d is the greatest (respectively least) term among the terms of the minimal weight.* We consider the case $d \in \mathbb{G}$, the case $d \in \mathbb{L}$ is similar. Note that by step (T2) there is no unary function symbol f such that $w_f \geq 0$ belongs to $\mathbb{D}^=$, therefore $w(f) > 0$ for all unary function symbols f. This implies that only constants may have the minimal weight. But by step (T1) for all constants c of the minimal weight we have $d \gg c$, and hence also $d \gg' c$.
4. *For every tuple inequality $\langle l_i, L_i \rangle > \langle r_i, R_i \rangle$ in $\mathbb{R}$ and every substitution σ minimal for $\mathbb{M}$ we have $\langle l_i\sigma, L_i\sigma \rangle \succ \langle r_i\sigma, R_i\sigma \rangle$.* In the proof we will use the fact that $w(e)$ is the minimal weight.
By our construction (step M3), the inequality $W(l_i, r_i)$ does not belong to $\mathbb{D}^=$ (otherwise $\langle l_i, L_i \rangle > \langle r_i, R_i \rangle$ would be removed at one of steps (M4)–(M8)). In Lemma 16 we proved that $|l_i\sigma| - |r_i\sigma|$ does not depend on σ, whenever σ is minimal for $\mathbb{M}$.
It follows from the definition of W that if $W(l_i, r_i) \in \mathbb{D} - \mathbb{D}^=$, then $|l_i\Theta_c| > |r_i\Theta_c|$. Therefore, $|l_i\sigma| > |r_i\sigma|$ for all substitutions σ minimal for $\mathbb{M}$.
5. *$\gg'$ extends $\gg$.* This follows immediately from our construction.

5.4 Time Complexity

Provided that we use a polynomial-time algorithm for solving homogeneous linear inequalities, and a polynomial-time algorithm for transitive closure, we can prove the following lemma (for a proof see [10]).

Lemma 22. *The algorithm runs in time polynomial of the size of the system of rewrite rules.*

5.5 A Simple Example

Let us consider how the algorithm works on the rewrite rule $g(x, a, b) \to g(b, b, a)$ of Example 1. Initially, $\mathbb{R}$ consists of one tuple inequality

$$\langle g(x, a, b) \rangle > \langle g(b, b, a) \rangle \tag{4}$$

and $\mathbb{D}$ consists of the following linear inequalities:

$$w_g \geq 0,\ w_e > 0,\ w_a - w_e \geq 0,\ w_b - w_e \geq 0.$$

At step (M1) we note that $n(x, g(x, a, b)) = 1 > n(x, g(b, b, a)) = 0$. Therefore, we add x to $\mathbb{M}$.

At step (M3) we add the linear inequality $w_e - w_b \geq 0$ to $\mathbb{D}$ obtaining

$$w_g \geq 0,\ w_e > 0,\ w_a - w_e \geq 0,\ w_b - w_e \geq 0,\ w_e - w_b \geq 0.$$

Now we compute $\mathbb{D}^=$. It consists of two equations $w_b - w_e \geq 0$ and $w_e - w_b \geq 0$, so we have to apply one of the steps (M4)–(M8), in this case the applicable step is (M5). We replace (4) by

$$\langle x, a, b \rangle > \langle b, b, a \rangle. \tag{5}$$

At the next iteration of step (M3) we should add to $\mathbb{D}$ the linear inequality $w_e - w_b \geq 0$, but this linear inequality is already a member of $\mathbb{D}$, and moreover a member of $\mathbb{D}^=$. So we proceed to step (M7). At this step we set $\mathbb{L} = \{b\}$ and replace (5) by

$$\langle a, b \rangle > \langle b, a \rangle. \tag{6}$$

Then at step (M2) we add $w_a - w_b \geq 0$ to $\mathbb{D}$ obtaining

$$w_g \geq 0, w_e > 0, w_a - w_e \geq 0, w_b - w_e \geq 0, w_e - w_b \geq 0, w_a - w_b \geq 0.$$

Now $w_a - w_b \geq 0$ does not belong to the degenerate subsystem of $\mathbb{D}$, so we proceed to TERMINATE. Steps (T1)–(T4) change neither $\mathbb{D}$ nor $\gg$, so we terminate with success.

Solutions extracted according to Lemma 21 will be any pairs $(w, \gg)$ such that $w(a) > w(b)$. Note that these are *not all* solutions. There are also solutions such that $w(a) = w(b)$ and $a \gg b$. However, if we try to find all solutions we cannot any more guarantee that the algorithm runs in polynomial time.

6 Main Results

Lemmas 9–21 guarantee that our algorithm is correct. Lemma 22 implies the algorithm runs in polynomial time. Hence we obtain the following theorem.

Theorem 1. *The problem of the existence of a KBO which orients a given rewrite rule systems can be solved in polynomial time.*

In [10], using a similar technique, we prove the following theorem.

Theorem 2. *The problem of solving a Knuth-Bendix constraint consisting of a single inequality can be solved in polynomial time.*

The *real-valued Knuth-Bendix order* is defined similar in the same way as above, except that the range of the weight function is the set of nonnegative real numbers. The real-valued KBO was introduced in [13]. Note that in view of the results of Section 3 on systems of homogeneous linear inequalities (Lemmas 3 and 4) the algorithm is also sound and complete for the real-valued orders. Therefore, we have

Theorem 3. *If a rewrite rule system is orientable using a real-valued KBO, then it is also orientable using an integer-valued KBO.*

Acknowledgments. The authors are partially supported by grants from EPSRC.

References

1. F. Baader and T. Nipkow. *Term Rewriting and All That.* Cambridge University press, Cambridge, 1998.
2. H. Comon. Solving symbolic ordering constraints. *International Journal of Foundations of Computer Science*, 1(4):387–411, 1990.
3. H. Comon and R. Treinen. Ordering constraints on trees. In S. Tison, editor, *Trees in Algebra and Programming: CAAP'94*, volume 787 of *Lecture Notes in Computer Science*, pages 1–14. Springer Verlag, 1994.
4. N. Dershowitz and D.A. Plaisted. Rewriting. In A. Robinson and A. Voronkov, editors, *Handbook of Automated Reasoning*, volume I, chapter 9, pages 533–608. Elsevier Science, 2001.
5. D. Detlefs and R. Forgaard. A procedure for automatically proving the termination of a set of rewrite rules. In J.-P. Jouannaud, editor, *Rewriting Techniques and Applications, First International Conference, RTA-85*, volume 202 of *Lecture Notes in Computer Science*, pages 255–270, Dijon, France, 1985. Springer Verlag.
6. J. Dick, J. Kalmus, and U. Martin. Automating the Knuth-Bendix ordering. *Acta Informatica*, 28(2):95–119, 1990.
7. D. Knuth and P. Bendix. Simple word problems in universal algebras. In J. Leech, editor, *Computational Problems in Abstract Algebra*, pages 263–297. Pergamon Press, Oxford, 1970.
8. K. Korovin and A. Voronkov. A decision procedure for the existential theory of term algebras with the Knuth-Bendix ordering. In *Proc. 15th Annual IEEE Symp. on Logic in Computer Science*, pages 291–302, Santa Barbara, California, June 2000.
9. K. Korovin and A. Voronkov. Knuth-Bendix constraint solving is NP-complete. Preprint CSPP-8, Department of Computer Science, University of Manchester, November 2000.
10. K. Korovin and A. Voronkov. Verifying orientability of rewrite rules using the knuth-bendix order. Preprint CSPP-11, Department of Computer Science, University of Manchester, March 2001.
11. M.S. Krishnamoorthy and P. Narendran. On recursive path ordering. *Theoretical Computer Science*, 40:323–328, 1985.
12. P. Lescanne. Term rewriting systems and algebra. In R.E. Shostak, editor, *7th International Conference on Automated Deduction, CADE-7*, volume 170 of *Lecture Notes in Computer Science*, pages 166–174, 1984.
13. U. Martin. How to choose weights in the Knuth-Bendix ordering. In *Rewriting Techniques and Applications*, volume 256 of *Lecture Notes in Computer Science*, pages 42–53, 1987.
14. P. Narendran, M. Rusinowitch, and R. Verma. RPO constraint solving is in NP. In G. Gottlob, E. Grandjean, and K. Seyr, editors, *Computer Science Logic, 12th International Workshop, CSL'98*, volume 1584 of *Lecture Notes in Computer Science*, pages 385–398. Springer Verlag, 1999.
15. R. Nieuwenhuis. Simple LPO constraint solving methods. *Information Processing Letters*, 47:65–69, 1993.
16. A. Schrijver. *Theory of Linear and Integer Programming.* John Wiley and Sons, 1998.

Relating Accumulative and Non-accumulative Functional Programs

Armin Kühnemann[1]*, Robert Glück[2]**, and Kazuhiko Kakehi[2]***

[1] Institute for Theoretical Computer Science, Department of Computer Science, Dresden University of Technology, D–01062 Dresden, Germany
kuehne@orchid.inf.tu-dresden.de

[2] Institute for Software Production Technology, Waseda University, 3-4-1 Okubo, Shinjuku, Tokyo 169-8555, Japan
glueck@acm.org and kaz@futamura.info.waseda.ac.jp

Abstract. We study the problem to transform functional programs, which intensively use append functions (like inefficient list reversal), into programs, which use accumulating parameters instead (like efficient list reversal). We give an (automatic) transformation algorithm for our problem and identify a class of functional programs, namely restricted 2-modular tree transducers, to which it can be applied. Moreover, since we get macro tree transducers as transformation result and since we also give the inverse transformation algorithm, we have a new characterization for the class of functions induced by macro tree transducers.

1 Introduction

Functional programming languages are very well suited to specify programs in a modular style, which simplifies the design and the verification of programs. Unfortunately, modular programs often have poor time- and space-complexity compared to other (sometimes less understandable) programs, which solve the same tasks. As running example we consider the following program p_{non} that contains functions *app* and *rev*, which append and reverse lists, respectively. For simplicity we only consider lists with elements A and B, where lists are represented by monadic trees. In particular, the empty list is represented by the symbol N.[1] The program p_{non} is a straightforward (but naive) solution for list reversal, since the definition of *rev* simply uses the function *app*.

$$\begin{aligned} rev\ (A\ x_1) &= app\ (rev\ x_1)\ (A\ N) & \quad app\ (A\ x_1)\ y_1 &= A\ (app\ x_1\ y_1)\\ rev\ (B\ x_1) &= app\ (rev\ x_1)\ (B\ N) & app\ (B\ x_1)\ y_1 &= B\ (app\ x_1\ y_1)\\ rev\ N &= N & app\ N\ y_1 &= y_1 \end{aligned}$$

* Research supported in part by the German Academic Exchange Service (DAAD).
** Research supported in part by the Japan Science and Technology Corporation (JST).
*** Research fellow of the Japan Society for the Promotion of Science (JSPS).

[1] Analogous functions for the usual list representation with (polymorphic) binary *Cons*-operator can be handled by our transformations just as well, but in order to describe such functions and transformations technically clean, an overhead to introduce types is necessary.

A. Middeldorp (Ed.): RTA 2001, LNCS 2051, pp. 154–168, 2001.

For list reversal, the program p_{non} has quadratic time-complexity in the length of an input list l, since it produces *intermediate results*, where the number and the maximum length of intermediate results depends on the length of l. Therefore we would prefer the following program p_{acc} with linear time-complexity in the length of an input list, which uses a binary auxiliary function *rev'*.

$$\begin{array}{ll} rev\ (A\ x_1) = rev'\ x_1\ (A\ N) & rev'\ (A\ x_1)\ y_1 = rev'\ x_1\ (A\ y_1) \\ rev\ (B\ x_1) = rev'\ x_1\ (B\ N) & rev'\ (B\ x_1)\ y_1 = rev'\ x_1\ (B\ y_1) \\ rev\ N \qquad = N & rev'\ N\ y_1 \qquad = y_1 \end{array}$$

Since p_{acc} reverses lists by accumulating their elements in the second argument of *rev'*, we call p_{acc} an *accumulative* program, whereas we call p_{non} *non-accumulative.* Already in [1] it is shown in the context of transforming programs into iterative form, how non-accumulative programs can be transformed (non-automatically) into their more efficient accumulative versions. An algorithm which removes append functions in many cases, e.g. also in p_{non}, was presented in [23]. In comparison to [23], our transformation technique is more general in three aspects (though so far we only have artificial examples to demonstrate these generalizations): we consider (i) arbitrary tree structures (instead of lists), (ii) functions defined by simultaneous recursion, and (iii) substitution functions on trees (instead of append) which may replace different designated symbols by different trees. On the other hand, our technique is restricted to unary functions (apart from substitutions), though also in [23] the only example program involving a non-unary function could not be optimized. Moreover, since we formally describe the two different program paradigms and since we also present a transformation of accumulative into non-accumulative programs, we obtain the equality of the classes of functions computed by the two paradigms.

Well-known techniques for eliminating intermediate results cannot improve p_{non}: (i) *deforestation* [24] and *supercompilation* [22,20] suffer from the phenomenon of the *obstructing function call* [2] and (ii) *shortcut deforestation* [14, 13] is hampered by the unknown number of intermediate results. In [3] an extension of shortcut deforestation was developed which is based on *type-inference* and splits function definitions into *workers* and *wrappers.* It successfully transforms p_{non} into p_{acc}, but is also less general in the above mentioned three aspects with respect to the transformation of non-accumulative into accumulative programs.

In [17,18] it was demonstrated that sometimes *composition* and *decomposition techniques* [8,11,6,12] for *attribute grammars* [16] and *tree transducers* [10] can help, when deforestation fails. For this purpose we have considered special functional programs as compositions of *macro tree transducers* (for short *mtts*) [5,4,6]. Every function f of an mtt is defined by a case analysis on the root symbol c of its first argument t. The right-hand side of the equation for f and c may only contain (*extended*) *primitive-recursive function calls*, i.e. the first argument of a function call has to be a variable that refers to a subtree of t. Under certain restrictions, compositions of mtts can be transformed into a single mtt. The function *app* in p_{non} is an mtt, whereas the function *rev* in p_{non} is not an mtt (since it calls *app* with a first argument that differs from x_1), such that these techniques cannot be applied directly.

In this paper we consider p_{non} as 2-*modular tree transducer* (for short 2-*modtt*) [7], where it is allowed that a function in module 1 (here *rev*) calls a function in module 2 (here *app*) non-primitive-recursively. Additionally, the two modules of p_{non} fulfill sufficient conditions (*rev* and *app* are so called top-down tree transducer- and substitution-modules, respectively), such that we can apply a decomposition step and two subsequent composition steps to transform p_{non} into the (more efficient) mtt p_{acc}. Since these constructions (called *accumulation*) transform every 2-modtt that fulfills our restrictions into an mtt, and since we also present inverse constructions (called *deaccumulation*) to transform mtts into the same class of restricted 2-modtts, we get a nice characterization of mtts in terms of restricted 2-modtts.

Besides this introduction, the paper contains five further sections. In Section 2 we fix elementary notions and notations. Section 3 introduces our functional language and tree transducers. Section 4 and Section 5 present accumulation and deaccumulation, respectively. Finally, Section 6 contains future research topics.

2 Preliminaries

We denote the set of natural numbers including 0 by $\mathbb{N}$ and the set $\mathbb{N} - \{0\}$ by $\mathbb{N}_+$. For every $m \in \mathbb{N}$, the set $\{1, \ldots, m\}$ is denoted by $[m]$. The cardinality of a set K is denoted by $card(K)$. We will use the sets $X = \{x_1, x_2, x_3, \ldots\}$, $Y = \{y_1, y_2, y_3, \ldots\}$, and $Z = \{z\}$ of *variables*. For every $n \in \mathbb{N}$, let $X_n = \{x_1, \ldots, x_n\}$ and $Y_n = \{y_1, \ldots, y_n\}$. In particular, $X_0 = Y_0 = \emptyset$.

Let $\Rightarrow$ be a binary relation on a set K. Then, $\Rightarrow^*$ denotes the transitive, reflexive closure of $\Rightarrow$. If $k \Rightarrow^* k'$ for $k, k' \in K$ and if there is no $k'' \in K$ such that $k' \Rightarrow k''$, then k' is called a *normal form of* k *with respect to* $\Rightarrow$, which is denoted by $nf(\Rightarrow, k)$, if it exists and if it is unique.

A *ranked alphabet* is a pair $(S, rank)$ where S is a finite set and *rank* is a mapping which associates with every symbol $s \in S$ a natural number called the *rank* of s. We simply write S instead of $(S, rank)$ and assume *rank* as implicitly given. The set of elements of S with rank n is denoted by $S^{(n)}$. The set of *trees over* S, denoted by T_S, is the smallest subset $T \subseteq (S \cup \{(,)\})^*$ such that $S^{(0)} \subseteq T$ and for every $s \in S^{(n)}$ with $n \in \mathbb{N}_+$ and $t_1, \ldots, t_n \in T$: $(s\, t_1 \ldots t_n) \in T$.

3 Language

We consider a simple first-order, constructor-based functional programming language P as source and target language for our transformations. Every program $p \in P$ consists of several modules and every module consists of several function definitions. The functions of a module are defined by a complete case analysis on the first argument (*recursion argument*) via pattern matching, where only flat patterns are allowed. The other arguments are called *context arguments*. If in the right-hand side of a function definition there is a call of a function that is defined in the same module, then the first argument of this function call has to be a subtree of the first argument in the corresponding left-hand side.

For simplicity we choose a unique ranked alphabet C_p of constructors, which is used to build up input trees and output trees of every function in p. In example programs and program transformations we relax the completeness of function definitions on T_{C_p} by leaving out those equations, which are not intended to be used in evaluations. Sometimes this leads to small technical difficulties, but avoids the overhead to introduce types.

Definition 1 Let C and F be ranked alphabets of *constructors* and *function symbols* (for short *functions*), respectively, such that $F^{(0)} = \emptyset$ and X, Y, C, and F are pairwise disjoint. We define the classes P, M, D, and R of *programs, modules, function definitions*, and *right-hand sides*, respectively, by the following grammar and the subsequent restrictions. We assume that p, m, d, r, c, and f (also equipped with indices) range over the sets P, M, D, R, C, and F, respectively.

$$\begin{array}{lll} p ::= m_1 \dots m_l & & \text{(program)} \\ m ::= d_1 \dots d_h & & \text{(module)} \\ d ::= f\ (c_1\ x_1 \dots x_{k_1})\ y_1 \dots y_n = r_1 & & \text{(function definition)} \\ \qquad\qquad \vdots & & \\ \quad\ \ f\ (c_q\ x_1 \dots x_{k_q})\ y_1 \dots y_n = r_q & & \\ r ::= x_i \mid y_j \mid c\, r_1 \dots r_k \mid f\ r_0\ r_1 \dots r_n & & \text{(right-hand side)} \end{array}$$

The sets of constructors, functions, and modules that occur in $p \in P$ are denoted by C_p, F_p, and M_p respectively. The set of functions that is defined in $m \in M_p$ is denoted by F_m. For every $i, j \in [l]$ with $i \neq j$: $F_{m_i} \cap F_{m_j} = \emptyset$. For every $i \in [l]$ and $f \in F_{m_i}$, the module m_i contains exactly one function definition for f. For every $i \in [l]$, $f \in F_{m_i}^{(n+1)}$, and $c \in C_p^{(k)}$ there is exactly one equation of the form

$$f\ (c\ x_1 \dots x_k)\ y_1 \dots y_n = rhs(f, c)$$

with $rhs(f,c) \in RHS(F_{m_i}, C_p \cup (F_p - F_{m_i}), X_k, Y_n)$, where for every $F' \subseteq F$, $C' \subseteq C \cup F$, and $k, n \in I\!N$, $RHS(F', C', X_k, Y_n)$ is the smallest set RHS with:

- For every $f \in F'^{(a+1)}$, $i \in [k]$, and $r_1, \dots, r_a \in RHS$: $(f\ x_i\ r_1 \dots r_a) \in RHS$.
- For every $c \in C'^{(a)}$ and $r_1, \dots, r_a \in RHS$: $(c\, r_1 \dots r_a) \in RHS$.
- For every $j \in [n]$: $y_j \in RHS$. □

Example 2

- $p_{non} \in P$ where $M_{p_{non}}$ contains two modules $m_{non,rev}$ and $m_{non,app}$ containing the definitions of *rev* and *app*, respectively.
- $p_{acc} \in P$ where $M_{p_{acc}}$ contains one module $m_{acc,rev}$ containing the definitions of *rev* and *rev'*.
- Let p_{fre} be the program

$$\begin{array}{ll} rev\ (A\ x_1) = APP\ (rev\ x_1)\ (A\ N) & app\ (A\ x_1)\ y_1 = A\ (app\ x_1\ y_1) \\ rev\ (B\ x_1) = APP\ (rev\ x_1)\ (B\ N) & app\ (B\ x_1)\ y_1 = B\ (app\ x_1\ y_1) \\ rev\ N \qquad\ = N & app\ N\ y_1 \qquad = y_1 \end{array}$$

$$\begin{array}{ll} int\ (APP\ x_1\ x_2) & = app\ (int\ x_1)\ (int\ x_2) \\ int\ (A\ x_1) & = A\ (int\ x_1) \\ int\ (B\ x_1) & = B\ (int\ x_1) \\ int\ N & = N \end{array}$$

Then, $p_{fre} \in P$ where $M_{p_{fre}}$ contains three modules $m_{fre,rev}$, $m_{fre,app}$, and $m_{fre,int}$ containing the definitions of *rev*, *app*, and *int*, respectively. □

Now we will introduce a modular tree transducer as hierarchy $(m_1, \ldots, m_u)$ of modules, where the functions in a module m_j may only call functions defined in the modules $m_j, \ldots, m_u$. Moreover, we will define interpretation-modules which interpret designated constructors as function calls, and substitution-modules which substitute designated constructors by context arguments.

Definition 3 Let $p \in P$.

- A sequence $(m_1, \ldots, m_u)$ with $u \in \mathbb{N}_+$ and $m_1, \ldots, m_u \in M_p$ is called *u-modular tree transducer* (for short *u-modtt*), iff $F_{m_1}^{(1)} \neq \emptyset$ and for every $i \in [u]$, $f \in F_{m_i}^{(n+1)}$, and $c \in C_p^{(k)}$: $rhs(f,c) \in RHS(F_{m_i}, C_p \cup \bigcup_{i+1 \leq j \leq u} F_{m_j}, X_k, Y_n)$. We call F_{m_i} and $\bigcup_{i+1 \leq j \leq u} F_{m_j}$ the set of *internal* and *external functions* (cf. also [9]) *of* m_i, respectively.[2]
- A 1-modtt (m_1), abbreviated by m_1, is called *macro tree transducer* (for short *mtt*).
- An mtt m with $F_m = F_m^{(1)}$ is called *top-down tree transducer* (for short *tdtt*) [19,21].
- A module $m \in M_p$ is also called *mtt-module.*[3]
- An mtt-module $m \in M_p$ with $F_m = F_m^{(1)}$ is called *tdtt-module.*
- A tdtt-module $m \in M_p$ is called *interpretation-module* (for short *int-module*), iff $card(F_m) = 1$, and there is $C_m \subseteq C_p$ such that m contains for $int \in F_m$ and for every $k \in \mathbb{N}$, $c \in C_m^{(k)}$ and for some $f_c \in (F_p - F_m)^{(k)}$ the equation

$$int\ (c\, x_1 \ldots x_k) = f_c\ (int\ x_1) \ldots (int\ x_k)$$

  and for $int \in F_m$ and for every $k \in \mathbb{N}$ and $c \in (C_p - C_m)^{(k)}$ the equation

$$int\ (c\, x_1 \ldots x_k) = c\ (int\ x_1) \ldots (int\ x_k).$$

- An mtt-module $m \in M_p$ is called *substitution-module* (for short *sub-module*), iff there are $n \in \mathbb{N}$ and distinct $\pi_1, \ldots, \pi_n \in C_p^{(0)}$ such that $card(F_m^{(n+1)}) = 1$, $card(F_m^{(i+1)}) = 0$ for every $i > n$, $card(F_m^{(i+1)}) \leq 1$ for every $i < n$, and m contains for every $i \in \mathbb{N}$, $sub_i \in F_m^{(i+1)}$, and $j \in [i]$ the equation

$$sub_i\ \pi_j\ y_1 \ldots y_i = y_j$$

  and for every $i \in \mathbb{N}$, $sub_i \in F_m^{(i+1)}$, $k \in \mathbb{N}$, and $c \in (C_p - \{\pi_1, \ldots, \pi_i\})^{(k)}$ the equation

$$sub_i\ (c\, x_1 \ldots x_k)\ y_1 \ldots y_i = c\ (sub_i\ x_1\ y_1 \ldots y_i) \ldots (sub_i\ x_k\ y_1 \ldots y_i).$$ □

[2] Our definition of modtts differs slightly from that in [7], since it allows a variable x_i in a right-hand side only as first argument of an internal function. Arbitrary occurrences of x_i can be achieved by applying an identity function to x_i, which is an additional internal function. Further note that the assumption $F_{m_1}^{(1)} \neq \emptyset$ only simplifies our presentation, but could be avoided.

[3] A module m is not necessarily an mtt, since it may call external functions.

Example 4

- $(m_{non,rev}, m_{non,app})$ is a 2-modtt, $m_{non,rev}$ is a tdtt-module, and $m_{non,app}$ is a sub-module (where $n = 1$, $\pi_1 = N$, $F^{(2)}_{m_{non,app}} = \{app\}$, and $F^{(1)}_{m_{non,app}} = \emptyset$).
- $m_{acc,rev}$ is an mtt-module and a 1-modtt, thus also an mtt.
- $m_{fre,rev}$ is a tdtt, $(m_{fre,int}, m_{fre,app})$ is a 2-modtt, $m_{fre,int}$ is an int-module (where $C_{m_{fre,int}} = \{APP\}$ and $f_{APP} = app$), and $m_{fre,app}$ is a sub-module.

□

We fix call-by-name semantics, i.e. for every program $p \in P$ we use a call-by-name reduction relation $\Rightarrow_p$ on $T_{C_p \cup F_p}$. It can be proved in analogy to [7] that for every program $p \in P$, u-modtt $(m_1, \ldots, m_u)$ with $m_1, \ldots, m_u \in M_p$, $f \in F^{(1)}_{m_1}$, and $t \in T_{C_p}$ the normal form $nf(\Rightarrow_p, (f\ t))$ exists. The proof is based on the result, that for every modtt the corresponding (nondeterministic) reduction relation is terminating (and confluent). This result can also be extended to normal forms of expressions of the form $(f_n\ (f_{n-1} \cdots (f_2\ (f_1\ t)) \ldots))$, where every f_i is a unary function of the first module of a modtt in p.

In the framework of this paper we would like to optimize the evaluation of expressions of the form $(f\ t)$, where t is a tree over constructors. Since the particular constructor trees are not relevant for the transformations, we abstract them by a variable z, i.e. we handle expressions of the form $(f\ z)$. The transformations will also deliver expressions of the form $(f_2\ (f_1\ z))$. All these expressions are initial expressions for programs, which are defined as follows.

Definition 5 Let $p \in P$ and let f range over $\{f \mid$ there is a modtt $(m_1, \ldots, m_u)$ with $m_1, \ldots, m_u \in M_p$ such that $f \in F^{(1)}_{m_1}\}$. The set of *initial expressions for p*, denoted by E_p, is defined as follows, where e ranges over E_p:

$e ::= f\ e \mid z$ (initial expression for a program) □

Example 6

- z, $(rev\ z)$, and $(rev\ (rev\ z))$ are initial expressions for p_{non} and for p_{acc}.
- z, $(rev\ z)$, $(int\ z)$, and $(int\ (rev\ z))$ are initial expressions for p_{fre}. □

Definition 7 Let $p \in P$ and $e \in E_p$. The function $\tau_{p,e} : T_{C_p} \longrightarrow T_{C_p}$ defined by $\tau_{p,e}(t) = nf(\Rightarrow_p, e[z/t])$ for every $t \in T_{C_p}$, where $e[z/t]$ denotes the substitution of z in e by t, is called the *tree transformation induced by p and e*. The *class of tree transformations* induced by all programs $p \in P$ and initial expressions $(f_n\ (f_{n-1} \cdots (f_2\ (f_1\ z)) \ldots)) \in E_p$ with $n \geq 1$, such that $p = m_1 \ldots m_l$ with $m_1, \ldots, m_l \in M_p$, and for every $i \in [n]$ there is $j_i \in [l]$ with $f_i \in F_{m_{j_i}}$, is denoted by

$$M_{j_1}\ ;\ M_{j_2}\ ; \ldots ;\ M_{j_n}, \quad \text{where } M_{j_i} \text{ is}$$

- T, if m_{j_i} is a tdtt,
- MT, if m_{j_i} is an mtt, and
- u-$ModT(M'_1, \ldots, M'_u)$, if there is a u-modtt $(m'_1, \ldots, m'_u)$ with $m'_1, \ldots, m'_u \in M_p$ and $m_{j_i} = m'_1$, where M'_j is

- T, if m'_j is a tdtt-module,
- INT, if m'_j is an int-module, and
- SUB, if m'_j is a sub-module. □

Example 8

- $\tau_{p_{acc},(rev\ z)} \in MT$
- $\tau_{p_{acc},(rev\ (rev\ z))} \in MT\ ;\ MT$
- $\tau_{p_{non},(rev\ z)} \in 2\text{-}ModT(T, SUB)$
- $\tau_{p_{fre},(int\ (rev\ z))} \in T\ ;\ 2\text{-}ModT(INT, SUB)$ □

Our program transformations should preserve the semantics, i.e. should transform pairs in $\{(p, e) \mid p \in P, e \in E_p\}$ into equivalent pairs. Since some transformations will introduce new constructors, the following definition of equivalence can later be instantiated, such that it is restricted to input trees over "old" constructors.

Definition 9 For every $i \in [2]$ let $(p_i, e_i) \in \{(p, e) \mid p \in P, e \in E_p\}$. Let $C' \subseteq C_{p_1} \cap C_{p_2}$. The pairs (p_1, e_1) and (p_2, e_2) are called *equivalent with respect to* C', denoted by $(p_1, e_1) \equiv_{C'} (p_2, e_2)$, if for every $t \in T_{C'}$: $\tau_{p_1,e_1}(t) = \tau_{p_2,e_2}(t)$. □

4 Accumulation

We would like to give an algorithm based on results from the theory of tree transducers, which translates non-accumulative programs of the kind "inefficient reverse" into accumulative programs of the kind "efficient reverse". According to our definitions in the previous section this means to compose the two modules of a 2-modtt to an mtt. A result in [7] shows that in general this is not possible, since there are 2-modtts, such that their induced tree transformations can even not be realized by any finite composition of mtts. Fortunately, there are sufficient conditions for the two modules, under which a composition to an mtt is possible, namely, if they are a tdtt-module and a sub-module, respectively.

Theorem 10 $2\text{-}ModT(T, SUB) \subseteq MT$.
Proof. Follows from Lemma 11, Corollary 15, and Lemma 17. □

4.1 Freezing

Surprisingly, in the first step we will decompose modtts. We use a technique, which is based on Lemma 5.1 of [7]: The first module of a modtt can be separated from the rest of the modtt by freezing the occurrences of external functions in right-hand sides, i.e. substituting them by new constructors. These new constructors are later activated by an interpretation function, which interprets every new constructor c as call of that function f_c, which corresponds to c by freezing.

Lemma 11 $2\text{-}ModT(T, SUB) \subseteq T\ ;\ 2\text{-}ModT(INT, SUB)$.
Proof. Let $p \in P$ and $e \in E_p$, such that $m_1, m_2 \in M_p$, (m_1, m_2) is a 2-modtt, m_1 is a tdtt-module, m_2 is a sub-module, and $e = (f\ z)$ for some $f \in F_{m_1}$. We construct $p' \in P$ from p by changing m_1 and by adding the int-module m_3:

1. For every $g \in F_{m_1}$ and $c \in C_p^{(k)}$, the equation $g\ (c\, x_1 \ldots x_k) = rhs(g,c)$ of m_1 is replaced by $g\ (c\, x_1 \ldots x_k) = \underline{freeze}(rhs(g,c))$, where $\underline{freeze}(rhs(g,c))$ is constructed from $rhs(g,c)$ by replacing every occurrence of $sub_i \in F_{m_2}^{(i+1)}$ by a new constructor $SUB_i \in (C - C_p)^{(i+1)}$. Thus, m_1 becomes a tdtt.
2. m_3 contains for a new function $int \in (F - F_p)^{(1)}$ and for every $c \in C_p^{(k)}$ the equation
$$int\ (c\, x_1 \ldots x_k) = c\ (int\ x_1) \ldots (int\ x_k)$$
and for every $sub_i \in F_{m_2}^{(i+1)}$ the equation
$$int\ (SUB_i\ x_1 \ldots x_{i+1}) = sub_i\ (int\ x_1) \ldots (int\ x_{i+1}).$$
Thus, (m_3, m_2) is a 2-modtt.

Let $e' = (int\ (f\ z))$, i.e. $e' \in E_{p'}$. Then, $(p, e) \equiv_{C_p} (p', e')$ (cf. [7]). □

Example 12 Freezing translates p_{non} with initial expression $(rev\ z)$ into p_{fre} with initial expression $(int\ (rev\ z))$. □

4.2 Integration

Now, having 2-modtts with int- and sub-module, we can compose the two modules to an mtt. The main idea is to integrate the behaviour of the interpretation into the substitution. This is best explained by means of our example: The equation $int\ (APP\ x_1\ x_2) = app\ (int\ x_1)\ (int\ x_2)$ is replaced by the new equation $app\ (APP\ x_1\ x_2)\ y_1 = app\ x_1\ (app\ x_2\ y_1)$, which interprets APP as function app and sends app into the subtrees of APP. Note that the new equation represents the associativity of app, if we interpret APP also on the left-hand side as app. The associativity of app, which is the function of a special sub-module (cf. Example 4), plays already in [1,23] an important role, and, for functions of general sub-modules, it will be proved as basis for our integration technique.

Lemma 13 $2\text{-}ModT(INT, SUB) \subseteq MT$.
Proof. Let $p \in P$ and $e \in E_p$, such that $m_1, m_2 \in M_p$, (m_1, m_2) is a 2-modtt, m_1 is an int-module, m_2 is a sub-module, and $e = (int\ z)$ with $F_{m_1} = \{int\}$. We construct $p' \in P$ by replacing m_1 and m_2 in p by the following mtt m:

1. Every equation $sub_i\ (c\, x_1 \ldots x_k)\ y_1 \ldots y_i = rhs(sub_i, c)$ with $sub_i \in F_{m_2}^{(i+1)}$ and $c \in (C_p - C_{m_1})^{(k)}$ (cf. Def. 3 for C_{m_1}) of m_2 is taken over to m, where every occurrence of $sub_i \in F_{m_2}^{(i+1)}$ is changed into $sub'_i \in F_m^{(i+1)}$.
2. If $F_{m_2}^{(1)} = \emptyset$, then for a new function $sub'_0 \in (F - F_p)^{(1)}$ and for every $c \in (C_p - C_{m_1})^{(k)}$ we add $sub'_0\ (c\, x_1 \ldots x_k) = c\ (sub'_0\ x1) \ldots (sub'_0\ x_k)$ to m.
3. For every $sub'_i \in F_m^{(i+1)}$ and $SUB_j \in C_{m_1}^{(j+1)}$, the equation
$$\begin{aligned} &sub'_i\ (SUB_j\ x_1\ x_2 \ldots x_{j+1})\ y_1 \ldots y_i \\ &= sub'_j\ x_1\ (sub'_i\ x_2\ y_1 \ldots y_i) \ldots (sub'_i\ x_{j+1}\ y_1 \ldots y_i) \end{aligned}$$
is added to m.

Let $e' = (sub'_0\ z)$, i.e. $e' \in E_{p'}$. Then $(p, e) \equiv_{C_p} (p', e')$, since for every $t \in T_{C_p}$ and $sub_i \in F_{m_2}^{(i+1)}$:

$$\begin{aligned}
& nf(\Rightarrow_p, (int\ t)) \\
&= nf(\Rightarrow_p, (sub_i\ (int\ t)\ \pi_1 \ldots \pi_i)) && (\text{"}\pi\text{s are substituted by }\pi\text{s" (Struct. Ind.)}) \\
&= nf(\Rightarrow_{p'}, (sub'_i\ t\ \pi_1 \ldots \pi_i)) && (*) \\
&= nf(\Rightarrow_{p'}, (sub'_0\ t)) && (\text{"}\pi\text{s are substituted by }\pi\text{s" (Struct. Ind.)})
\end{aligned}$$

The statement $(*)$ For every $sub_i \in F_{m_2}^{(i+1)}$, and $t, t_1, \ldots, t_i \in T_{C_p}$:
$$nf(\Rightarrow_p, (sub_i\ (int\ t)\ t_1 \ldots t_i)) = nf(\Rightarrow_{p'}, (sub'_i\ t\ t_1 \ldots t_i)).$$

is proved by structural induction on $t \in T_{C_p}$. We only show the interesting case $t = (SUB_j\ t'_0\ t'_1 \ldots t'_j)$ with $SUB_j \in C_{m_1}^{(j+1)}$ and $t'_0, \ldots, t'_j \in T_{C_p}$:

$$\begin{aligned}
& nf(\Rightarrow_p, (sub_i\ (int\ (SUB_j\ t'_0\ t'_1 \ldots t'_j))\ t_1 \ldots t_i)) \\
&= nf(\Rightarrow_p, (sub_i\ (sub_j\ (int\ t'_0)\ (int\ t'_1) \ldots (int\ t'_j))\ t_1 \ldots t_i)) && (\text{Def. } int) \\
&= nf(\Rightarrow_p, (sub_j\ (int\ t'_0)\ (sub_i\ (int\ t'_1)\ t_1 \ldots t_i) \ldots \\
&\qquad (sub_i\ (int\ t'_j)\ t_1 \ldots t_i))) && (**) \\
&= nf(\Rightarrow_p, (sub_j\ (int\ t'_0)\ nf(\Rightarrow_p, (sub_i\ (int\ t'_1)\ t_1 \ldots t_i)) \ldots \\
&\qquad nf(\Rightarrow_p, (sub_i\ (int\ t'_j)\ t_1 \ldots t_i)))) && (\text{"Split } nf\text{"}) \\
&= nf(\Rightarrow_{p'}, (sub'_j\ t'_0 \quad nf(\Rightarrow_{p'}, (sub'_i\ t'_1\ t_1 \ldots t_i)) \ldots \\
&\qquad nf(\Rightarrow_{p'}, (sub'_i\ t'_j\ t_1 \ldots t_i)))) && (\text{Ind. Hyp. } (*)) \\
&= nf(\Rightarrow_{p'}, (sub'_j\ t'_0\ (sub'_i\ t'_1\ t_1 \ldots t_i) \ldots (sub'_i\ t'_j\ t_1 \ldots t_i))) && (\text{"Collect } nf\text{"}) \\
&= nf(\Rightarrow_{p'}, (sub'_i\ (SUB_j\ t'_0\ t'_1 \ldots t'_j)\ t_1 \ldots t_i)) && (\text{Def. } sub'_i)
\end{aligned}$$

The associativity of substitutions

$(**)$ For every $sub_j \in F_{m_2}^{(j+1)}$, $sub_i \in F_{m_2}^{(i+1)}$, $s_0, s_1, \ldots, s_j \in T_{C_p - C_{m_1}}$, and $t_1, \ldots, t_i \in T_{C_p}$:
$$\begin{aligned}
& nf(\Rightarrow_p, (sub_i\ (sub_j\ s_0\ s_1 \ldots s_j)\ t_1 \ldots t_i)) \\
&= nf(\Rightarrow_p, (sub_j\ s_0\ (sub_i\ s_1\ t_1 \ldots t_i) \ldots (sub_i\ s_j\ t_1 \ldots t_i))).
\end{aligned}$$

is also proved by structural induction on $s_0 \in T_{C_p - C_{m_1}}$. We only show the "central" case $s_0 = \pi_k$ with $k \in [j]$:

$$\begin{aligned}
& nf(\Rightarrow_p, (sub_i\ (sub_j\ \pi_k\ s_1 \ldots s_j)\ t_1 \ldots t_i)) \\
&= nf(\Rightarrow_p, (sub_i\ s_k\ t_1 \ldots t_i)) \\
&= nf(\Rightarrow_p, (sub_j\ \pi_k\ (sub_i\ s_1\ t_1 \ldots t_i) \ldots (sub_i\ s_j\ t_1 \ldots t_i)))
\end{aligned}$$

□

Example 14 Integration translates p_{fre} with initial expression $(int\ z)$ into the following program p_{int} with initial expression $(app'_0\ z)$:

$$\begin{array}{ll}
rev\ (A\ x_1) = APP\ (rev\ x_1)\ (A\ N) & app'_0\ (APP\ x_1\ x_2) = app'\ x_1\ (app'_0\ x_2) \\
rev\ (B\ x_1) = APP\ (rev\ x_1)\ (B\ N) & app'_0\ (A\ x_1) = A\ (app'_0\ x_1) \\
rev\ N = N & app'_0\ (B\ x_1) = B\ (app'_0\ x_1) \\
 & app'_0\ N = N \\
 & app'\ (APP\ x_1\ x_2)\ y_1 = app'\ x_1\ (app'\ x_2\ y_1) \\
 & app'\ (A\ x_1)\ y_1 = A\ (app'\ x_1\ y_1) \\
 & app'\ (B\ x_1)\ y_1 = B\ (app'\ x_1\ y_1) \\
 & app'\ N\ y_1 = y_1
\end{array}$$

□

Let $p, p' \in P$, $(f_1\ z), (f_2\ z), (f_2\ (f_1\ z)) \in E_p$, $(f_1\ z), (f_2'\ z), (f_2'\ (f_1\ z)) \in E_{p'}$, and $C' = C_p \cap C_{p'}$. If $(p, (f_1\ z)) \equiv_{C'} (p', (f_1\ z))$ and $(p, (f_2\ z)) \equiv_{C'} (p', (f_2'\ z))$, then $(p, (f_2\ (f_1\ z))) \equiv_{C'} (p', (f_2'\ (f_1\ z)))$. Thus in particular we get from Lemma 13:

Corollary 15 $T\,;\, 2\text{-}ModT(INT, SUB) \subseteq T\,;\, MT$. □

Example 16 Integration translates p_{fre} with initial expression $(int\ (rev\ z))$ into p_{int} with initial expression $(app_0'\ (rev\ z))$. □

4.3 Composition

In [19] it was shown that the composition of two tdtts can be simulated by only one tdtt. This result was generalized in [6], where in particular a composition technique is presented which constructs an mtt m for the composition of a tdtt m_1 with an mtt m_2. The central idea is the observation that, roughly speaking, intermediate results are built up from right-hand sides of m_1. Thus, instead of translating intermediate results by m_2, right-hand sides of m_1 are translated by m_2 to get the equations of m. For this purpose, m uses $F_{m_1} \times F_{m_2}$ as function set. In the following, we abbreviate every pair $(f, g) \in F_{m_1} \times F_{m_2}$ by $\overline{fg}$.

In our construction it will be necessary to extend the call-by-name reduction relation to expressions containing variables (they are handled like 0-ary constructors) and to restrict the call-by-name reduction relation to use only equations of a certain mtt m, which will be denoted by $\Rightarrow_m$.

Lemma 17 $T\,;\, MT \subseteq MT$.
Proof. Let $p \in P$ and $e \in E_p$, such that $m_1, m_2 \in M_p$, m_1 is a tdtt, m_2 is an mtt, and $e = (f_2\ (f_1\ z))$ for some $f_1 \in F_{m_1}$ and $f_2 \in F_{m_2}^{(1)}$. We construct $p' \in P$ by replacing m_1 and m_2 in p by the following mtt m:

1. From m_2 we construct an mtt $\bar{m}_2$ which is able to translate right-hand sides of equations of m_1. Note that $\bar{m}_2$ is not part of p'.
   - $\bar{m}_2$ contains the equations of m_2.
   - For every $g \in F_{m_2}^{(n+1)}$ and $f \in F_{m_1}$ we add the following equation to $\bar{m}_2$:
   $$g\ (f\ x_1)\ y_1 \ldots y_n = \overline{fg}\ x_1\ y_1 \ldots y_n$$
   where every $f \in F_{m_1}$ and $\overline{fg}$ with $f \in F_{m_1}$ and $g \in F_{m_2}^{(n+1)}$ is viewed as additional unary and $(n+1)$-ary constructor, respectively.[4]
2. Let $F_m^{(n+1)} = \{\overline{fg} \mid f \in F_{m_1}, g \in F_{m_2}^{(n+1)}\}$. For every $g \in F_{m_2}^{(n+1)}$, $f \in F_{m_1}$, and $c \in C_p^{(k)}$, such that $f\ (c\ x_1 \ldots x_k) = rhs(f, c)$ is an equation in m_1, m contains the equation
   $$\overline{fg}\ (c\ x_1 \ldots x_k)\ y_1 \ldots y_n = nf(\Rightarrow_{\bar{m}_2}, g\ (rhs(f, c))\ y_1 \ldots y_n).$$

[4] In a strong sense, $\overline{fg}\ x_1\ y_1 \ldots y_n$ is not a legal right-hand side, since x_1 does not occur as first argument of a function symbol, but of a constructor. Again, an additional identity function would solve the problem formally.

Let $e' = (\overline{f_1 f_2}\ z)$, i.e. $e' \in E_{p'}$. Then, $(p, e) \equiv_{C_p} (p', e')$. We omit the proof and only mention that in [6] another construction was used, which first splits the mtt into a tdtt and a so called *yield function*, which handles parameter substitutions (cf. also Subsection 5.1), then composes the two tdtts, and finally joins the resulting tdtt to the yield function. We get the same transformation result in one step by avoiding the explicit splitting and joining (cf. also [18]). □

Example 18 Composition translates p_{int} with initial expression $(app'_0\ (rev\ z))$ into a program and an initial expression, which can be obtained from p_{acc} and $(rev\ z)$, respectively, by a renaming of functions:

Let $M_{p_{int}}$ contain the tdtt $m_{int,rev}$ and the mtt $m_{int,app}$ containing the definitions of rev and of app'_0 and app', respectively. The mtt $\bar{m}_{int,app}$ is given by:

$$
\begin{array}{llll}
app'_0\,(APP\,x_1\,x_2) & = app'\,x_1\,(app'_0\,x_2) & app'\ (APP\,x_1\,x_2)\,y_1 & = app'\,x_1\,(app'\,x_2\,y_1)\\
app'_0\,(A\ x_1) & = A\ (app'_0\ x_1) & app'\ (A\ x_1)\ y_1 & = A\ (app'\ x_1\ y_1)\\
app'_0\,(B\ x_1) & = B\ (app'_0\ x_1) & app'\ (B\ x_1)\ y_1 & = B\ (app'\ x_1\ y_1)\\
app'_0\,N & = N & app'\ N\ y_1 & = y_1\\
app'_0\,(rev\ x_1) & = \overline{revapp}'_0\ x_1 & app'\ (rev\ x_1)\ y_1 & = \overline{revapp}'\ x_1\ y_1
\end{array}
$$

The new program contains the following equations with underlined left- and right-hand sides:

$$
\begin{array}{ll}
\underline{\overline{revapp}'\ (A\ x_1)\ y_1} & \underline{\overline{revapp}'\ (B\ x_1)\ y_1}\\
= nf(\Rightarrow_{\bar{m}_{int,app}}, app'\ (rhs(rev, A))\ y_1) & = \ldots = \underline{\overline{revapp}'\ x_1\ (B\ y_1)}\\
= nf(\Rightarrow_{\bar{m}_{int,app}}, app'\ (APP\ (rev\ x_1)\ (A\ N))\ y_1) & \\
= nf(\Rightarrow_{\bar{m}_{int,app}}, app'\ (rev\ x_1)\ (app'\ (A\ N)\ y_1)) & \underline{\overline{revapp}'\ N\ y_1}\\
= nf(\Rightarrow_{\bar{m}_{int,app}}, \overline{revapp}'\ x_1\ (app'\ (A\ N)\ y_1)) & = \ldots = \underline{y_1}\\
= \underline{\overline{revapp}'\ x_1\ (A\ y_1)} &
\end{array}
$$

$$
\begin{array}{ll}
\underline{\overline{revapp}'_0\ (A\ x_1)} & \underline{\overline{revapp}'_0\ (B\ x_1)}\\
= nf(\Rightarrow_{\bar{m}_{int,app}}, app'_0\ (rhs(rev, A))) & = \ldots = \underline{\overline{revapp}'\ x_1\ (B\ N)}\\
= nf(\Rightarrow_{\bar{m}_{int,app}}, app'_0\ (APP\ (rev\ x_1)\ (A\ N))) & \\
= nf(\Rightarrow_{\bar{m}_{int,app}}, app'\ (rev\ x_1)\ (app'_0\ (A\ N))) & \underline{\overline{revapp}'_0\ N}\\
= nf(\Rightarrow_{\bar{m}_{int,app}}, \overline{revapp}'\ x_1\ (app'_0\ (A\ N))) & = \ldots = \underline{N}\\
= \underline{\overline{revapp}'\ x_1\ (A\ N)} &
\end{array}
$$

The new initial expression is $(\overline{revapp}'_0\ z)$. Note that by a result in [18] we could have used also deforestation to get the same transformation result. □

5 Deaccumulation

This section shows that results from the theory of tree transducers can also be applied to translate accumulative programs of the kind "efficient reverse" into non-accumulative programs of the kind "inefficient reverse", thereby proving that the classes 2-$ModT(T, SUB)$ and MT are equal. So far we consider this transformation direction as purely theoretical, but there may be cases, in which the non-accumulative programs are more efficient than their related accumulative versions (see Section 6).

Theorem 19 $MT \subseteq 2\text{-}ModT(T, SUB)$.
Proof. Follows from Lemma 21 and Lemma 23. □

From Theorems 10 and 19 we get a new characterization of MT:

Corollary 20 $MT = 2\text{-}ModT(T, SUB)$. □

5.1 Decomposition

We have already mentioned that an mtt can be simulated by the composition of a tdtt with a yield function [5,6]. In this paper we will use a construction for this result, which is based on the proof of Lemma 5.5 in [7]. The key idea is to simulate the task of an $(n+1)$-ary function g by a unary function g'.[5] Since g' does not know the current values of its context arguments, it uses a new constructor π_j, wherever g uses its j-th context argument. For this purpose, every variable y_j in the right-hand sides of equations for g is replaced by π_j. The current context arguments themselves are integrated into the calculation by replacing every occurrence of the form $(g\ x_i \ldots)$ in a right-hand side by $(SUB_n\ (g'\ x_i) \ldots)$, where SUB_n is a new constructor. Roughly speaking, an int-module interprets every occurrence of SUB_n as substitution, which replaces every occurrence of π_j in the first subtree of SUB_n by the j-th context argument.

Lemma 21 $MT \subseteq T;\, 2\text{-}ModT(INT, SUB)$.
Proof. Let $p \in P$ and $e \in E_p$, such that $m \in M_p$ is an mtt and $e = (f\ z)$ for some $f \in F_m^{(1)}$. We construct $p' \in P$ by replacing m in p by the following tdtt m_1, int-module m_2, and sub-module m_3, where (m_2, m_3) is a 2-modtt. Let $A = \{n \in I\!N \mid F_m^{(n+1)} \neq \emptyset\}$ and mx be the maximum of A.

1. For every $a \in A - \{0\}$ let $SUB_a \in (C - C_p)^{(a+1)}$ and for every $j \in [mx]$ let $\pi_j \in (C - C_p)^{(0)}$ be distinct new constructors.
   For every $g \in F_m^{(n+1)}$, $c \in C_p^{(k)}$, and equation $g\ (c\ x_1 \ldots x_k)\ y_1 \ldots y_n = rhs(g, c)$ in m, an equation $g\ (c\ x_1 \ldots x_k) = \underline{tr}(rhs(g, c))$ with $g \in F_{m_1}^{(1)}$ is contained in m_1, where $\underline{tr} : RHS(F_m, C_p, X_k, Y_n) \longrightarrow RHS(F_{m_1}, C_p \cup \{SUB_a \mid a \in A - \{0\}\} \cup \{\pi_1, \ldots, \pi_n\}, X_k, Y_0)$ is defined by:

   For every $a \in I\!N_+, h \in F_m^{(a+1)}, i \in [k]$,
   and $r_1, \ldots, r_a \in RHS(F_m, C_p, X_k, Y_n)$:
   $\underline{tr}(h\ x_i\ r_1 \ldots r_a) = SUB_a\ (h\ x_i)\ \underline{tr}(r_1) \ldots \underline{tr}(r_a)$.
   For every $h \in F_m^{(1)}$ and $i \in [k]$:
   $\underline{tr}(h\ x_i) = (h\ x_i)$.
   For every $a \in I\!N, c' \in C_p^{(a)}$, and $r_1, \ldots, r_a \in RHS(F_m, C_p, X_k, Y_n)$:
   $\underline{tr}(c'\ r_1 \ldots r_a) = c'\ \underline{tr}(r_1) \ldots \underline{tr}(r_a)$.
   For every $j \in [n]$:
   $\underline{tr}(y_j) = \pi_j$.

[5] In the formal construction g is not renamed.

2. m_2 contains for a new function $int \in (F - F_p)^{(1)}$ and for every $c \in (C_p \cup \{\pi_1, \ldots, \pi_{mx}\})^{(k)}$ the equation

$$int\ (c\ x_1 \ldots x_k) = c\ (int\ x_1) \ldots (int\ x_k)$$

and for every $a \in A - \{0\}$ the equation

$$int\ (SUB_a\ x_1 \ldots x_{a+1}) = sub_a\ (int\ x_1) \ldots (int\ x_{a+1}).$$

3. m_3 contains for every $a \in A - \{0\}$[6], for every new function $sub_a \in (F - F_p)^{(a+1)}$, and for every $c \in C_p^{(k)}$ the equation

$$sub_a\ (c\ x_1 \ldots x_k)\ y_1 \ldots y_a = c\ (sub_a\ x_1\ y_1 \ldots y_a) \ldots (sub_a\ x_k\ y_1 \ldots y_a)$$

and for every $j \in [a]$ the equation

$$sub_a\ \pi_j\ y_1 \ldots y_a = y_j.$$

Let $e' = (int\ (f\ z))$, i.e. $e' \in E_{p'}$. Then, $(p, e) \equiv_{C_p} (p', e')$ (cf. [7]). □

Example 22 Decomposition translates p_{acc} with initial expression $(rev\ z)$ into the following program p_{dec} with initial expression $(int\ (rev\ z))$:

$$\begin{array}{ll} rev\ (A\ x_1) = SUB_1\ (rev'\ x_1)\ (A\ N) & rev'\ (A\ x_1) = SUB_1\ (rev'\ x_1)\ (A\ \pi_1) \\ rev\ (B\ x_1) = SUB_1\ (rev'\ x_1)\ (B\ N) & rev'\ (B\ x_1) = SUB_1\ (rev'\ x_1)\ (B\ \pi_1) \\ rev\ N \quad = N & rev'\ N \quad = \pi_1 \end{array}$$

$$\begin{array}{ll} int\ (SUB_1\ x_1\ x_2) = sub_1\ (int\ x_1)\ (int\ x_2) & sub_1\ (A\ x_1)\ y_1 = A\ (sub_1\ x_1\ y_1) \\ int\ (A\ x_1) \quad = A\ (int\ x_1) & sub_1\ (B\ x_1)\ y_1 = B\ (sub_1\ x_1\ y_1) \\ int\ (B\ x_1) \quad = B\ (int\ x_1) & sub_1\ N\ y_1 \quad = N \\ int\ N \quad = N & sub_1\ \pi_1\ y_1 \quad = y_1 \\ int\ \pi_1 \quad = \pi_1 & \end{array}$$

□

5.2 Thawing

We use an inverse construction as in the proof of Lemma 11 to thaw "frozen functions" of a tdtt, i.e. we substitute occurrences of those constructors SUB_i, which the interpretation replaces by functions sub_i, by occurrences of sub_i.

Lemma 23 T; 2-$ModT(INT, SUB) \subseteq$ 2-$ModT(T, SUB)$.
Proof. Let $p \in P$ and $e \in E_p$, such that $m_1, m_2, m_3 \in M_p$, m_1 is a tdtt, (m_2, m_3) is a 2-modtt, m_2 is an int-module, m_3 is a sub-module, and $e = (int\ (f\ z))$ with $F_{m_2} = \{int\}$ and for some $f \in F_{m_1}$. We construct $p' \in P$ from p by dropping m_2 and by changing m_1:

For every $g \in F_{m_1}$ and $c \in C_p^{(k)}$, the equation $g\ (c\ x_1 \ldots x_k) = rhs(g, c)$ of m_1 is replaced by $g\ (c\ x_1 \ldots x_k) = \underline{thaw}(rhs(g, c))$, where $\underline{thaw}(rhs(g, c))$

[6] If $A = \{0\}$, i.e. m was already a tdtt, then we construct a "dummy function" sub_0.

is constructed from $rhs(g, c)$ by replacing every occurrence of $SUB_i \in C_{m_2}^{(i+1)}$ (cf. Def. 3 for C_{m_2}) by $sub_i \in F_{m_3}^{(i+1)}$, iff the equation $int\ (SUB_i\ x_1 \ldots x_{i+1}) = sub_i\ (int\ x_1) \ldots (int\ x_{i+1})$ is in m_2. Thus, m_1 becomes a tdtt-module and (m_1, m_3) a 2-modtt.

Let $e' = (f\ z)$, i.e. $e' \in E_{p'}$. Then, $(p, e) \equiv_{C_p - C_{m_2}} (p', e')$, since

$$(*)\quad \text{For every } g \in F_{m_1} \text{ and } t \in T_{C_p - C_{m_2}}\text{: } nf(\Rightarrow_p, (int\ (g\ t))) = nf(\Rightarrow_{p'}, (g\ t)).$$

is proved by structural induction on $t \in T_{C_p - C_{m_2}}$. This proof requires another structural induction on $r \in RHS(F_{m_1}, C_p, X_k, Y_0)$ to prove

$$(**)\quad \text{For every } k \in I\!N, r \in RHS(F_{m_1}, C_p, X_k, Y_0), \text{and } t_1, \ldots, t_k \in T_{C_p - C_{m_2}}\text{:}$$
$$nf(\Rightarrow_p, (int\ r[x_1/t_1, \ldots, x_k/t_k])) = nf(\Rightarrow_{p'}, \underline{thaw}(r)[x_1/t_1, \ldots, x_k/t_k]),$$

where $[x_1/t_1, \ldots, x_k/t_k]$ denotes the substitution of every occurrence of x_i in r and $\underline{thaw}(r)$, respectively, by t_i. □

Example 24 Thawing translates p_{dec} with initial expression $(int\ (rev\ z))$ into the following program p'_{non} with initial expression $(rev\ z)$:

$$\begin{array}{ll}
rev\ (A\ x_1) = sub_1\ (rev'\ x_1)\ (A\ N) & rev'\ (A\ x_1) = sub_1\ (rev'\ x_1)\ (A\ \pi_1) \\
rev\ (B\ x_1) = sub_1\ (rev'\ x_1)\ (B\ N) & rev'\ (B\ x_1) = sub_1\ (rev'\ x_1)\ (B\ \pi_1) \\
rev\ N \quad\quad\ = N & rev'\ N \quad\quad\ = \pi_1
\end{array}$$

$$\begin{array}{ll}
sub_1\ (A\ x_1)\ y_1 & = A\ (sub_1\ x_1\ y_1) \\
sub_1\ (B\ x_1)\ y_1 & = B\ (sub_1\ x_1\ y_1) \\
sub_1\ N\ y_1 & = N \\
sub_1\ \pi_1\ y_1 & = y_1
\end{array}$$

It is surprising that p'_{non} is very similar to p_{non}: sub_1 in p'_{non} corresponds to app in p_{non}, but substitutes the symbol π_1 instead of N. rev' in p'_{non} corresponds to rev in p_{non}, but uses sub_1 and π_1 instead of app and N. The additional rev in p'_{non} achieves that π_1 does not occur in the output. □

6 Future Work

In this paper we have always considered non-accumulative functional programs as less efficient than their related accumulative versions, like we have considered in [17] the pure elimination of intermediate results as success. This assumption is true for so far studied example programs, but may be wrong in general. It is necessary to find out sufficient conditions for our source programs, under which we can guarantee that the accumulation technique (and maybe also the deaccumulation technique, respectively) does not deteriorate the efficiency. In particular, is the linearity of programs a sufficient condition for accumulation (like for deforestation [24] and tree transducer composition [18,15])?

We have presented sufficient conditions, such that we can compose the modules of a 2-modtt. Is it possible to extend the applicability of the accumulation technique by relaxing these conditions or by using other conditions? Additionally, it would be interesting to analyze u-modtts with $u > 2$ in this context.

References

1. R. M. Burstall and J. Darlington. A transformation system for developing recursive programs. *J. Assoc. Comput. Mach.*, 24:44–67, 1977.
2. W.-N. Chin. Safe fusion of functional expressions II: Further improvements. *Journal of Functional Programming*, 4:515–555, 1994.
3. O. Chitil. Type-inference based short cut deforestation (nearly) without inlining. In *IFL'99, Lochem, The Netherlands, Proceedings, September 1999*, volume 1868 of *LNCS*, pages 19–36. Springer-Verlag, April 2000.
4. B. Courcelle and P. Franchi–Zannettacci. Attribute grammars and recursive program schemes. *Theor. Comp. Sci.*, 17:163–191, 235–257, 1982.
5. J. Engelfriet. Some open questions and recent results on tree transducers and tree languages. In R.V. Book, editor, *Formal language theory; perspectives and open problems*, pages 241–286. New York, Academic Press, 1980.
6. J. Engelfriet and H. Vogler. Macro tree transducers. *J. Comp. Syst. Sci.*, 31:71–145, 1985.
7. J. Engelfriet and H. Vogler. Modular tree transducers. *Theor. Comp. Sci.*, 78:267–304, 1991.
8. Z. Fülöp. On attributed tree transducers. *Acta Cybernetica*, 5:261–279, 1981.
9. Z. Fülöp, F. Herrmann, S. Vágvölgyi, and H. Vogler. Tree transducers with external functions. *Theor. Comp. Sci.*, 108:185–236, 1993.
10. Z. Fülöp and H. Vogler. *Syntax-directed semantics — Formal models based on tree transducers.* Monographs in Theoretical Computer Science. Springer-Verlag, 1998.
11. H. Ganzinger. Increasing modularity and language–independency in automatically generated compilers. *Science of Computer Programming*, 3:223–278, 1983.
12. R. Giegerich. Composition and evaluation of attribute coupled grammars. *Acta Informatica*, 25:355–423, 1988.
13. A. Gill. *Cheap deforestation for non-strict functional languages.* PhD thesis, University of Glasgow, 1996.
14. A. Gill, J. Launchbury, and S.L. Peyton Jones. A short cut to deforestation. In *FPCA'93, Copenhagen, Denmark, Proceedings*, pages 223–231. ACM Press, 1993.
15. M. Höff. Vergleich von Verfahren zur Elimination von Zwischenergebnissen bei funktionalen Programmen. Master's thesis, Dresden University of Technology,1999.
16. D.E. Knuth. Semantics of context–free languages. *Math. Syst. Th.*, 2:127–145, 1968. Corrections in *Math. Syst. Th.*, 5:95-96, 1971.
17. A. Kühnemann. Benefits of tree transducers for optimizing functional programs. In *FST & TCS'98, Chennai, India, Proceedings*, volume 1530 of *LNCS*, pages 146–157. Springer-Verlag, December 1998.
18. A. Kühnemann. Comparison of deforestation techniques for functional programs and for tree transducers. In *FLOPS'99, Tsukuba, Japan, Proceedings*, volume 1722 of *LNCS*, pages 114–130. Springer-Verlag, November 1999.
19. W.C. Rounds. Mappings and grammars on trees. *Math. Syst.Th.*, 4:257–287, 1970.
20. M. H. Sørensen, R. Glück, and N. D. Jones. A positive supercompiler. *Journal of Functional Programming*, 6:811–838, 1996.
21. J.W. Thatcher. Generalized^2 sequential machine maps. *J. Comp. Syst. Sci.*, 4:339–367, 1970.
22. V. F. Turchin. The concept of a supercompiler. *ACM TOPLAS*, 8:292–325, 1986.
23. P. Wadler. The concatenate vanishes. Note, University of Glasgow, December 1987 (Revised, November 1989).
24. P. Wadler. Deforestation: Transforming programs to eliminate trees. *Theor. Comp. Sci.*, 73:231–248, 1990.

Context Unification and Traversal Equations*

Jordi Levy[1] and Mateu Villaret[2]

[1] IIIA, CSIC, Campus de la UAB, Barcelona, Spain.
`http://www.iiia.csic.es/~levy`
[2] IMA, UdG, Campus de Montilivi, Girona, Spain.
`http://www.ima.udg.es/~villaret`

Abstract. Context unification was originally defined by H. Comon in ICALP'92, as the problem of finding a unifier for a set of equations containing first-order variables and *context variables*. These context variables have arguments, and can be instantiated by *contexts*. In other words, they are second-order variables that are restricted to be instantiated by linear terms (a linear term is a λ-expression $\lambda x_1 \cdots \lambda x_n . t$ where every x_i occurs exactly once in t).
In this paper, we prove that, if the so called *rank-bound conjecture* is true, then the context unification problem is decidable. This is done reducing context unification to solvability of *traversal equations* (a kind of word unification modulo certain permutations) and then, reducing traversal equations to *word equations with regular constraints*.

1 Introduction

Context unification is defined as the problem of finding a unifier for a finite set of equations where, in addition to first-order variables, we also consider *context variables*. These variables are applied to terms, and can be instantiated by contexts, i.e. by *linear second-order terms*. A linear second-order term is a λ-expression $\lambda x_1 \cdots \lambda x_n . t$ where $x_1,\ldots,x_n$ are first-order bound variables and occur *exactly once* in t. Therefore, context unification can be considered as a variant of second-order unification where possible instances of second-order variables are restricted to be linear. Sometimes, context variables are required to be unary. However, this restriction does not help to prove the decidability of the problem, and it will not be used in this paper. Given an instance of the problem, if it has a solution considered as a context unification problem, then it has also a solution as second-order unification problem. Obviously, the converse is not true.

The context unification problem was originally formulated by H. Comon in [Com92,Com98]. There it is proved that context unification is decidable when, for any context variable, all its occurrences have the same argument. Later, it was proved [SS96,SS98,SS99b] that the problem is also decidable when context variables are stratified, i.e. when, for any variable, the list of context variables

* This work has been partially supported by the CICYT research projects DENOC (BFM 2000-1054-C02-01) and MODELOGOS (TIC 97-0579-C02-01).

A. Middeldorp (Ed.): RTA 2001, LNCS 2051, pp. 169–184, 2001.

we find going from the root of the term to any occurrence of this variable is always the same. It was also proved [Lev96] that a generalization of the problem –the *linear second-order unification problem*, where third-order constants are also allowed– is decidable when no variable occurs more than twice. Recently, it has been proved [SSS99] that context unification is also decidable for problems containing no more than two context variables. The relationship between the context unification problem and the linear second-order unification problem is studied in [LV00b].

Decidability of context unification would have important consequences in different research areas. For instance, some partial decidability results are used in [Com92] to prove decidability of membership constraints, in [SS96] to prove decidability of distributive unification, in [LA96] to define a completion procedure for bi-rewriting systems. In [ENRX98] it is proved that *parallelism constraints* –a kind of partial description of trees– are equivalent to context unification. *Dominance constraints* are a subset of parallelism constraints, and their solvability is decidable [KNT98]. Other application areas of context unification include computational linguistics [NPR97b]. The common assumption is that context unification is decidable. This is because the various restrictions that make context unification decidable, when they are applied to second-order unification, they do not make it decidable [Lev98,LV00a].

In [Lev96] there is a description of a sound and complete context unification procedure, based on Pietrzykowski's procedure [Pie73] for second-order unification. Like Pietrzykowski's procedure, this procedure does not always terminate. The linearity restriction makes some trivially solvable second-order unification problems, like $X(a) \stackrel{?}{=} X(b)$, unsolvable when we only consider context unifiers. Notice that this problem has only one unifier $[X \mapsto \lambda x \,.\, Y]$ which is not linear because x does not occur once in Y. In particular, flexible-flexible pairs, which are always solvable in second-order unification, now are not necessarily solvable.

The *bounded second-order unification problem* is another variant of second-order unification, similar to context unification. There, instances of second-order variables are required to use their arguments a bounded number of times. We can easily reduce any k-bounded second-order unification problem, like $X(Y(a,b)) \stackrel{?}{=} Y(X(a),b)$, to a context unification problem, like

$$X(Y(a,\overset{q}{\dots},a,b,\overset{r}{\dots},b),\overset{p}{\dots},Y(a,\overset{q}{\dots},a,b,\overset{r}{\dots},b)) \stackrel{?}{=}$$
$$\stackrel{?}{=} Y(X(a,\overset{p}{\dots},a),\overset{q}{\dots},X(a,\overset{p}{\dots},a),b,\overset{r}{\dots},b)$$

nondeterministically, for any possible choice of $p, q, r \le k$ satisfying the bound. The converse reduction does not seem easy to find. The bounded second-order unification problem has recently been proved decidable [SS99a].

The relationship between context unification and word unification [Mak77] was originally suggested in [Lev96]. In [SSS98] it is proved that the exponent of periodicity lemma also holds for context unification. We can easily reduce word unification to context unification by encoding any word unification problem, like $F\,a\,G \stackrel{?}{=} G\,a\,F$, as a monadic context unification problem $F(a(G(b))) \stackrel{?}{=} G(a(F(b)))$, where b is a new constant. This paper suggests that the opposite reduction may also be possible. In the following Section we motivate this statement using a naive reduction. Although it does not work, we will see in the rest of the paper how it could be adapted properly.

2 A Naive Reduction

Given a signature where every symbol has a fixed arity, we can encode a term using its *pre-order traversal* sequence. We can use this fact to encode a context unification problem, like the following one

$$X(Y(a,b)) \stackrel{?}{=} Y(X(a),b) \tag{1}$$

as the following word unification problem

$$X_0\, Y_0\, a\, Y_1\, b\, Y_2\, X_1 \stackrel{?}{=} Y_0\, X_0\, a\, X_1\, Y_1\, b\, Y_2 \tag{2}$$

We can prove easily that *if* the context unification problem (1) is solvable, *then* its corresponding word unification problem (2) is also solvable. In our example, the solution corresponding to the following unifier

$$\begin{array}{l} X \mapsto \lambda x\,.\, f(f(x,b),b) \\ Y \mapsto \lambda x\,.\,\lambda y\,.\, f(f(f(x,b),y),b) \end{array} \tag{3}$$

is

$$\begin{array}{ll} X_0 \mapsto f\,f & Y_0 \mapsto f\,f\,f \\ X_1 \mapsto b\,b & Y_1 \mapsto b \\ & Y_2 \mapsto b \end{array}$$

Unfortunately, the converse is not true. We can find a solution of the word unification problem which does not correspond to the pre-order traversal of any instantiation of the original context unification problem (consider the unifier that instantiates X_0, X_1, Y_0, Y_1 and Y_2 by the empty word). Word unification is decidable [Mak77], and given a solution of the word unification problem we can check if it corresponds to a solution of the context unification problem. Unfortunately, word unification is also infinitary, and we can not repeat this test for infinitely many word unifiers.

The idea to overcome this difficulty comes from the notion of *rank of a term.* In figure 1 there are some examples of terms (trees) with different ranks. Notice that terms with rank bounded by zero are isomorphic to words, and those with rank bounded by one are caterpillars. For signatures of binary symbols, the rank of a term can be defined as follows

$$\begin{array}{l} \mathrm{rank}(a) = 0 \\ \mathrm{rank}(f(t_1,t_2)) = \begin{cases} 1 + \mathrm{rank}(t_1) & \text{if } \mathrm{rank}(t_1) = \mathrm{rank}(t_2) \\ \max\{\mathrm{rank}(t_1), \mathrm{rank}(t_2)\} & \text{if } \mathrm{rank}(t_1) \neq \mathrm{rank}(t_2) \end{cases} \end{array}$$

Alternatively, the rank of a binary tree can also be defined as the depth of the greatest complete binary tree that is embedded in the tree, using the standard embedding of trees.

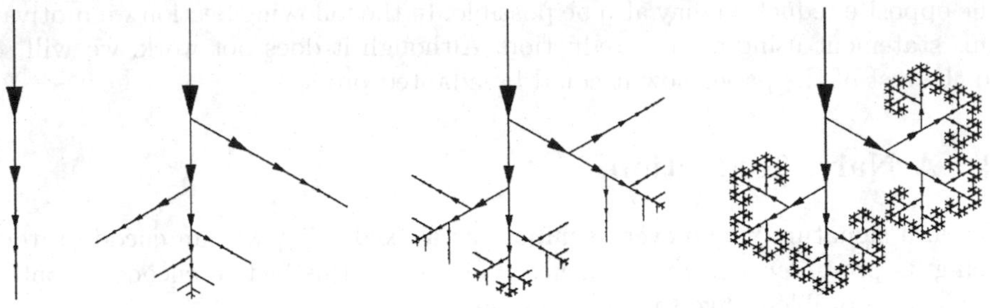

Fig. 1. Examples of trees with ranks equal to 0, 1, 2 and ∞.

We conjecture that there is a computable function Φ such that, for every solvable context unification problem $t \stackrel{?}{=} u$, there exists a ground unifier σ, such that the rank of $\sigma(t)$ is bounded by the size of the problem: $\text{rank}(\sigma(t)) \leq \Phi(\text{size}(t \stackrel{?}{=} u))$.

The other idea is to generalise pre-order traversal sequences to a more general notion of *traversal sequence*, by allowing subterms to be traversed in different orders. Then, any rank-bounded term has a traversal sequence belonging to a regular language. We also introduce a new notion of *traversal equation*, noted $t \equiv u$, and meaning t and u are traversal sequences of the same term. We prove that a variant of these constraints can be reduced to word equations with regular constraints, that are decidable [Sch91].

The rest of this paper proceeds as follows. In Section 3 we introduce basic notation. In Section 4 we define the notions of traversal sequence, rank of a traversal sequence, rank of a term, and normal traversal sequence. Traversal equations are introduced in Section 5. There, we prove that solvability of rank- and permutation-bounded traversal equations is decidable, by reducing the problem to solvability of word equations with regular constraints. In Section 6, we state the rank-bound conjecture. Finally, in Section 7 we show how, if the conjecture is true, context unification could be reduced to rank- and permutation-bounded traversal systems.

3 Preliminary Definitions

In this section, we introduce some definitions and notations. Most of them are standard and can be skipped.

We define terms over a *second-order signature* $\langle \Sigma, \mathcal{X} \rangle$ of constants $\Sigma = \bigcup_{i \geq 0} \Sigma_i$ and variables $\mathcal{X} = \bigcup_{i \geq 0} \mathcal{X}_i$, where any constant $f \in \Sigma_i$ or variable $X \in \mathcal{X}_j$ has a fixed arity: $\text{arity}(f) = i$, $\text{arity}(X) = j$. Constants from Σ_0 are called *first-order constants* whereas constants from $\Sigma \backslash \Sigma_0$ are called second-order constants or *function symbols.* Similarly, variables from $\mathcal{X}_0$ are *first-order variables*, and those from $\mathcal{X} \backslash \mathcal{X}_0$ are *context variables. First-order terms* $T^1(\Sigma, \mathcal{X})$ and *second-order terms* $T^2(\Sigma, \mathcal{X})$ are defined as usual. The set of *free variables* of a term t is denoted by Var(t). The *size* of a first-order term is defined inductively by $\text{size}(f(t_1,..., t_n)) = 1 + \sum_{i\in[1..n]} \text{size}(t_i)$ being f either a n-ary constant or variable. The arity of a ($\beta\eta$-normalised) second-order term is defined by $\text{arity}(\lambda x_1 \cdots \lambda x_n . t) = n$. A second-order term $\lambda x_1 \cdots \lambda x_n . t$ is said to be *linear* if any bound variable x_i occurs exactly once in t. As far as first-order terms do not contain bound variables, any first-order term is linear.

A *position* within a term is defined, using Dewey decimal notation, as a sequence of integers $i_1 \cdots i_n$, being λ the empty sequence. The concatenation of two sequences is denoted by $p_1 \cdot p_2$. The concatenation of an integer and a sequence is also denoted by $i \cdot p$, standing $i, j,...$ for integers and $p, q,...$ for sequences. The subterm of t at position p is denoted by $t|_p$. By $t[u]_p$ we denote the term t where the subterm at position p has been replaced by u.

The group of permutations of n elements is denoted by Π_n. A permutation ρ of n elements is denoted as a sequence of integers $[\rho(1),..., \rho(n)]$.

A *context unification problem* is a finite sequence of equations $\{t_i \stackrel{?}{=} u_i\}_{i\in[1..n]}$, being an *equation* $t \stackrel{?}{=} u$ a pair of first-order terms $t, u \in T^1(\Sigma, \mathcal{X})$. The *size of a problem* is defined by $\text{size}(\{t_i \stackrel{?}{=} u_i\}_{i\in[1..n]}) = \sum_{i\in[1..n]}(\text{size}(t_i) + \text{size}(u_i))$ A position within a problem or an equation is defined by

$$
\begin{aligned}
\{t_i \stackrel{?}{=} u_i\}_{i\in[1..n]}|_{j\cdot p} &= (t_j \stackrel{?}{=} u_j)|_p \\
(t \stackrel{?}{=} u)|_{1\cdot p} &= t|_p \\
(t \stackrel{?}{=} u)|_{2\cdot p} &= u|_p
\end{aligned}
$$

A *second-order substitution* is a finite sequence of pairs of variables and terms $\sigma = [X_1 \mapsto s_1,..., X_m \mapsto s_m]$, where X_i and s_i are restricted to have the same arity. A *context substitution* is a second-order substitution where the s_i's are linear terms. A substitution $\sigma = [X_1 \mapsto s_1,..., X_n \mapsto s_n]$ defines a mapping from terms to terms. A substitution σ_1 is said to be *more general* than another σ_2, if there exist another substitution ρ such that $\sigma_2 = \rho \circ \sigma_1$.

Given a context unification problem $\{t_i \stackrel{?}{=} u_i\}_{i\in[1..n]}$, a context [second-order] substitution $\sigma = [X_1 \mapsto s_1,..., X_m \mapsto s_m]$, is said to be a *context [second-order] unifier* if $\sigma(t_i) = \sigma(u_i)$, for any $i \in [1..n]$. A unifier σ is said to be *most general*, m.g.u. for short, if no other unifier is strictly more general than it. It is said to be *ground* if $\sigma(t_i)$ does not contain variables, for any $i \in [1..n]$. A context unification problem is said to be *solvable* if it has a context unifier.

The *context unification problem* is defined as the problem of deciding if, given context unification problem, does it have a context unifier or not.

Without loss of generality, we can assume that the unification problem only contains just one equation $t \stackrel{?}{=} u$. We will also assume that the signature Σ is finite, and that it contains, at least, a first-order constant, and a binary function symbol. This ensures that any solvable context unification problem has a ground unifier, and we can guess constant symbols in non-deterministic computations. If nothing is said, the signature of a problem is the set of symbols occurring in the problem, plus a first-order and a binary constant, if required.

In the appendix we include a variant of the sound and complete context unification procedure described in [Lev96], and adapted to our actual settings.

This procedure can be used to find most general unifiers, and a variant of it, to find minimal ground unifiers.

4 Terms and Traversal Sequences

The solution to the problems pointed out in the introduction comes from generalising the definition of pre-order traversal sequence. It will allow us to traverse the branches of a tree, i.e. the arguments of a function, in any possible order. In order to reconstruct the term from the traversal sequence, we have to annotate the permutation we have used in this particular traversal sequence. For this purpose, we define a new signature Σ_Π containing $n!$ symbols f^ρ for each n-ary symbol $f \in \Sigma$, where $\rho \in \Pi_n$ and Π_n is the group of permutations of n elements.

Definition 1. *Given a signature $\Sigma = \bigcup_{i\geq 0} \Sigma_i$, we define the* extended signature

$$\Sigma_\Pi = \{f^\rho \mid f \in \Sigma \ \wedge \ \rho \in \Pi_{\text{arity}(f)}\}$$

where Π_n is the group of permutations over n elements.
For any $f^\rho \in \Sigma_\Pi$, and its corresponding $f \in \Sigma$, we define $\text{arity}(f^\rho) = \text{arity}(f)$.
A sequence $s \in (\Sigma_\Pi)^*$ is said to be a traversal sequence *of a term $t \in T(\Sigma)$ if:*

1. *$t \in \Sigma_0$, and $s = t$ (the permutation is omitted for first-order constants); or*
2. *$t = f(t_1,\ldots,t_n)$, for any $i \in [1..n]$, there exists a sequence s_i such that it is a traversal sequences of t_i, and there exists a permutation $\rho \in \Pi_n$ such that $s = f^\rho s_{\rho(1)} \cdots s_{\rho(n)}$.*

Definition 2. *Given a sequence of symbols $a_1 \cdots a_n \in (\Sigma_\Pi)^*$, we define its* width *as*

$$\text{width}(a) = \text{arity}(a) - 1$$
$$\text{width}(a_1 \cdots a_n) = \textstyle\sum_{i\in[1..n]} \text{width}(a_i)$$

This definition can be used to characterize *traversal sequences.*

Lemma 3. *A sequence of symbols $a_1 \cdots a_n \in (\Sigma_\Pi)^*$ is a* traversal sequence, *of some term $t \in T(\Sigma)$, if, and only if,*

$$\text{width}(a_1 \cdots a_n) = -1, \textit{ and}$$
$$\text{width}(a_1 \cdots a_i) \geq 0, \textit{ for any } i \in [1..n-1].$$

Now we define the rank of a traversal sequence, and by extension, the rank of a term as the minimal rank of its traversal sequences. This definition coincides with the definition given in the introduction for the rank of a term for binary signatures.

Definition 4. *Given a sequence of symbols $a_1 \cdots a_n \in (\Sigma_\Pi)^*$, we define its* rank *as*

$$\text{rank}(a_1 \cdots a_n) = \max\{\text{width}(a_i \cdots a_j) \mid i,j \in [1..n]\}$$

Given a term $t \in T(\Sigma)$, we define its rank *as*

$$\text{rank}(t) = \min\{\text{rank}(w) \mid w \textit{ is a traversal of } t\}$$

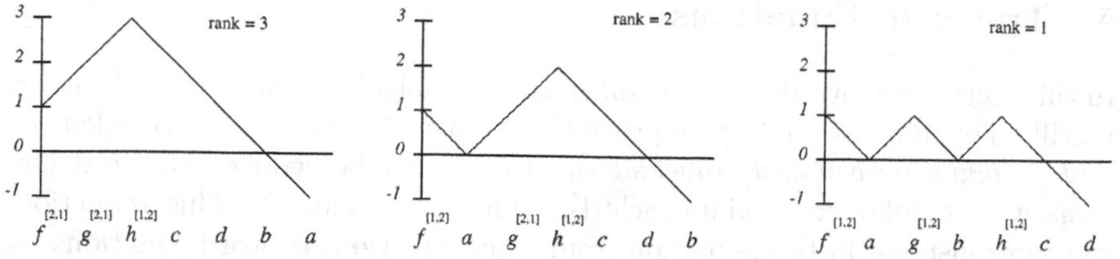

Fig. 2. Representations of the function $f(i) = \text{width}(a_1 \cdots a_i)$, for some traversal sequences of $f(a, g(b, h(c, d)))$.

In general, a term has more than one traversal sequence associated. The rank of the term is always smaller or equal to the rank of its traversals, and for at least one of them we have equality. These rank-minimal traversals are relevant for us, and we choose one of them as the *normal traversal sequence.* In figure 2, the third traversal sequence $f^{[1,2]}\, a\, g^{[1,2]}\, b\, h^{[1,2]}\, c\, d$ is the normal one.

Definition 5. *Given a term t, its* normal traversal sequence $\text{NF}(t)$ *is defined recursively as follows:*

1. *If* $t = a$ *then* $\text{NF}(t) = a$.
2. *If* $t = f(t_1, ..., t_n)$ *then let* $\rho \in \Pi_n$ *be the permutation satisfying*

$$i < j \Rightarrow \begin{cases} \text{rank}(t_{\rho(i)}) < \text{rank}(t_{\rho(j)}) \\ \vee \\ \text{rank}(t_{\rho(i)}) = \text{rank}(t_{\rho(j)}) \wedge \rho(i) < \rho(j) \end{cases}$$

Then, $\text{NF}(t) = f^{\rho}\ \text{NF}(t_{\rho(1)}) \ \cdots\ \text{NF}(t_{\rho(n)})$.

Lemma 6. *For any term, its normal traversal sequence has minimal rank, i.e.* $\text{rank}(t) = \text{rank}(\text{NF}(t))$.

Rank-upper bounded traversal sequences define a regular language. The construction of associated automata can be found in [LV00b].

Lemma 7. *Given an extended signature* Σ_Π *and a constant* k*, the following set is a regular language.*

$$R^k_\Sigma = \{s \in (\Sigma_\Pi)^* \mid \text{rank}(s) \le k\ \wedge\ s \textit{ is a traversal}\}$$

Proof. We can define R^k_Σ inductively as follows:

$$R^0_\Sigma = (\Sigma_1)^*\ \Sigma_0$$
$$R^k_\Sigma = R^{k-1}_\Sigma \cup \Big(\bigcup_{n \ge 1} \Sigma_n\ R^{k-n+1}_\Sigma \cdots R^{k-1}_\Sigma \Big)^*\ \Sigma_0$$

5 Traversal Equations

In this section we introduce *traversal equations*. Solvability of traversal equations is still an open question, but we prove that a variant of them (the so called *rank- and permutation-bounded traversal equations*) can be reduced to word equations with regular constraints [Sch91], which are decidable. This reduction is somehow inspired in the reduction from trace equations to word equations used in [DMM97] to prove decidability of trace equations. Later in Section 7, we will reduce context unification to solvability of traversal equations. We need the rank-bound conjecture to prove that the reduction can be done to rank- and permutation-bounded traversal equations.

Definition 8. *A* traversal system *over an extended signature with word variables* $(\Sigma_\Pi, \mathcal{W})$ *is a conjunction of literals, where every literal has the form* $w_1 \stackrel{?}{=} w_2$ (word equation), $w_1 \equiv w_2$ (traversal equation) *or* $w \in R$ (regular constraint), *being* $w_i \in (\Sigma_\Pi \cup \mathcal{W})^*$ *words with variables and* $R \subseteq (\Sigma_\Pi)^*$ *a regular language.*
A solution *of a traversal system is a word substitution* $\sigma : \mathcal{W} \to (\Sigma_\Pi)^*$ *such that*

1. $\sigma(w_1) = \sigma(w_2)$ *for any word equation* $w_1 \stackrel{?}{=} w_2$*,*
2. $\sigma(w_1)$ *and* $\sigma(w_2)$ *are both traversal sequences of the same term, for any traversal equation* $w_1 \equiv w_2$*,*
3. *and* $\sigma(w)$ *belongs to* R*, for any regular constraint* $w \in R$*.*

Definition 9. *A traversal system is said to be* rank-bounded *if, for every traversal equation* $w_1 \equiv w_2$*, there exist two constants* k_1 *and* k_2*, and two regular constraints* $w_1 \in R^{k_1}_\Sigma$ *and* $w_2 \in R^{k_2}_\Sigma$ *in the system, where* R^k_Σ *is the (regular) set of* k*-bounded traversal sequences.*

We can transform rank-bounded traversal systems into equivalent traversal systems using the following transformation rules.

Definition 10. *The following rules define a non-deterministic translation procedure from rank-bounded traversal systems into word equations with regular constraints.*

Rule 1: *For some* n*-ary symbol* $f \in \Sigma$ *and permutations* $\rho_1, \rho_2 \in \Pi_n$*, we replace the traversal equation* $w_1 \equiv w_2$ *and the corresponding regular constraints* $w_1 \in R^{k_1}_\Sigma$ *and* $w_2 \in R^{k_2}_\Sigma$ *by*

$$
\begin{array}{l}
w_1 \equiv w_2 \\
w_1 \in R^{k_1}_\Sigma \\
w_2 \in R^{k_2}_\Sigma
\end{array}
\Longrightarrow
\begin{array}{l}
w_1 \in R^{k_1}_\Sigma \\
w_2 \in R^{k_2}_\Sigma \\
w_1 \stackrel{?}{=} X_1\ f^{\rho_1}\ Y_{\rho_1(1)} \cdots Y_{\rho_1(n)}\ X_2 \\
w_2 \stackrel{?}{=} X_1\ f^{\rho_2}\ Y'_{\rho_2(1)} \cdots Y'_{\rho_2(n)}\ X_2 \\
\left.\begin{array}{l}
Y_i \equiv Y'_i \\
Y_{\rho_1(i)} \in R^{k_1-n+i}_\Sigma \\
Y'_{\rho_2(i)} \in R^{k_2-n+i}_\Sigma
\end{array}\right\} \text{for any } i \in [1..n]
\end{array}
$$

where X_1*,* X_2 *and* $\{Y_i, Y'_i\}_{i \in [1..n]}$ *are fresh word variables.*

Rule 2: *We replace the traversal equation* $w_1 \equiv w_2$ *and the corresponding regular constraints* $w_1 \in R_\Sigma^{k_1}$ *and* $w_2 \in R_\Sigma^{k_2}$ *by*

$$\begin{array}{l} w_1 \equiv w_2 \\ w_1 \in R_\Sigma^{k_1} \\ w_2 \in R_\Sigma^{k_2} \end{array} \Longrightarrow \begin{array}{l} w_1 \stackrel{?}{=} w_2 \\ w_1 \in R_\Sigma^{\min\{k_1,k_2\}} \end{array}$$

If the rank of a traversal sequence $f^\rho\ w_1 \cdots w_n$ is bounded by k_1, then, for any $i \in [1..n]$, the rank of w_i is bounded by $k_1 - n + i$. These are the values of the exponents used in the regular restrictions of the right-hand side of Rule 1. Rank-boundness is crucial in order to ensure soundness of Rule 2. For instance, the traversal equation $X\,a\,a\,Y \equiv Y\,a\,a\,X$ has no solution, whereas the word equation $X\,a\,a\,Y \stackrel{?}{=} Y\,a\,a\,X$ is solvable. Notice that some substitutions, like $X, Y \mapsto a$, give equal sequences, but they are not traversal sequences.

Theorem 11. *The rules of Definition 10 describe a sound and complete decision procedure for rank-bounded traversal systems. In other words, for any rank-bounded traversal system* S,

1. *if* $S \Longrightarrow^* S'$ *and the substitution* σ *is a solution of* S', *then* σ *is also a solution of* S, *and*
2. *if the substitution* σ *is a solution of* S, *then there exists a word unification problem with regular constraints* S', *a transformation sequence* $S \Longrightarrow^* S'$, *and an extension* σ' *of* σ, *such that* σ' *is a solution of* S'.

Unfortunately, this nondeterministic transformation procedure does not always terminate. Notice that we can have $\rho_1(n) = \rho_2(n) = r$, and in such case we obtain a traversal equation $Y_r \equiv Y_r'$ with the same bounds $Y_r \in R_\Sigma^{k_1}$ and $Y_r' \in R_\Sigma^{k_2}$ as the original one. However, these transformation rules can be used to find solutions σ of equations $w_1 \equiv w_2$, such that $\sigma(w_1)$ and $\sigma(w_2)$ are traversal sequences for the same term, and they are "similar", where "similar" means that they only differ in a bounded number of permutations.

Definition 12. *Given two traversal sequences* v *and* w *over* Σ_Π, *we say that they differ in* n *permutations if, either*

1. $v = f^\rho\, r_1 \cdots r_m$ *and* $w = f^\rho\, s_1 \cdots s_m$, *for any* $i \in [1..m]$, r_i *and* s_i *differ in* n_i *permutations, and* $\sum_{i=1}^m n_i = n$, *or*
2. $v = f^\rho\, r_{\rho(1)} \cdots r_{\rho(m)}$ *and* $w = f^\tau\, s_{\tau(1)} \cdots s_{\tau(m)}$, *where* $\rho \neq \tau$, *for any* $i \in [1..m]$, r_i *and* s_i *differ in* n_i *permutations, and* $\sum_{i=1}^n n_i = n - 1$.

Definition 13. *A* permutation-bounded traversal equation, *noted* $w_1 \equiv_k w_2$, *is a tuple of two words with variables* w_1 *and* w_2, *and an integer* k.

A substitution σ *is said to be a solution of a permutation-bounded traversal equation* $w_1 \equiv_k w_2$ *if* $\sigma(w_1)$ *and* $\sigma(w_2)$ *are both traversal sequences of the same term, and they only differ in at most* k *permutations.*

A permutation- and rank-bounded traversal system *is a rank-bounded traversal system where all traversal equations are permutation-bounded.*

Theorem 14. *Solvability of permutation- and rank-bounded traversal systems is decidable.*

Proof. We can reduce the problem to an equivalent word unification problem with regular constraints using a variant of the rules of Definition 10 for permutation-bounded equations, finitely many times.

When we apply Rule 1 with $\rho_1 = \rho_2$, we transform $w_1 \equiv_k w_2$ into $\{Y_i \equiv_{k_i} Y_i'\}_{i\in[1..n]}$ where $\sum_{i=1}^n k_i = k$. We can require the existence of $i, j \in [1..n]$, such that $i \neq j$, $k_i \neq 0$ and $k_j \neq 0$ without loosing completeness. When we apply this rule with $\rho_1 \neq \rho_2$, we transform $w_1 \equiv_k w_2$ into $\{Y_i \equiv_{k_i} Y_i'\}_{i\in[1..n]}$ where $\sum_{i=1}^n k_i = k - 1$.

Rule 2 can be applied to transform $w_1 \equiv_k w_2$ into $w_1 \stackrel{?}{=} w_2$, for any k.

It is easy to prove that this transformation process always terminates using a multiset ordering on the multisets of bounds of the traversal equations.

6 The Rank-Bound Conjecture

In this section we introduce the rank-bound conjecture. This is the base of the reduction of context unification to permutation- and rank-bounded traversal systems that we describe in the next section. As we will see, this conjecture is essential in order to prove that the traversal equations that we find in the reduction are both permutation-bounded and rank-bounded.

Conjecture 15 (Rank-Bound Conjecture). *There exists a computable function Φ such that, for any solvable context unification problem $t \stackrel{?}{=} u$ there exists a ground unifier σ satisfying*

$$\mathrm{rank}(\sigma(t)) \leq \Phi(\mathrm{size}(t \stackrel{?}{=} u))$$

The validity of the conjecture is still an open question. In fact, we think that the conjecture is true, not only for *just one* ground unifier, but for *any* most general unifier. This stronger version of the conjecture is not true for second-order unification, because we can have most general second-order unifiers with arbitrarily large rank, as the following example shows.

Example 16. The second-order unification problem

$$F(f(a,a)) \stackrel{?}{=} f(F(a), F(a))$$

has only one context unifier $\sigma = [F \mapsto \lambda x \,.\, x]$. However, it has infinitely many second-order unifiers which are not context unifiers, like

$$\sigma = [F \mapsto \lambda x \,.\, f(f(f(x,x), f(x,x)), f(f(x,x), f(x,x)))]$$

For any $n \geq 0$, there is a second-order unifier where bound variable x occurs 2^n many times in the body of the function, and the rank of $\sigma(F(f(a,a)))$ is equal to $n+1$. This term $\sigma(F(f(a,a)))$ can be represented as follows for $n = \infty$.

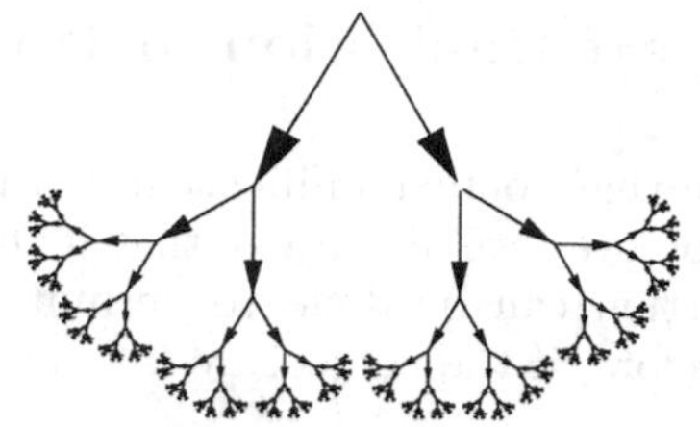

In the following Lemma we prove that the conjecture is true for first-order unification.

Lemma 17. *Given a solvable first-order unification problem* $t \stackrel{?}{=} u$*, its m.g.u.* σ *satisfies*

$$\text{rank}(\sigma(t)) \leq \text{size}(t) + \text{size}(u)$$

Proof. Suppose we have an unification problem $t \stackrel{?}{=} u$ like

$$g(f(a,b), f(X,X), f(Y,Y)) \stackrel{?}{=} g(X,Y,Z)$$

We can represent it by a directed acyclic graph (DAG) where we have two initial nodes (one for each side of the equation), and a unique node per variable. We can solve the unification problem by re-addressing the arrows pointing to a variable, when this variable is instantiated. Therefore we can represent $\sigma(t)$ by means of a DAG D, where $\text{size}(D) \leq \text{size}(t) + \text{size}(s)$, being the size of a DAG its number of arrows. This is the representation of the DAG corresponding to our example (where, for simplicity, we have added a thick arrow instead of re-addressing arrows pointing to variables):

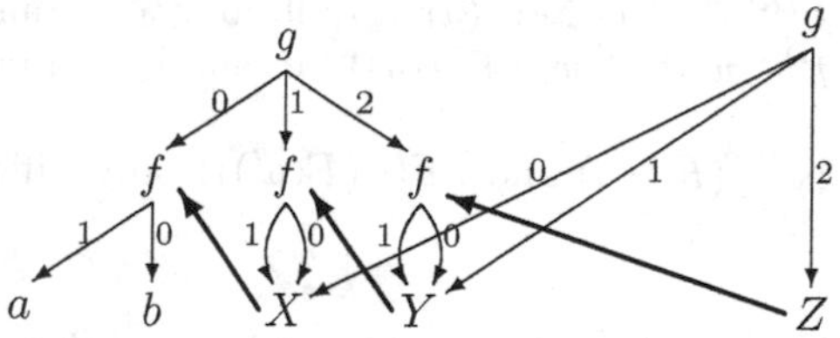

For any labelling of the original DAG, the same labels in the DAG resulting from instantiation represent a traversal sequence of $\sigma(t)$ and a traversal sequence of $\sigma(u)$. Defining the rank of a node as the addition of the label in the path from the root to this node, the rank of the traversal sequence will be the maximal of the rank of all leaves. In our example, this rank is 5 and it is obtained from the following path

$$g \xrightarrow{2} f \xrightarrow{1} f \xrightarrow{1} f \xrightarrow{1} a$$

The rank of a path never exceeds the number of arrows of the DAG, i.e. its size, because, to avoid occur check, we can not repeat nodes in a path. Therefore, when we use an arrow with label n, there are at least n other arrows (the ones with the same origin) that can not be contained in the same path. We can conclude that the traversal sequence of $\sigma(t)$ represented in the path satisfies $\text{rank}(s) \leq \text{size}(t) + \text{size}(u)$, thus $\text{rank}(\sigma(t)) \leq \text{size}(t) + \text{size}(u)$.

7 Reducing Context Unification to Traversal Equations

In this section we prove that context unification can be reduced to solvability of traversal systems. Moreover, we also prove that if the rank-bound conjecture is true, then this reduction can be done to permutation- and rank-bounded traversal systems. Therefore, if the conjecture is true, then context unification is decidable.

The reduction is very similar to the naive reduction described in Section 2: First-order variables X are encoded as word variables X' such that, if σ is a solution of the context unification problem, and σ' is the corresponding solution of the equivalent word unification problem, then $\sigma'(X') = \mathrm{NF}(\sigma(X))$.

For every n-ary context variable F, we would need $n+1$ word variables $F'_0,\dots,F'_n$, such that $\sigma'(F'_0\, a\, F'_1\, a \cdots F'_{n-1}\, a\, F'_n) = \mathrm{NF}(\sigma(F(a,\dots,a)))$. However, this simple translation does not work. If a term t contains two occurrences of a first-order variable X, then $\mathrm{NF}(\sigma(t))$ will contain two occurrences of $\mathrm{NF}(\sigma(X))$. However, two different occurrences of a context variable can have different arguments, and this means that the context $\sigma(F)$ can be traversed in different ways, depending on the arguments. Notice that, in general, even if $\mathrm{NF}(t[a]) = w_0\, a\, w_1$, we can have $\mathrm{NF}(t[u]) \neq w_0\, \mathrm{NF}(u)\, w_1$. Fortunately, the different ways in which the occurrences of $\sigma(F)$ are traversed in the normal form of $\sigma(t)$ are not very different, i.e. they differ in at most a bounded number of permutations.

Example 18. Let $\sigma(F) = \lambda x\,.\,f(f(x,t_1),t_2)$, where $\mathrm{rank}(t_1) < \mathrm{rank}(t_2)$, and $w_i = \mathrm{NF}(t_i)$, for $i = 1,2$. Depending on the argument u, we have

$$\mathrm{NF}(\sigma(F(u))) = \begin{cases} f^{[1,2]}\, f^{[1,2]}\, \mathrm{NF}(\sigma(u))\, w_1\, w_2 & \text{if } \mathrm{rank}(\sigma(u)) \leq \mathrm{rank}(t_1) \\ f^{[1,2]}\, f^{[2,1]}\, w_1\, \mathrm{NF}(\sigma(u))\, w_2 & \text{if } \mathrm{rank}(t_1) < \mathrm{rank}(\sigma(u)) \leq \mathrm{rank}(t_2) \\ f^{[2,1]}\, w_2\, f^{[2,1]}\, w_1\, \mathrm{NF}(\sigma(u)) & \text{if } \mathrm{rank}(t_2) < \mathrm{rank}(\sigma(u)) \end{cases}$$

For any u and u', $\mathrm{NF}(\sigma(F(u)))$ and $\mathrm{NF}(\sigma(F(u')))$ only differ in at most 2 permutations.

Lemma 19. *Let F be a context variable and σ a substitution. For any two terms $F(t_1,\dots,t_n)$ and $F(u_1,\dots,u_n)$, there exist sequences $v_0,\dots,v_n, w_0,\dots,w_n$ and permutations $\rho,\tau \in \Pi_n$, such that*

$$\begin{aligned} \mathrm{NF}(\sigma(F(t_1,\dots,t_n))) &= v_0\, \mathrm{NF}(\sigma(t_{\rho(1)}))\, v_1 \cdots v_{n-1}\, \mathrm{NF}(\sigma(t_{\rho(n)}))\, v_n \\ \mathrm{NF}(\sigma(F(u_1,\dots,u_n))) &= w_0\, \mathrm{NF}(\sigma(u_{\tau(1)}))\, w_1 \cdots w_{n-1}\, \mathrm{NF}(\sigma(u_{\tau(n)}))\, w_n \end{aligned}$$

and, for any sequence of constants $\{a_i\}_{i\in[1..n]}$,

$$\begin{gathered} v_0\, a_{\rho(1)}\, v_1 \cdots v_{n-1}\, a_{\rho(n)}\, v_n \\ w_0\, a_{\tau(1)}\, w_1 \cdots w_{n-1}\, a_{\tau(n)}\, w_n \end{gathered}$$

are both traversal sequences of $\sigma(F(a_1,\dots,a_n))$, and they only differ in at most $n \cdot \mathrm{rank}(\sigma(F(a_1,\dots,a_n)))$ permutations.

Notice that we need the rank-bound conjecture in order to bound the value of $\text{rank}(\sigma(F(a_1,...,a_n)))$, i.e. to prove that these two traversal sequences differ in a bounded number of permutations.

In the rest we describe how a context unification problem could be effectively translated into an equivalent system of traversal equations.

Theorem 20. *Context unification can be reduced to solvability of traversal systems.*
If the Rank-Bound Conjecture is true, then context unification can be reduced to solvability of permutation- and rank-bounded traversal systems.

Proof. Let $t \stackrel{?}{=} u$ be the original context unification problem, and $(\Sigma, \mathcal{X})$ be the original signature. We assume that Σ is finite, and contains at least $2 \cdot n$ distinct first-order constants $a_1,...,a_n,b_1,...,b_n$, where $n = \max\{\text{arity}(F) \mid F \in \text{Var}(t \stackrel{?}{=} u)\}$, and a binary symbol f, and that $a_1,...,a_n,b_1,...,b_n$ do not occur in $t \stackrel{?}{=} u$. Therefore, if a problem is solvable, it has a ground unifier.

First step. The order of the arguments in F and in $\sigma(F)$ are not necessarily the same. In this first step we guess a permutation $\rho_F \in \Pi_{\text{arity}(F)}$ for any context variable and transform $t \stackrel{?}{=} u$ into $\sigma_0(t) \stackrel{?}{=} \sigma_0(u)$ where

$$\sigma_0 = \bigcup_{F \in \text{Var}(t \stackrel{?}{=} u)} [F \mapsto \lambda x_1 \cdots x_n \,.\, F'(x_{\rho_F(1)},...,x_{\rho_F(n)})]$$

Now, we can assume that F' and its instance have the arguments in the same order. Moreover, as far as σ_0 is simply a renaming substitution, $t \stackrel{?}{=} u$ and $\sigma_0(t) \stackrel{?}{=} \sigma_0(u)$ are equivalent problems.

Second step. We introduce a word variable $X' \in \mathcal{W}$ *for every first order variable* $X \in \mathcal{X}$, and $\text{arity}(F) + 1$ many word variables $F_0^p,...,F_{\text{arity}(F)}^p \in \mathcal{W}$ *for every occurrence* p *of a context variable* F in the problem (notice that in this case we use different word variables for every occurrence).

We guess a permutation ρ_p for any occurrence of a constant function f or of a context variable F, with arity greater or equal than two, in a position p of the problem.

We define the following translating function $\mathcal{T}$ that given a subterm $t \in T^1(\Sigma, \mathcal{X})$ of the problem, and its position p, returns its translation in terms of words with variables $w \in (\Sigma_\Pi \cup \mathcal{W})^*$.

For any first-order constant a, or variable X,

$$\mathcal{T}(a,p) = a$$
$$\mathcal{T}(X,p) = X'$$

For every n-ary function symbol f, or context variable F, occurring at position p, let $w_i = \mathcal{T}(t_i, p \cdot i)$, and ρ_p be the permutation conjectured for this position, then

$$\mathcal{T}(f(t_1,...,t_n),p) = f^{\rho_p}\, w_{\rho_p(1)} \cdots w_{\rho_p(n)}$$
$$\mathcal{T}(F(t_1,...,t_n),p) = F_0^p\, w_{\rho_p(1)}\, F_1^p \cdots F_{n-1}^p\, w_{\rho_p(n)}\, F_n^p$$

Finally, the traversal system will contain the following equations:

1. A word equation for the original problem $t \stackrel{?}{=} u$

$$\mathcal{T}(t,1) \stackrel{?}{=} \mathcal{T}(u,2)$$

2a. For any two occurrences $F(t_1,\ldots,t_n)$ and $F(u_1,\ldots,u_n)$ of a context variable F at positions p and q, we introduce the following traversal equations and regular constraints:[1]

$$\begin{array}{ll} \mathcal{T}(F(a_1,\ldots,a_n),p) \equiv_k \mathcal{T}(F(a_1,\ldots,a_n),q) & \mathcal{T}(F(b_1,\ldots,b_n),p) \equiv_k \mathcal{T}(F(b_1,\ldots,b_n),q) \\ \mathcal{T}(F(a_1,\ldots,a_n),p) \in R^{k_1}_{\Sigma\Pi} & \mathcal{T}(F(b_1,\ldots,b_n),p) \in R^{k_1}_{\Sigma\Pi} \\ \mathcal{T}(F(a_1,\ldots,a_n),q) \in R^{k_2}_{\Sigma\Pi} & \mathcal{T}(F(b_1,\ldots,b_n),q) \in R^{k_2}_{\Sigma\Pi} \end{array}$$

where $k = \text{arity}(F) \cdot \Phi(\text{size}(t \stackrel{?}{=} u))$
$k_1 = k_2 = \Phi(\text{size}(t \stackrel{?}{=} u))$
and Φ is the computable function introduced in the rank-bound conjecture.

2b. In case we want to reduce context unification to (non-bounded) traversal systems, we will introduce

$$\mathcal{T}(F(a_1,\ldots,a_n),p) \equiv \mathcal{T}(F(a_1,\ldots,a_n),q)$$
$$\mathcal{T}(F(b_1,\ldots,b_n),p) \equiv \mathcal{T}(F(b_1,\ldots,b_n),q)$$

In this second case, we do not need the conjecture to fix k, k_1 and k_2.

The duplication of traversal equations with distinct constants a_i and b_i ensures that these constants occur in the place of the arguments. Otherwise, if we only introduce a traversal equation $X_0\, a\, X_1 \equiv X'_0\, a\, X'_1$, we can get solutions like $\sigma = [X_0 \mapsto f^{[1,2]}\, a][X_1 \mapsto \lambda][X'_0 \mapsto f^{[1,2]}][X'_1 \mapsto a]$, that do not satisfy $\sigma(X_0\, b\, X_1) \equiv \sigma(X'_0\, b\, X'_1)$, and leads to incompatible definitions of $\sigma(F) = \lambda x\,.\, f(a,x)$ and $\sigma(F) = \lambda x\,.\, f(x,a)$.

Corollary 21. *If the Rank-Bound Conjecture is true, then Context Unification is decidable.*

Example 22. To conclude, let's see how problem $X(Y(a,b)) \stackrel{?}{=} Y(X(a),b)$ could be translated into a traversal system.

We guess σ_0 equals to identity in the first step. In second step, we introduce the word variables X_0, X_1, X'_0, X'_1 for the two occurrences of X, and $Y_0, Y_1, Y_2, Y'_0, Y'_1, Y'_2$ for Y. For both occurrences of Y, the only symbol with arity 2 or greater, we guess the same permutation $\rho_{1\cdot 1} = \rho_2 = [2,1]$.

The translation of the unification problem results then into:

$$X_0\, Y_0\, b\, Y_1\, a\, Y_2\, X_1 \stackrel{?}{=} Y'_0\, b\, Y'_1\, X'_0\, a\, X'_1\, Y'_2$$

$$\begin{array}{ll} X_0\, a_1\, X_1 \equiv_k X'_0\, a_1\, X'_1 & X_0\, b_1\, X_1 \equiv_k X'_0\, b_1\, X'_1 \\ X_0\, a_1\, X_1 \in R^k_{\Sigma\Pi} & X_0\, b_1\, X_1 \in R^k_{\Sigma\Pi} \\ X'_0\, a_1\, X'_1 \in R^k_{\Sigma\Pi} & X'_0\, b_1\, X'_1 \in R^k_{\Sigma\Pi} \end{array}$$

[1] We can avoid to introduce a context variable occurrence in more than two traversal equation. If we have $p_1,\ldots,p_n$ occurrences of F, we can introduce an equation relating p_1 and p_2, p_2 and $p_3,\ldots, p_{n-1}$ and p_n.

$$Y_0\, a_2\, Y_1\, a_1\, Y_2 \equiv_{2k} Y_0'\, a_2\, Y_1'\, a_1\, Y_2' \qquad Y_0\, b_2\, Y_1\, b_1\, Y_2 \equiv_{2k} Y_0'\, b_2\, Y_1'\, b_1\, Y_2'$$
$$Y_0\, a_2\, Y_1\, a_1\, Y_2 \in R^k_{\Sigma_\Pi} \qquad Y_0\, b_2\, Y_1\, b_1\, Y_2 \in R^k_{\Sigma_\Pi}$$
$$Y_0'\, a_2\, Y_1'\, a_1\, Y_2' \in R^k_{\Sigma_\Pi} \qquad Y_0'\, b_2\, Y_1'\, b_1\, Y \in R^k_{\Sigma_\Pi}$$

where $k = \Phi(8)$, and Φ is the function introduced by the rank-bound conjecture.

8 Conclusions and Further Work

In this paper we prove that, if the rank-bound conjecture is true, then context unification is decidable. The decidability of context unification is still an open question, and a positive answer would have important implications in very different research areas. Additionally, we define *traversal equations* and *rank- and permutation-bounded traversal equations*, and prove that solvability of the second ones is decidable.

We are currently trying to prove the rank-bound conjecture, and finding a reduction from traversal equations to context unification, to prove the equivalence of both problems.

Acknowledgements. We are in debt with M. Schmidt-Schauß and K.U. Schulz who carefully read a preliminary version of this paper and found some errors. We also acknowledge M.L. Bonet for her valuable comments and suggestions, and all the anonymous referees.

References

[Com92] Hubert Comon. Completion of rewrite systems with membership constraints. In *Int. Coll. on Automata, Languages and Programming, ICALP'92*, volume 623 of *LNCS*, Vienna, Austria, 1992.

[Com98] Hubert Comon. Completion of rewrite systems with membership constraints. *Journal of Symbolic Computation*, 25(4):397–453, 1998.

[DMM97] Volker Diekert, Yuri Matiyasevich, and Anca Muscholl. Solving trace equations using lexicographical normal forms. In *Int. Colloquium on Automata, Languages and Programming, ICALP'97*, pages 336–346, Bologna, Italy, 1997.

[ENRX98] Markus Egg, Joachim Niehren, Peter Ruhrberg, and Feiyu Xu. Constraints over lambda-structures in semantic underspecification. In *Proceedings of the 36th Annual Meeting of the Association for Computational Linguistics and the 17th International Conference on Computational Linguistics (ACL'98)*, pages 353–359, Montreal, Quebec, Canada, 1998.

[KNT98] Alexander Koller, Joachim Niehren, and Ralf Treinen. Dominance constraints: Algorithms and complexity. In *Proceedings of the Third Conf. on Logical Aspects of Computational Linguistics*, Grenoble, France, 1998.

[LA96] Jordi Levy and Jaume Agustí. Bi-rewrite systems. *Journal of Symbolic Computation*, 22(3):279–314, 1996.

[Lev96] Jordi Levy. Linear second-order unification. In *Proceedings of the 7th Int. Conf. on Rewriting Techniques and Applications (RTA'96)*, volume 1103 of *LNCS*, pages 332–346, New Brunswick, New Jersey, 1996.

[Lev98] Jordi Levy. Decidable and undecidable second-order unification problems. In *Proceedings of the 9th Int. Conf. on Rewriting Techniques and Applications (RTA'98)*, volume 1379 of *LNCS*, pages 47–60, Tsukuba, Japan, 1998.

[LV98] Jordi Levy and Margus Veanes. On unification problems in restricted second-order languages. In *Annual Conf. of the European Ass. of Computer Science Logic (CSL98)*, Brno, Czech Republic, 1998.

[LV00a] Jordi Levy and Margus Veanes. On the undecidability of second-order unification. *Information and Computation*, 159:125–150, 2000.

[LV00b] Jordi Levy and Mateu Villaret. Linear second-order unification and context unification with tree-regular constraints. In *Proceedings of the 11th Int. Conf. on Rewriting Techniques and Applications (RTA'00)*, volume 1833 of *LNCS*, pages 156–171, Norwich, UK, 2000.

[Mak77] G. S. Makanin. The problem of solvability of equations in a free semigroup. *Math. USSR Sbornik*, 32(2):129–198, 1977.

[NPR97a] Joachim Niehren, Manfred Pinkal, and Peter Ruhrberg. On equality up-to constraints over finite trees, context unification, and one-step rewriting. In *Proceedings of the 14th Int. Conference on Automated Deduction (CADE-14)*, volume 1249 of *LNCS*, pages 34–48, Townsville, North Queensland, Australia, 1997.

[NPR97b] Joachim Niehren, Manfred Pinkal, and Peter Ruhrberg. A uniform approach to underspecification and parallelism. In *Proceedings of the 35th Annual Meeting of the Association for Computational Linguistics and the 8th Conference of the European Chapter of the Association for Computational Linguistics (ACL'97)*, pages 410–417, Madrid, Spain, 1997.

[NTT00] Joachim Niehren, Sophie Tison, and Ralf Treinen. On rewrite constraints and context unification. *Information Processing Letters*, 74(1-2):35–40, 2000.

[Pie73] Tomasz Pietrzykowski. A complete mechanization of second-order logic. *J. of the ACM*, 20(2):333–364, 1973.

[Sch91] Klaus U. Schulz. Makanin's algorithm, two improvements and a generalization. Technical Report CIS-Bericht-91-39, Centrum für Informations und Sprachverarbeitung, Universität München, 1991.

[SS96] Manfred Schmidt-Schauß. An algorithm for distributive unification. In *Proceedings of the 7th Int. Conf. on Rewriting Techniques and Applications (RTA'96)*, volume 1103 of *LNCS*, pages 287–301, New Jersey, USA, 1996.

[SS98] Manfred Schmidt-Schauß. A decision algorithm for distributive unification. *Theoretical Computer Science*, 208:111–148, 1998.

[SS99a] Manfred Schmidt-Schauß. Decidability of bounded second-order unification. Technical Report Frank-report-11, FB Informatik, J.W. Goethe Universität Frankfurt, 1999.

[SS99b] Manfred Schmidt-Schauß. A decision algorithm for stratified context unification. Technical Report Frank-report-12, FB Informatik, J.W. Goethe Universität Frankfurt, 1999.

[SSS98] Manfred Schmidt-Schauß and Klaus U. Schulz. On the exponent of periodicity of minimal solutions of context equations. In *Proceedings of the 9th Int. Conf. on Rewriting Techniques and Applications (RTA'98)*, volume 1379 of *LNCS*, pages 61–75, Tsukuba, Japan, 1998.

[SSS99] Manfred Schmidt-Schauß and Klaus U. Schulz. Solvability of context equations with two context variables is decidable. In *Proceedings of the 16th Int. Conf. on Automated Deduction (CADE-16)*, LNAI, pages 67–81, 1999.

Weakly Regular Relations and Applications

Sébastien Limet[1], Pierre Réty[1], and Helmut Seidl[2]

[1] LIFO, Université d'Orléans, France, {limet,rety}@lifo.univ-orleans.fr
[2] Dept. of Computer Science, University of Trier, Germany seidl@psi.uni-trier.de

Abstract. A new class of tree-tuple languages is introduced: the weakly regular relations. It is an extension of the regular case (regular relations) and a restriction of tree-tuple synchronized languages, that has all usual nice properties, except closure under complement. Two applications are presented: to unification modulo a rewrite system, and to one-step rewriting.

1 Introduction

Several classes of tree-tuple languages (also viewed as tree relations), have been defined by means of automata or grammars. In particular, a simple one, the *Regular Relations (RR)*, consists in defining regularity as being the one of the tree language (over the product-alphabet) obtained by overlapping the tuple components.

A more sophisticated class, the *Tree-Tuple Synchronized Languages (TTSL)*, is obtained by extending RRs thanks to synchronization constraints between independent branches. TTSLs have first been introduced by means of *Tree-Tuple Synchronized Grammars (TTSG)* and have been applied to equational unification [7], to logic program validation [10], and to one-step rewriting theory [8]. They have next been reformulated in a simpler way by means of *Constraint Systems (CS)*, and applied to rewriting again and to concurrency [4].

RRs have all usual nice properties[1], but expressiveness is poor. It is just the opposite for TTSLs. In particular, CSs are not closed under intersection[2], nor under complement.

In this paper we define an intermediate class, called *Weakly Regular Relations (WRR)*, between RRs and synchronized languages, that has all RRs properties, except closure under complement. Thanks to its expressiveness greater than RRs, WRRs enable to prove new decidability results:

- on unification modulo a rewrite system. Unlike [7], a non-linear goal is allowed. This result is obtained by using the general method of [11] for deciding unifiability with the help of a tree-tuple language.

[1] Except closure under iteration (transitive closure).
[2] Horizontal TTSGs are closed under intersection, provided a more complicated control is used [8], which makes great difficulties. In particular, we do not know if they are then still closed under projection.

A. Middeldorp (Ed.): RTA 2001, LNCS 2051, pp. 185–200, 2001.

– on the existential one-step rewriting theory. Unlike [8], non-linear rewrite rules are allowed.

All missing proofs and details can be found in [9]

2 Regular Relations and Synchronized Languages

Let Σ be an alphabet, i.e. a set of symbols with fixed arities. T_Σ denotes the set of ground terms over Σ. For a position p in $t \in T_\Sigma$, $t|_p$ denotes the subterm of t at position p, and $t(p)$ denotes the symbol occurring in t at position p.

Given a language S of l-tuples, and T of n-tuples, and for $i \in \{1, \dots, l\}$, $j \in \{1, \dots, n\}$ the *i, j-join* of S and T is a $l + n - 1$-tuple language, denoted $S \bowtie_{i,j} T$, and defined by:

$$\{(s_1, \dots, s_l, t_1, \dots, t_{j-1}, t_{j+1}, \dots, t_n) \mid (s_1, \dots, s_l) \in S \wedge (t_1, \dots, t_n) \in T \wedge s_i = t_j\}$$

$S \bowtie T$ stands for $S \bowtie_{l,1} T$. For $i_1, \dots, i_k \in \{1, \dots, l\}$, the *projection* of S on components $i_1, \dots, i_k$ is the k-tuple language, denoted $\Pi_{i_1,\dots,i_k}(S)$, defined by:

$$\Pi_{i_1,\dots,i_k}(S) = \{(s_{i_1}, \dots, s_{i_k}) \mid \forall j \neq i_1, \dots, i_k, \exists s_j, (s_1, \dots, s_l) \in S\}$$

For tuples $s \in S$, $t \in T$, $s\,t$ denotes the $l + n$-tuple obtained by *concatenation* of s and t.

2.1 Regular Relations

We define regularity for n-ary relations as in [2], i.e. a relation R is regular iff the tree language over the product-alphabet obtained by overlapping the tuple-components is regular.

Formally, let $\Sigma_\oplus$ be the product alphabet defined by $\Sigma_\oplus = (\Sigma \cup \{\perp\}) \times (\Sigma \cup \{\perp\}) - \{\perp\perp\}$ where $\perp$ is a new constant. For $s, t \in T_\Sigma$ we recursively define $s \oplus t \in T_{\Sigma_\oplus}$ by:

$$f(s_1, \dots, s_n) \oplus g(t_1, \dots, t_m) = \begin{cases} fg(s_1 \oplus t_1, \dots, s_n \oplus t_n, \perp \oplus t_{n+1}, \dots, \perp \oplus t_m) \\ \quad \text{if } n < m \\ fg(s_1 \oplus t_1, \dots, s_m \oplus t_m, s_{m+1} \oplus \perp, \dots, s_n \oplus \perp) \\ \quad \text{otherwise} \end{cases}$$

For instance $f(a, g(b)) \oplus f(f(a, a), b)$ is $ff(af(\perp a, \perp a), gb(b\perp))$. This definition trivially extends to product of k terms $t_1 \oplus \dots \oplus t_k$.

A n-ary relation $R \subseteq T_\Sigma^n$ is regular iff the tree language $\{t_1 \oplus \dots \oplus t_n \mid (t_1, \dots, t_n) \in R\}$ is regular. *RR* stands for *Regular Relation*. For example, $\{(t, t) \mid t \in T_\Sigma\}$ and for given symbols $f, g \in \Sigma$ of same arity, $\{(t, t[f \leftarrow g]) \mid t \in T_\Sigma\}$ are RRs. On the other hand $\{(t, t_{sym}) \mid t \in T_\Sigma,\ t_{sym}$ is the symmetric tree of $t\}$ is not a RR if Σ is not monadic.

2.2 Constraint Systems for Synchronized Languages

Synchronized languages refer to the class of languages defined by means of TTSG with bounded synchronizations of [7]. The exact definition is rather technical and will not be given here. The aim of this section is to define a more uniform way to recognize this class of languages. We use constraint systems (CS), and we consider their least fix-point solutions. A CS can be viewed as a grammar for which a bottom-up point of view is adopted. We will sometimes identify non-terminals with the sets they generate.

Example 1. In the signature $\Sigma = \{f^{\backslash 2}, b^{\backslash 0}\}$ let $Id_2 = \{(t,t) \mid t \in T_\Sigma\}$ be the set of pairs of identical terms. Id_2 can be defined by the following CS:

$$Id_2 \supseteq (b,b)$$
$$Id_2 \supseteq (f(1_1, 2_1),\ f(1_2, 2_2))\ (Id_2, Id_2)$$

where $1_1, 2_1, 1_2, 2_2$ abbreviate pairs (for readability). For example 2_1 means $(2,1)$, which denotes the first component of the second argument (the second Id_2). Note that since 1_1 and 1_2 come from the same Id_2, they represent two identical terms, in other words they are linked (*synchronized*), whereas for example 1_1 and 2_1 are independent.

Example 2. Now if we consider the slightly different CS :

$$X_{sym} \supseteq (b,b)$$
$$X_{sym} \supseteq (f(1_1, 2_1),\ f(2_2, 1_2))\ (X_{sym}, X_{sym})$$

we get the set $L_{sym} = \{(t, t_{sym}) \mid t_{sym} \text{ is the symmetric tree of } t\}$.

Example 3. In the signature $\Sigma = \{s^{\backslash 1}, b^{\backslash 0}\}$ let $L_{dble} = \{(s^n(b), s^{2n}(b))\}$. It can be defined by the CS :

$$X_{dble} \supseteq (b,b)$$
$$X_{dble} \supseteq (s(1_1),\ s(s(1_2)))\ X_{dble}$$

General Formalization

Assume we are given a (universal) index set N for tuple components. For $I \subseteq N$ and any set M, the set of I-tuples $a : I \to M$ is denoted by M^I. Often, we also write $a = (a_i)_{i \in I}$ provided $a(i) = a_i$ which for $I = \{1, \ldots, k\} \subseteq \mathbb{N}$, is also written as $a = (a_1, \ldots, a_k)$.

Different tree tuple languages may refer to tuples of different length, or, to different index sets. Our constraint variables represent tree tuple languages. Consequently, they have to be equipped with the intended index set. Such an assignment is called *classification*. Accordingly, a *classified* set of tuple variables (over N) is a pair $(\mathcal{X}, \rho)$ where $\rho : \mathcal{X} \to 2^N$ assigns to each variable X a subset of indices. This subset is called the *class* of X. For convenience and whenever ρ

is understood, we omit ρ and denote the classified set $(\mathcal{X}, \rho)$ by $\mathcal{X}$. The maximal cardinality of the classes in $\mathcal{X}$ is also called the *width* of $\mathcal{X}$. In particular, in example 1, $N = \{1,2\}$, $\mathcal{X} = \{Id_2\}$, $\rho(Id_2) = \{1,2\}$, and the width of $\mathcal{X}$ is 2.

A *constraint system* (*CS*) for tree tuple languages consists of a classified set $(\mathcal{X}, \rho)$ of constraint variables, together with a finite set $\mathcal{E}$ of inequations of the form

$$X \supseteq \Box(X_1, \ldots, X_k) \quad (1)$$

where $X, X_1, \ldots, X_k \in \mathcal{X}$ and $\Box$ is an operator mapping the concatenation of tuples for the variables X_i to tuples for X. More precisely, let

$$J = \{(i,x) \mid 1 \leq i \leq k, x \in \rho(X_i)\} \quad (2)$$

denote the *disjoint union* of the index sets corresponding to the variables X_i (in example 1, $J = \{(1,1), (1,2), (2,1), (2,2)\}$ abbreviated into $\{1_1, 1_2, 2_1, 2_2\}$). Then $\Box$ denotes a mapping $T_\Sigma^J \to T_\Sigma^{\rho(X)}$. Each component of this mapping is specified through a tree expression t which may access the components of the argument tuple and apply constructors from signature Σ. Thus, t can be represented as an element of $T_\Sigma(J)$ where $T_\Sigma(J)$ denotes all trees over Σ which additionally may contain nullary symbols from the index set J.

Consider, e.g., the second constraint in example 1. There, the first component of the operator is given by $t = f(1_1, 2_1)$.

The mapping induced by such a tree t then is defined by

$$t(s_j)_{j \in J} = t\{j \mapsto s_j\}_{j \in J}$$

for every $(s_j)_{j \in J} \in T_\Sigma^J$. Accordingly, $\Box$ is given by a tuple $\Box \in T_\Sigma(J)^{\rho(X)}$.

Let us collect a set of useful special forms of constraint systems. The constraint (1) is called

- *non-copying* iff no index $j \in J$ occurs twice in $\Box$;
- *horizontal* iff for any components $\Box_x$, $\Box_y$ and any positions p, q, $\Box_x|_p = (i, x')$, $\Box_y|_q = (i, y')$ implies p and q are positions at the same depth.
- *regular* iff each component of $\Box$ is a single constructor application of the form $\Box_x = a_x((1,x), \ldots, (n,x))$ for each $x \in \rho(X)$ (a_x being of arity n). Note that regularity implies horizontality.

The whole constraint system is called non-copying, (horizontal, regular) iff each constraint in $\mathcal{E}$ is so.

For example, the CSs that define Id_2 and L_{sym} are non-copying, and horizontal. Moreover Id_2 is regular. On the other hand, L_{dble} is not horizontal.

The class of non-copying CSs corresponds to the class of TTSGs of [7], so recognizes synchronized languages. In the rest of the paper, we only consider *non-copying* CSs even it is not explicitly written.

Proposition 1. *[4] The class of non-copying CSs is closed under union, projection, cartesian product. Moreover membership and emptiness are decidable.*

Proposition 2. *RRs are exactly the languages defined by regular CSs.*

Proof. If top symbols of all components of $\square$ have the same arity, i.e. $\forall x, y \in \rho(X)$, $arity(a_x) = arity(a_y)$, then it is obvious since overlapping tuple-components amounts to synchronize identical positions together. Otherwise, a non-regular CS may define a RR, like for example $\{X \supseteq (s(1_1), f(1_2, 2_1))(X, Y),\ X \supseteq (b, b),\ Y \supseteq b\}$. But it can be transformed into a regular CS by changing the index set of Y into $\rho(Y) = \{2\}$. The CS becomes $\{X \supseteq (s(1_1), f(1_2, 2_2))(X, Y),\ X \supseteq (b, b),\ Y \supseteq b\}$, which is regular. However, for simplicity, we always consider in examples that for any X, $\rho(X) = \{1, 2, \ldots\}$.

A variable assignment is a mapping σ assigning to each variable $X \in \mathcal{X}$ a subset of $T_\Sigma^{\rho(X)}$ (i.e. a set of tuples of ground terms). Variable assignment σ satisfies the constraint (1) iff

$$\sigma(X) \supseteq \{\square(t_1, \ldots, t_k) \mid t_i \in \sigma(X_i)\} \tag{3}$$

Note that the $\square$-operator is applied to the cartesian product of the argument sets $\sigma(X_i)$. In particular, the tuples inside the argument sets are kept "synchronized" while tuples from different argument sets may be arbitrarily combined.

The variable assignment σ is a *solution* of the constraint system iff it satisfies all constraints in the system. Since, the operator application in (3) is monotonic, even continuous (w.r.t. set inclusion of tuple languages) we conclude that each constraint system has a unique least solution.

3 Study of Intersection of Synchronized Languages

Finding a CS that recognizes the intersection of two synchronized languages is difficult, precisely because of synchronizations. The first example exhibits the first difficulty: deadlocks.

Example 4. Let L_1 and L_2 recognized respectively by the variables X_1 and X_2 of the following constraint systems:

L_1	L_2
$X_1 \supseteq (1_1, g(1_2))Y_1$	$X_2 \supseteq (f(1_1), 1_2)Y_2$
$Y_1 \supseteq (f(1_1), f(1_2))Z_1$	$Y_2 \supseteq (g(1_1), g(1_2))Z_2$
$Z_1 \supseteq (g(b), b)$	$Z_2 \supseteq (b, f(b))$

Clearly $L_1 = L_2 = \{(f(g(b)), g(f(b)))\}$, So $L_1 \cap L_2$ is not empty but in L_1 occurrences 1 (in the first component) and 2.1 (in the second component) are synchronized whereas occurrences 2 and 1.1 are synchronized in L_2. This produces a deadlock when trying to run the two CSs in parallel to recognize $L_1 \cap L_2$.

More generally, it is possible to encode the Post Correspondence Problem by testing emptiness of the intersection of two synchronized languages.

Theorem 1. *Emptiness of the intersection of synchronized languages is undecidable.*

Proof. Let A and C be two alphabets and ϕ, ϕ' be morphisms from A^* to C^*. Let us consider the two synchronized tree languages $L = \{(\alpha, \phi(\alpha)) | \alpha \in A^+\}$ and $L' = \{(\alpha, \phi'(\alpha)) | \alpha \in A^+\}$. Let us write $\phi(a_i) = c_{i,1} \dots c_{i,p_i}$ and $\phi'(a_i) = c'_{i,1} \dots c'_{i,p'_i}$ for each $a_i \in A$. Then L and L' are recognized by the following CSs[3]:

L	L'	
$X \supseteq (a_i(1_1),\, c_{i,1} \dots c_{i,p_i}(1_2))\, X$	$X' \supseteq (a_i(1_1),\, c'_{i,1} \dots c'_{i,p'_i}(1_2))\, X'$	$\forall a_i \in A$
$X \supseteq (a_i(\bot),\, c_{i,1} \dots c_{i,p_i}(\bot))$	$X' \supseteq (a_i(\bot),\, c'_{i,1} \dots c'_{i,p'_i}(\bot))$	$\forall a_i \in A$

So deciding whether $L \cap L'$ is empty amounts to decide the existence of $\alpha \in A^+$ such that $\phi(\alpha) = \phi'(\alpha)$, i.e. solving the Post correspondence problem.

Therefore, since emptiness of synchronized languages is decidable [4], this class is not effectively closed under intersection.

Considering the two previous examples, it seems unavoidable to consider a subclass of synchronized languages to get closure under intersection. The first idea is to avoid deadlocks by forbidding leaning synchronizations (i.e. imposing that synchronization points are always at the same depth): from now we consider only horizontal CSs.

Example 5. Consider again languages Id_2 and L_{sym} as defined in Examples 1 and 2. So $Id_2 \cap L_{sym}$ is the set of pairs of terms (t, t') such that $t = t'$ and t is the symmetric tree of t'. This means that t is a self-symmetric tree. So $Id_2 \cap L_{sym}$ is the set $\{(b, b)\} \cup \{(f(t, t_{sym}), f(t, t_{sym}))\}$ where t_{sym} denotes the symmetric tree of t. Let $X = \{(t, t_{sym}, t, t_{sym})\}$, $Id_2 \cap L_{sym}$ can be defined by

$$Id_2 \cap L_{sym} \supseteq (b, b) | (f(1_1, 1_2), f(1_3, 1_4))X$$
$$X \supseteq (f(1_1, 2_1), f(2_2, 1_2), f(1_3, 2_3), f(2_4, 1_4))(X, X) | (b, b, b, b)$$

It seems that horizontality is a good criterion to get closure under intersection, but unfortunately it is not enough as shown by the following example.

Example 6. Let L_{pb} be the language defined by the following constraints:

$$L_{pb} \supseteq (f(1_1, 1_2), f(2_1, 2_2))(Id_2, L_{pb}) \quad L_{pb} \supseteq (b, b)$$

The following picture gives an intuition on how L_{pb} looks like:

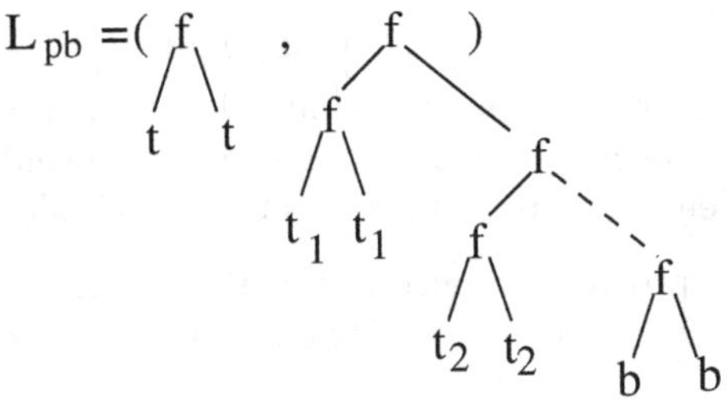

3 ⊥ just marks word ends.

The intersection of L_{pb} with Id_2 gives the language of pairs of identical balanced terms (i.e. terms all branches of which have the same length). To express it with a non-copying CS, we have to synchronize all occurrences of the same depth together, which requires wider and wider tuples, then infinitely many intermediate languages. This is impossible because a CS is always supposed to be finite. That is why the definition of WRRs needs restrictions stronger than horizontality.

4 Weakly Regular Relations

A horizontal constraint system $\mathcal{C} = (\mathcal{X}, \mathcal{E})$ is *weakly regular* iff there is a rank function $r : \mathcal{X} \to \mathbb{N}$ s.t. for every constraint c given by $X \supseteq \square(X_1, \ldots, X_k)$:

- $r(X) \geq r(X_1) + \ldots + r(X_k)$; and
- $r(X) = r(X_1) + \ldots + r(X_k)$ only provided c is regular.

In essence, this definition implies that each tree tuple of any variable X in the system is constructed by using at most $r(X)$ horizontal but non-regular constraints.

Accordingly, a (set of tree tuples or a) tree relation is called *weakly regular* (*WRR* stands for Weakly Regular Relation) iff it is defined by a weakly regular constraint system.

Example 7. Let $\Sigma = \{f^{\backslash 2}, g^{\backslash 2}, a^{\backslash 0}\}$, consider the rewrite system $R = \{f(x, y) \to g(y, x)\}$, and let $S = \{(t_1, t_2) \mid t_1 \to_R t_2\} = \{C[f(x, y)\sigma], C[g(y, x)\sigma]\}$. S is a WRR:

$$S \supseteq (f(1_1, 2_1), f(1_2, 2_2))(S, Id_2) \qquad S \supseteq (g(1_1, 2_1), g(1_2, 2_2))(S, Id_2)$$
$$S \supseteq (f(1_1, 2_1), f(1_2, 2_2))(Id_2, S) \qquad S \supseteq (g(1_1, 2_1), g(1_2, 2_2))(Id_2, S)$$
$$S \supseteq (f(1_1, 2_1), g(2_2, 1_2))(Id_2, Id_2)$$
$$Id_2 \supseteq (f(1_1, 2_1), f(1_2, 2_2))(Id_2, Id_2) \quad Id_2 \supseteq (g(1_1, 2_1), g(1_2, 2_2))(Id_2, Id_2)$$

with $r(Id_2) = 0, \;\; r(S) = 1$.

Fact 1 – *If the constraint system $\mathcal{C}$ is weakly regular, then it is weakly regular for a rank function with maximal rank $2^{|\mathcal{C}|}$ (where $|\mathcal{C}|$ denotes the number of variables of $\mathcal{C}$.*

– *It can be decided in linear time whether or not a constraint system is weakly regular.*

There is another structural characterization of weakly regular constraint systems. For variable X, let us denote by $\mathcal{C}_X$ the restriction of the constraint system $\mathcal{C}$ to all variables possibly influencing X. Let us call a constraint $X \supseteq \square(X_1, \ldots, X_k)$ *recursive* iff some variable X_j on the right-hand side depends on X, i.e., X occurs in $\mathcal{C}_{X_j}$. Then we have:

Theorem 2. *A horizontal constraint system $\mathcal{C}$ is weakly regular iff every recursive constraint $X \supseteq \Box(X_1, \ldots, X_k)$ is regular where furthermore the constraint systems $\mathcal{C}_{X_i}$ are regular for all i except at most one.*

Example 7 defines a WRR indeed, since every recursive rule is regular and Id_2 is a RR. On the other hand, L_{sym} (Example 2) is not a WRR.

Our main new result is:

Theorem 3. *Weakly regular relations are closed under union, projection, cartesian product and intersection.*

The first three closure properties of WRRs stated in theorem 3 are obtained simply by considering the corresponding constructions for arbitrary tree tuple constraint systems and verifying that these preserve the WRR property. In the sequel, we therefore concentrate on the proof of closure under intersection. We need the following auxiliary notion.

A constraint system $\mathcal{C}$ is *single-permuting*, iff every non-regular constraint of $\mathcal{C}$ has only one variable on the right-hand side, i.e., is of the form

$$X \supseteq \Box(X_1)$$

The proof of theorem 3 is based onto the following two auxiliary lemmas:

Lemma 1. *Let $\mathcal{C}$ be a weakly regular constraint system. Then an equivalent (up to further auxiliary variables) weakly regular constraint system $\mathcal{C}'$ can be constructed which is single-permuting.*

If $\mathcal{C}$ has maximal rank r, maximal size of classes d, at most a variables in right-hand sides and size n, then $\mathcal{C}'$ has size $\mathcal{O}(n^{(a+1)^r})$ and classes of size $\mathcal{O}(d \cdot a^r)$ where neither the rank nor the number of variables in right-hand sides has increased. Moreover, the constraint system $\mathcal{C}'$ can be constructed in double-exponential time.

Lemma 2. *Assume that the tree-tuple language L of class I is defined by a single-permuting constraint system $\mathcal{C}$ and $\sim \subseteq I \times I$ is an equivalence relation. Then a single-permuting constraint system $\mathcal{C}'$ can be constructed for the language*

$$L^{\sim} = \{t \in L \mid \forall (i,j) \in \sim : t_i = t_j\}$$

In particular, if $\mathcal{C}$ was weakly regular, then so is $\mathcal{C}'$. If $\mathcal{C}$ is of size n and has classes of size at most d, then $\mathcal{C}$ can be constructed in time $d^{\mathcal{O}(d)} \cdot n$.

Using lemmas 1 and 2, the proof of theorem 3 proceeds as follows. Assume we are given languages L_1 and L_2 both defined by weakly regular constraint systems $\mathcal{C}_i$, $i = 1, 2$. By closure under cartesian product, the language $L = \{t^{(1)}t^{(2)} \mid t^{(i)} \in L_i\}$ is defined by a weakly regular constraint system $\mathcal{C}$. By lemma 1, we also replace $\mathcal{C}$ with a single-permuting (weakly regular) constraint system $\mathcal{C}'$. Now consider the equivalence relation $\sim$ on the index set of L which equates the corresponding components from L_1 and L_2. By lemma 2, we can construct from C' a constraint system C'' which defines $L^{\sim}$. Finally, we may (due to closure under projection) construct a weakly regular constraint system for the language

$$\{t^{(1)} \mid \exists t^{(2)} : t^{(1)}t^{(2)} \in L^{\sim}\} = L_1 \cap L_2$$

and we are done. Calculating the costs of the individual construction steps, we furthermore find that the overall construction can be implemented in double-exponential time.

As an immediate corollary of theorem 3, we obtain:

Corollary 1. *Weakly regular relations are closed under joins.*

Proof. Assume L_1 (resp. L_2) is a l-tuple (resp. n-tuple) language.

$$\begin{aligned} L_1 \bowtie_{i,j} L_2 = & \Pi_{1,\ldots,l+j-1,l+j+1,\ldots,l+n}(L_1 \times L_2) \\ & \cap\{(t_1,\ldots,t_i,\ldots,t_l,t_{l+1},\ldots,t_{l+j-1},t_i,t_{l+j+1},\ldots,t_{l+n}) \mid \forall k, t_k \in T_\Sigma\} \end{aligned}$$

Non-closure under Complement

Consider the balanced full binary trees. Balanced means that all leaves appear at the same depth, and it is well known that a full binary tree t is balanced iff for all non-leaf position v in t, $t|_{v.1} = t|_{v.2}$. Therefore t is unbalanced iff there is a non-leaf position v and a position u s.t. $t(v.1.u) \neq t(v.2.u)$.

Full binary trees are simulated by terms (with fixed arities) over the signature $\Sigma = \{f^{\backslash 2}, a^{\backslash 0}\}$. Thus the set of unbalanced full binary trees is the set of unbalanced terms over Σ, which is generated by variable V of the following constraint system:

$$\begin{aligned} & V \supseteq f(1_1, 2_1)(V, Id) \\ & V \supseteq f(1_1, 2_1)(Id, V) \\ & V \supseteq f(1_1, 1_2)\, U \end{aligned} \qquad \begin{aligned} & U \supseteq (f(1_1, 2_1),\, f(1_2, 2_2))(U, Id^2) \\ & U \supseteq (f(1_1, 2_1),\, f(1_2, 2_2))(Id^2, U) \\ & U \supseteq (a,\, f(Id, Id)) \\ & U \supseteq (f(Id, Id),\, a) \end{aligned}$$

where $Id = T_\Sigma$ and $Id^2 = Id \times Id$. It is a WRR since Id is regular and the only non-regular constraint is $V \supseteq f(1_1, 1_2)U$, which is not recursive.

The complement of unbalanced terms, i.e. the balanced ones, cannot be defined by a WRR since, as shown in Section 3, they cannot be defined by a non-copying constraint system.

5 Application to *R*-Unification

This section addresses the problem of unification modulo a confluent constructor-based rewrite system. The goal is to decide the existence of data-unifiers (unifiability), and to express ground data-unifiers by a tree-tuple language.

Under some restrictions, a decidability result has been established using TTSGs [7]. Next the method has been generalized [11]: any class of tree-tuple language can be used, provided:

1. it is closed under join,
2. emptiness is decidable,
3. it can express the tuple-set $N_f = \{(r, \sigma x_1, \ldots, \sigma x_n) \mid f(x_1, \ldots, x_n) \rightsquigarrow^*_{[\sigma]} r$ and $r, \sigma x_1, \ldots, \sigma x_n$ are ground data-terms$\}$ for each defined function f,
4. For each constructor c, it can express the tuple-set

$$N_c = \{(c(t_1, \ldots, t_n), t_1, \ldots, t_n) \mid t_i \text{ are ground data-terms}\}$$

If in addition it is closed under intersection[4] and projection, the goal to be unified may be non-linear.

The unification method presented in [11] has been used with TTSGs and with primal grammars [5] providing some decidable subclasses of R-unification problems. In this section, we present the unification method of [11] which can be used with WRRs getting a new subclass of decidable R-unification problems. The restrictions are those needed for TTSGs, except that the goal and the non-recursive rewrite rules may be non-linear[5], additional technical restrictions are needed otherwise unifiability is undecidable as shown in [7].

Let us first recall the principle of the general method, using a simple example:

Example 8.

$$R = \{f(s(x)) \overset{r_1}{\to} p(f(x)),\quad f(p(x)) \overset{r_2}{\to} s(f(x)),\quad f(0) \overset{r_3}{\to} 0\}$$

where $s, p, 0$ are constructors, and consider the linear goal $p(f(x)) \doteq f(s(f(x')))$. We assume that we have a tree-tuple language that satisfies the above properties. In particular it can express N_f, N_s, N_p.

The method consists in simulating the innermost narrowing derivations issued from the goal. We compute:

$$\begin{aligned} N = N_p \bowtie N_f &= \{(s_1, s_2, t) \mid (s_1, s_2) \in N_p \wedge (s_2, t) \in N_f\} \\ &= \{(s_1, s_2, t) \mid p(f(x)) \rightsquigarrow^*_{[x/t]} p(s_2) = s_1\} \end{aligned}$$

$$\begin{aligned} N' &= N_f \bowtie N_s \bowtie N_f \\ &= \{(s'_1, s'_2, s'_3, t') \mid (s'_1, s'_2) \in N_f \wedge (s'_2, s'_3) \in N_s \wedge (s'_3, t') \in N_f\} \\ &= \{(s'_1, s'_2, s'_3, t') \mid f(s(f(x'))) \rightsquigarrow^*_{[x'/t']} f(s(s'_3)) = f(s'_2) \to^* s'_1\} \end{aligned}$$

From narrowing properties [3], there exists a data-unifier iff there exist instances t, t' such that $s_1 = s'_1$, i.e. such that $N \bowtie_{1,1} N' \neq \emptyset$. Moreover $N \bowtie_{1,1} N'$ expresses the solutions thanks to t, t'.

Now if the goal to be unified is not linear, like $p(f(x)) \doteq f(s(f(x)))$, t must be in addition equal to t'. By projection, we keep only t, t', s_1, s'_1, and force equalities by intersection. Thus, there exists a data-unifier iff $\Pi_{1,3}(N) \cap \Pi_{1,4}(N') \neq \emptyset$.

For each narrowing step occurring within a narrowing derivation $f(x) \rightsquigarrow^*_{[\sigma]} r$ where r is a ground data-term, either s is added in σ and p in r (using r_1), or

[4] This implies the closure under join.

[5] WRRs are closed under intersection and projection.

the opposite (using r_2), and the derivation ends by adding 0 in both (using r_3). So N_f can be described by the expression:

$$N_f \equiv ((p,s) \cup (s,p))^*.(0,0)$$

On the other hand

$$N_s = \{(s(t),t)\} \equiv (s,\epsilon).((s,s) \cup (p,p))^*.(0,0)$$
$$N_p = \{(p(t),t)\} \equiv (p,\epsilon).((s,s) \cup (p,p))^*.(0,0)$$

where ϵ is the empty string. Consequently,

$$N = (p,\epsilon,\epsilon).((s,s,p) \cup (p,p,s))^*.(0,0,0)$$
$$N_f \bowtie N_s = (p,s,\epsilon).((p,s,s) \cup (s,p,p))^*.(0,0,0)$$
$$N' = (N_f \bowtie N_s) \bowtie N_f = (p,s,\epsilon,\epsilon).((p,s,s,p) \cup (s,p,p,s))^*.(0,0,0,0)$$

$$\Pi_{1,3}(N) \cap \Pi_{1,4}(N') = (p,\epsilon).((s,p) \cup (p,s))^*.(0,0) \bigcap (p,\epsilon).((p,p) \cup (s,s))^*.(0,0) = (p(0),0)$$

The solutions are given by the second component. So there is one solution: $x/0$.

This is correct: if x/t is a data-unifier of $p(f(x)) \doteq f(s(f(x)))$, necessarily $p(f(t)) = p(f(f(t)))$, then $f(t) = f(f(t))$, then $f(t) = t$ because $f(f(t)) = t$. Therefore $t = 0$.

Using WRRs

$\boldsymbol{N_c}$ **:** unfortunately, the sets $N_c = \{(c(t_1,\dots,t_n),t_1,\dots,t_n)\}$ cannot be expressed by WRRs (nor by RRs), because generating two copies of t_i needs synchronization between them, and one occurs on top and the other at depth 1. So this language is not horizontal.

We slightly modify the method so that N_c is horizontal. Now, using an extra symbol $\natural$, we define $N_c = \{(c(t_1,\dots,t_n),\natural t_1,\dots,\natural t_n)\}$ for each constructor c, and $\natural L = \{(\natural t_1,\dots,\natural t_n) \mid (t_1,...,t_n) \in L\}$ for any language L. Their constraint systems are:

$$N_c \supseteq (c(1_1,\dots,n_1),\natural 1_2,\dots,\natural n_2)(Id_2,\dots,Id_2)$$
$$\natural L \supseteq (\natural 1_1,\dots,\natural 1_n)(L)$$

where $Id_2 = \{(t,t) \mid t \in T_C\}$ and T_C is the set of constructor-terms. These constraints are not recursive and Id_2 is regular. So N_c is a WRR and if L is regular, $\natural L$ is a WRR.

It is however necessary to slightly modify the computation of N and N'.

Example 9. Consider the previous example again. Now:

$$N = N_p \bowtie \natural N_f = \{(r_1,\natural r_2,\natural t) \mid (r_1,\natural r_2) \in N_p \wedge (\natural r_2,\natural t) \in \natural N_f\}$$
$$= \{(r_1,\natural r_2,\natural t) \mid p(f(x)) \rightsquigarrow^*_{[x/t]} p(r_2) = r_1\}$$

$$N' = N_f \bowtie N_s \bowtie \natural N_f$$
$$= \{(r'_1, r'_2, \natural r'_3, \natural t') \mid (r'_1, r'_2) \in N_f \wedge (r'_2, \natural r'_3) \in N_s \wedge (\natural r'_3, \natural t') \in \natural N_f\}$$
$$= \{(r'_1, r'_2, \natural r'_3, \natural t') \mid f(s(f(x'))) \rightsquigarrow^*_{[x'/t']} f(s(r'_3)) = f(r'_2) \rightarrow^* r'_1\}$$

If the goal is linear, we check $N \bowtie_{1,1} N' \neq \emptyset$ as previously. Otherwise we can still force $t = t'$ by computing $\Pi_{1,3}(N) \cap \Pi_{1,4}(N')$, because the number of $\natural$ above t in N and t' in N' are the same (one). This comes from the fact that the number of constructors appearing above the two occurrences of x in the goal is the same (one).

Definition 1. *For a term t and $u \in Pos(t)$, let $\|u\|$ denote the number of constructors appearing in t above u. The term t is* weak-horizontal *if*

$$\forall u, u' \in Pos(t),\ t(u) = t(u') = x \in Var(t) \implies \|u\| = \|u'\|$$

In this case we also define $\|x\| \stackrel{def}{=} \|u\|$. The goal $t \doteq t'$ (resp. the rewrite rule $l \rightarrow r$) is weak-horizontal *if $t \doteq t'$ (resp. $l \rightarrow r$) is a weak-horizontal term, considering $\doteq$ (resp. $\rightarrow$) is a binary symbol.*

Lemma 3. *If the goal is weak-horizontal and x is a variable occurring several times in the goal, then every component in N, N' that gives the instances of an occurrence of x contains exactly $\|x\|$ times $\natural$.*

Proof. When using n-ary symbols, the intermediate language N^u corresponding to position u ($N^\epsilon = N$) is

$$N^u = (((N_c \bowtie_{2,1} \natural N^{u.1}) \bowtie_{3,1} \natural N^{u.2}) \ldots \bowtie_{n+1,1} \natural N^{u.n}$$

if $t(u)$ is a constructor, and

$$N^u = (((N_f \bowtie_{2,1} N^{u.1}) \bowtie_{3,1} N^{u.2}) \ldots \bowtie_{n+1,1} N^{u.n}$$

otherwise. If x occurs below u, by induction we get that the number of $\natural$ in the component giving the instances of x is exactly the number of constructors occurring along the path from u to x.

Thus, if the goal is weak-horizontal the modified method works, provided N_f can be expressed.

$\boldsymbol{N_f}$ **:** let us explain how to transform a rewrite system into a constraint system that expresses the sets N_f. We give an example before the general algorithm.

Example 10. Consider the rewrite rule $f(c(x, y)) \rightarrow d(f(y), x)$. By narrowing, we get $f(x) \rightsquigarrow_{[\sigma = x/c(x,y)]} t' = d(f(y), x)$. So $(t', \sigma) = (d(f(y), x), c(x, y))$. To get ground data-terms, x should be instantiated by something and $f(y)$ should be narrowed further. Then the corresponding constraint is:

$$F \supseteq (d(1_1, 2_1),\ c(2_2, 1_2))(F, Id_2)$$

Definition 2. *A function position p in a term t is* shallow *if $t|_p = f(x_1, \ldots, x_n)$ where $x_1, \ldots, x_n$ are variables.*

Definition 3. *Given a rewrite system R, we define an ordering on defined function symbols by $f > g$ if $f(t_1, \ldots, t_n) \to r \in R$ and g occurs in r. $f \equiv g$ means $f > g \wedge g > f$. The rewrite rule $f(t_1, \ldots, t_n) \to r \in R$ is* recursive *if there is a function g in r s.t. $f \equiv g$.*

Algorithm: we assume that function calls in rhss are shallow, and recursive rewrite rules are linear. Let us write rewrite rules as follows: $f(t_1, \ldots, t_n) \to C[u_1, \ldots, u_m]$ where C is a constructor context containing no variables and each u_i is either a variable or a function call of the form $f_i(x_1^i, \ldots, x_{k_i}^i)$. For each rewrite rule, we create the constraint

$$F \supseteq (C[1_1, \ldots, m_1], \theta t_1, \ldots, \theta t_n)(X_1, \ldots, X_m)$$

where $X_i = Id_2$ if u_i is a variable and $X_i = F_i$ if $u_i = f_i(x_1^i, \ldots, x_{k_i}^i)$ and θ is the substitution $\{u_i/i_2 \mid u_i \text{ is a variable}\}$.

If this rewrite rule is not linear and not recursive, the F_is are defined independently of F. So we can make some variables equal by computing the intersection[6] of the constraint argument $(X_1, \ldots, X_n)$ with the regular relation

$$\{(s_1, \ldots s_k, t, s_{k+2}, \ldots s_p, t, s_{p+2}, \ldots, s_q) | s_i, t \in T_C\}$$

However, the resultant constraint system is not necessarily a WRR: in the previous example, the constraint is recursive and not regular. The non-regularity comes from the fact that when going through the leaves from left-to-right, we get x, y for the lhs, and y, x for the rhs: there is a variable permutation. If we remove the permutation by considering the rule $f(c(x, y)) \to d(x, f(y))$, the resultant constraint is regular. But the absence of variable permutation does not ensure regularity.

Example 11. Let $f(a, c(x, y)) \to s(f(x, y))$.
The corresponding constraint $F \supseteq (s(1_1), a, c(1_2, 1_3))\, F$ is not regular because there is an internal synchronization in the third component.

And the presence of permutation does not imply non-regularity.

Example 12. Let $f(c(x, y),\ s(z)) \to c(f(x, z),\ y)$.
The corresponding constraint $F \supseteq (c(1_1, 2_1),\ c(1_2, 2_2),\ s(1_3))(F, Id_2)$ is regular.

This is why we introduce the following definition:

Definition 4. *A constructor-based rewrite system is* weak-regular *if function positions in rhs's are shallow, its recursive rules are linear, and the corresponding constraint system generated by the above algorithm is a WRR.*

---

[6] if one of the F_is is not a WRR, the algorithm fails.

Note that weak-regularity implies in particular weak-horizontality of rewrite rules. Thanks to decidability of emptiness for WRRs, we get:

Theorem 4. *The unifiability of a weak-horizontal goal modulo a weak-regular confluent constructor-based rewrite system is decidable. Moreover, the unifiers can be expressed by a WRR.*

6 Application to One-Step-Rewriting

Given a signature Σ, the theory of one-step rewriting for a finite rewrite system R is the first order theory over the universe of ground Σ-terms that uses the only predicate symbol $\rightarrow$, where $x \rightarrow y$ means x rewrites into y by one step.

It has been shown undecidable in [16]. Sharper undecidability results have been obtained for some subclasses of rewrite systems, about the $\exists^*\forall^*$-fragment [15,12] and the $\exists^*\forall^*\exists^*$-fragment [17].

It has been shown decidable in the case of unary signatures [6], in the case of linear rewrite systems whose left and right members do not share any variables [2][7], for the positive existential fragment [13], for the whole existential fragment in the case of quasi-shallow[8] rewrite systems [1] and also in the case of linear, non-overlapping, non-ϵ-left-right-overlapping[9] rewrite systems [8].

Thanks to WRRs, we get a new result about the existential fragment.

Definition 5. *A rewrite system R is* ϵ-left-right-clashing *if for all rewrite rule $l \rightarrow r$, $l(\epsilon) \neq r(\epsilon)$. R is* horizontal *if in each rewrite rule, all occurrences of the same variable appear at the same depth.*

Note that ϵ-left-right-clashing excludes collapsing rules.

Theorem 5. *The existential one-step rewriting theory is decidable in the case of ϵ-left-right-clashing horizontal rewrite systems.*

Since quasi-shallowness is a particular case of horizontality, our result extends that of [1], except that we assume in addition ϵ-left-right-clashing. Compared to [8], rewrite rules may now be non-linear and overlapping, but they must be horizontal.

Consider a finite rewrite system $R = \{ru_1, \ldots, ru_n\}$ and an existential formula in the prenex form. Our decision procedure consists in the following steps:

1. Since the symbols of Σ are not allowed in formulas, every atom is of the form $x \rightarrow x$ or $x \rightarrow y$. Because of ϵ-left-right-clashing, $x \rightarrow x$ has no solutions and is replaced by the predicate without solutions $\perp$, and $x \rightarrow y$ is replaced by the equivalent proposition $(x \overset{?}{\rightarrow}_{[ru_1]} y \vee \ldots \vee x \overset{?}{\rightarrow}_{[ru_n]} y) \wedge x \neq y$, where $\overset{?}{\rightarrow}_{[ru_i]}$ is the rewrite relation in zero or one step with rule ru_i. Next, the formula

[7] Even the theory of several-step rewriting is decidable.

[8] All variables in the rewrite rules occur at depth one.

[9] I.e. no left-hand-side overlaps on top with the corresponding right-hand-side.

is transformed into a disjunction of conjunctions of (possibly) negations of atoms of the form $x \overset{?}{\rightarrow}_{[ru_i]} y$ or $x = y$. We show that the set of solutions of each atom, and of its negation, is a WRR.

2. The solutions of a conjunctive factor are obtained by making cartesian products with the set T_Σ of all ground terms (which is a particular WRR), as well as intersections. For instance let $C = x \overset{?}{\rightarrow}_{[ru_1]} y \wedge \neg(y \overset{?}{\rightarrow}_{[ru_2]} z)$. The solutions of $x \overset{?}{\rightarrow}_{[ru_1]} y$, denoted $SOL(x \overset{?}{\rightarrow}_{[ru_1]} y))$, and those of $\neg(y \overset{?}{\rightarrow}_{[ru_2]} z)$, denoted $SOL(\neg(y \overset{?}{\rightarrow}_{[ru_2]} z))$, are WRRs of pairs, then we can compute $SOL(C) = SOL(x \overset{?}{\rightarrow}_{[ru_1]} y)) \times T_\Sigma \ \cap \ T_\Sigma \times SOL(\neg(y \overset{?}{\rightarrow}_{[ru_2]} z))$, which is still a WRR (of triples).
3. The validity of the formula is tested by applying the WRR emptiness test on every disjunctive factor.

$SOL(x = y)$ and $SOL(x \neq y)$ are trivially WRRs since they are RRs.

Lemma 4. *$SOL(x \overset{?}{\rightarrow}_{[ru_i]} y)$ and $SOL(\neg(x \overset{?}{\rightarrow}_{[ru_i]} y))$ are WRRs.*

7 Further Work and Conclusion

Computing descendants through a rewrite system may give rise to several applications. [14] shows that the set of descendants of a regular tree language through a rewrite system is still regular, assuming some restrictions. Using non-regular languages, like WRRs, still closed under intersection (for applications), could extend the result of [14] by weakening the restrictions.

Compared to automata with (dis)equality constraints [2], WRRs can define more constraints than only (dis)equality, but they cannot define the balanced terms. However, WRRs and the subclass of reduction automata have something in common: when deriving (recognizing) a term, the number of non-regular constraints (of (dis)equality constraints) applied is supposed to be bounded.

As shown in Example 5, the intersection of some horizontal CSs that are not necessarily WRRs, is still a horizontal CS. So, is there a subclass of horizontal CSs, larger than WRRs, and closed under intersection?

References

1. A.C. Caron, F. Seynhaeve, S. Tison, and M. Tommasi. Deciding the Satisfiability of Quantifier Free Formulae on One-Step Rewriting. In *Proceedings of 10th Conference RTA, Trento (Italy)*, volume 1631 of *LNCS*. Springer-Verlag, 1999.
2. H. Comon, M. Dauchet, R. Gilleron, D. Lugiez, S. Tison, and M. Tommasi. *Tree Automata Techniques and Applications (TATA)*. http://l3ux02.univ-lille3.fr/tata.
3. L. Fribourg. SLOG: A Logic Programming Language Interpreter Based on Clausal Superposition and Rewriting. In *proceedings IEEE Symposium on Logic Programming*, pages 172–185, Boston, 1985.

4. V. Gouranton, P. Réty, and H. Seidl. Synchronized Tree Languages Revisited and New Applications. In *Proceedings of FoSSaCs*, volume to appear of *LNCS*. Springer-Verlag, 2001.
5. M. Hermann and R. Galbavý. Unification of Infinite Sets of Terms Schematized by Primal Grammars. *Theoretical Computer Science*, 176, 1997.
6. F. Jacquemard. *Automates d'Arbres et Réécriture de Termes*. Thèse de Doctorat d'Université, Université de Paris-sud, 1996. In French.
7. S. Limet and P. Réty. E-Unification by Means of Tree Tuple Synchronized Grammars. In *Proceedings of 6th Colloquium on Trees in Algebra and Programming*, volume 1214 of *LNCS*, pages 429–440. Springer-Verlag, 1997. Full version in DMTCS (http://dmtcs.loria.fr/), volume 1, pages 69-98, 1997.
8. S. Limet and P. Réty. A New Result about the Decidability of the Existential One-step Rewriting Theory. In *Proceedings of 10th Conference on Rewriting Techniques and Applications, Trento (Italy)*, volume 1631 of *LNCS*. Springer-Verlag, 1999.
9. S. Limet, P. Réty, and H. Seidl. Weakly Regular Relations and Applications. Research Report RR-LIFO-00-17, LIFO, 2000. http://www.univ-orleans.fr/SCIENCES/LIFO/Members/rety/publications.html.
10. S. Limet and F. Saubion. On partial validation of logic programs. In M. Johnson, editor, *proc of the 6th Conf. on Algebraic Methodology and Software Technology, Sydney (Australia)*, volume 1349 of *LNCS*, pages 365–379. Springer Verlag, 1997.
11. S. Limet and F. Saubion. A general framework for *R*-unification. In C. Palamidessi, H. Glaser, and K. Meinke, editors, *proc of PLILP-ALP'98*, volume 1490 of *LNCS*, pages 266–281. Springer Verlag, 1998.
12. J. Marcinkowski. Undecidability of the First-order Theory of One-step Right Ground Rewriting. In *Proceedings 8th Conference RTA, Sitges (Spain)*, volume 1232 of *LNCS*, pages 241–253. Springer-Verlag, 1997.
13. J. Niehren, M. Pinkal, and P. Ruhrberg. On Equality up-to Constraints over Finite Trees ,Context Unification and One-step Rewriting. In W. Mc Cune, editor, *Proc. of CADE'97, Townsville (Australia)*, volume 1249 of *LNCS*, pages 34–48, 1997.
14. P. Réty. Regular Sets of Descendants for Constructor-based Rewrite Systems. In *Proceedings of the 6th international conference on Logic for Programming and Automated Reasoning (LPAR), Tbilisi (Republic of Georgia)*, Lecture Notes in Artificial Intelligence. Springer-Verlag, 1999.
15. F. Seynhaeve, M. Tommasi, and R. Treinen. Grid Structures and Undecidable Constraint Theories. In *Proceedings of 6th Colloquium on Trees in Algebra and Programming*, volume 1214 of *LNCS*, pages 357–368. Springer-Verlag, 1997.
16. R. Treinen. The First-order Theory of One-step Rewriting is Undecidable. *Theoretical Computer Science*, 208:179–190, 1998.
17. S. Vorobyov. The First-order Theory of One-step Rewriting in Linear Noetherian Systems is Undecidable. In *Proceedings 8th Conference RTA, Sitges (Spain)*, volume 1232 of *LNCS*, pages 241–253. Springer-Verlag, 1997.

On the Parallel Complexity of Tree Automata

Markus Lohrey

Universität Stuttgart, Institut für Informatik
Breitwiesenstr. 20–22, 70565 Stuttgart, Germany
lohreyms@informatik.uni-stuttgart.de

Abstract. We determine the parallel complexity of several (uniform) membership problems for recognizable tree languages. Furthermore we show that the word problem for a fixed finitely presented algebra is in DLOGTIME-uniform NC^1.

1 Introduction

Tree automata are a natural generalization of usual word automata to terms. Tree automata were introduced in [11,12] and [27] in order to solve certain decision problems in logic. Since then they were successfully applied to many other decision problems in logic and term rewriting, see e.g. [7]. These applications motivate the investigation of decision problems for tree automata like emptiness, equivalence, and intersection nonemptiness. Several complexity results are known for these problems, see [28] for an overview. Another important decision problem is the membership problem, i.e, the problem whether a given tree automaton accepts a given term. It is easily seen that this problem can be solved in deterministic polynomial time [7], but up to now no precise bounds on the complexity are known.

In this paper we investigate the complexity of several variants of membership problems for tree automata. In Section 3 we consider the membership problem for a fixed tree automaton, i.e, for a fixed tree automaton $\mathcal{A}$ we ask whether a given input term is accepted by $\mathcal{A}$. We prove that this problem is contained in the parallel complexity class DLOGTIME-uniform NC^1, and furthermore that there exists a fixed tree automaton for which this problem is complete for DLOGTIME-uniform NC^1. Using these results, in Section 4 we prove that the word problem for a fixed finitely presented algebra is in DLOGTIME-uniform NC^1. This result nicely contrasts a result of Kozen that the uniform word problem for finitely presented algebras is P-complete [21]. Finally in Section 5 we investigate uniform membership problems for tree automata. In these problems the input consists of a tree automaton $\mathcal{A}$ from some fixed class $\mathcal{C}$ of tree automata and a term t, and we ask whether $\mathcal{A}$ accepts t. For the class $\mathcal{C}$ we consider the class of all deterministic top-down, deterministic bottom-up, and nondeterministic (bottom-up) tree automata, respectively. The complexity of the corresponding uniform membership problem varies between the classes log-space and LOGCFL, which is the class of all languages that can be reduced in log-space to a context free language. Again we prove several completeness results. Table 1 at the end of this paper summarizes the presented complexity results for membership problems.

A. Middeldorp (Ed.): RTA 2001, LNCS 2051, pp. 201–215, 2001.

2 Preliminaries

In the following let Σ be a finite alphabet. The empty word is denoted by ϵ. The set of all finite words over Σ is Σ^*. We set $\Sigma^+ = \Sigma^* \backslash \{\epsilon\}$. For $\Gamma \subseteq \Sigma$ we denote by $|s|_\Gamma$ the number of occurrences of symbols from Γ in s. We set $|s| = |s|_\Sigma$. For a binary relation $\to$ on some set we denote by $\stackrel{+}{\to}$ ($\stackrel{*}{\to}$) the transitive (reflexive and transitive) closure of $\to$. *Context-free grammars* are defined as usual. If $G = (N, \Sigma, S, P)$ is a context-free grammar then N is the set of *non-terminals*, Σ is the set of *terminals*, $S \in N$ is the *initial non-terminal*, and $P \subseteq N \times (N \cup \Sigma)^*$ is the finite set of *productions*. With $\to_G$ we denote the derivation relation of G. The language generated by G is denoted by $L(G)$. A context-free grammar is *ϵ-free* if it does not contain productions of the form $A \to \epsilon$.

We assume that the reader is familiar with the basic concepts of computational complexity, see for instance [23]. We just recall a few definitions concerning parallel complexity theory, see [30] for more details. It is not necessary to be familiar with this field in order to understand the constructions in this paper. L denotes deterministic logarithmic space. The definition of DLOGTIME-uniformity and DLOGTIME-reductions can be found in [2]. An important subclass of L is DLOGTIME-uniform NC^1, briefly uNC^1. More general, for $k \geq 1$ the class uNC^k contains all languages K such that there exists a DLOGTIME-uniform family $(\mathcal{C}_n)_{n \geq 0}$ of Boolean circuits with the following properties: (i) for some constant c the depth of the circuit $\mathcal{C}_n$ is bounded by $c \cdot \log(n)^k$, (ii) for some polynomial $p(n)$ the size of $\mathcal{C}_n$, i.e., the number of gates in $\mathcal{C}_n$, is bounded by $p(n)$, (iii) all gates in $\mathcal{C}_n$ have fan-in at most two, and (iv) the circuit $\mathcal{C}_n$ recognizes exactly the set of all words in K of length n. By [25] uNC^1 is equal to ALOGTIME.

An important subclass of uNC^1 is DLOGTIME-uniform-TC^0, briefly uTC^0. A language K is in uTC^0 if there exists a DLOGTIME-uniform family $(\mathcal{C}_n)_{n \geq 0}$ of circuits built up from Boolean gates and majority gates (or equivalently arbitrary threshold-gates) with the following properties: (i) for some constant c the depth of the circuit $\mathcal{C}_n$ is bounded by c, (ii) for some polynomial $p(n)$ the size of $\mathcal{C}_n$ is bounded by $p(n)$, (iii) all gates in $\mathcal{C}_n$ have unbounded fan-in, and (iv) the circuit $\mathcal{C}_n$ recognizes exactly the set of all words in K of length n. For more details see [2]. In this paper we will use a more convenient characterization of uTC^0 using first-order formulas with majority quantifiers, briefly FOM-formulas. Let Σ be a fixed finite alphabet of symbols. An FOM-formula is built up from the unary predicate symbols Q_a ($a \in \Sigma$) and the binary predicate symbols $<$ and BIT, using Boolean operators, first-order quantifiers, and the majority quantifier M. Such formulas are interpreted over words from Σ^+. Let $w = a_1 \cdots a_m$, where $m \geq 1$ and $a_i \in \Sigma$ for $i \in \{1, \ldots, m\}$. If we interpret an FOM-formula over w then all variables range over the interval $\{1, \ldots, m\}$, $<$ is interpreted by the usual order on this interval, $\mathrm{BIT}(n, i)$ is true if the i-th bit in the binary representation of n is one (we will not need this predicate any more), and $Q_a(x)$ is true if $a_x = a$. Boolean connectives and first-order quantifiers are interpreted as usual. Finally the formula $Mx\,\varphi(x)$ evaluates to true if $\varphi(x)$ is true for at least half of all $x \in \{1, \ldots, m\}$. The language defined by an FOM-sentence φ is the set of words from Σ^+ for which the FOM-sentence φ evaluates to true. For instance the FOM-sentence

$$Mx\, Q_a(x) \wedge Mx\, Q_b(x) \wedge \forall x, y\, \{x < y \rightarrow \neg(Q_b(x) \wedge Q_a(y))\}$$

defines the language $\{a^n b^n \mid n \geq 1\}$. It is well-known that uTC^0 is the set of languages that can be defined by an FOM-sentence [2].

In FOM-formulas we will often use constants and relations that can be easily defined in FOM, like for instance the equality of positions or the constants 1 and max, which denote the first and last position in a word, respectively. Furthermore by [2, Lemma 10.1] also the predicates $x+y = z$ and $x = \#y\ \varphi(y)$, i.e, the number of positions y that satisfy the FOM-formula $\varphi(y)$ is exactly x, can be expressed in FOM. Finally let us mention that uNC^1 also has a logical characterization similar to uTC^0. The only difference is that instead of majority quantifiers so called group quantifiers for a non-solvable group are used, see [2] for the details. Of course the resulting logic is at least as expressive as FOM.

In this paper we also use reductions between problems that can be defined within FOM. Formally let $f : \Sigma^+ \to \Gamma^+$ be a function such that for some constant k we have $|f(w)| \leq k \cdot |w|$ for all $w \in \Sigma^+$. [1] Then we say that f is *FOM-definable* if there exist formulas $\phi(x)$ and $\phi_a(x)$ for $a \in \Gamma$ such that when interpreted over a word w and $i \in \{1, \dots, k \cdot |w|\}$ then $\phi(i)$ evaluates to true if and only if $i = |f(w)|$, and $\phi_a(i)$ evaluates to true if and only if the i-th symbol in $f(w)$ is a (here also all quantified variables in ϕ and ϕ_a range over the interval $\{1, \dots, |w|\}$). We say that ϕ and ϕ_a $(a \in \Gamma)$ define f.

Lemma 1. *Let $f : \Sigma^+ \to \Gamma^+$ be FOM-definable and let $L \subseteq \Sigma^+$, $K \subseteq \Gamma^+$ such that $w \in L$ if and only if $f(w) \in K$ (in this case we say that L is FOM-reducible to K). If K is in uTC^0 (resp. uNC^1) then also L is in uTC^0 (resp. uNC^1).*

Proof. Let ϕ, ϕ_a $(a \in \Gamma)$ be FOM-formulas that define the function f and let K be in uTC^0, i.e., it can be defined by an FOM-sentence ψ. Let $|f(w)| \leq k \cdot |w|$ for all $w \in \Sigma^+$. In the following we restrict to the case $k = 2$, the generalization to an arbitrary k is obvious. In principle we can define the language L by the sentence that results from ψ by replacing every subformula $Q_a(x)$ by the formula $\phi_a(x)$. The only problem is that if we interpret this sentence over a word w then the variables quantified in ψ have to range over the interval $\{1, \dots, |f(w)|\}$. Hence we define L by the FOM-sentence $\exists z\ \{(\phi(z) \wedge \psi^{z,0}) \vee (\phi(\max + z) \wedge \psi^{z,1})\}$, where the sentence $\psi^{z,i}$ is inductively defined as follows:

- $(\exists x\ \varphi(x))^{z,i} \equiv \exists x\ \{(x \leq i \cdot \max \wedge \varphi(x)^{z,i}) \vee (x \leq z \wedge \varphi(x + i \cdot \max)^{z,i})\}$

- $(Mx\ \varphi(x))^{z,i} \equiv \exists x_1, x_2 \left\{ \begin{array}{l} x_1 = \#y\ (y \leq i \cdot \max \wedge \varphi(y)^{z,i}) \wedge \\ x_2 = \#y\ (y \leq z \wedge \varphi(y + i \cdot \max)^{z,i}) \wedge \\ \exists y\ \{2(x_1 + x_2) = y - 1 + z + i \cdot \max\} \end{array} \right\}$

- $Q_a(x)^{z,i} \equiv \phi_a(x)$ and $Q_a(x + \max)^{z,i} \equiv \phi_a(x + \max)$

If K belongs to uNC^1 the arguments are similar using the logical characterization of uNC^1. □

[1] This linear length-bound may be replaced by a polynomial bound, but this is not necessary for this paper.

LOGCFL (respectively LOGDCFL) is the class of all languages that are log-space reducible to a context-free language (respectively deterministic context-free language) [26]. In [26] it was shown that LOGCFL is the class of all languages that can be recognized in polynomial time on a log-space bounded auxiliary push-down automaton, whereas the deterministic variants of these machines precisely recognize all languages in LOGDCFL. The following inclusions are well-known and it is conjectured that they are all proper.

$$\mathrm{uTC}^0 \subseteq \mathrm{uNC}^1 = \mathrm{ALOGTIME} \subseteq \mathrm{L} \subseteq \mathrm{LOGDCFL} \subseteq \mathrm{LOGCFL} \subseteq \mathrm{uNC}^2 \subseteq \mathrm{P}$$

A *ranked alphabet* is a pair $(\mathcal{F}, \text{arity})$ where $\mathcal{F}$ is a finite set of function symbols and arity is a function from $\mathcal{F}$ to $\mathbb{N}$ which assigns to each $\alpha \in \mathcal{F}$ its arity arity(α). A function symbol a with arity$(a) = 0$ is called a *constant*. In all examples we will use function symbols a and f, where arity$(a) = 0$ and arity$(f) = 2$. Mostly we omit the function arity in the description of a ranked alphabet. With $\mathcal{F}_i$ we denote the set of all function symbols in $\mathcal{F}$ of arity i. In FOM-formulas we use $Q_{\mathcal{F}_i}(x)$ as an abbreviation for $\bigvee_{\alpha \in \mathcal{F}_i} Q_\alpha(x)$. Let $\mathcal{X}$ be a countably infinite set of *variables*. Then $T(\mathcal{F}, \mathcal{X})$ denotes the set of *terms* over $\mathcal{F}$ and $\mathcal{X}$, it is defined as usual. The word *tree* is used as a synonym for term. We use the abbreviation $T(\mathcal{F}, \emptyset) = T(\mathcal{F})$, this set is called the set of *ground terms* over $\mathcal{F}$. We identify the set $T(\mathcal{F})$ with the corresponding *free term algebra* over the signature $\mathcal{F}$. In computational problems terms will be always represented by their prefix-operator notation, which is a word over the alphabet $\mathcal{F}$. The set of all these words is known as the Lukasiewicz-language $\text{Ł}(\mathcal{F})$ for the ranked alphabet $\mathcal{F}$. For instance $ffafaafaa \in \text{Ł}(\mathcal{F})$ but $fafaf \notin \text{Ł}(\mathcal{F})$. When we write terms we will usually use the prefix-operator notation including brackets and commas in order to improve readability.

Lemma 2. *For every ranked alphabet $\mathcal{F}$ the language $\text{Ł}(\mathcal{F}) \subseteq \mathcal{F}^+$ is in uTC^0.*

A similar result for Dyck-languages was shown in [1].

Proof. Let $m = \max\{\text{arity}(\alpha) \mid \alpha \in \mathcal{F}\}$. For $s \in \mathcal{F}^+$ define

$$\|s\| = \sum_{i=0}^{m} (i-1) \cdot |s|_{\mathcal{F}_i}.$$

Then for $s \in \mathcal{F}^+$ it holds $s \in \text{Ł}(\mathcal{F})$ if and only if $\|s\| = -1$ and $\|t\| \geq 0$ for every prefix $t \neq s$ of s, see [17, p 323]. This characterization can be easily converted into an FOM-sentence:

$$\exists x_0, \dots, x_m \{ \bigwedge_{i=0}^{m} x_i = \#z\, Q_{\mathcal{F}_i}(z) \wedge \sum_{i=0}^{m} (i-1) \cdot x_i = -1 \} \wedge$$

$$\forall y < \max \exists x_0, \dots, x_m \{ \bigwedge_{i=0}^{m} x_i = \#z\, (Q_{\mathcal{F}_i}(z) \wedge z \leq y) \wedge \sum_{i=0}^{m} (i-1) \cdot x_i \geq 0 \}$$

□

If the ranked alphabet $\mathcal{F}$ is clear from the context then in the following we will always write Ł instead of Ł$(\mathcal{F})$. From the formula above it is straight forward to construct a formula Ł(i, j) which evaluates to true for a word $\alpha_1 \cdots \alpha_n \in \mathcal{F}^+$ and two positions $i, j \in \{1, \ldots, n\}$ if and only if $i \leq j$ and $\alpha_i \cdots \alpha_j \in$ Ł. The height height(t) of the term $t \in T(\mathcal{F})$ is inductively defined by $\text{height}(\alpha(t_1, \ldots, t_n)) = 1 + \max\{\text{height}(t_1), \ldots, \text{height}(t_n)\}$, where $\text{arity}(\alpha) = n \geq 0$ and $t_1, \ldots, t_n \in T(\mathcal{F})$ (here $\max(\emptyset) = 0$).

A *term rewriting system*, briefly TRS, over a ranked alphabet $\mathcal{F}$ is a finite set $\mathcal{R} \subseteq T(\mathcal{F}, \mathcal{X}) \times T(\mathcal{F}, \mathcal{X})$ such that for all $(s, t) \in \mathcal{R}$ every variable that occurs in t also occurs in s and furthermore $s \notin \mathcal{X}$. With a TRS $\mathcal{R}$ the *one-step rewriting relation* $\rightarrow_{\mathcal{R}}$ over $T(\mathcal{F}, \mathcal{X})$ is associated as usual, see any text on term rewriting like for instance [10]. A *ground term rewriting system* $\mathcal{P}$ is a finite subset of $T(\mathcal{F}) \times T(\mathcal{F})$, i.e., the rules only contain ground terms. The symmetric, transitive, and reflexive closure of the one-step rewriting relation $\rightarrow_{\mathcal{P}}$ of a ground TRS is the smallest congruence relation on the free term algebra $T(\mathcal{F})$ that contains all pairs in $\mathcal{P}$, it is denoted by $\equiv_{\mathcal{P}}$. The corresponding quotient algebra $T(\mathcal{F})/\equiv_{\mathcal{P}}$ is denoted by $A(\mathcal{F}, \mathcal{P})$, it is a *finitely presented algebra.*

For a detailed introduction into the field of tree automata see [14,7]. A *top-down tree automaton*, briefly TDTA, is a tuple $\mathcal{A} = (Q, \mathcal{F}, q_0, \mathcal{R})$, where Q is a finite set of states, $Q \cup \mathcal{F}$ is a ranked alphabet with $\text{arity}(q) = 1$ for all $q \in Q$, $q_0 \in Q$ is the initial state, and $\mathcal{R}$ is a TRS such that all rules of $\mathcal{R}$ have the form $q(\alpha(x_1, \ldots, x_n)) \rightarrow \alpha(q_1(x_1), \ldots, q_n(x_n))$, where $q, q_1, \ldots, q_n \in Q$, $x_1, \ldots, x_n \in \mathcal{X}$, $\alpha \in \mathcal{F}$, and $\text{arity}(\alpha) = n$. $\mathcal{A}$ is a *deterministic TDTA* if there are no two rules in $\mathcal{R}$ with the same left-hand side. The language that is accepted by a TDTA $\mathcal{A}$ is defined by

$$T(\mathcal{A}) = \{t \in T(\mathcal{F}) \mid q_0(t) \overset{*}{\rightarrow}_{\mathcal{R}} t\}.$$

A *bottom-up tree automaton*, briefly BUTA, is a tuple $\mathcal{A} = (Q, \mathcal{F}, q_f, \mathcal{R})$, where Q is a finite set of states, $Q \cup \mathcal{F}$ is a ranked alphabet with $\text{arity}(q) = 1$ for all $q \in Q$, $q_f \in Q$ is the final state, and $\mathcal{R}$ is a TRS such that all rules of $\mathcal{R}$ have the form $\alpha(q_1(x_1), \ldots, q_n(x_n)) \rightarrow q(\alpha(x_1, \ldots, x_n))$, where $q, q_1, \ldots, q_n \in Q$, $x_1, \ldots, x_n \in \mathcal{X}$, $\alpha \in \mathcal{F}$, and $\text{arity}(\alpha) = n$. $\mathcal{A}$ is a *deterministic BUTA* if there are no two rules in $\mathcal{R}$ with the same left-hand side. The language that is accepted by a BUTA $\mathcal{A}$ is defined by

$$T(\mathcal{A}) = \{t \in T(\mathcal{F}) \mid t \overset{*}{\rightarrow}_{\mathcal{R}} q_f(t)\}.$$

It is well known that TDTAs, BUTAs, and deterministic BUTAs, respectively, all recognize the same subsets of $T(\mathcal{F})$. These subsets are called *recognizable tree languages* over $\mathcal{F}$. On the other hand deterministic TDTAs cannot recognize all recognizable tree languages.

As already remarked, if a term is part of the input for a Turing machine then the term will be encoded by its corresponding word from Ł$(\mathcal{F})$, where the symbols from $\mathcal{F}$ are binary coded. A tree automaton will be encoded by basically listing its rules, we omit the formal details. The *membership problem* for a fixed TDTA $\mathcal{A}$, defined over a ranked alphabet $\mathcal{F}$, is the following decision problem:

INPUT: A term $t \in T(\mathcal{F})$.
QUESTION: Does $t \in T(\mathcal{A})$ hold?
If the TDTA $\mathcal{A}$ is also part of the input we speak of the *uniform membership problem* for TDTAs. It is the following decision problem:
INPUT: A TDTA $\mathcal{A} = (Q, \mathcal{F}, q_0, \mathcal{R})$ and a term $t \in T(\mathcal{F})$.
QUESTION: Does $t \in T(\mathcal{A})$ hold?
(Uniform) membership problems for other classes of automata or grammars are defined analogously. Note that the uniform membership problem for TDTAs can be reduced trivially to the uniform membership problem for BUTAs (and vice versa) by reversing the rules. Thus these two problems have the same computational complexity.

3 Membership Problems

In this section we will study the membership problem for a fixed recognizable tree language. First we need some preliminary results.

A *parenthesis grammar* is a context-free grammar $G = (N, \Sigma, S, P)$ that contains two distinguished terminal symbols (and) such that all productions of G are of the form $A \to (s)$, where $A \in N$ and $s \in (N \cup \Sigma \backslash \{(,)\})^*$. A language that is generated by a parenthesis grammar is called a *parenthesis language*. Parenthesis languages where first studied in [22]. In [5] it was shown that every parenthesis language is in uNC^1.

Lemma 3. *Every recognizable tree language is FOM-reducible to a parenthesis language. Furthermore the uniform membership problem for TDTAs is log-space reducible to the uniform membership problem for parenthesis grammars.*

Proof. Let $\mathcal{A} = (Q, \mathcal{F}, q_0, \mathcal{R})$ be a TDTA. Let G be the parenthesis grammar $G = (Q, \mathcal{F} \cup \{(,)\}, q_0, P)$ where

$$P = \{q \to (f q_1 \cdots q_m) \mid q(f(x_1, \ldots, x_m)) \to f(q_1(x_1), \ldots, q_m(x_m)) \in \mathcal{R}\}.$$

Let us define a function $\beta : \text{Ł}(\mathcal{F}) \to (\mathcal{F} \cup \{(,)\})^+$ inductively by $\beta(f t_1 \cdots t_m) = (f \beta(t_1) \cdots \beta(t_m))$ for $f \in \mathcal{F}_m$ and $t_1, \ldots, t_m \in \text{Ł}(\mathcal{F})$. Then we have $t \in T(\mathcal{A})$ if and only if $\beta(t) \in L(G)$. Thus by Lemma 1 it suffices to show that the function β is FOM-definable. Let $t = \alpha_1 \cdots \alpha_n$, where $\alpha_j \in \mathcal{F}$. Then in order to construct $\beta(t)$ from t, an opening bracket has to be inserted in front of every symbol in t. Furthermore for $j \in \{1, \ldots, n\}$ the number of closing brackets following α_j in $\beta(t)$ is precisely the number of positions $i \leq j$ such that $\alpha_i \cdots \alpha_j \in \text{Ł}$. Hence β can be defined by the following formulas, where $\alpha \in \mathcal{F}$:

$$\begin{aligned}
\phi(x) &\equiv x = 3 \cdot \max \\
\phi_\alpha(x) &\equiv \exists y, z \, \{ \, Q_\alpha(y) \wedge z = \#i(\exists j (j < y \wedge \text{Ł}(i,j))) \wedge x = 2y + z \} \\
\phi_((x) &\equiv \bigvee_{\alpha \in \mathcal{F}} \phi_\alpha(x+1) \\
\phi_)(x) &\equiv \neg \phi_((x) \wedge \bigwedge_{\alpha \in \mathcal{F}} \neg \phi_\alpha(x)
\end{aligned}$$

For the second statement note that in the uniform case all constructions can be easily done in log-space. □

Theorem 1. *Let T be a fixed recognizable tree language. Then the membership problem for T is in uNC^1. Furthermore there exists a fixed deterministic TDTA $\mathcal{A}$ such that the membership problem for $T(\mathcal{A})$ is uNC^1-complete under DLOGTIME-reductions.*

Proof. The first statement follows from Lemma 3 and the results of [5]. For the hardness part let $L \subseteq \Sigma^*$ be a fixed regular word language, whose membership problem is uNC^1-complete under DLOGTIME-reductions. By [2, Proposition 6.4] such a language exists. If we define $\mathrm{arity}(a) = 1$ for all $a \in \Sigma$ and let $\# \notin \Sigma$ be a constant then we can identify a word $a_1 a_2 \cdots a_n \in \Sigma^*$ with the ground term $a_1 a_2 \cdots a_n \# \in T(\Sigma \cup \{\#\})$, and the language L can be recognized by a fixed deterministic TDTA. □

4 Word Problems for Finitely Presented Algebras

In this section we present an application of Theorem 1 to the word problem for a finitely presented algebra. The *uniform word problem* for finitely presented algebras is the following problem:

INPUT: A ranked alphabet $\mathcal{F}$, a ground TRS $\mathcal{P}$ over $\mathcal{F}$, and $t_1, t_2 \in T(\mathcal{F})$.
QUESTION: Does $t_1 \equiv_{\mathcal{P}} t_2$ hold?

In [21] it was shown that the uniform word problem for finitely presented algebras is P-complete. Here we will study the *word problem* for a fixed finitely presented algebra $A(\mathcal{F}, \mathcal{P})$, where $\mathcal{F}$ is a fixed ranked alphabet, and $\mathcal{P}$ is a fixed ground TRS over $\mathcal{F}$:

INPUT: Two ground terms $t_1, t_2 \in T(\mathcal{F})$.
QUESTION: Does $t_1 \equiv_{\mathcal{P}} t_2$ hold?

For the rest of this section let us fix two ground terms $t_1, t_2 \in T(\mathcal{F})$. We want to decide whether $t_1 \equiv_{\mathcal{P}} t_2$. The following definition is taken from [9]. Let Ω be a new constant. Let $\Delta = (\mathcal{F} \cup \{\Omega\}) \times (\mathcal{F} \cup \{\Omega\}) \backslash \{(\Omega, \Omega)\}$ and define the arity of $[\alpha, \beta] \in \Delta$ by $\max\{\mathrm{arity}(\alpha), \mathrm{arity}(\beta)\}$. We define the function $\sigma : T(\mathcal{F}) \times T(\mathcal{F}) \to T(\Delta)$ inductively by

$$\sigma(f(u_1, \dots, u_m), g(v_1, \dots, v_n)) = \\ [f, g](\sigma(u_1, v_1), \dots, \sigma(u_n, v_n), \sigma(u_{n+1}, \Omega), \dots, \sigma(u_m, \Omega))$$

if $m \geq n$ plus the symmetric rules for the case $m < n$. The term $\sigma(t_1, t_2)$ is a kind of parallel superposition of t_1 and t_2.

Example 1. Let $t_1 = faffafaaa$, $t_2 = fffaafaafaa$. Then $\sigma(t_1, t_2)$ is the term $[f, f][a, f][\Omega, f][\Omega, a][\Omega, a][\Omega, f][\Omega, a][\Omega, a][f, f][f, a][a, \Omega][f, \Omega][a, \Omega][a, \Omega][a, a]$.

In [9] the was shown that the set $T_{\mathcal{P}} = \{\sigma(t_1, t_2) \mid t_1, t_2 \in T(\mathcal{F}), t_1 \equiv_{\mathcal{P}} t_2\}$ is recognizable. Since $\mathcal{P}$ is a fixed ground TRS, $T_{\mathcal{P}}$ is also a fixed recognizable tree language. Thus by Theorem 1 we can decide in uNC^1 whether a term $t \in T(\Delta)$ belongs to $T_{\mathcal{P}}$. Therefore in order to put the word problem for $A(\mathcal{F}, \mathcal{P})$ into uNC^1 it suffices by Lemma 1 to prove the following lemma:

Lemma 4. *The function σ is FOM-definable.*

Proof. We assume that the input for σ is given as $t_1 t_2$. Let $x \in \{1, \ldots, |t_i|\}$. Then the x-th symbol of t_i corresponds to a node in the tree associated with t_i, and we denote the sequence of numbers that labels the path from the root to this node by $p_i(x)$, where each time we descend to the k-th child we write k. For instance if $t_1 = faffafaaa$ then $p_1(7) = 2121$. This sequence can be also constructed as follows. Let us fix some constant $a \in \mathcal{F}_0$. Let s be the prefix of t_i of length $x - 1$. Now we replace in s an arbitrary subword which belongs to $Ł\backslash\{a\}$ by a and repeat this as long as possible. Formally we define a function Π inductively by $\Pi(s) = s$ if $s \notin \mathcal{F}^*(Ł\backslash\{a\})\mathcal{F}^*$ and $\Pi(vtw) = \Pi(vaw)$ if $t \in Ł\backslash\{a\}$. We have for instance $\Pi(faff) = faff$ and $\Pi(fffaafaafa) = fafa$. Then it is easy to see that $p_i(x) = k_1 \cdots k_m$ if and only if $\Pi(s) = f_1 a^{k_1 - 1} \cdots f_m a^{k_m - 1}$ where $\mathrm{arity}(f_j) > 0$ for $j \in \{1, \ldots, m\}$.

First we construct an FOM-formula $c(x_1, x_2)$ that evaluates to true for two positions $x_1 \in \{1, \ldots, |t_1|\}$ and $x_2 \in \{1, \ldots, |t_2|\}$ if and only if $p_1(x_1) = p_2(x_2)$. For this we formalize the ideas above in FOM. In the following formulas we use the constants $o_1 = 0$ and o_2, where o_2 is uniquely defined by the formula $Ł(1, o_1)$. Thus, if interpreted over the word $t_1 t_2$, we have $o_2 = |t_1|$ and $\max - o_2 = |t_2|$. Furthermore let $I_1 = \{1, \ldots, o_2\}$ and $I_2 = \{1, \ldots, \max - o_2\}$. Quantification over these intervals can be easily done in FOM. If $t_i = \alpha_1 \cdots \alpha_n$ and $1 \leq x \leq n$ then the formula $\varphi_i(\ell, r, x)$ evaluates to true if $r < x$, $\alpha_\ell \cdots \alpha_r \in Ł$, and the interval between the positions ℓ and r is maximal with these two properties. The formula $\pi_i(u, x)$ evaluates to true if $u = |\Pi(\alpha_1 \cdots \alpha_{x-1})|$ and finally $f_i(u, x)$ evaluates to true if the u-th symbol of $\Pi(\alpha_1 \cdots \alpha_{x-1})$ has a nonzero arity. Formally for $i \in \{1, 2\}$ we define:

$$\varphi_i(\ell, r, x) \equiv \left\{ \begin{array}{l} r < x \wedge Ł(\ell + o_i, r + o_i) \wedge \\ \neg \exists y, z \in I_i \{ y < \ell \wedge r \leq z < x \wedge Ł(y + o_i, z + o_i) \} \end{array} \right\}$$

$$\pi_i(u, x) \equiv x = u + 1 + \#z\, (\exists \ell, r \in I_i \{ \varphi_i(\ell, r, x) \wedge \ell < z \leq r \})$$

$$f_i(u, x) \equiv \exists z \{ z < x \wedge \neg \exists \ell, r \in I_i (\varphi_i(\ell, r, x) \wedge \ell \leq z \leq r) \wedge \pi_i(u - 1, z) \}$$

$$c(x_1, x_2) \equiv \exists u \left\{ \begin{array}{l} \pi_1(u, x_1) \wedge \pi_2(u, x_2) \wedge \\ \forall y\, (1 \leq y \leq u \rightarrow (f_1(y, x_1) \leftrightarrow f_2(y, x_2))) \end{array} \right\}$$

Finally we can define the functions σ by the following formulas, where $\alpha, \beta \in \mathcal{F}$ (the formulas $\phi_{[\Omega,\alpha]}(x)$ and $\phi_{[\alpha,\Omega]}(x)$ can be defined similarly to $\phi_{[\alpha,\beta]}(x)$):

$$\phi(x) \equiv \max = x + \#y \in I_1\, (\exists y \in I_2\, (c(y_1, y_2)))$$

$$\phi_{[\alpha,\beta]}(x) \equiv \exists y \in I_1, z \in I_2 \left\{ \begin{array}{l} c(y, z) \wedge Q_\alpha(y) \wedge Q_\beta(z + o_2) \wedge \\ y + z = x + \#y'\, (y' \leq y \wedge \exists z' \leq z\, (c(y', z'))) \end{array} \right\}$$

□

Example 2. Let t_1, t_2 be from Example 1. In the following picture two positions satisfy the formula $c(y, z)$ if they are connected by a line. If $x = 15$ then the formula $\phi_{[a,a]}(x)$ is satisfied if we choose $y = 9$ and $z = 11$. Indeed, the 15-th symbol of $\sigma(t_1, t_2)$ is $[a, a]$.

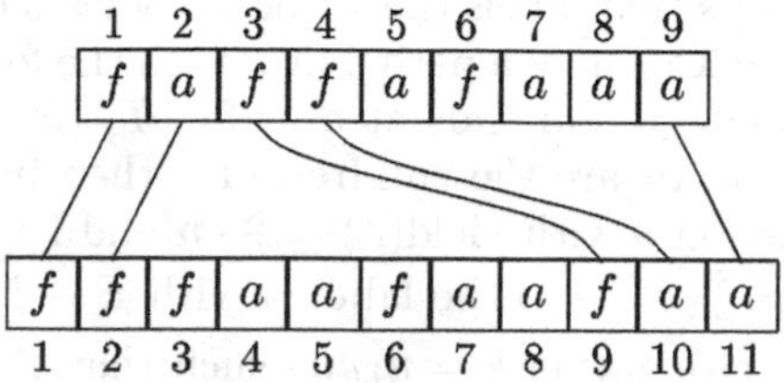

Corollary 1. *For every finitely presented algebra the word problem is in uNC^1.*

Clearly there are also finitely presented algebras whose word problems are uNC^1-complete, like for instance the Boolean algebra $(\{0, 1\}, \wedge, \vee)$ [5]. An interesting open problem might be to find criteria for a finitely presented algebra $A(\mathcal{F}, \mathcal{P})$ which imply that the word problem is uNC^1-complete. For similar work in the context of finite groupoids see [4].

We should also say a few words concerning the input representation. In Theorem 1 and Corollary 1 we represent the input terms as strings over the alphabet $\mathcal{F}$. This is in fact crucial for the uNC^1-upper bounds. If we would represent input terms by their pointer representations then the problems considered would be in general L-complete. For instance if Boolean expressions are represented by their pointer representations then the expression evaluation problem becomes L-complete [3]. For other problems on trees for which it is crucial whether the string or the pointer representation is chosen see [6,19]. For the uniform membership problems in the next section the encoding of the input terms is not crucial for the complexity since these problems are at least L-hard regardless of the chosen encoding.

5 Uniform Membership Problems

In this section we will investigate uniform membership problems for TDTAs. First we need some preliminary results.

Remark 1. The uniform membership problem for the class of all ϵ-free context-free grammars is in LOGCFL.

This fact seems to be folklore. In fact the usual algorithm for recognizing a context-free language on a push-down automaton can be implemented on a log-space bounded auxiliary push-down automaton also if the context-free grammar is part of the input. Furthermore if the grammar does not contain ϵ-productions then this automaton runs in polynomial time. Thus Remark 1 follows from [26]. The next lemma is stated in a similar form in [24, Lemma 3].

Lemma 5. *Let $G = (N, \Sigma, S, P)$ be a context-free grammar in Chomsky normal form. Assume that $A \xrightarrow{*}_G s$, where $A \in N$, $s \in (N \cup \Sigma)^*$, and $|s| > 2$. Then there exist a factorization $s = u_1 v u_2$ and $B \in N$ such that $A \xrightarrow{*}_G u_1 B u_2$, $B \xrightarrow{*}_G v$ and $|v|, |u_1 B u_2| \leq \frac{8}{9} \cdot |s|$.*

Proof. Consider a derivation tree T for the derivation $A \xrightarrow{*}_G s$ and let $n = |s|$. Since G is in Chomsky normal form T is a binary tree. For a node ν of T let yield(ν) be the factor of s that labels the sequence of leafs from left to right of the subtree of T rooted at ν. Consider a path p in T with the following two properties: (i) p starts at the root of T and ends at a leaf of T. (ii) If an edge (ν, ν_1) of T belongs to p and ν_1 and ν_2 are the children of ν then $|\text{yield}(\nu_1)| \geq |\text{yield}(\nu_2)|$. Let ν be the first node on p with yield$(\nu) \leq \frac{8}{9} \cdot n$ and let ν' be the parent node of ν. Thus yield$(\nu') > \frac{8}{9} \cdot n$. Let ν be labeled with $B \in N$ and let yield$(\nu) = v$. Thus there exists a factorization $s = u_1 v u_2$ such that $A \xrightarrow{*}_G u_1 B u_2$, $B \xrightarrow{*}_G v$, and $|v| \leq \frac{8}{9} \cdot |s|$. Furthermore since $n > 2$ and $|v| \geq |\text{yield}(\nu')|/2 > \frac{4}{9} \cdot n$ we also have $|u_1 B u_2| = n - |v| + 1 < n - \frac{4}{9} \cdot n + 1 \leq \frac{8}{9} \cdot n$. □

Theorem 2. *The uniform membership problem for the class of all TDTAs is LOGCFL-complete under log-space reductions.*

Proof. By the second statement from Lemma 3 and Remark 1 the uniform membership problem for TDTAs is in LOGCFL. It remains to show LOGCFL-hardness. For this we will make use of a technique from [24, Proof of Theorem 2]. Let $G = (N, \Sigma, P, S)$ be an arbitrary fixed context-free grammar [2] and let $w \in \Sigma^*$. We may assume that G is in Chomsky normal form and that $\epsilon \notin L(G)$. Let $|w| = n$ and $\mathcal{F} = \{a, f\}$, where arity$(a) = 0$ and arity$(f) = 2$. We will construct a TDTA $\mathcal{A} = (Q, \mathcal{F}, q_0, \mathcal{R})$ and a term $t \in T(\mathcal{F})$ such that $t \in T(\mathcal{A})$ if and only if $w \in L(G)$. Furthermore $\mathcal{A}$ and t can be computed in log-space from w. Let

$$W = \{w_1 A_1 w_2 \cdots w_i A_i w_{i+1} \mid 0 \leq i \leq 3,\ A_1, \ldots, A_i \in N,\\ w \in \Sigma^* w_1 \Sigma^* w_2 \ldots w_i \Sigma^* w_{i+1} \Sigma^*\}.$$

Thus W is the set of all $s \in (N \cup \Sigma)^*$ with $|s|_N \leq 3$ such that a subword of w can be obtained by substituting terminal words for the non-terminals in s. Note that $|W|$ is bounded polynomially in $|w| = n$, more precisely $|W| \in O(n^8)$. The set Q of states of $\mathcal{A}$ is $Q = \{\langle A, s\rangle \mid A \in N, s \in W\}$. The state $\langle A, s\rangle$ may be seen as the assertion that $A \xrightarrow{*}_G s$ holds. The initial state q_0 is $\langle S, w\rangle$. Finally the set $\mathcal{R}$ contains all rules of the following form, where $A \in N$, $s \in W$, and $v, u_1, u_2 \in (N \cup \Sigma)^*$ such that $u_1 v u_2 \in W$.

(1) $\langle A, s\rangle(a) \to a$ if $(A, s) \in P$
(2) $\langle A, s\rangle(f(x, y)) \to f(\langle A, s\rangle(x), \langle A, s\rangle(y))$ if $(A, s) \in P$
(3) $\langle A, u_1 v u_2\rangle(f(x, y)) \to f(\langle A, u_1 B u_2\rangle(x), \langle B, v\rangle(y))$ if $|u_1 v u_2|_N < 3$ or $(|v|_N = 2$ and $|u_1 v u_2|_N = 3)$.

Note that in (3), if $u_1 v u_2$ contains three non-terminals then we must choose a factorization $u_1 v u_2$ such that v contains exactly two non-terminals. Then also $u_1 B u_2$ contains exactly two non-terminals. On the other hand if $u_1 v u_2$ contains less then three non-terminals in (3) then we may choose any factorization. In

[2] In fact we may choose for G Greibach's hardest context-free grammar [16].

this case both v and u_1Bu_2 contain at most three non-terminals. This concludes the description of $\mathcal{A}$. Note that $\mathcal{A}$ can be constructed in log-space from w. For the definition of the term t we need some further notations. In the following let $\gamma = 9/8 > 1$ and for $m > 0$ let $g_m = 2 \cdot \lceil \log_\gamma(m) \rceil + 2$. Furthermore for $m > 0$ let $\mathrm{bal}(m) \in T(\{a, f\})$ be a fully balanced term of height m, i.e., if $m = 1$ then $\mathrm{bal}(m) = a$, otherwise $\mathrm{bal}(m) = f(\mathrm{bal}(m-1), \mathrm{bal}(m-1))$. Now let $t = \mathrm{bal}(g_n)$. Since g_n is logarithmic in $n = |w|$, the size of t is polynomially bounded in n and t can be constructed from w in log-space. We claim that $w \in L(G)$ if and only if $t \in T(\mathcal{A})$. For the if-direction it suffices to prove the following more general claim for all $A \in N$, $s \in W$, and $t' \in T(\{a, f\})$:

$$\text{If } \langle A, s\rangle(t') \xrightarrow{*}_{\mathcal{R}} t' \text{ then } A \xrightarrow{*}_G s.$$

This statement can be shown easily by an induction on the structure of the term t'. For the (only if)-direction we will first show the following two statements, where $A \in N$, $s \in W$, $A \xrightarrow{*}_G s$, and $|s| = m > 0$ (note that $\epsilon \notin L(G)$).

(1) If $|s|_N < 3$ then $\langle A, s\rangle(t') \xrightarrow{*}_{\mathcal{R}} t'$ for some t' with $\mathrm{height}(t') \leq g_m - 1$.
(2) If $|s|_N = 3$ then $\langle A, s\rangle(t') \xrightarrow{*}_{\mathcal{R}} t'$ for some t' with $\mathrm{height}(t') \leq g_m$.

We prove these two statements simultaneously by an induction on the length of the derivation $A \xrightarrow{*}_G s$. If $m = 1$ then the derivation $A \xrightarrow{*}_G s$ has length one. Choose $t' = a$, then $\langle A, s\rangle(t') \rightarrow_{\mathcal{R}} t'$ and $\mathrm{height}(t') = 1 = g_1 - 1$. If $m = 2$ then since G is in Chomsky normal form the derivation $A \xrightarrow{*}_G s$ has length at most 3. Let $t' = f(f(a, a), a)$. Then $\langle A, s\rangle(t') \xrightarrow{*}_{\mathcal{R}} t'$ and $\mathrm{height}(t') = 3 \leq g_2 - 1$. Now assume $m \geq 3$. First let $|s|_N = 3$. Then there exist a factorization $s = u_1 v u_2$ and $B \in N$ with $A \xrightarrow{*}_G u_1Bu_2$, $B \xrightarrow{*}_G v$, and $|v|_N = |u_1Bu_2|_N = 2$. Let $m_1 = |u_1Bu_2|$ and $m_2 = |v|$, thus $m = m_1 + m_2 - 1$. W.l.o.g. assume $m_1 \geq m_2$. Now the induction hypothesis implies $\langle A, u_1Bu_2\rangle(t_1) \xrightarrow{*}_{\mathcal{R}} t_1$ and $\langle B, v\rangle(t_2) \xrightarrow{*}_{\mathcal{R}} t_2$ for some terms t_1 and t_2 with $\mathrm{height}(t_1) \leq g_{m_1} - 1$ and $\mathrm{height}(t_2) \leq g_{m_2} - 1$. It follows $\langle A, s\rangle(f(t_1, t_2)) \xrightarrow{*}_{\mathcal{R}} f(t_1, t_2)$ where $\mathrm{height}(f(t_1, t_2)) = 1 + \mathrm{height}(t_1) \leq g_{m_1} \leq g_m$ since $m_1 = m + 1 - m_2 \leq m$. Finally assume that $|s|_N < 3$. By Lemma 5 there exist a factorization $s = u_1 v u_2$ and $B \in N$ such that $A \xrightarrow{*}_G u_1Bu_2$, $B \xrightarrow{*}_G v$, and $v, u_1Bu_2 \in W$, and $|u_1Bu_2|, |v| \leq m/\gamma$. Let $m_1 = |u_1Bu_2|$ and $m_2 = |v|$. The induction hypothesis implies $\langle A, u_1Bu_2\rangle(t_1) \xrightarrow{*}_{\mathcal{R}} t_1$ and $\langle B, v\rangle(t_2) \xrightarrow{*}_{\mathcal{R}} t_2$ for some terms t_1 and t_2 such that $\mathrm{height}(t_1) \leq g_{m_1} \leq 2 \cdot \lceil \log_\gamma(\frac{m}{\gamma}) \rceil + 2 = g_m - 2$ and similarly $\mathrm{height}(t_2) \leq g_m - 2$. It follows $\langle A, s\rangle(f(t_1, t_2)) \xrightarrow{*}_{\mathcal{R}} f(t_1, t_2)$ where $\mathrm{height}(f(t_1, t_2)) \leq g_m - 1$. This concludes the proof of the statements (1) and (2). Now assume that $w \in L(G)$. Then $\langle S, w\rangle(t') \xrightarrow{*}_{\mathcal{R}} t'$ for some t' with $\mathrm{height}(t') \leq g_n$. But then the first two groups of rules of $\mathcal{A}$ imply $\langle S, w\rangle(\mathrm{bal}(g_n)) \xrightarrow{*}_{\mathcal{R}} \mathrm{bal}(g_n)$, i.e, $t = \mathrm{bal}(g_n) \in T(\mathcal{A})$. This concludes the proof of the theorem. □

Remark 1, Theorem 2 and the second statement from Lemma 3 immediately imply the following corollary.

Corollary 2. *The uniform membership problem for the class of all parenthesis grammars is LOGCFL-complete under log-space reductions.*

In [15] it was shown that the problem of evaluating acyclic Boolean conjunctive queries is LOGCFL–complete. In order to show LOGCFL–hardness, in [15] Venkateswaran's characterization [29] of LOGCFL in terms of semi–unbounded circuits is used. In fact the method from [15] may be modified in order to prove Theorem 2. On the other hand our proof does not use Venkateswaran's result and seems to be more elementary.

It should be also noted that since directed reachability in graphs is NL-complete [20], the uniform membership problem for usual nondeterministic word automata is NL-complete. Thus, since NL is supposed to be a proper subset of LOGCFL, for the nondeterministic case the complexity seems to increase when going from words to trees. The next theorem shows that this is not the case for the deterministic case if we restrict to TDTAs.

Theorem 3. *The uniform membership problem for the class of all deterministic TDTAs is L-complete under DLOGTIME-reductions.*

Proof. Hardness follows from the fact that the uniform membership problem for deterministic word automata is L-complete under DLOGTIME-reductions, see e.g. [8] and the remark in [18, Theorem 15]. For the upper bound we will use an idea that appeared in a similar form in [13, Section 4] in the context of tree walking automata with pebbles. Let $\mathcal{A} = (Q, \mathcal{F}, q_0, \mathcal{R})$ be a deterministic TDTA and let $t \in T(\mathcal{F})$. We will outline a high-level description of a deterministic log-space Turing machine that decides whether $t \in T(\mathcal{A})$. We use a result of [14], which roughly speaking says that in order to check whether a tree is accepted by a deterministic TDTA it suffices to check each path from the root to a leaf separately.

We assume that the input word $t \in \text{Ł}$ is stored on the input tape starting at position 1. In the following we will identify a position $i \in \{1, \ldots, |t|\}$ on the input tape with the corresponding node of the tree t. For the term $t = ffaaffaaa$ for instance, position 1 is the root of t and $3, 4, 7, 8$, and 9 are the leafs of t. In the high-level description we will use the following variables:

- $h_i \in \{1, \ldots, |t|\}$ $(i \in \{1, 2\})$ is a position on the input tape. With h_1 we visit all nodes of t. Each time h_1 visits a leaf of t, with h_2 we walk down the path from the root 1 to h_1 and check whether it is accepted by $\mathcal{A}$.
- $f_i \in \mathcal{F}$ $(i \in \{1, 2\})$ is the label of node h_i.
- $q \in Q$ is the state to which node h_2 evaluates under the automaton $\mathcal{A}$.

All these variables only need logarithmic space. We use the following routines:

- brother(h) returns the position of the right brother of h, or undefined if h does not have a right brother. This value can be calculated in log-space by counting, using the characterization of Ł from [17], see the proof of Lemma 2.
- $\delta(f, q, i)$, where $f \in \mathcal{F}$, $q \in Q$, and $i \in \{1, \ldots, \text{arity}(f)\}$, returns the state q' such that if $q(f(x_1, \ldots, x_n)) \to f(q_1(x_1), \ldots, q_n(x_n)) \in \mathcal{R}$ then $q' = q_i$.

For instance for the term t above we have brother(2) = 5. Finally we present the algorithm. It is clear that this algorithm runs in log-space.

```
for h1 := 1 to |t| do
    if arity(f1) = 0 then
        q := q0; h2 := 1;
        while h2 < h1 do
            f := f2; i := 1; h2 := h2 + 1;
            while brother(h2) is defined and brother(h2) ≤ h1 do
                i := i + 1; h2 := brother(h2)
            endwhile
            q := δ(f, q, i)
        endwhile
        if (q(f2) → f2) ∉ R then reject
endfor
accept
```

□

Finally we consider deterministic BUTAs. Note that the uniform membership problem for nondeterministic BUTAs was implicitly considered in Theorem 2, since the uniform membership problems for nondeterministic BUTAs and nondeterministic TDTAs, respectively, can be directly translated into each other.

Theorem 4. *The uniform membership problem for the class of all deterministic BUTAs is in LOGDCFL.*

Proof. Let $\mathcal{A} = (Q, \mathcal{F}, q_f, \mathcal{R})$ be a deterministic BUTA and let $t \in T(\mathcal{F})$. Let $\# \notin \{0,1\}$ be an additional symbol. By [26] it suffices to outline a deterministic log-space bounded auxiliary push-down automaton $\mathcal{M}$ that checks in polynomial time whether $t \in T(\mathcal{A})$. The input word t is scanned from from right to left. A sequence of the form $\#\text{bin}(q_1)\#\text{bin}(q_2)\cdots\#\text{bin}(q_m)$ is stored on the push-down, where $\text{bin}(q_i)$ is the binary coding of the state $q_i \in Q$ and the top-most push-down symbol corresponds to the right-most symbol in this word. The length of this coding is bounded logarithmically in the input length. If $\mathcal{M}$ reads the symbol f from the input, where $\text{arity}(f) = n$, then $\mathcal{M}$ replaces the sequence $\#\text{bin}(q_1)\#\text{bin}(q_2)\cdots\#\text{bin}(q_n)$ by the sequence $\#\text{bin}(q)$ on top of the push-down, where $f(q_1(x_1), \ldots, q_n(x_n)) \to q(f(x_1, \ldots, x_n))$ is a rule in $\mathcal{R}$. The auxiliary tape is used for storing binary coded states. □

The precise complexity of the uniform membership problem for deterministic BUTAs remains open. For the lower bound we can only prove L-hardness. This problem has also an interesting reformulation in terms of finite algebras. A deterministic BUTA $\mathcal{A}$ corresponds in a straight-forward way to a finite algebra A. The carrier set of A is the set Q of states of $\mathcal{A}$ and every function symbol f of arity n is interpreted as an n-ary function on Q. Now the question whether a term t is accepted by $\mathcal{A}$ is equivalent to the question whether the expression t evaluates in the algebra A to a distinguished element q (namely the final state of $\mathcal{A}$). Thus the uniform membership problem for deterministic BUTAs is equivalent to the uniform expression evaluation problem for finite algebras. In the case of a fixed groupoid, the complexity of the expression evaluation problem was considered in [4].

Table 1 summarizes the complexity results for tree automata shown in this paper.

Table 1. Complexity results for tree automata

	det. TDTA	det. BUTA	TDTA (BUTA)
membership	uNC^1-complete	uNC^1-complete	uNC^1-complete
uniform membership	L-complete	LOGDCFL	LOGCFL-complete

Acknowledgments. I would like to thank the referees for valuable comments.

References

1. D. A. M. Barrington and J. Corbet. On the relative complexity of some languages in NC^1. *Information Processing Letters*, 32:251–256, 1989.
2. D. A. M. Barrington, N. Immerman, and H. Straubing. On uniformity within NC^1. *Journal of Computer and System Sciences*, 41:274–306, 1990.
3. M. Beaudry and P. McKenzie. Circuits, matrices, and nonassociative computation. *Journal of Computer and System Sciences*, 50(3):441–455, 1995.
4. J. Berman, A. Drisko, F. Lemieux, C. Moore, and D. Thérien. Circuits and expressions with non–associative gates. In *Proceedings of the 12th Annual IEEE Conference on Computational Complexity, Ulm (Germany)*, pages 193–203. IEEE Computer Society Press, 1997.
5. S. R. Buss. The Boolean formula value problem is in ALOGTIME. In *Proceedings of the 19th Annual Symposium on Theory of Computing (STOC 87)*, pages 123–131. ACM Press, 1987.
6. S. R. Buss. Alogtime algorithms for tree isomorphism, comparison, and canonization. In *Kurt Gödel Colloquium 97*, pages 18–33, 1997.
7. H. Comon, M. Dauchet, R. Gilleron, F. Jacquemard, D. Lugiez, S. Tison, and M. Tommasi. Tree automata techniques and applications. Available on: `http://www.grappa.univ-lille3.fr/tata`, 1997.
8. S. A. Cook and P. McKenzie. Problems complete for deterministic logarithmic space. *Journal of Algorithms*, 8:385–394, 1987.
9. M. Dauchet and S. Tison. The theory of ground rewrite systems is decidable. In *Proceedings of the 5th Annual IEEE Symposium on Logic in Computer Science (LICS '90)*, pages 242–256. IEEE Computer Society Press, 1990.
10. N. Dershowitz and J.-P. Jouannaud. Rewriting systems. In J. van Leeuwen, editor, *Handbook of Theoretical Computer Science*, pages 243–320. Elsevier Publishers, Amsterdam, 1990.
11. J. E. Doner. Decidability of the weak second-order theory of two successors. *Notices Amer. Math. Soc.*, 12:365–468, 1965.
12. J. E. Doner. Tree acceptors and some of their applications. *Journal of Computer and System Sciences*, 4:406–451, 1970.

13. J. Engelfriet and H. J. Hoogeboom. Tree-walking pebble automata. In J. Karhumäki, H. Maurer, G. Paun, and G. Rozenberg, editors, *Jewels are Forever, Contributions on Theoretical Computer Science in Honor of Arto Salomaa*, pages 72–83. Springer, 1999.
14. F. Gécseg and M. Steinby. *Tree automata.* Akadémiai Kiadó, 1984.
15. G. Gottlob, N. Leone, and F. Scarcello. The complexity of acyclic conjunctive queries. In *Proceedings of the 39th Annual Symposium on Foundations of Computer Science (FOCS 98, Palo Alto, California, USA)*, pages 706–715. IEEE Computer Society Press, 1998.
16. S. Greibach. The hardest context-free language. *SIAM Journal on Computing*, 2(4):304–310, 1973.
17. M. A. Harrison. *Introduction to Formal Language Theory.* Addison-Wesley, 1978.
18. M. Holzer and K.-J. Lange. On the complexities of linear LL(1) and LR(1) grammars. In Z. Ésik, editor, *Proceedings of the 9th International Symposium on Fundamentals of Computation Theory (FCT'93, Szeged, Hungary)*, number 710 in Lecture Notes in Computer Science, pages 299–308. Springer, 1993.
19. B. Jenner, P. McKenzie, and J. Torán. A note on the hardness of tree isomorphism. In *Proceedings of the 13th Annual IEEE Conference on Computational Complexity*, pages 101–105. IEEE Computer Society Press, 1998.
20. N. D. Jones. Space-bounded reducibility among combinatorial problems. *Journal of Computer and System Sciences*, 11(1):68–85, 1975.
21. D. C. Kozen. Complexity of finitely presented algebras. In *9th Annual Symposium on Theory of Computing (STOC 77)*, pages 164–177. ACM Press, 1977.
22. R. McNaughton. Parenthesis grammars. *Journal of the Association for Computing Machinery*, 14(3):490–500, 1967.
23. C. H. Papadimitriou. *Computational Complexity.* Addison Wesley, 1994.
24. W. L. Ruzzo. Tree–size bounded alternation. *Journal of Computer and System Sciences*, 21:218–235, 1980.
25. W. L. Ruzzo. On uniform circuit complexity. *Journal of Computer and System Sciences*, 22:365–383, 1981.
26. I. H. Sudborough. On the tape complexity of deterministic context–free languages. *Journal of the Association for Computing Machinery*, 25(3):405–414, 1978.
27. J. W. Thatcher and J. B. Wright. Generalized finite automata with an application to a decision problem of second order logic. *Mathematical Systems Theory*, 2:57–82, 1968.
28. M. Veanes. On computational complexity of basic decision problems of finite tree automata. Technical Report 133, Uppsala Computing Science Department, 1997.
29. H. Venkateswaran. Properties that characterize LOGCFL. *Journal of Computer and System Sciences*, 43:380–404, 1991.
30. H. Vollmer. *Introduction to Circuit Complexity.* Springer, 1999.

Transfinite Rewriting Semantics for Term Rewriting Systems*

Salvador Lucas

Departamento de Sistemas Informáticos y Computación
Universidad Politécnica de Valencia
Camino de Vera s/n, E-46022 Valencia, Spain
slucas@dsic.upv.es

Abstract. We provide some new results concerning the use of transfinite rewriting for giving semantics to rewrite systems. We especially (but not only) consider the computation of possibly infinite constructor terms by transfinite rewriting due to their interest in many programming languages. We reconsider the problem of compressing transfinite rewrite sequences into shorter (possibly finite) ones. We also investigate the role that (finitary) confluence plays in transfinite rewriting. We consider different (quite standard) rewriting semantics (mappings from input terms to sets of reducts obtained by –transfinite– rewriting) in a unified framework and investigate their algebraic structure. Such a framework is used to formulate, connect, and approximate different properties of TRSs.

1 Introduction

Rewriting that considers infinite terms and reduction sequences of any ordinal length is called *transfinite rewriting*; rewriting sequences of length ω are often called *infinitary*. The motivation to distinguish between them is clear: reduction sequences of length of at most ω seem more adequate for real applications (but transfinite rewriting is suitable for modeling rewriting with finite cyclic graphs). There are two main frameworks for transfinite rewriting: [DKP91] considers standard *Cauchy convergent* rewriting sequences; [KKSV95] only admits *strongly convergent* sequences which are Cauchy convergent sequences in which redexes are contracted at deeper and deeper positions. Cauchy convergent sequences are more powerful than strongly convergent ones w.r.t. their *computational strength*, i.e., the ability to compute canonical forms of terms (normal forms, values, etc.).

Example 1. Consider the TRS (see [KKSV95]):

$$\begin{array}{llll} \mathtt{f(x,g)} \to \mathtt{f(c(x),g)} & & \mathtt{g} \to \mathtt{a} \\ \mathtt{f(x,a)} \to \mathtt{c(x)} & & \mathtt{h} \to \mathtt{c(h)} \end{array}$$

and the derivation of length $\omega + 2$:

$$\underline{\mathtt{f(a,g)}} \to \underline{\mathtt{f(c(a),g)}} \to \cdots\ \mathtt{f(c^{\omega},\underline{g})} \to \underline{\mathtt{f(c^{\omega},a)}} \to \mathtt{c^{\omega}}$$

No strongly convergent reduction rewrites $\mathtt{f(a,g)}$ into the infinite term $\mathtt{c^{\omega}}$.

* This work has been partially supported by CICYT TIC 98-0445-C03-01.

A. Middeldorp (Ed.): RTA 2001, LNCS 2051, pp. 216–230, 2001.

Transfinite, strongly convergent sequences can be *compressed* into infinitary ones when dealing with left-linear TRSs [KKSV95]. This is not true for Cauchy convergent sequences: e.g., there is no Cauchy convergent *infinitary* sequence reducing `f(a,g)` into c^ω in Example 1. Because of this, Kennaway et al. argue that strongly convergent transfinite rewriting is the best basis for a theory of transfinite rewriting [KKSV95]. However, Example 1 shows that the restriction to strongly convergent sequences may lose computational power, since many possibly useful sequences are just disallowed. Thus, from the semantic point of view, it is interesting to compare them further.

In this paper, we especially consider the computation of possibly infinite constructor terms by (both forms of) transfinite rewriting. In algebraic and functional languages, constructor terms play the role of completely meaningful pieces of information that (defined) functions take as their input and produce as the outcome. We prove that every infinitary rewrite sequence leading to a constructor term is strongly convergent. We prove that, for left-linear TRSs, transfinite rewrite sequences leading to finite terms can always be compressed into *finite* rewrite sequences and that infinite terms obtained by transfinite rewrite sequences can always be finitely (but arbitrarily) approximated by finitary rewrite sequences. We have investigated the role of *finitary* confluence in transfinite rewriting. We prove that for left-linear, (finitary) confluent TRSs, Cauchy convergent transfinite rewritings leading to infinite constructor terms can always be compressed into infinitary ones. We also prove that finitary confluence ensures the uniqueness of infinite constructor terms obtained by infinitary rewriting. We use our results to define and compare *rewriting semantics*. By a rewriting semantics we mean a mapping from input terms to sets of reducts obtained by finite, infinitary, or transfinite rewriting. We study different rewriting semantics and their appropriateness for computing different kinds of interesting semantic values in different classes of TRSs. We also investigate two orderings between semantics that provide an algebraic framework for approximating semantic properties of TRSs. We motivate this framework using some well-known problems in term rewriting.

Section 2 introduces transfinite rewriting. Section 3 deals with compression. Section 4 investigates the role of finitary confluence in transfinite rewriting computations. Section 5 introduces the semantic framework and Section 6 discusses its use in approximating properties of TRSs. Section 7 compares different semantics studied in the paper. Section 8 discusses related work.

2 Transfinite Rewriting

Terms are viewed as labelled trees in the usual way. An (infinite) term on a signature Σ is a finite (or infinite) ordered tree such that each node is labeled by a symbol $f \in \Sigma$ and has a tuple of descendants, and the size of such a tuple is equal to $ar(f)$ (see [Cou83] for a formal definition). When considering a denumerable set of variables $\mathcal{X}$, we obtain terms with variables in the obvious way. The set of (ground) infinite terms is denoted by $\mathcal{T}^\omega(\Sigma,\mathcal{X})$ (resp. $\mathcal{T}^\omega(\Sigma)$) and $\mathcal{T}(\Sigma,\mathcal{X})$

(resp. $\mathcal{T}(\Sigma)$) is the set of (resp. ground) finite terms. Notations $\mathcal{T}^\infty(\Sigma,\mathcal{X})$ or $\mathcal{T}^\infty(\Sigma)$ are more frequent; by using ω, we emphasize that transfinite terms are not considered. This is more consistent with the use of ω and ∞ in the paper.

By $\mathcal{P}os(t)$, we denote the set of positions of a term t, and $|p|$ is the length of a position p. By Λ, we denote the empty chain. The height d_t of $t \in \mathcal{T}(\Sigma,\mathcal{X})$ is given by $\mathsf{d}_t = 1 + max(\{|p| \mid p \in \mathcal{P}os(t)\})$. We use the following distance d among terms: $d(t,s) = 0$ if $t = s$; otherwise, $d(t,s) = 2^{-p(t,s)}$ (where $p(t,s)$ is the length $|p|$ of the shortest position $p \in \mathcal{P}os(t) \cap \mathcal{P}os(s)$ such that $root(t|_p) \neq root(s|_p)$ [Cou83]). Therefore, $(\mathcal{T}^\omega(\Sigma,\mathcal{X}), d)$ is a metric space. Note that, if $t \in \mathcal{T}(\Sigma)$ and $\epsilon \leq 2^{-\mathsf{d}_t}$, then $\forall s \in \mathcal{T}^\omega(\Sigma)$, $d(t,s) < \epsilon \Leftrightarrow t = s$. A *substitution* is a mapping $\sigma : \mathcal{X} \to \mathcal{T}^\omega(\Sigma,\mathcal{X})$ which we homomorphically extend to a mapping $\sigma : \mathcal{T}^\omega(\Sigma,\mathcal{X}) \to \mathcal{T}^\omega(\Sigma,\mathcal{X})$ by requiring that it be continuous w.r.t. the metric d. A rewrite rule is an ordered pair (l,r), written $l \to r$, with $l, r \in \mathcal{T}(\Sigma,\mathcal{X})$, $l \notin \mathcal{X}$ and $\mathcal{V}ar(r) \subseteq \mathcal{V}ar(l)$. The left-hand side (*lhs*) of the rule is l and r is the right-hand side[1] (*rhs*). A TRS is a pair $\mathcal{R} = (\Sigma, R)$ where R is a set of rewrite rules. Given $\mathcal{R} = (\Sigma, R)$, we consider Σ as the disjoint union $\Sigma = \mathcal{C} \uplus \mathcal{F}$ of symbols $c \in \mathcal{C}$, called *constructors* and symbols $f \in \mathcal{F}$, called *defined functions*, where $\mathcal{F} = \{f \mid f(\bar{l}) \to r \in R\}$ and $\mathcal{C} = \Sigma - \mathcal{F}$. Then, $\mathcal{T}(\mathcal{C},\mathcal{X})$ (resp. $\mathcal{T}(\mathcal{C}), \mathcal{T}^\omega(\mathcal{C})$) is the set of (resp. ground, possibly infinite, ground) constructor terms. A term $t \in \mathcal{T}^\omega(\Sigma,\mathcal{X})$ is a redex if there exist a substitution σ and a rule $l \to r$ such that $t = \sigma(l)$. A term $t \in \mathcal{T}^\omega(\Sigma,\mathcal{X})$ rewrites to s (at position p), written $t \to_{\mathcal{R}} s$ (or just $t \to s$), if $t|_p = \sigma(l)$ and $s = t[\sigma(r)]_p$, for some rule $\rho : l \to r \in R$, $p \in \mathcal{P}os(t)$ and substitution σ. This can eventually be detailed by writing $t \stackrel{[p,\rho]}{\to} s$ (substitution σ is uniquely determined by $t|_p$ and l). A normal form is a term without redexes. By $\mathsf{NF}_{\mathcal{R}}$ ($\mathsf{NF}^\omega_{\mathcal{R}}$) we denote the set of finite (resp. possibly infinite) ground normal forms of $\mathcal{R}$. A term t is root-stable (or a head normal form) if $\forall s$, if $t \to^* s$, then s is not a redex. By $\mathsf{HNF}_{\mathcal{R}}$ ($\mathsf{HNF}^\omega_{\mathcal{R}}$) we denote the set of ground root-stable finite (resp. possibly infinite) terms of $\mathcal{R}$. In the following, we are mainly concerned with ground terms.

A *transfinite* rewrite sequence of a TRS $\mathcal{R}$ is a mapping A whose domain is an *ordinal* α such that A maps each $\beta < \alpha$ to a reduction step $A_\beta \to A_{\beta+1}$ [KKSV95]. If α is a limit ordinal, A is called *open*; otherwise, it is *closed.* If $\alpha \in \{\omega, \omega+1\}$, we will say that A is *infinitary*, rather than transfinite. The *length* of the sequence is α if α is a limit ordinal; otherwise, it is $\alpha - 1$. For limit ordinals $\beta < \alpha$, the previous definition of transfinite rewrite sequence does not stipulate any relationship between A_β and the earlier terms in the sequence. Thus, the following notion is helpful [KKSV95]: Given a distance d on terms, a rewriting sequence A is said to be Cauchy continuous if for every ordinal limit $\lambda < \alpha$, $\forall \epsilon > 0, \exists \beta < \lambda, \forall \gamma\ (\beta < \gamma < \lambda \Rightarrow d(A_\gamma, A_\lambda) < \epsilon)$. Given a reduction sequence A, let $\mathfrak{d}_\beta$ be the length of the position of the redex reduced in the step from A_β to $A_{\beta+1}$. A Cauchy continuous closed sequence A is called *Cauchy convergent.* A Cauchy convergent sequence is called *strongly continuous* if for every limit ordinal $\lambda < \alpha$, the sequence $(\mathfrak{d}_\beta)_{\beta<\lambda}$ tends to infinity. If A is strongly

[1] In [KKSV95], infinite right-hand sides are also allowed.

continuous and closed, then S is strongly convergent. We write: $A : t \hookrightarrow^{\alpha} s$ (resp. $A : t \to^{\alpha} s$) for a Cauchy (resp. strongly) convergent sequence of length α starting from t and ending at s. We write: $A : t \hookrightarrow^{\leq\alpha} s$ (resp. $A : t \to^{\leq\alpha} s$) for a Cauchy (resp. strongly) convergent sequence starting from t and ending at s whose length is less than or equal to α; moreover, we often take the initial term t of A as its name and write t_β for an ordinal $\beta < \alpha$ rather than A_β. We write $t \hookrightarrow^{\infty} s$ (resp. $t \to^{\infty} s$) if we do not wish to explicitly indicate the length of the sequence.

3 Compression of Transfinite Rewrite Sequences

The following result shows that, for computing *constructor terms*, Cauchy convergent and strongly convergent *infinitary* rewriting are equivalent.

Theorem 1. *Let $\mathcal{R}$ be a TRS, $t \in \mathcal{T}^{\omega}(\Sigma)$ and $\delta \in \mathcal{T}^{\omega}(\mathcal{C})$. Every Cauchy convergent infinitary rewrite sequence from t to δ is strongly convergent.*

Proof. Let $t \hookrightarrow^{\omega} \delta$. Since no rule may overlap any constructor prefix of δ, $\mathfrak{d}_\beta$ for redexes contracted in each step $t_\beta \to t_{\beta+1}$ must tend to infinity as the normal form does.

Example 1 shows that, in general, Theorem 1 does not hold for transfinite sequences. Moreover, Theorem 1 does not hold for arbitrary normal forms.

Example 2. Consider the orthogonal (infinite) TRS [KKSV95]:

$$\begin{array}{lll} \mathtt{f(g}^n\mathtt{(c))} & \to \mathtt{f(g}^{n+1}\mathtt{(c))} & \text{for } n \geq 0 \\ \mathtt{a} & \to \mathtt{g(a)} & \end{array}$$

Thus, $\mathtt{f(c)} \hookrightarrow^{\omega} \mathtt{f(g}^{\omega}\mathtt{)} \in \mathsf{NF}^{\omega}_{\mathcal{R}}$, but $\mathtt{f(c)} \not\to^{\omega} \mathtt{f(g}^{\omega}\mathtt{)}$.

Kennaway et al. proved the following:

Theorem 2. [KKSV95] *Let $\mathcal{R}$ be a left-linear TRS and $t, s \in \mathcal{T}^{\omega}(\Sigma)$. If $t \to^{\infty} s$, then $t \to^{\leq\omega} s$.*

Theorem 2 does not hold for Cauchy convergent reductions.

Remark 1. The compression of Cauchy convergent sequences into infinitary ones has been studied[2] in [DKP91], where *top-termination* (i.e., no infinitary reduction sequence performs infinitely many rewrites at Λ), is required to achieve it. Kennaway et al. noticed that it implies that "*every reduction starting from a finite term is strongly convergent*" thus arising as a consequence of Theorem 2.

For Cauchy convergent sequences, we prove several restricted compression properties.

Proposition 1. *Let $\mathcal{R}$ be a TRS, $t \in \mathcal{T}^{\omega}(\Sigma)$ and s be a finite term. If $t \hookrightarrow^{\lambda} s$ for a limit ordinal λ, then $t \hookrightarrow^{\beta} s$ for some $\beta < \lambda$.*

[2] In the terminology of [DKP91], ω-closure.

Proof. If $t \hookrightarrow^{\lambda} s$, then $\forall \epsilon > 0, \exists \beta_\epsilon, \forall \gamma, \beta_\epsilon < \gamma < \lambda$, $d(t_\gamma, s) < \epsilon$. Since s is finite, we let $\epsilon = 2^{-\mathsf{d}_s}$. Then, $\forall \gamma, \beta_\epsilon < \gamma < \lambda$, it must be $t_\gamma = s$. In particular, $t \hookrightarrow^{\beta} s$ for $\beta = \beta_\epsilon + 1 < \lambda$ (since λ is a limit ordinal).

In Theorem 3 below, we prove that, for left-linear TRSs, finite pieces of information obtained from transfinite rewriting can always be obtained from finitary sequences. In the following auxiliary result, we use $\mathsf{d}_\rho = max(\mathsf{d}_l, \mathsf{d}_r)$ for a rule $\rho : l \to r$ (note that d_ρ only makes sense for rules whose *rhs*'s are finite).

Proposition 2. *Let $\mathcal{R}$ be a left-linear TRS. Let $t, s \in \mathcal{T}^\omega(\Sigma)$ be such that $t = t_1 \stackrel{[p_1,\rho_1]}{\to} t_2 \to \cdots \to t_n \stackrel{[p_n,\rho_n]}{\to} t_{n+1} = s$. For all $\kappa \in]0,1]$, and $t' \in \mathcal{T}^\omega(\Sigma)$ such that $d(t',t) < 2^{-(\sum_{i=1}^{n}|p_i| + \mathsf{d}_{\rho_i}) + log_2(\kappa)}$, there exists $s' \in \mathcal{T}^\omega(\Sigma)$ such that $t' \to^* s'$ and $d(s', s) < \kappa$.*

Proof. By induction on n. If $n = 0$, then $t = s$ and $d(t', s) < 2^{log_2(\kappa)} = \kappa$. We let $s' = t'$ and the conclusion follows. If $n > 0$, then, since $0 < \kappa \leq 1$, we have $d(t',t) < 2^{-(\sum_{i=1}^{n}|p_i| + \mathsf{d}_{\rho_i}) + log_2(\kappa)} \leq 2^{-|p_1| - \mathsf{d}_{l_1}}$. Since $\mathcal{R}$ is left-linear, it follows that $t'|_{p_1}$ is a redex of ρ_1. Thus, $t' \to t'[\sigma_1'(r_1)]_{p_1}$ and for all $x \in \mathcal{V}ar(l_1)$, $d(\sigma_1'(x), \sigma_1(x)) < 2^{-(\sum_{i=2}^{n}|p_i| + \mathsf{d}_{\rho_i}) + log_2(\kappa)}$. Hence, since $\mathcal{V}ar(r_1) \subseteq \mathcal{V}ar(l_1)$ and $\mathsf{d}_{\rho_1} = max(\mathsf{d}_{l_1}, \mathsf{d}_{r_1})$, we have $d(t'[\sigma_1'(r_1)]_{p_1}, t_2) \leq 2^{-(\sum_{i=2}^{n}|p_i| + \mathsf{d}_{\rho_i}) - |p_1| - \mathsf{d}_{r_1} + log_2(\kappa)}$ $< 2^{-(\sum_{i=2}^{n}|p_i| + \mathsf{d}_{\rho_i}) + log_2(\kappa)}$. By the induction hypothesis, $t'[\sigma_1'(r_1)]_{p_1} \to^* s'$ and $d(s', s) < \kappa$; hence $t' \to^* s'$ and the conclusion follows.

Proposition 3. *Let $\mathcal{R}$ be a left-linear TRS, $t, s \in \mathcal{T}^\omega(\Sigma)$, and $t' \in \mathcal{T}^\omega(\Sigma) - \mathcal{T}(\Sigma)$. If $t \hookrightarrow^{\lambda} t' \to^* s$ for a limit ordinal λ, then for all $\kappa \in]0,1]$, there exist $s' \in \mathcal{T}^\omega(\Sigma)$ and $\beta < \lambda$ such that $t \hookrightarrow^{\beta} s'$ and $d(s', s) < \kappa$.*

Proof. Since λ is a limit ordinal and t' is infinite, for all $\epsilon > 0$, there exists $\beta_\epsilon < \lambda$ such that, for all γ, $\beta_\epsilon < \gamma < \lambda$, $d(t_\gamma, t') < \epsilon$. Let $t' = t_1 \stackrel{[p_1,\rho_1]}{\to} t_2 \to \cdots \to t_n \stackrel{[p_n,\rho_n]}{\to} t_{n+1} = s$, $\epsilon = 2^{-(\sum_{i=1}^{n}|p_i| + \mathsf{d}_{\rho_i}) + log_2(\kappa)}$, and $\beta_0 = \beta_\epsilon + 1$. Thus, $d(t_{\beta_0}, t') < \epsilon$ and, by Proposition 2, $t_{\beta_0} \to^* s'$ and $d(s', s) < \kappa$. Thus, $t \hookrightarrow^{\beta} s'$ for $\beta = \beta_0 + n + 1 < \lambda$ and the conclusion follows.

Theorem 3. *Let $\mathcal{R}$ be a left-linear TRS and $t, s \in \mathcal{T}^\omega(\Sigma)$. If $t \hookrightarrow^{\infty} s$, then for all $\kappa \in]0,1]$, there exists $s' \in \mathcal{T}^\omega(\Sigma)$ such that $t \to^* s'$ and $d(s', s) < \kappa$.*

Proof. Let $t \hookrightarrow^{\alpha} s$ for an arbitrary ordinal α. We proceed by transfinite induction. The finite case $\alpha < \omega$ (including the base case $\alpha = 0$) is immediate since $t \to^* s$ and $d(s, s) = 0 < \kappa$, for all $\kappa \in]0,1]$. Assume $\alpha \geq \omega$. We can write $t \hookrightarrow^{\lambda} t' \to^* s$ for some limit ordinal $\lambda \leq \alpha$. If t' is finite, by Proposition 1, $\exists \beta < \lambda$ such that $t \hookrightarrow^{\beta} t'$. Thus, we have $t \hookrightarrow^{\beta+n} s$ where n is the length of the finite derivation $t' \to^* s$. Since λ is a limit ordinal, $\beta < \lambda$, and $n \in \mathbb{N}$, we have $\beta + n < \lambda$ and by the induction hypothesis, the conclusion follows. If t' is not finite, by Proposition 3, there exist $s' \in \mathcal{T}^\omega(\Sigma)$ and $\beta < \lambda$ such that $t \hookrightarrow^{\beta} s'$ and $d(s', s) < \kappa$. Since $\beta < \alpha$, by the induction hypothesis, $t \to^* s'$ and the conclusion follows.

Corollary 1. *Let $\mathcal{R}$ be a left-linear TRS, $t \in \mathcal{T}^\omega(\Sigma)$, and s be a finite term. If $t \hookrightarrow^\infty s$, then $t \to^* s$.*

The following example shows the need for left-linearity to ensure Corollary 1.

Example 3. Consider the TRS [DKP91]:

```
a → g(a)          b → g(b)          f(x,x) → c
```

Note that $\mathtt{f(a,b)} \to^\omega \mathtt{f(g^\omega,g^\omega)} \to \mathtt{c}$, but $\mathtt{f(a,b)} \not\to^* \mathtt{c}$.

4 Finitary Confluence and Transfinite Rewriting

A TRS $\mathcal{R}$ is (finitary) confluent if $\to$ is confluent, i.e., for all $t, s, s' \in \mathcal{T}(\Sigma, \mathcal{X})$, if $t \to^* s$ and $t \to^* s'$, there exists $u \in \mathcal{T}(\Sigma, \mathcal{X})$ such that $s \to^* u$ and $s' \to^* u$. Finitary confluence of $\to$ for finite terms extends to infinite terms.

Proposition 4. *Let $\mathcal{R}$ be a confluent TRS and $t, s, s' \in \mathcal{T}^\omega(\Sigma, \mathcal{X})$. If $t \to^* s$ and $t \to^* s'$, then there exists $u \in \mathcal{T}^\omega(\Sigma, \mathcal{X})$ such that $s \to^* u \;{}^*\!\!\leftarrow s'$.*

Proof. (sketch) If t is finite, then it is obvious. If t is infinite, we consider the derivations $t = t_1 \overset{[p_1,\rho_1]}{\to} t_2 \to \cdots \to t_n \overset{[p_n,\rho_n]}{\to} t_{n+1} = s$ and $t = t'_1 \overset{[p'_1,\rho'_1]}{\to} t'_2 \to \cdots \to t'_m \overset{[p'_m,\rho'_m]}{\to} t'_{m+1} = s'$. Let $P = \frac{1}{2} + \sum_{i=1}^n |p_i| + \mathsf{d}_{\rho_i} + \sum_{i=1}^m |p'_i| + \mathsf{d}_{\rho'_i}$ and $\widehat{t}$ be the finite term obtained by replacing in t all subterms at positions $q_1, \ldots, q_k$ such that $|q_i| = P$, for $1 \leq i \leq k$, by new variables $x_1, \ldots, x_\ell$, $\ell \leq k$. Let $\nu : \{1, \ldots, k\} \to \{1, \ldots, \ell\}$ be a surjective mapping that satisfies $\nu(i) = \nu(j) \Leftrightarrow t|_{p_i} = t|_{p_j}$ for $1 \leq i \leq j \leq k$. Thus, each variable $x \in \{x_1, \ldots, x_\ell\}$ will name equal subterms at possibly different positions. It is not difficult to show that $\exists \widehat{s}, \widehat{s'} \in \mathcal{T}(\Sigma, \mathcal{X})$ such that $\widehat{t} \to^* \widehat{s}$, $\widehat{t} \to^* \widehat{s'}$, and $s = \sigma(\widehat{s})$, $s' = \sigma(\widehat{s'})$ for σ defined by $\sigma(x_{\nu(i)}) = t|_{q_i}$ for $1 \leq i \leq k$. By confluence, $\exists \widehat{u} \in \mathcal{T}(\Sigma, \mathcal{X})$ such that $\widehat{s} \to^* \widehat{u} \;{}^*\!\!\leftarrow \widehat{s'}$ and by stability of rewriting, $s = \sigma(\widehat{s}) \to^* \sigma(\widehat{u}) \;{}^*\!\!\leftarrow \sigma(\widehat{s'}) = s'$. Thus, $u = \sigma(\widehat{u})$ is the desired (possibly infinite) term.

Proposition 4 could fail for TRSs whose rhs's are infinite (see [KKSV95], Counterexample 6.2). Concerning the computation of infinite constructor terms, we prove that, for left-linear, confluent TRSs, infinitary sequences suffice. First we need an auxiliary result.

Proposition 5. *Let $\mathcal{R}$ be a left-linear, confluent TRS, $t \in \mathcal{T}^\omega(\Sigma)$, and $\delta \in \mathcal{T}^\omega(\mathcal{C}) - \mathcal{T}(\mathcal{C})$. If $t \hookrightarrow^\infty \delta$, then (1) $\nexists s \in \mathsf{NF}_\mathcal{R}$ such that $t \to^* s$, (2) $\nexists \delta' \in \mathcal{T}^\omega(\mathcal{C})$, $\delta \neq \delta'$ such that $t \hookrightarrow^\infty \delta'$.*

Proof. (1) Assume that $t \to^* s \in \mathsf{NF}_\mathcal{R}$ and let $\kappa = 2^{-\mathsf{d}_s - 1}$. Since $t \hookrightarrow^\infty \delta$, by Theorem 3, there exists s' such that $t \to^* s'$ and $d(s', \delta) < \kappa$. Since $s \in \mathsf{NF}_\mathcal{R}$, by confluence, $s' \to^* s$ which is not possible as s' has a constructor prefix whose height is greater than the height of s. (2) Assume that $t \hookrightarrow^\infty \delta'$ for some $\delta' \in \mathcal{T}^\omega(\mathcal{C})$, $\delta \neq \delta'$. Let $\kappa = d(\delta, \delta')/2$. By Theorem 3, $\exists s, s'$ such that $t \to^* s$, $t \to^* s'$, $d(s, \delta) < \kappa$, and $d(s', \delta') < \kappa$. By Proposition 4, there exists u such that $s \to^* u \;{}^*\!\!\leftarrow s'$. However, this is not possible since there exists $p \in \mathcal{P}os(s) \cap \mathcal{P}os(s')$ such that $root(s|_p) = c \neq c' = root(s'|_p)$ and $c, c' \in \mathcal{C}$.

Theorem 4. *Let $\mathcal{R}$ be a left-linear, confluent TRS, $t \in \mathcal{T}^\omega(\Sigma)$, and $\delta \in \mathcal{T}^\omega(\mathcal{C})$. If $t \hookrightarrow^\infty \delta$, then $t \hookrightarrow^{\leq\omega} \delta$.*

Proof. If $\delta \in \mathcal{T}(\mathcal{C})$, it follows by Corollary 1. Let $\delta \in \mathcal{T}^\omega(\mathcal{C}) - \mathcal{T}(\mathcal{C})$. Assume that $t \not\hookrightarrow^{\leq\omega} \delta$. Then, by Proposition 5(2), there is no other infinite constructor term reachable from t. Thus, we can associate a number $M_\mathcal{C}(A) \in \mathbb{N}$ to every (open or closed) infinitary sequence A starting from t, as follows: $M_\mathcal{C}(A) = max(\{m_\mathcal{C}(s) \mid s \in A\})$ where $m_\mathcal{C}(s)$ is the minimum length of maximal sub-constructor positions of s (i.e., positions $p \in \mathcal{P}os(s)$ such that $\forall q < p, root(s|_q) \in \mathcal{C}$). Note that, since we assume that no infinitary derivation A converges to δ, $M_\mathcal{C}(A)$ is well defined. We also associate a number $n(A) \in \mathbb{N}$ to each infinite derivation A to be $n(A) = min(\{i \in \mathbb{N} \mid m_\mathcal{C}(A_i) = M_\mathcal{C}(A)\})$, i.e., from term $A_{n(A)}$ on, terms of derivation A does not increase its $m_\mathcal{C}$ number. We claim that $\mathsf{M}_\mathcal{C}(t) = \{M_\mathcal{C}(A) \mid A \text{ is an infinitary derivation starting from } t\}$ is bounded by some $\mathsf{N} \in \mathbb{N}$. In order to prove this, we use a kind of 'diagonal' argument. If $\mathsf{M}_\mathcal{C}(t)$ is not bounded, there would exist infinitely many classes $\mathcal{A}$ of infinitary derivations starting from t such that $\forall A, A' \in \mathcal{A}$, $M_\mathcal{C}(A) = M_\mathcal{C}(A')$ (that we take as $M_\mathcal{C}(\mathcal{A})$ for the class $\mathcal{A}$) that can be ordered by $\mathcal{A} < \mathcal{A}'$ iff $M_\mathcal{C}(\mathcal{A}) < M_\mathcal{C}(\mathcal{A}')$. Thus, classes of infinite derivations starting from t can be enumerated by $0, 1, \ldots$ according to their $M_\mathcal{C}(\mathcal{A})$. Without losing generality, we can assume $M_\mathcal{C}(\mathcal{A}_0) > m_\mathcal{C}(t)$. Consider an arbitrary $A \in \mathcal{A}_i$ for some $i \geq 0$. By confluence and Proposition 5(1), there exist a least $j > i$ and a derivation $A' \in \mathcal{A}_j$ such that the first $n(A)$ steps of A and A' coincide. We say that A *continues into* A'. Obviously, since $M_\mathcal{C}(\mathcal{A}_i) < M_\mathcal{C}(\mathcal{A}_j)$, and the first $n(A)$ steps of A and A' coincide, we have $n(A) < n(A')$. Once $A^0 \in \mathcal{A}_0$ has been fixed, by induction, we can define an infinite sequence $\mathsf{A} : A^0, A^1, \ldots, A^n, \ldots$ of infinitary derivations such that, for all $n \in \mathbb{N}$, A^n continues into A^{n+1}. For such a sequence A, we let $\iota : \mathbb{N} \to \mathbb{N}$ be as follows: $\iota(n) = min(\{m \mid n < n(A^m)\})$. We define an infinite derivation A starting from t as follows: $\forall \beta < \omega, A^{\iota(\beta)}_\beta \to A^{\iota(\beta)}_{\beta+1}$. It is not difficult to see that A is well defined and, by construction, $M_\mathcal{C}(A)$ is not finite thus contradicting that $t \not\hookrightarrow^{\leq\omega} \delta$. Thus, $\mathsf{M}_\mathcal{C}(t) = \{M_\mathcal{C}(A) \mid A \text{ is an infinitary derivation starting from } t\}$ is bounded by some $\mathsf{N} \in \mathbb{N}$. Let $\kappa = 2^{-\mathsf{N}-1}$. By Theorem 3, there exists s such that $t \to^* s$ and $d(s, \delta) < \kappa$. Since Proposition 5(1) ensures that every finite sequence $t \to^* s$ can be extended into an infinitary one, it follows that $A : t \to^* s \hookrightarrow^\omega \cdots$ satisfies $M_\mathcal{C}(A) > \mathsf{N}$, thus contradicting that N is an upper bound of $\mathsf{M}_\mathcal{C}(t)$.

Since $\mathcal{R}$ in Example 1 is not confluent, it shows the need for confluence in order to ensure Theorem 4. Theorem 4 does not generalize to arbitrary normal forms.

Example 4. Consider the following left-linear, confluent TRS:

```
h(f(a(x),y)) → h(f(x,b(y)))        f(a(x),b(y)) → g(f(x,y))
h(f(x,b(y))) → h(f(a(x),y))        g(f(x,y))    → f(a(x),b(y))
```

The transfinite rewrite sequence

$$\texttt{h(f(a}^\omega\texttt{,y))} \hookrightarrow^\omega \texttt{h(f(a}^\omega\texttt{,b}^\omega\texttt{))} \hookrightarrow^\omega \texttt{h(g}^\omega\texttt{)}$$

cannot be compressed into an infinitary one.

With regard to Remark 1, we note that the conditions of Theorem 4 do not forbid 'de facto' non-strongly convergent transfinite sequences (as in [DKP91]).

Example 5. Consider TRS $\mathcal{R}$ of Example 2 plus the rule `h(x)` $\to$ `b(h(x))`, and the Cauchy convergent reduction sequence of length $\omega \cdot 2$:

$$\mathtt{h(\underline{f(c)})} \to \mathtt{h(\underline{f(g(c))})} \to \mathtt{h(f(g(g(c))))} \hookrightarrow^{\omega} \mathtt{h(f(g^{\omega}))} \to^{\omega} \mathtt{b}^{\omega}$$

which is not strongly convergent. Theorem 4 ensures the existence of an (obvious) Cauchy convergent infinitary sequence leading to $\mathtt{b}^{\omega}$ (which, by Theorem 1, is strongly convergent). However, results in [DKP91] do not apply here since the TRS is not top-terminating.

A TRS $\mathcal{R} = (\Sigma, R)$ is left-bounded if the set $\{\mathsf{d}_l \mid l \to r \in R\}$ is bounded.

Theorem 5. [KKSV95] *Let $\mathcal{R}$ be a left-bounded, orthogonal TRS, $t \in \mathcal{T}^{\omega}(\Sigma)$ and $s \in \mathsf{NF}^{\omega}_{\mathcal{R}}$. If $t \hookrightarrow^{\infty} s$, then $t \to^{\infty} s$.*

By Theorem 2, we conclude that if $t \hookrightarrow^{\infty} s \in \mathsf{NF}^{\omega}_{\mathcal{R}}$, then $t \to^{\leq\omega} s$. Thus, for TRSs with finite rhs's, Theorem 4 along with Theorem 1 improves Theorem 5 with regard to normal forms of a special kind: constructor terms. In particular, note that Theorem 5 along with Theorem 2 does not apply to Example 5.

The uniqueness of infinite normal forms obtained by transfinite rewriting is a consequence of the confluence of transfinite rewriting. Even though orthogonality does *not* imply confluence of transfinite rewriting, Kennaway et al. proved that it implies uniqueness of (possibly infinite) normal forms obtained by strongly convergent reductions. We prove that finitary confluence implies the uniqueness of *constructor* normal forms obtained by infinitary rewriting.

Theorem 6. *Let $\mathcal{R}$ be a confluent TRS, $t \in \mathcal{T}^{\omega}(\Sigma)$, and $\delta, \delta' \in \mathcal{T}^{\omega}(\mathcal{C})$. If $t \hookrightarrow^{\leq\omega} \delta$ and $t \hookrightarrow^{\leq\omega} \delta'$, then $\delta = \delta'$.*

Proof. If $t \to^* \delta$ and $t \to^* \delta'$, it follows by Proposition 4. Assume that $t \hookrightarrow^{\omega} \delta$ and $\delta \neq \delta'$. In this case, δ is necessarily infinite. Let $\epsilon = d(\delta, \delta')$ and $p \in \mathcal{P}os(\delta)$ be such that $|p| = -log_2(\epsilon)$. Since $t \hookrightarrow^{\omega} \delta$, there exists $n_\epsilon \in \mathbb{N}$ such that, $\forall m$, $n_\epsilon < m < \omega$, $d(t_m, \delta) < \epsilon$. Let $m = n_\epsilon + 1$; thus, $t \to^* t_m$. If $t \to^* \delta'$, then, by Proposition 4, $t_m \to^* \delta'$. However, since $root(t_m|_p) = root(\delta|_p) \in \mathcal{C}$, and $root(\delta|_p) \neq root(\delta'|_p)$, rewriting steps issued from t_m cannot rewrite t_m into δ', thus leading to a contradiction. On the other hand, if $t = t' \hookrightarrow^{\omega} \delta'$, we consider a term $t'_{m'}$ such that $d(t'_{m'}, \delta') < \epsilon$ whose existence is ensured by the convergence of $t = t'$ into δ'. The confluence of $t'_{m'}$ and t_m into a common reduct s is also impossible for similar reasons.

Theorem 6 does not hold for arbitrary normal forms.

Example 6. Consider the (ground) TRS $\mathcal{R}$:

`f(a)` $\to$ `a`   `f(a)` $\to$ `f(f(a))`

Since $t \to^* \mathtt{a}$ for every ground term t, $\mathcal{R}$ is ground confluent and hence confluent (see [BN98]). We have $\mathtt{\underline{f(a)}} \to \mathtt{f(\underline{f(a)})} \to \mathtt{f(f(f(a)))} \to^{\omega} \mathtt{f}^{\omega} \in \mathsf{NF}^{\omega}_{\mathcal{R}}$, but also $\mathtt{\underline{f(a)}} \to \mathtt{a} \in \mathsf{NF}^{\omega}_{\mathcal{R}}$.

For left-linear TRSs, Theorem 6 generalizes to transfinite rewriting.

Theorem 7. *Let $\mathcal{R}$ be a left-linear, confluent TRS, $t \in \mathcal{T}^\omega(\Sigma)$, and $\delta, \delta' \in \mathcal{T}^\omega(\mathcal{C})$. If $t \hookrightarrow^\infty \delta$ and $t \hookrightarrow^\infty \delta'$, then $\delta = \delta'$.*

5 Term and Rewriting Semantics

Giving semantics to a programming language (e.g., TRSs) is the first step for discussing the properties of programs. Relating semantics of (the same or different TRSs) is also essential in order to effectively analyze properties (via approximation). We provide a notion of semantics which can be used with term rewriting, and investigate two different orderings between semantics aimed at approximating semantics and (hence) at properties of programs. Our notion of semantics is aimed at couching both operational and denotational aspects in the style of *computational*, *algebraic*, or *evaluation* semantics [Bou85,Cou90,Pit97]. Since many definitions and relationships among our semantics do not depend on any computational mechanism, we consider rewriting issues later (Section 5.1).

Definition 1. *A (ground)* term semantics *for a signature Σ is a mapping* $\mathsf{S} : \mathcal{T}(\Sigma) \to \mathcal{P}(\mathcal{T}^\omega(\Sigma))$.

A trivial example of term semantics is empty given by $\forall t \in \mathcal{T}(\Sigma), \mathsf{empty}(t) = \varnothing$. We say that a semantics S is *deterministic* (resp. *defined*) if $\forall t \in \mathcal{T}(\Sigma), |\mathsf{S}(t)| \leq 1$ (resp. $|\mathsf{S}(t)| \geq 1$). Partial order $\sqsubseteq$ among semantics is the pointwise extension of partial order $\subseteq$ on sets of terms: $\mathsf{S} \sqsubseteq \mathsf{S}'$ if and only if $\forall t \in \mathcal{T}(\Sigma), \mathsf{S}(t) \subseteq \mathsf{S}'(t)$. Another partial order $\preceq$ is defined: $\mathsf{S} \preceq \mathsf{S}'$ if there exists $T \subseteq \mathcal{T}^\omega(\Sigma)$ such that, for all $t \in \mathcal{T}(\Sigma)$, $\mathsf{S}(t) = \mathsf{S}'(t) \cap T$. Given semantics S and S' such that $\mathsf{S} \preceq \mathsf{S}'$, any set $T \subseteq \mathcal{T}^\omega(\Sigma)$ satisfying $\mathsf{S}(t) = \mathsf{S}'(t) \cap T$ is called a *window set* of S' w.r.t. S. Clearly, $\mathsf{S} \preceq \mathsf{S}'$ implies $\mathsf{S} \sqsubseteq \mathsf{S}'$ (but not vice versa). Note that $\mathsf{empty} \preceq \mathsf{S}$ for all semantics S. The *range* $W_\mathsf{S} = \bigcup_{t \in \mathcal{T}(\Sigma)} \mathsf{S}(t)$ of a semantics S suffices to compare it with all the others.

Theorem 8. *Let S, S' be semantics for a signature Σ. Then, $\mathsf{S} \preceq \mathsf{S}'$ if and only if $\forall t \in \mathcal{T}(\Sigma)$, $\mathsf{S}(t) = \mathsf{S}'(t) \cap W_\mathsf{S}$.*

Proposition 6. *Let S, S' be semantics for a signature Σ. (1) If $\mathsf{S} \sqsubseteq \mathsf{S}'$, then $W_\mathsf{S} \subseteq W_{\mathsf{S}'}$. (2) If $\mathsf{S} \prec \mathsf{S}'$, then $W_\mathsf{S} \subset W_{\mathsf{S}'}$.*

Remark 2. As W_S is the union of possible outputs of semantics S, it collects the 'canonical values' we are interested in. Hence, it provides a kind of computational reference. S' is 'more powerful' than S whenever $\mathsf{S} \sqsubseteq \mathsf{S}'$. However, if $\mathsf{S} \sqsubseteq \mathsf{S}'$ but $\mathsf{S} \not\preceq \mathsf{S}'$, there will be input terms t for which S is not able to compute interesting values (according to W_S!) which, in turn, will be available by using S'. Thus, $\mathsf{S} \preceq \mathsf{S}'$ ensures that (w.r.t. semantic values in W_S) we do *not* need to use S' to compute them.

The following results that further connect $\preceq$ and $\sqsubseteq$ will be used later.

Proposition 7. *Let* $\mathsf{S}_1, \mathsf{S}'_1, \mathsf{S}_2, \mathsf{S}'_2$ *be semantics for a signature* Σ. *If* $\mathsf{S}_1 \not\preceq \mathsf{S}_2$, $W_{\mathsf{S}_1} \subseteq W_{\mathsf{S}_2}$, $\mathsf{S}_1 \preceq \mathsf{S}'_1$, *and* $\mathsf{S}_2 \preceq \mathsf{S}'_2$, *then* $\mathsf{S}'_1 \not\preceq \mathsf{S}'_2$.

Proposition 8. *Let* $\mathsf{S}, \mathsf{S}', \mathsf{S}''$ *be semantics for a signature* Σ. *If* $\mathsf{S} \preceq \mathsf{S}'$ *and* $\mathsf{S} \sqsubseteq \mathsf{S}'' \sqsubseteq \mathsf{S}'$, *then* $\mathsf{S} \preceq \mathsf{S}''$.

5.1 Rewriting Semantics

Definition 1 does not indicate how to associate a set of terms $\mathsf{S}(t)$ to each term t. Rewriting can be used for setting up such an association.

Definition 2. *A (ground)* rewriting semantics *for a TRS* $\mathcal{R} = (\Sigma, R)$ *is a term semantics* S *for* Σ *such that for all* $t \in \mathcal{T}(\Sigma)$ *and* $s \in \mathsf{S}(t)$, $t \hookrightarrow^{\infty} s$

A *finitary* rewriting semantics S only considers reducts reached by means of a finite number of rewriting steps, i.e., $\forall t \in \mathcal{T}(\Sigma), s \in \mathsf{S}(t) \Rightarrow t \to^* s$. Given a TRS $\mathcal{R}$, semantics red, hnf, nf, and eval (where $\forall t \in \mathcal{T}(\Sigma)$, $\mathsf{red}(t) = \{s \mid t \to^* s\}$, $\mathsf{hnf}(t) = \mathsf{red}(t) \cap \mathsf{HNF}_{\mathcal{R}}$, $\mathsf{nf}(t) = \mathsf{hnf}(t) \cap \mathsf{NF}_{\mathcal{R}}$, and $\mathsf{eval}(t) = \mathsf{nf}(t) \cap \mathcal{T}(\mathcal{C})$) are the most interesting finitary rewriting semantics involving reductions to arbitrary (finitely reachable) reducts, head-normal forms, normal forms and values, respectively. In general, if no confusion arises, we will not make the underlying TRS $\mathcal{R}$ explicit in the notations for these semantics (by writing $\mathsf{red}_{\mathcal{R}}$, $\mathsf{hnf}_{\mathcal{R}}$, $\mathsf{nf}_{\mathcal{R}}, \ldots$). Concerning non-finitary semantics, the corresponding (Cauchy convergent) infinitary counterparts are: $\omega\mathsf{-red}$, $\omega\mathsf{-hnf}$, $\omega\mathsf{-nf}$, and $\omega\mathsf{-eval}$ using $\hookrightarrow^{\leq\omega}$, $\mathsf{HNF}^{\omega}_{\mathcal{R}}$, $\mathsf{NF}^{\omega}_{\mathcal{R}}$, and $\mathcal{T}^{\omega}(\mathcal{C})$. For a given TRS $\mathcal{R}$, we have $\mathsf{eval} \preceq \mathsf{nf} \preceq \mathsf{hnf} \preceq \mathsf{red}$ and $\omega\mathsf{-eval} \preceq \omega\mathsf{-nf} \preceq \omega\mathsf{-hnf} \preceq \omega\mathsf{-red}$ with $W_{\mathtt{red}} = \{s \in \mathcal{T}(\Sigma) \mid \exists t \in \mathcal{T}(\Sigma), t \to^* s\}$, $W_{\mathtt{hnf}} = \mathsf{HNF}_{\mathcal{R}}$, $W_{\mathtt{nf}} = \mathsf{NF}_{\mathcal{R}}$, and $W_{\mathtt{eval}} = \mathcal{T}(\mathcal{C})$. Also, $W_{\omega\mathtt{-red}} = \{s \in \mathcal{T}^{\omega}(\Sigma) \mid \exists t \in \mathcal{T}(\Sigma), t \hookrightarrow^{\leq\omega} s\}$, $W_{\omega\mathtt{-hnf}} = W_{\omega\mathtt{-red}} \cap \mathsf{HNF}^{\omega}_{\mathcal{R}}$, $W_{\omega\mathtt{-nf}} = W_{\omega\mathtt{-red}} \cap \mathsf{NF}^{\omega}_{\mathcal{R}}$, and $W_{\omega\mathtt{-eval}} = W_{\omega\mathtt{-red}} \cap \mathcal{T}^{\omega}(\mathcal{C})$. We also consider the strongly convergent versions $\omega\mathsf{-Sred}, \omega\mathsf{-Shnf}, \omega\mathsf{-Snf}$, and $\omega\mathsf{-Seval}$ using $\to^{\leq\omega}$. As a simple consequence of Theorem 1, we have:

Theorem 9. *For all TRS,* $\omega\mathsf{-eval} = \omega\mathsf{-Seval}$.

We consider the following (Cauchy convergent) *transfinite semantics*: $\infty\mathsf{-red}(t) = \{s \in \mathcal{T}^{\omega}(\Sigma) \mid t \hookrightarrow^{\infty} s\}$, $\infty\mathsf{-hnf}(t) = \infty\mathsf{-red}(t) \cap \mathsf{HNF}^{\omega}_{\mathcal{R}}$, $\infty\mathsf{-nf}(t) = \infty\mathsf{-hnf}(t) \cap \mathsf{NF}^{\omega}_{\mathcal{R}}$, and $\infty\mathsf{-eval}(t) = \infty\mathsf{-nf}(t) \cap \mathcal{T}^{\omega}(\mathcal{C})$ (note that these definitions implicitly say that $\infty\mathsf{-eval} \preceq \infty\mathsf{-nf} \preceq \infty\mathsf{-hnf} \preceq \infty\mathsf{-red}$). The '*strongly convergent*' versions are $\infty\mathsf{-Sred}, \infty\mathsf{-Shnf}, \infty\mathsf{-Snf}$, and $\infty\mathsf{-Seval}$ using $\to^{\infty}$ instead of $\hookrightarrow^{\infty}$.

6 Semantics, Program Properties, and Approximation

Semantics of programming languages and partial orders between semantics can be used to express and analyze properties of programs. For instance, definedness of (term) semantics is obviously monotone w.r.t. $\sqsubseteq$ (i.e., if S is defined and

$\mathsf{S} \sqsubseteq \mathsf{S}'$, then S' is defined); on the other hand, determinism is antimonotone (i.e., if S' is deterministic and $\mathsf{S} \sqsubseteq \mathsf{S}'$, then S is deterministic). Given a TRS $\mathcal{R}$, definedness of $\mathsf{nf}_{\mathcal{R}}$ is usually known as ($\mathcal{R}$) being *normalizing* [BN98]; definedness of $\mathsf{eval}_{\mathcal{R}}$ corresponds to the standard notion of 'completely defined'; definedness of $\omega{-}\mathsf{eval}_{\mathcal{R}}$ is 'sufficient completeness' of [DKP91]. Determinism of $\mathsf{nf}_{\mathcal{R}}$ is the standard *unique normal form (w.r.t. reductions)* property ($UN^{\rightarrow}$); determinism of $\infty{-}\mathsf{Snf}_{\mathcal{R}}$ is the *unique normal form (w.r.t. strongly convergent transfinite reductions)* property of [KKSV95].

A well-known property of TRSs that we can express in our framework is neededness [HL91]. In [DM97], Durand and Middeldorp define neededness (for normalization) without using the notion of residual (as in [HL91]). In order to formalize it, the authors use an augmented signature $\Sigma \cup \{\bullet\}$ (where $\bullet$ is a new constant symbol). TRSs over a signature Σ are also used to reduce terms in $\mathcal{T}(\Sigma \cup \{\bullet\})$. According to this, neededness for normalization can be expressed as follows (we omit the proof which easily follows using the definitions in [DM97]):

Theorem 10. *Let $\mathcal{R} = (\Sigma, R)$ be an orthogonal TRS. A redex $t|_p$ in $t \in \mathcal{T}(\Sigma)$ is needed for normalization if and only if* $\mathsf{nf}_{\mathcal{R}}(t[\bullet]_p) \subseteq \mathcal{T}(\Sigma \cup \{\bullet\}) - \mathcal{T}(\Sigma)$.

Theorem 10 suggests the following *semantic* definition of neededness.

Definition 3. *Let S be a term semantics for the signature $\Sigma \cup \{\bullet\}$. Let $t \in \mathcal{T}(\Sigma) - W_{\mathsf{S}}$ and $p \in \mathcal{P}os(t)$. Subterm $t|_p$ is S-needed in t if $\mathsf{S}(t[\bullet]_p) \subseteq \mathcal{T}^{\omega}(\Sigma \cup \{\bullet\}) - \mathcal{T}^{\omega}(\Sigma)$.*

We do not consider any TRS in our definition but only a term semantics; thus, we do not require that $t|_p$ be a redex but just a subterm of t. The restriction $t \in \mathcal{T}(\Sigma) - W_{\mathsf{S}}$ in Definition 3 is natural since terms in W_{S} already have complete semantic meaning according to S. Our notion of $\mathsf{nf}_{\mathcal{R}}$-neededness coincides with the standard notion of neededness for normalization when considering an orthogonal TRS $\mathcal{R}$ and we restrict the attention to redexes within terms. S-neededness applies in other cases.

Example 7. Consider the TRS $\mathcal{R}$:

$$\mathtt{g(x)} \rightarrow \mathtt{c(g(x))} \qquad\qquad \mathtt{a} \rightarrow \mathtt{b}$$

Redex $\mathtt{a}$ in $\mathtt{g(a)}$ is $\mathsf{nf}_{\mathcal{R}}$-needed, since $\mathsf{nf}_{\mathcal{R}}(\mathtt{g}(\bullet)) = \varnothing$. However, it is *not* $\omega{-}\mathsf{Seval}_{\mathcal{R}}$-needed, since $\omega{-}\mathsf{Seval}_{\mathcal{R}}(\mathtt{g}(\bullet)) = \{\mathtt{c}^{\omega}\}$.

Neededness of redexes for infinitary or transfinite normalization has been studied in [KKSV95] for strongly convergent sequences. As remarked by the authors, their definition is closely related to standard Huet and Lévy's finitary one. In fact, it is not difficult to see (considering Theorems 4 and 1), that, for every orthogonal TRS $\mathcal{R}$, $\omega{-}\mathsf{Seval}_{\mathcal{R}}$-needed redexes are needed in the sense of [KKSV95].

Unfortunately, S-neededness does not always 'naturally' coincide with other well-known notions of neededness such as, e.g., root-neededness [Mid97].

Example 8. Consider the TRS $\mathcal{R}$:

$$\mathtt{a} \to \mathtt{b} \qquad \mathtt{c} \to \mathtt{b} \qquad \mathtt{f(x,b)} \to \mathtt{g(x)}$$

and $t = \mathtt{f(a,c)}$ which is not root-stable (hence, $\mathtt{f(a,c)} \notin W_{\mathtt{hnf}}$). Since derivation $\mathtt{f(a,\underline{c})} \to \underline{\mathtt{f(a,b)}} \to \mathtt{g(a)}$ does not reduce redex $\mathtt{a}$ in t, $\mathtt{a}$ is not root-needed. However, $\mathtt{f(\bullet,\underline{c})} \to \underline{\mathtt{f(\bullet,b)}} \to \mathtt{g(\bullet)}$, which means that $\mathtt{a}$ is hnf-needed.

This kind of problem has already been noticed in [DM97]. A deep comparison between semantic neededness of Definition 3 and other kinds of neededness is out of the scope of this paper. However, it allows us to show the use of $\sqsubseteq$ in *approximation*. For instance, S-neededness is antimonotone with regard to $\sqsubseteq$.

Theorem 11. *Let* S, S' *be term semantics for a signature* Σ *such that* $\mathsf{S} \sqsubseteq \mathsf{S}'$. *If* $t|_p$ *is* S'*-needed in* $t \in \mathcal{T}(\Sigma) - W_{\mathsf{S}'}$, *then* $t|_p$ *is* S*-needed in* t.

Theorem 11 suggests using $\sqsubseteq$ for approximating the neededness of a TRS, either by using semantics for *other* TRSs or other semantics for the *same* TRS. Given TRSs $\mathcal{R}$ and $\mathcal{S}$ over the same signature, we say that $\mathcal{S}$ *approximates* $\mathcal{R}$ if $\to^*_{\mathcal{R}} \subseteq \to^*_{\mathcal{S}}$ and $\mathsf{NF}_{\mathcal{R}} = \mathsf{NF}_{\mathcal{S}}$ [DM97,Jac96]. An approximation of TRSs is a mapping α from TRSs to TRSs with the property that $\alpha(\mathcal{R})$ approximates $\mathcal{R}$ [DM97]. By using approximations of TRSs we can *decide* a property of $\alpha(\mathcal{R})$ (e.g., neededness) which is valid (but often undecidable) for the 'concrete' TRS $\mathcal{R}$. In [DM97], four approximations, namely $\mathsf{s}, \mathsf{nv}, \mathsf{sh}$, and g, have been studied[3] and neededness proved decidable w.r.t. these approximations. Since $\mathsf{nf}_{\mathcal{R}} \sqsubseteq \mathsf{nf}_{\alpha(\mathcal{R})}$, by Theorem 11 $\mathsf{nf}_{\alpha(\mathcal{R})}$ correctly approximates $\mathsf{nf}_{\mathcal{R}}$-neededness (as proved in [DM97]).

A final example is *redundancy* of arguments. Given a term semantics S for the signature Σ, the i-th argument of $f \in \Sigma$ is redundant w.r.t. S if, for all context $C[\,]$, $t \in \mathcal{T}(\Sigma)$ such that $root(t) = f$, and $s \in \mathcal{T}(\Sigma)$, $\mathsf{S}(C[t]) = \mathsf{S}(C[t[s]_i])$. Redundancy is antimonotone with regard to $\preceq$ but *not* w.r.t. $\sqsubseteq$ [AEL00].

7 Relating Transfinite, Infinitary, and Finitary Semantics

Following our previous discussion about the usefulness of relating different semantics, in this section, we investigate orders $\sqsubseteq$ and $\preceq$ among semantics of Section 5.1. Strongly convergent sequences are Cauchy convergent; thus, $\infty{-}\mathsf{S}\varphi \sqsubseteq \infty{-}\varphi$ for $\varphi \in \{\mathsf{red}, \mathsf{hnf}, \mathsf{nf}, \mathsf{eval}\}$. In general, these inequalities are strict: consider $\mathcal{R}$ in Example 1. We have $\infty{-}\mathsf{Seval} \not\preceq \infty{-}\mathsf{eval}$; otherwise, $\infty{-}\mathsf{Seval}(\mathtt{f(a,g)}) = \infty{-}\mathsf{eval}(\mathtt{f(a,g)}) \cap W_{\infty-\mathsf{Seval}}$. Since $\mathsf{c}^\omega \in \infty{-}\mathsf{eval}(\mathtt{f(a,g)})$ but we have that $\mathsf{c}^\omega \notin \infty{-}\mathsf{Seval}(\mathtt{f(a,g)})$, it follows that $\mathsf{c}^\omega \notin W_{\infty-\mathsf{Seval}}$. However, $\mathsf{c}^\omega \in \infty{-}\mathsf{Seval}(\mathtt{h}) = \infty{-}\mathsf{eval}(\mathtt{h}) \cap W_{\infty-\mathsf{Seval}}$ contradicts the latter. Thus, in general, $\infty{-}\mathsf{Seval} \not\preceq \infty{-}\mathsf{eval}$, i.e., $\infty{-}\mathsf{Seval} \sqsubset \infty{-}\mathsf{eval}$. By Proposition 6, $W_{\infty-\mathsf{Seval}} \subseteq W_{\infty-\mathsf{eval}}$; thus, by Proposition 7 we also have $\infty{-}\mathsf{S}\varphi \not\preceq \infty{-}\varphi$, and hence $\infty{-}\mathsf{S}\varphi \sqsubset \infty{-}\varphi$ for $\varphi \in \{\mathsf{red}, \mathsf{hnf}, \mathsf{nf}, \mathsf{eval}\}$. By Theorem 5, for left-bounded, orthogonal TRSs, $\infty{-}\mathsf{nf} = \infty{-}\mathsf{Snf}$ and $\infty{-}\mathsf{eval} = \infty{-}\mathsf{Seval}$.

With regard to *infinitary semantics*, the situation is similar to the transfinite case (but consider Theorem 9). We also have the following:

[3] Names s, nv, sh, and g correspond to *strong*, *NV*, *shallow*, and *growing* approximations, respectively.

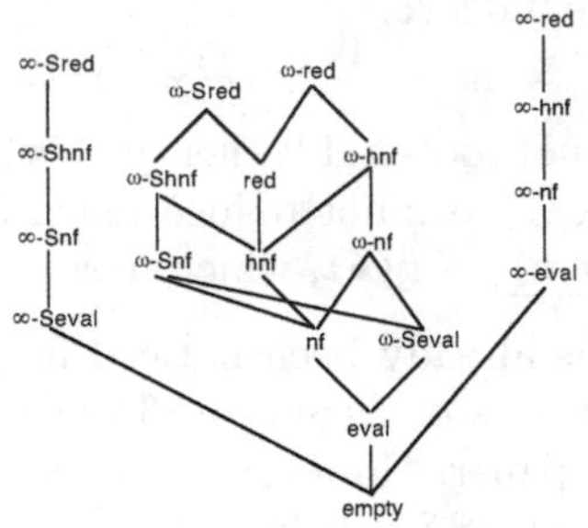

Fig. 1. Semantics for a TRS ordered by $\preceq$

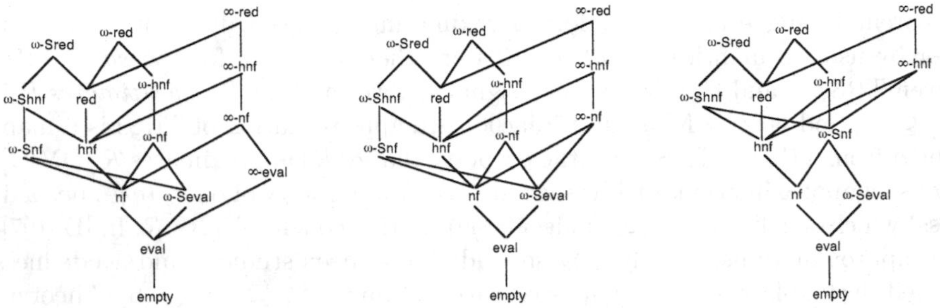

Fig. 2. Semantics for a left-linear / left-linear confluent / left-bounded orthogonal TRS (ordered by $\preceq$)

Proposition 9. *For all TRS,* $\varphi \preceq \omega{-}\varphi$ *for* $\varphi \in \{\mathsf{red}, \mathsf{hnf}, \mathsf{nf}, \mathsf{eval}\}$.

By Propositions 9 and 8, $\varphi \preceq \omega{-}\mathsf{S}\varphi$ for $\varphi \in \{\mathsf{red}, \mathsf{hnf}, \mathsf{nf}, \mathsf{eval}\}$. However (consider the TRS in Example 3), in general $\mathsf{eval} \not\preceq \infty{-}\mathsf{Seval}$ (and $\mathsf{eval} \not\preceq \infty{-}\mathsf{eval}$)! By Proposition 7, no comparison (with $\preceq$) is possible between (considered here) transfinite semantics and infinitary or finitary ones (except empty). Figure 1 shows the hierarchy of semantics for a TRS ordered by $\preceq$.

By Theorem 2, for *left-linear* TRSs, we have $\infty{-}\mathsf{S}\varphi = \omega{-}\mathsf{S}\varphi$ for $\varphi \in \{\mathsf{red}, \mathsf{hnf}, \mathsf{nf}, \mathsf{eval}\}$. Example 1 shows that $\omega{-}\mathsf{eval} \not\preceq \infty{-}\mathsf{eval}$ and, since $\omega{-}\mathsf{eval} \sqsubseteq \infty{-}\mathsf{eval}$, by Propositions 6 and 7, $\omega{-}\varphi \not\preceq \infty{-}\varphi$ for $\varphi \in \{\mathsf{red}, \mathsf{hnf}, \mathsf{nf}, \mathsf{eval}\}$. As a simple consequence of Corollary 1, for every left-linear TRS, $\varphi \preceq \infty{-}\varphi$ for $\varphi \in \{\mathsf{red}, \mathsf{hnf}, \mathsf{nf}, \mathsf{eval}\}$. By additionally requiring (finitary) *confluence*, Theorem 4 entails that, for every left-linear, confluent TRS, $\infty{-}\mathsf{eval} = \omega{-}\mathsf{eval}$. By using Theorem 2, we have that, for every *left-bounded, orthogonal* TRS, $\infty{-}\mathsf{nf} = \omega{-}\mathsf{Snf}$. Diagrams of Figure 2 summarize these facts. Table 1 shows the appropriateness of different semantics for computing different kinds of interesting semantic values (see Remark 2). Left-linear, confluent TRSs provide the best framework for computing constructor terms, since infinitary, strongly convergent sequences suffice and determinism of computations is guaranteed.

Table 1. Semantics for computing different canonical forms; (!) means determinism

	$\mathcal{T}(\mathcal{C})$	$\mathtt{NF}_\mathcal{R}$	$\mathcal{T}^\omega(\mathcal{C})$	$\mathtt{NF}^\omega_\mathcal{R}$
arbitrary TRSs	∞−eval	∞−nf	∞−eval	∞−nf
left-linear TRSs	eval	nf	∞−eval	∞−nf
left-linear, confluent TRSs	eval(!)	nf (!)	ω−Seval(!)	∞−nf
left-bounded, orthogonal TRSs	eval(!)	nf (!)	ω−Seval(!)	ω−Snf (!)

8 Related Work

Our rewriting semantics are related to other (algebraic) approaches to semantics of recursive program schemes [Cou90] and TRSs [Bou85]. For instance, a *computational semantics*, $Comp_{\langle\mathcal{R},\mathcal{A}\rangle}(t)$, of a ground term t in a TRS $\mathcal{R} = (\Sigma, R)$ is given in [Bou85] as the collection of lub's of increasing partial information obtained along maximal computations starting from t (see [Bou85], page 212). The partial information associated to each component of a rewrite sequence is obtained by using an interpretation $\mathcal{A}$ of $\mathcal{R}$ which is a complete Σ-algebra[4] satisfying $l_\mathcal{A} = \bot$ for all rules $l \to r \in R$. Consider $\mathcal{R}$ in Example 6 and let $\mathcal{A}$ be a Σ-algebra interpreting $\mathcal{R}$. Since $\mathtt{f(a)} \to \mathtt{a}$ is a rule of $\mathcal{R}$, then $f(a) = \bot$, where f and a interpret $\mathtt{f}$ and $\mathtt{a}$, respectively. Since $\bot \sqsubseteq a$ and f is monotone, we have $f(f(a)) = f(\bot) \sqsubseteq f(a) = \bot$, i.e., $f(f(a)) = \bot$. In general, $f^n(a) = \bot$, for $n \geq 1$, and thus $Comp_{\langle\mathcal{R},\mathcal{A}\rangle}(\mathtt{f(a)}) = \{\bot, a\}$. However, ω−$\mathsf{Snf}(\mathtt{f(a)}) = \{\mathtt{a}, \mathtt{f}^\omega\}$. Moreover, since $\mathcal{R}$ in Example 6 is confluent, according to [Bou85], we can provide another semantics $Val_{\langle\mathcal{R},\mathcal{A}\rangle}(t)$ which is the lub of the interpretations of all finite reducts of t. In particular, $Val_{\langle\mathcal{R},\mathcal{A}\rangle}(\mathtt{f(a)}) = \bigsqcup\{\bot, a\} = a$, which actually corresponds to $\mathsf{nf}(\mathtt{f(a)})$, since there is no reference to the infinitary term $\mathtt{f}^\omega$ obtained from $\mathtt{f(a)}$ by strongly convergent infinitary rewriting.

Closer to ours, in [HL01,Luc00], an *observable semantics* is given to (computations of) terms without referring its meaning to any external semantic domain. An observation mapping is a lower closure operator on Ω-terms ordered by a partial order $\sqsubseteq$ on them. An *adequate* observation mapping $(\!|\ |\!)$ permits us to describe computations which are issued from a term t as the set of all observations $(\!|s|\!)$ of finitary reducts s of t. An observation mapping $(\!|\ |\!)$ is *adequate* for observing rewriting computations if the observations of terms according to $(\!|\ |\!)$ are refined as long as the computation proceeds. For instance, the observation mapping $(\!|\ |\!)_\omega$ that yields the normal form of a term w.r.t. Huet and Lévy's Ω-reductions [HL91] is adequate for observing rewriting computations [Luc00]. However, using $(\!|\ |\!)_\omega$ to observe computations issued from $\mathtt{f(a)}$ in the TRS of Example 6 does not provide any information about $\mathtt{f}^\omega$, since only the set $\{\Omega, \mathtt{a}\}$ is obtained.

In [DKP91], continuous Σ-algebras (based on quasi-orders) are used to provide models for ω-canonical TRSs, i.e., ω-confluent and ω-normalizing (that is,

[4] By a complete Σ-algebra we mean $\mathcal{A} = (D, \sqsubseteq, \bot, \{f_\mathcal{A} \mid f \in \Sigma\})$ where $(D, \sqsubseteq, \bot)$ is a *cpo* with least element $\bot$ and each $f_\mathcal{A}$ is continuous (hence monotone) [Bou85].

every term has a normal form[5] which is reachable in at most ω-steps) TRSs. Models are intended to provide *inequality* of left- and right-hand sides rather than equality, as usual. Also, in contrast to [Bou85,HL01,Luc00], semantics is given to symbols (via interpretations) rather than terms (via computations). This semantic framework does not apply to $\mathcal{R}$ in Example 6 since it is not ω-canonical (it lacks ω-confluence).

Acknowledgements. I thank the anonymous referees for their remarks. Example 4 was suggested by a referee.

References

[AEL00] M. Alpuente, S. Escobar, and S. Lucas. Redundancy Analyses in Term Rewriting. In *Proc of WFLP'2000*, pages 309-323, 2000.

[BN98] F. Baader and T. Nipkow. Term Rewriting and All That. Cambridge University Press, 1998.

[Bou85] G. Boudol. Computational semantics of term rewriting systems. In M. Nivat and J. Reynolds, editors, *Algebraic Methods in Semantics*, pages 169-236, Cambridge University Press, Cambridge, 1985.

[Cou83] B. Courcelle. Fundamental Properties of Infinite Trees. *Theoretical Computer Science* 25:95-169, 1983.

[Cou90] B. Courcelle. Recursive Applicative Program Schemes. In *Handbook of Theoretical Computer Science*, pages 459-492, Elsevier, 1990.

[DKP91] N. Dershowitz, S. Kaplan, and D. Plaisted. Rewrite, rewrite, rewrite, rewrite, rewrite. *Theoretical Computer Science* 83:71-96, 1991.

[DM97] I. Durand and A. Middeldorp. Decidable Call by Need Computations in Term Rewriting (Extended Abstract). In *Proc. of CADE'97*, LNAI 1249:4-18, 1997.

[HL01] M. Hanus and S. Lucas. An Evaluation Semantics for Narrowing-Based Functional Logic Languages. *Journal of Functional and Logic Programming, to appear*, 2001.

[HL91] G. Huet and J.J. Lévy. Computations in orthogonal term rewriting systems. In *Computational logic: essays in honour of J. Alan Robinson*, pages 395-414 and 415-443. The MIT Press, Cambridge, MA, 1991.

[Jac96] F. Jacquemard. Decidable approximations to Term Rewriting Systems. In *Proc. of RTA'96*, LNCS 1103:362-376, 1996.

[KKSV95] R. Kennaway, J.W. Klop, M.R. Sleep, and F.-J. de Vries. Transfinite reductions in Orthogonal Term Rewriting Systems. *Information and Computation* 119(1):18-38, 1995.

[Luc00] S. Lucas. Observable Semantics and Dynamic Analysis of Computational Processes. Technical Report LIX/RR/00/02, École Polytechnique, Palaiseau, France, 2000.

[Mid97] A. Middeldorp. Call by Need Computations to Root-Stable Form. In *Proc. of POPL'97*, pages 94-105, ACM Press, 1997.

[Pit97] A. Pitts. Operationally-Based Theories of Program Equivalence. In *Semantics and Logics of Computation*, pages 241-298, Cambridge University Press, 1997.

[5] Notice that the notion of normal form of [DKP91] (a term t such that $t = t'$ whenever $t \to t'$) differs from the standard one.

Goal-Directed E-Unification

Christopher Lynch and Barbara Morawska

Department of Mathematics and Computer Science Box 5815, Clarkson University, Potsdam, NY 13699-5815, USA, clynch@clarkson.edu,morawskb@clarkson.edu**

Abstract. We give a general goal directed method for solving the E-unification problem. Our inference system is a generalization of the inference rules for Syntactic Theories, except that our inference system is proved complete for any equational theory. We also show how to easily modify our inference system into a more restricted inference system for Syntactic Theories, and show that our completeness techniques prove completeness there also.

1 Introduction

E-unification [1] is a problem that arises in several areas of computer science, including automated deduction, formal verification and type inference. The problem is, given an equational theory E and a goal equation $u \approx v$, to find the set of all substitutions θ such that $u\theta$ and $v\theta$ are identical modulo E. In practice, it is not necessary to find all such substitution. We only need to find a set from which all such substitutions can be generated, called a *complete set of E-unifiers.*

The decision version of E-unification (Does an E-unifier exist?) is an undecidable problem, even for the simpler *word problem* which asks if all substitutions θ will make $u\theta$ and $v\theta$ equivalent modulo E. However there are procedures which are *complete* for the problem. Complete, in this sense, means that each E-unifier in a complete set will be generated eventually. However, because of the undecidability, the procedure may continue to search for an E-unifier forever, when no E-unifier exists.

One of the most successful general methods for solving the E-unification problem has been Knuth-Bendix Completion[12] (in particular, Unfailing Completion[2]) plus Narrowing[7]. This procedure deduces new equalities from E. If the procedure ever halts, it solves the word problem. However, because of the undecidability, Knuth-Bendix Completion cannot always halt.

Our goal in this paper is to develop an alternative E-unification procedure. Why do we want an alternative to Knuth-Bendix Completion? There are several reasons. First, there are simple equational theories for which Completion does not halt. An example is the equational theory $E = \{f(g(f(x))) \approx g(f(x))\}$. So then it is impossible to decide any word problem in this theory, even a simple example like $a \approx b$, which is obviously not true. Using our method, examples like this will quickly halt and say there is no solution.

** This work was supported by NSF grant number CCR-9712388.

A. Middeldorp (Ed.): RTA 2001, LNCS 2051, pp. 231–245, 2001.

A related deficiency of Completion is that it is difficult to identify classes of equational theories where the procedure halts, and to analyze the complexity of solving those classes. That is our main motivation for this line of research. We do not pursue that subject in this paper, since we first need to develop a complete inference system. That subject is addressed in [14,15], where we deal classes of equations where the E-unification is decidable in an inference system similar to the one given in this paper.

Another aspect of Completion is that it is insensitive to the goal. It is possible to develop heuristics based on the goal, but problems like the example above still exist, because of the insensitivity to the goal. The method we develop in this paper is goal directed, in the sense that every inference step is a step backwards from the goal, breaking the given goal into separate subgoals. Therefore we call our method a goal directed inference system for equational reasoning. This quality of goal-directedness is especially important when combining an equational inference system with another inference system. Most of the higher order inference systems used for formal verification have been goal directed inference systems. Even most inference systems for first order logic, like OTTER, are often run with a set of support strategy. For things like formal verification, we need equality inference systems that can be added as submodules of previously existing inference systems. We believe that the best method for achieving this is to have a goal directed equality inference system.

We do not claim that our procedure is the first goal directed equational inference system. Our inference system is similar to the inference system Syntactic Mutation first developed by Claude Kirchner [8,10]. That inference system applies to a special class of equational theories called Syntactic Theories. In such theories, any true equation has an equational proof with at most one step at the root. The problem of determining if an equational theory is syntactic is undecidable[11]. In the Syntactic Mutation inference system, it is possible to determine which inference rule to apply next by looking at the root symbols on the two sides of a goal equation. This restricts which inference rules can be applied at each point, and makes the inference system more efficient than a blind search.

Our inference system applies to every equational theory, rather than just Syntactic Theories. Therefore, it would be incomplete for us to examine the root symbol at both sides of a goal equation. However, we do prove that we may examine the root symbol of one side of an equation to decide which inference rule to apply. Other than that, our inference system is similar to Syntactic Mutation. We prove that our inference system is complete. The Syntactic Mutation rules were never proved to be complete. In [9], it is stated that there is a problem proving completeness because the Variable Elimination rule (called "Replacement" there) does not preserve the form of the proof. We think we effectively deal with that problem.

There is still an open problem of whether the Variable Elimination rule can be applied eagerly.[1] We have not solved that problem. But we have avoided those problems as much as possible. The inefficiency of the procedure comes from cases

[1] See [16] for a discussion of the problem and a solution for a very specific case.

where one side of a goal equation is a variable. We prove that any equation where both sides are variables may be ignored without losing completeness. We also orient equations so that inference rules are applied to the nonvariable side of an equation. This gives some of the advantages of Eager Variable Elimination.

Other similar goal-directed inference procedures are given in [4,5]. The inference system from [4] is called BT. The one in [5] is called **Trans**. The main difference between our results and the results in those papers all of our inference rules involve a root symbol of a term in the goal. This limits the number of inference rules that can be applied at any point. For BT and **Trans** there are inference rules that only involve variables in the goal. These rules allow an explosion of inferences at each step, which expands the search space. This is similar to the situation in Paramodulation completeness proofs required paramodulation into variables until Brand[3] proved that this was not necessary for completeness. We believe that the completeness results in this paper are analogous to the results of Brand, but for goal-directed E-unification. In the case of Paramodulation, the results of Brand prove essential in practice. Another difference between our results and BT and **Trans** is that those papers require Variable Elimination, while ours do not. Gallier and Snyder[4] pointed out the problem of inference rules involving variables. However, their solution was to design a different inference system, called T, that allows inferences below the root. Our results solve this problem without requiring inferences below the root. The problem of Eager Variable Elimination was first presented in [4].

The format of the paper is to first give some preliminary definitions. Then present our inference system. After a discussion of normal form, we present soundness results. In order to prove completeness, we first give a bottom-up method for deducing ground equations, then use that method to prove completeness of our goal-directed method. After that we show how our completeness technique can be applied to Syntactic Theories to show completeness of a procedure similar to Syntactic Mutation. Finally, we conclude the paper. All missing proofs are in [13].

2 Preliminaries

We assume we are given a set of variables and a set of uninterpreted function symbols of various arities. An arity is a non-negative integer. *Terms* are defined recursively in the following way: each variable is a term, and if $t_1, \cdots, t_n$ are terms, and f is of arity $n \geq 0$, then $f(t_1, \cdots, t_n)$ is a term, and f is the symbol at the *root* of $f(t_1, \cdots, t_n)$. A term (or any object) without variables is called *ground*. We consider equations of the form $s \approx t$, where s and t are terms. Please note that throughout this paper these equations are considered to be oriented, so that $s \approx t$ is a different equation that $t \approx s$. Let E be a set of equations, and $u \approx v$ be an equation, then we write $E \models u \approx v$ (or $u =_E v$) if $u \approx v$ is true in any model containing E. If G is a set of equations, then $E \models G$ means that $E \models e$ for all e in G.

A *substitution* is a mapping from the set of variables to the set of terms, such that it is almost everywhere the identity. We identify a substitution with its homomorphic extension. If θ is a substitution then $Dom(\theta) = \{x \mid x\theta \neq x\}$. A substitution θ is an *E-unifier* of an equation $u \approx v$ if $E \models u\theta \approx v\theta$. θ is an *E-unifier* of a set of equations G if θ is an E-unifier of all equations in G.

If σ and θ are substitutions, then we write $\sigma \leq_E \theta[Var(G)]$ if there is a substitution ρ such that $E \models x\sigma\rho \approx x\theta$ for all x appearing in G. If G is a set of equations, then a substitution θ is a *most general unifier of* G, written $\theta = mgu(G)$ if θ is an E unifier of G, and for all E unifiers σ of G, $\theta \leq_E \sigma[Var(G)]$. A complete set of E-unifiers of G, is a set of E-unifiers Θ of G such that for all E-unifiers σ of G, there is a θ in Θ such that $\theta \leq_E \sigma[Var(G)]$.

3 The Goal Directed Inference Rules

In this section, we will give a set of inference rules for finding a complete set of E-unifiers of a goal G, and in the following sections we prove that, for every goal G and substitution θ such that $E \models G\theta$, G can be converted into a *normal form* (see Section 4), which determines a substitution which is more general than θ. The inference rules decompose an equational proof by choosing a potential step in the proof and leaving what is remaining when that step is removed.

We define two special kinds of equations appearing in the goal G. An equation of the form $x \approx y$ where x and y are both variables is called a *variable-variable* equation. An equation $x \approx t$ appearing in G where x only appears once in G is called *solved.*

As in Logic Programming, we can have a selection rule for goals. For each goal G, we don't-care nondeterministically select an equation $u \approx v$ from G, such that $u \approx v$ is not a variable-variable equation and $u \approx v$ is not solved. We say that $u \approx v$ is *selected in* G. If there is no such equation $u \approx v$ in the goal, then nothing is selected. We will prove that if nothing is selected, then the goal is in normal form and a most general-E unifier can be easily determined.

There is a Decomposition rule.

Decomposition

$$\frac{\{f(s_1,\cdots,s_n) \approx f(t_1,\cdots,t_n)\} \cup G}{\{s_1 \approx t_1,\cdots,s_n \approx t_n\} \cup G}$$

where $f(s_1,\cdots,s_n) \approx f(t_1,\cdots,t_n)$ is selected in the goal.

This is just an application of the Congruence Axiom, in a goal-directed way. If f is of arity 0 (a constant) then this is a goal-directed application of Reflexivity.

We additionally add a second inference rule that is applied when one side of an equation is a variable.

Variable Decomposition

$$\frac{\{x \approx f(t_1,\cdots,t_n)\} \cup G}{\{x \approx f(x_1,\cdots,x_n)\} \cup (\{x_1 \approx t_1,\cdots,x_n \approx t_n\} \cup G)[x \mapsto f(x_1,\cdots,x_n)]}$$

where x is a variable, and $x \approx f(t_1,\cdots,t_n)$ is selected in the goal.

This is similar to the Variable Elimination rule for syntactic equalities. It can be considered a gradual form of Variable Elimination, since it is done one step at a time. This rule is the same as the rule that is called Imitation in **Trans** and Root Imitation in BT. We have chosen our name to emphasize its relationship with the Decomposition rule.

Now we add a rule called Mutate. We call it Mutate, because it is very similar to the inference rule Mutate that is used in the inference procedure for syntactic theories. Mutate is a kind of goal-directed application of Transitivity, but only transitivity steps involving equations from the theory.

Mutate

$$\frac{\{u \approx f(v_1, \cdots, v_n)\} \cup G}{\{u \approx s, t_1 \approx v_1, \cdots, t_n \approx v_n\} \cup G}$$

where $u \approx f(v_1, \cdots, v_n)$ is selected in the goal, and $s \approx f(t_1, \cdots, t_n) \in E$. [2] [3]

This rule assumes that there is an equational proof of the goal equation at the root of the equation (see Section 7). If one of the equations in this proof is $s \approx t$ then that breaks up the proof at the root into two separate parts. We have performed a Decomposition on one of the two equations that is created. Contrast this with the procedure for Syntactic Theories[8] which allows a Decomposition on both of the newly created equations. However, that procedure only works for Syntactic Theories, whereas our procedure is complete for any equational theory. The names of our inference rules are chosen to coincide with the names from [8]. In **Trans** the Mutate rule is called Lazy Narrowing, and in BT it is called Root Rewriting.

Next we give a Mutate rule for the case when one side of the equation from E is a variable.

Variable Mutate

$$\frac{\{u \approx f(v_1, \cdots, v_n)\} \cup G}{\{u \approx s\}[x \mapsto f(x_1, \cdots, x_n)] \cup \{x_1 \approx v_1, \cdots, x_n \approx v_n\} \cup G}$$

where $s \approx x \in E$, x is a variable, and $u \approx f(v_1, \cdots, v_n)$ is selected in the goal. This is called Application of a Trivial Clause in **Trans**, and it is a special case of Root Rewriting in BT.

We will write $G \longrightarrow G'$ to indicate that G goes to G' by one application of an inference rule. Then $\stackrel{*}{\longrightarrow}$ is the reflexive, transitive closure of $\longrightarrow$.

When an inference is performed, we may eagerly reorient any new equations in the goal. The way they are reoriented is don't-care nondeterministic, except that any equation of the form $t \approx x$, where t is not a variable and x is a variable, must be reoriented to $x \approx t$. This way there is never an equation with a nonvariable on the left hand side and a variable on the right hand side.

---

[2] For simplicity, we assume that E is closed under symmetry.

[3] $s \approx f(t_1, \cdots, t_n)$ is actually a variant of an equation in E such that it has no variables in common with the goal. We assume this throughout the paper.

We will prove that the above inference rules solve a goal G by transforming it into normal forms representing a complete set of E-unifiers of G. There are two sources of non-determinism involved in the procedure defined by the inference rules. The first is "don't-care" non-determinism in deciding which equation to select in the goal, and in deciding which way to orient equations with non-variable terms on both sides. The second is "don't-know" non-determinism in deciding which rule to apply. Not all paths of inference steps will lead us to the normal form, and we do not know beforehand which ones do.

4 Normal Form

Notice that there are no inference rules that apply to an equation $x \approx y$, where x and y are both variables.[4] In fact, such an equation can never be selected. The reason is that so many Mutate and Variable Decomposition inferences could possibly apply to variable-variable pairs (as in BT) that we have designed the system to avoid them. That changes the usual definition of normal form, as in Standard Unification, and shows that inferences with variable-variable pairs are unnecessary.

Let G be a goal of the form $\{x_1 \approx t_1, \cdots, x_n \approx t_n, y_1 \approx z_1, \cdots, y_m \approx z_m\}$, where all x_i, y_i and z_i are variables, the t_i are not variables, and for all i and j,

1. $x_i \notin Var(t_j)$,
2. $x_i \neq y_j$ and
3. $x_i \neq z_j$.

Then G is said to be in *normal form*. Let σ_G be the substitution $[x_1 \mapsto t_1, \cdots x_n \mapsto t_n]$. Let τ_G be a most general (syntactic) unifier of $y_1 = z_1, \cdots, y_m = z_m$, with no new variables, such as what is calculated by a syntactic unification procedure. We know an mgu of only variable-variable equations must exist. Any such unifier effectively divides the variables into equivalence classes such that for each class E, there is some variable z in E such that $y\tau_G = z$ for all $y \in E$. Then we write $\hat{y} = z$. Note that for any E-unifier θ of G, $y\theta =_E \hat{y}\theta$. Finally, define θ_G to be the substitution $\sigma_G\tau_G$.

Proposition 1. *A goal with nothing selected is in normal form.*

Proof. Let G be a goal with nothing selected. Then all equations in G have a variable on the left hand side. So G is of the form $x_1 \approx t_1, \cdots, x_n \approx t_n, y_1 \approx z_1, \cdots, y_m \approx z_m$. Since nothing is selected, each equation $x_1 \approx t_1$ must be solved. So each x_i appears only once in G. Therefore the three conditions of normal form are satisfied. □

Now we will prove that the substitution represented by a goal in normal form is a most general E-unifier of that goal.

[4] This is similar to the flex-flex pairs for higher order unification in [6].

Lemma 1. *Let G be a set of equations in normal form. Then θ_G is a most general E-unifier of G.*

Proof. Let G be the goal $\{x_1 \approx t_1, \cdots, x_n \approx t_n, y_1 \approx z_1, \cdots, y_m \approx z_m\}$, such that for all i and j, $x_i \notin t_j$, $x_i \neq y_j$ and $x_i \neq z_j$. Let $\sigma_G = [x_1 \mapsto t_1, \cdots, x_n \mapsto t_n]$. Let $\tau_G = mgu(y_1 = z_1, \cdots, y_m = z_m)$. Let $\theta_G = \sigma_G \tau_G$. We will prove that θ_G is a most general E unifier of G.

Let i and j be integers such that $1 \leq i \leq n$ and $1 \leq j \leq n$. First we need to show that θ_G is a unifier of G, i.e. that $x_i\theta_G = t_i\theta_G$ and $y_j\theta_G = z_j\theta_G$. In other words, prove that $x_i\sigma_G\tau_G = t_i\sigma_G\tau_G$ and $y_j\sigma_G\tau_G = z_j\sigma_G\tau_G$. Since t_i, y_j and z_j are not in the domain of σ, this is equivalent to $t_i\tau_G = t_i\tau_G$ and $y_j\tau_G = z_j\tau_G$, which is trivially true, since τ_G is mgu of $\{y_1 \approx z_1, \cdots, y_m \approx z_m\}$.

Next we need to show that θ_G is more general than all other unifiers of G. So let θ be an E-unifier of G. In other words, $x_i\theta =_E t_i\theta$ and $y_j\theta =_E z_j\theta$. We need to show that $\theta_G \leq_E \theta[Var(G)]$. In particular, we will show that $G\theta_G\theta =_E G\theta$.

Then $x_i\theta_G\theta = x_i\sigma_G\tau_G\theta = t_i\tau_G\theta =_E t_i\theta =_E x_i\theta$. The only step that needs justification is the fact that $t_i\tau_G\theta =_E t_i\theta$. This can be verified by examining the variables of t_i. So let w be a variable in t_i. If $w \notin Dom(\tau_G)$ then obviously $w\tau_G\theta = w\theta$. If $w \in Dom(\tau_G)$ then w is some y_k. Note that $y_k\tau_G\theta = \hat{y_k}\theta =_E y_k\theta$. So $t_i\tau_G\theta =_E t_i\theta$.

Also, $y_j\theta_G\theta = y_j\sigma_G\tau_G\theta = y_j\tau_G\theta = \hat{y_j}\theta =_E y_j\theta$. Similarly for z_j. □

5 An Example

Here is an example of the procedure. (The selected equations are underlined.)

Example 1. Let $E = E_0 = \{ffx \approx gfx\}$, $G = G_0 = \{\underline{fgfy \approx ggfz}\}$.
By rule Mutate applied to G_0 we have
$G_1 = \{\underline{fgfy \approx ffx_1}, fx_1 \approx gfz\}$.
After Decomposition,
$G_2 = \{\underline{gfy \approx fx_1}, fx_1 \approx gfz\}$.
After Mutate,
$G_3 = \{\underline{gfy \approx gfx_2}, x_1 \approx fx_2, fx_1 \approx gfz\}$
After Decomposition is used 2 times on G_3,
$G_4 = \{y \approx x_2, \underline{x_1 \approx fx_2}, fx_1 \approx gfz\}$.
Variable Decomposition:
$G_5 = \{y \approx x_2, x_1 \approx fx_3, x_3 \approx x_2, \underline{ffx_3 \approx gfz}\}$.
Mutate:
$G_6 = \{y \approx x_2, x_1 \approx fx_3, x_3 \approx x_2, \underline{ffx_3 \approx ffx_4}, fx_4 \approx fz\}$.
2× Decomposition:
$G_7 = \{y \approx x_2, x_1 \approx fx_3, x_3 \approx x_2, x_3 \approx x_4, \underline{fx_4 \approx fz}\}$.
Decomposition:
$G_8 = \{y \approx x_2, x_1 \approx fx_3, x_3 \approx x_2, x_3 \approx x_4, x_4 \approx z\}$.

The extended θ' that unifies the goal G_0 is equal to: $[x_1 \mapsto fx_3][y \mapsto z, x_3 \mapsto z, x_2 \mapsto z, x_4 \mapsto z]$. θ' is equivalent on the variables of G to θ equal to: $[y \mapsto z]$.

6 Soundness

Theorem 1. *The above procedure is sound, i.e. if* $G' \xrightarrow{*} G$ *and* G *is in normal form, then* $E \models G'\theta_G$.

7 A Bottom Up Inference System

In order to prove the completeness of this procedure, we first define a bottom-up equational proof using Congruence and Equation Application rules. We prove that this equational proof is equivalent to the usual definition of equational proof for ground terms, which involves Reflexivity, Symmetry, Transitivity and Congruence.

$$\text{Congruence:} \quad \frac{s_1 \approx t_1 \cdots s_n \approx t_n}{f(s_1, \cdots, s_n) \approx f(t_1, \cdots, t_n)}$$

$$\text{Equation Application:} \quad \frac{u \approx s \quad t \approx v}{u \approx v},$$

if $s \approx t$ is a ground instance of an equation in E.

We define $E \vdash u \approx v$ if there is a proof of $u \approx v$ using the Congruence and Equation Application rules. If π is a proof, then $|\pi|$ is the number of steps in the proof. $|u \approx v|_E$ is the number of steps in the shortest proof of $u \approx v$.

We need to prove that $\{u \approx v \,|\, E \vdash u \approx v\}$ is closed under Reflexivity, Symmetry and Transitivity. First we prove Reflexivity.

Lemma 2. *Let* E *be an equational theory. Then* $E \vdash u \approx u$ *for all ground* u.

Next we prove closure under symmetry.

Lemma 3. *Let* E *be an equational theory such that* $E \vdash u \approx v$ *and* $|u \approx v|_E = n$. *Then* $E \vdash v \approx u$, *and* $|v \approx u|_E = n$.

Next we show closure under Transitivity.

Lemma 4. *Let* E *be an equational theory such that* $E \vdash s \approx t$ *and* $E \vdash t \approx u$. *Suppose that* $|s \approx t|_E = m$ *and* $|t \approx u|_E = n$. *Then* $E \vdash s \approx u$, *and* $|s \approx u|_E \leq m + n$.

Closure under Congruence is trivial. Now we put these lemmas together to show that anything true under the semantic definition of Equality is also true under the syntactic definition given here.

Theorem 2. *If* $E \models u \approx v$, *then* $E \vdash u \approx v$, *for all ground* u *and* v.

We can restrict our proofs to only certain kinds of proofs. In particular, if the root step of a proof tree is an Equation Application, then we can show there is a proof such that the proof step of the right child is not an Equation Application.

Lemma 5. *Let π be a proof of $u \approx v$ in E, which is derived by Equation Application, and whose right child is also derived by Equation Application. Then there is a proof π' of $u \approx v$ in E such that the root of π' is Equation Application but the right child is derived by Congruence, and $|\pi'| = |\pi|$.*

Proof. Let π be a proof of $u \approx v$ in E such that the step at the top is Equation Application, and the step at the right child is also Equation Application. We will show that there is another proof π' of $u \approx v$ in E such that $|\pi'| = |\pi|$, and the size of the right subtree of π' is smaller than the size of the right subtree of π. So this proof is an induction on the size of the right subtree of the proof.

Suppose $u \approx v$ is at the root of π and $u \approx s$ labels the left child n_1. Suppose the right child n_2 is labeled with $t \approx v$. Further suppose that the left child of n_2 is labeled with $t \approx w_1$ and the right child of n_2 is labeled with $w_2 \approx v$. Then $s \approx t$ and $w_1 \approx w_2$ must be ground instances of members of E.

$$\dfrac{\begin{array}{c}\pi_1\\ \vdots\\ n_1\colon u \approx s\end{array} \qquad \dfrac{\begin{array}{c}\pi_2\\ \vdots\\ t \approx w_1\end{array} \qquad \begin{array}{c}\pi_3\\ \vdots\\ w_2 \approx v\end{array}}{n_2\colon t \approx v}\ \text{Eq. App.}}{u \approx v}\ \text{Eq. App.}$$

Then we can let π' be the proof whose root is labeled with $u \approx v$, whose left child n_3 is labeled with $u \approx w_1$. Let the left child of n_3 be labeled with $u \approx s$ and the right child of n_3 be labeled with $t \approx w_1$. Also let the right child of the root n_4 be labeled with $w_2 \approx v$.

$$\text{Eq. App.}\ \dfrac{\text{Eq. App.}\ \dfrac{\begin{array}{c}\pi_1\\ \vdots\\ u \approx s\end{array} \qquad \begin{array}{c}\pi_2\\ \vdots\\ t \approx w_1\end{array}}{n_3\colon u \approx w_1} \qquad \begin{array}{c}\pi_3\\ \vdots\\ n_4\colon w_2 \approx v\end{array}}{u \approx v}$$

By induction, π' is a proof of $u \approx v$ of the same size as π. □

8 Completeness of the Goal-Directed Inference System

Now we finally get to the main theorem of this paper, which is the completeness of the inference rules given in section 3. But first we need to define a measure on the equations in the goal.

Definition 1. *Let E be an equational theory and G be a goal. Let θ be a substitution such that $E \models G\theta$. We will define a measure μ, parameterized by θ and G. Define $\mu(G, \theta)$ as the multiset $\{|u\theta \approx v\theta|_E \mid u \approx v$ is an unsolved equation in $G\}$.*

The intension of the definition is that the measure of an equation in a goal is the number of steps it takes to prove that equation. However, solved equations are ignored.

Now, finally, the completeness theorem:

Theorem 3. *Suppose that E is an equational theory, G is a set of goal equations, and θ is a ground substitution. If $E \models G\theta$ then there exists a goal H in normal form such that $G \overset{*}{\longrightarrow} H$ and $\theta_H \leq_E \theta[Var(G)]$.*

Proof. Let G be a set of goal equations, and θ a ground substitution such that $E \models G\theta$. Let $\mu(G, \theta) = M$. We will prove by induction on M that there exists a goal H such that $G \overset{*}{\longrightarrow} H$ and $\theta_H \leq_E \theta[Var(G)]$.

If nothing is selected in G, then G must be in normal form, by Proposition 1. By Lemma 1, θ_G is the most general E-unifier of G, so $\theta_G \leq_E \theta[Var(G)]$.

If some equation is selected in G, we will prove that there is a goal G' and a substitution θ' such that $G \longrightarrow G'$, $\theta' \leq_E \theta[Var(G)]$, and $\mu(G', \theta') \leq \mu(G, \theta)$.

So assume that some equation $u \approx v$ is selected in G. Then G is of the form $\{u \approx v\} \cup G_1$. We assume that v is not a variable, because any term-variable equation $t \approx x$ is immediately reoriented to $x \approx t$. By Lemma 3, $|v\theta \approx u\theta|_E = |u\theta \approx v\theta|_E$. Also, according to our selection rule, a variable-variable equation is never selected. Since v is not a variable, it is in the form $f(v_1, \cdots, v_n)$. Let $|u\theta \approx v\theta|_E = m$.

Consider the rule used at the root of the smallest proof tree that $E \vdash u\theta \approx v\theta$. This was either an application of Congruence or Equation Application.

Case 1: Suppose the rule at the root of the proof tree of $E \vdash u\theta \approx v\theta$ is an Equation Application. Then there exists an extension θ' of θ and a ground instance $s\theta' \approx t\theta'$ of an equation $s \approx t$ in E, such that $E \vdash u\theta' \approx s\theta'$ and $E \vdash t\theta' \approx v\theta'$. Let $|u\theta' \approx s\theta'|_E = p$. Let $|t\theta' \approx v\theta'|_E = q$. Then $m = p+q+1$. We now consider two subcases, depending on whether or not t is a variable.

Case 1A: Suppose that t is not a variable. Then, we can assume that the rule at the root of the proof tree of $E \vdash t\theta' \approx v\theta'$, is Congruence. Otherwise, by Lemma 5, it could be converted into one, without making the proof any longer. So then t is of the form $f(t_1, \cdots, t_n)$, and the previous nodes of the proof tree are labeled with $t_1\theta' \approx v_1\theta', \cdots, t_n\theta' \approx v_n\theta'$. And, for each i, $|t_i\theta' \approx v_i\theta'|_E = q_i$ such that $1 + \Sigma_{1 \leq i \leq n} q_i = q$.
Therefore, there is an application of Mutate that can be applied to $u \approx v$, resulting in the new goal $G' = \{u \approx s, t_1 \approx v_1, \cdots, t_n \approx v_n\} \cup G_1$. Then $|u\theta' \approx s\theta'|_E = p$, and $|t_i\theta' \approx v_i\theta'|_E = q_i$ for all i, so $\mu(G'\theta') < \mu(G, \theta)$. By the induction assumption there is an H such that $G' \overset{*}{\longrightarrow} H$ with $\theta_H \leq_E \theta'[Var(G')]$. This implies that $G \overset{*}{\longrightarrow} H$. Also, $\theta_H \leq_E \theta'[Var(G)]$, since the variables of G are a subset of the variables of G'. Since $G\theta' = G\theta$, we know that $\theta_H \leq_E \theta[Var(G)]$.

Case 1B: Suppose that t is a variable. Then, by Lemma 5, we can assume that the rule at the root of the proof tree of $E \vdash t\theta' \approx v\theta'$ is Congruence. So then $t\theta'$ is of the form $f(t_1, \cdots, t_n)$, and the previous nodes of the proof tree are labeled with $t_1 \approx v_1\theta', \cdots, t_n \approx v_n\theta'$. And, for each i, $|t_i \approx v_i\theta'|_E = q_i$ such that $1 + \Sigma_{1\le i\le n} q_i = q$.

Therefore, there is an application of Variable Mutate that can be applied to $u \approx v$, resulting in the new goal $G' = \{u \approx s[t \mapsto f(x_1, \cdots, x_n)], x_1 \approx v_1, \cdots, x_n \approx v_n\} \cup G_1\}$. We will extend θ' so that $x_i\theta' = t_i$ for all i. Then $|u\theta' \approx s\theta'|_E = p$, and $|x_i\theta' \approx v_i\theta'|_E = q_i$ for all i, so $\mu(G'\theta') < \mu(G, \theta)$. By the induction assumption there is an H such that $G' \xrightarrow{*} H$ with $\theta_H \le_E \theta'[Var(G')]$. This implies that $G \xrightarrow{*} H$. Also, $\theta_H \le_E \theta'[Var(G)]$, since the variables of G are a subset of the variables of G'. Since $G\theta' = G\theta$, we know that $\theta_H \le_E \theta[Var(G)]$.

Case 2: Now suppose that the rule at the root of the proof tree of $E \vdash u\theta \approx v\theta$ is an application of Congruence. There are two cases here: u is a variable or u is not a variable.

Case 2A: First we will consider the case where u is not a variable. Then $u = f(u_1, \cdots, u_n)$, $v = f(v_1, \cdots, v_n)$ and $E \vdash u_i\theta \approx v_i\theta$ for all i. There is an application of Decomposition that can be applied to $u \approx v$, resulting in the new goal $G' = \{u_1 \approx v_1, \cdots, u_n \approx v_n\} \cup G_1$. Then $|u_i\theta \approx v_i\theta|_E < |u\theta \approx v\theta|$ for all i, so $\mu(G', \theta) < \mu(G, \theta)$. By the induction assumption there is an H such that $G' \xrightarrow{*} H$ with $\theta_H \le_E \theta[Var(G')]$. This implies that $G \xrightarrow{*} H$ and $\theta_H \le_E \theta[Var(G)]$.

Case 2B: Now we consider the final case, where u is a variable and the rule at the root of the proof tree of $E \vdash u\theta \approx v\theta$ is an application of Congruence. Let $u\theta = f(u_1, \cdots, u_n)$. Then, for each i, $E \vdash u_i \approx v_i\theta$, and $|u_i \approx v_i\theta|_E < |u\theta \approx v\theta|_E$. There is an application of Variable Decomposition that can be applied to $u \approx v$, resulting in the new goal $G' = \{u \approx f(x_1, \cdots, x_n)\} \cup (\{x_1 \approx v_1, \cdots, x_n \approx v_n\} \cup G_1)[u \mapsto f(x_1, \cdots, x_n)]$. Let θ' be the substitution $\theta \cup [x_1 \mapsto u_1, \cdots, x_n \mapsto u_n]$. Then $u \approx f(x_1, \cdots, x_n)$ is solved in G'. Also $|x_i\theta' \approx v_i\theta'|_E < |u\theta \approx v\theta|_E$ for all i. Therefore $\mu(G, \theta) < \mu(G', \theta')$. By the induction assumption there is an H such that $G' \xrightarrow{*} H$ with $\theta_H \le_E \theta'[Var(G')]$. This implies that $G \xrightarrow{*} H$. Also, $\theta_H \le_E \theta'[Var(G)]$, since the variables of G are a subset of the variables of G'. Since $G\theta' = G\theta$, we know that $\theta_H \le_E \theta[Var(G)]$.

□

The fact that we required θ to be ground in the theorem does not limit our results. This implies that any substitution will work

Corollary 1. *Suppose that E is an equational theory, G is a set of goal equations, and θ is any substitution. If $E \models G\theta$ then there exists a goal H such that $G \xrightarrow{*} H$ and $\theta_H \le_E \theta[Var(G)]$.*

Proof. Let θ' be a skolemized version of θ, i.e., θ' is the same as θ except that every variable in the range of θ is replaced by a new constant. Then θ' is ground, so by Theorem 3 there exists a goal H such that $G \xrightarrow{*} H$ and $\theta_H \le_E \theta'[Var(G)]$. Then θ_H cannot contain any of the new constants, so $\theta_H \le_E \theta[Var(G)]$. □

9 *E*-Unification for Syntactic Theories

In this section we will show how we can restrict our inference rules further to get a set of inference rules that resembles the Syntactic Mutation rules of Kirchner. Then we prove that that set of inference rules is complete for syntactic theories.

The definition of a syntactic theory is in terms of equational proofs. The definition of a proof is as follows.

Definition 2. *An* equational proof of $u \approx v$ from E *is a sequence* $u_0 \approx u_1 \approx u_w \approx \cdots \approx u_n$, *for* $n \geq 0$ *such that* $u_0 = u$, $u_n = v$ *and for all* $i \geq 0$, $u_i = u_i[s\theta]$ *and* $u_{i+1} = u_i[t\theta]$ *for some* $s \approx t \in E$ *and some substitution* θ.

Now we give Kirchner's definition of *syntactic theory.*

Definition 3. *An equational theory* E *is* resolvent *if every equation* $u \approx v$ *with* $E \models u \approx v$ *has an equational proof such that there is at most one step at the root. A theory is* syntactic *if it has an equivalent finite resolvent presentation.*

From now on, when we discuss a Syntactic Theory E, we will assume that E is the resolvent presentation of that theory.

In this paper, we are considering bottom-up proofs instead of equational replacement proofs. We will call a bottom-up proof *resolvent* if whenever an equation appears as a result of Equation Application, then its left and right children must have appeared as a result of an application of Congruence at the root. We will call E *bottom-up resolvent* if every ground equation $u \approx v$ implied by E has a bottom-up resolvent proof. Now we show that the definition of resolvent for equational proofs is equivalent to the definition of resolvent for bottom-up proofs.

Theorem 4. *E is resolvent if and only if E is bottom-up resolvent.*

Proof. We need to show how to transform a resolvent equational proof into a resolvent bottom-up proof and vice versa.

Case 1: First consider transforming a resolvent equational proof into a resolvent bottom-up proof. We will prove this can be done by induction on the the lexicographic combination of the number of steps in the equational proof and the number of symbols appearing in the equation.

Case 1A: Suppose $u \approx v$ has an equational proof with no steps at the root. Then $u \approx v$ is of the form $f(u_1, \cdots, u_n) \approx f(v_1, \cdots, v_n)$, and there are equational proofs of $u_i \approx v_i$ for all i. Since each equation $u_i \approx v_i$ has fewer symbols than $u \approx v$ and does not have a longer proof, then, by the induction argument there is a resolvent bottom-up proof of each $u_i \approx v_i$, and by adding one more congruence step to all the $u_i \approx v_i$, we get a resolvent bottom-up proof of $u \approx v$.

Case 1B: Now suppose $u \approx v$ has an equational proof with one step at the root. Then there is a ground instance $s \approx t$ of something in E such that the proof of $u \approx v$ is a proof of $u \approx s$ with no steps at the top, followed by a replacement of s with t, followed by a proof of $t \approx v$ with no steps at the root.

By induction, each child in the proof of $u \approx s$ has a resolvent bottom-up proof. Therefore $u \approx s$ has a resolvent bottom-up proof with a Congruence step at the root. Similarly, $t \approx v$ has a resolvent bottom-up proof with a Congruence step at the root. If we apply Equation Application to those two proofs, we get a bottom-up resolvent proof of $u \approx v$.

Case 2: Now we will transform a resolvent bottom-up proof of $u \approx v$ to an equational proof of $u \approx v$, by induction on $|u \approx v|_E$.

Case 2A: Suppose $u \approx v$ has a bottom-up resolvent proof with an application of Congruence at the root. Then $u \approx v$ is of the form $f(u_1, \cdots, u_n) \approx f(v_1, \cdots, v_n)$, and there are bottom-up resolvent proofs of $u_i \approx v_i$ for all i. Since each equational proof of $u_i \approx v_i$ is shorter than the proof of $u \approx v$, then, by the induction argument there is a resolvent equational proof of each $u_i \approx v_i$, and they can be combined to give a resolvent equational proof of $u \approx v$.

Case 2B: Now suppose $u \approx v$ has a resolvent bottom-up proof with one Equation Application step at the root. Then there is some $s \approx t$ in E such that the proof of $u \approx v$ is a proof of $u \approx s$ with a Congruence step at the root, and a proof of $t \approx v$ with a Congruence step at the root, then an Equation Application using the equation $s \approx t$ from E. By induction, the corresponding equalities of subterms of $u \approx s$ have resolvent equational proofs. So $u \approx s$ has a resolvent equational proof with no steps at the root. Similarly, $t \approx v$ also has a resolvent equational proof with no steps at the root. So $u \approx v$ has a resolvent equational proof with one step at the root.

□

Now we give the inference rules for solving E-unification problems in Syntactic Theories. The rules for Decomposition and Variable Decomposition remain the same, but Mutate becomes more restrictive. We replace Mutate and Variable Mutate with one rule that covers several cases.

Mutate

$$\frac{\{u \approx v\} \cup G}{\{Dec(u \approx s), Dec(v \approx t)\} \cup G}$$

where $u \approx v$ is selected in the goal, $s \approx t \in E$, v is not a variable, if both u and s are not variables then they have the same root symbol, and if t is not a variable then v and t have the same root symbol. We also introduce a function *Dec*, which when applied to an equation indicates that the equation should be decomposed eagerly according to the following rules:

$$\frac{\{Dec(f(u_1, \cdots, u_n) \approx f(s_1, \cdots, s_n))\} \cup G}{\{u_1 \approx s_1, \cdots, u_n \approx s_n\} \cup G}$$

$$\frac{\{Dec(x \approx f(s_1, \cdots, s_n)\} \cup G}{\{x \approx f(x_1, \cdots, x_n)\} \cup G[x \mapsto f(x_1, \cdots, x_n)] \cup \{x_1 \approx s_1, \cdots, x_n \approx s_n\}}$$

where the x_i are fresh variables.

$$\frac{\{Dec(x \approx y)\} \cup G}{\{x \approx y\} \cup G}$$

$$\frac{\{Dec(f(s_1, \cdots, s_n) \approx x)\} \cup G}{G[x \mapsto f(x_1, \cdots, x_n)] \cup \{x_1 \approx s_1, \cdots, x_n \approx s_n\}}$$

where the x_i are fresh variables.

Now we prove a completeness theorem for this new set of inference rules, which is Decomposition, Variable Decomposition, and the Mutate rule given above.

Theorem 5. *Suppose that E is a resolvent presentation of an equational theory, G is a set of goal equations, and θ is a ground substitution. If $E \models G\theta$ then there exists a goal H in normal form such that $G \stackrel{*}{\longrightarrow} H$ and $\theta_H \leq_E \theta[Var(G)]$.*

Proof. The proof is the same as the proof of Theorem 3, except for Case 1. In this case, we can show that one of the forms of the Mutate rules from this section is applicable. Here, instead of using Lemma 5 to say that an Equation Application must have a Congruence as a right child, we instead use the definition of bottom-up resolvent to say that an Equation Application has a Congruence as both children. The full proof is in [13]. □

10 Conclusion

We have given a new goal-directed inference system for E-unification. We are interested in goal-directed E-unification for two reasons. One is that many other inferences systems for which E-unification would be useful are goal directed, and so a goal-directed inference system will be easier to combine with other inference systems. The second reason is that we believe this particular inference system is such that we can use it to find some decidable classes of equational theories for E-unification and analyze their complexity. We have already made progress in this direction in [14,15].

Our inference system is an improvement over the inference systems BT of [4] and **Trans** of [5] for Equational Unification. There are two important differences between our inference system an those other two. The first is that those inference systems require the Variable Elimination rule. This blows up the search space, because, for an equation $x \approx t$, both Variable Elimination and (Root) Imitation will be applicable. We do not require Variable Elimination. The second difference is that both of those inference systems require an inference with a variable in the goal. In BT, Root Rewriting inferences are performed on variable-variable pairs. This blows up the search space, because everything unifies with a variable. Similarly, in BT, Root Imitation inferences are performed on variable-variable pairs. That blows up the search space because it must be attempted for every function symbol and constant. In **Trans**, there is a rule called Paramodulation at Variable Occurence. This is like a Mutate (Lazy Paramodulation) inference applied to a variable x in a goal equation $x \approx t$. Again, every equation will unify with x, so the search space will blow up. Gallier and Snyder recognize the above-mentioned problems of $\mathcal{BT}$. There solution is to create another inference system called $\mathcal{T}$, but that one is different because Root Rewriting inferences are now allowed at non-root positions.

The inference system we have given is similar to the Syntactic Mutation inference system of [9]. The difference is that our inference system can be applied to all equational theories, not just Syntactic Theories as in their case. Also, we show how our results are easily adapted to give an inference similar to the Syntactic Mutation rules of [9]. While the rules in [9] have not been proved complete, we prove that ours are complete.

References

1. F. Baader and T. Nipkow. *Term Rewriting and All That.* Cambridge, 1998.
2. L. Bachmair, N. Dershowitz, D. Plaisted. Completion without failure. In *Resolution of Equations in Algebraic Structures*, ed. H. Aït-Kaci, M. Nivat, vol. 2, 1-30, Academic Press, 1989.
3. D. Brand. Proving theorems with the modification method. in *SIAM J. Computing* 4, 412-430, 1975.
4. J. Gallier and W. Snyder. Complete sets of transformations for general E-unification. In *TCS*, vol. 67, 203-260, 1989.
5. S. Hölldobler. *Foundations of Equational Logic Programming* Lecture Notes in Artificial Intelligence 353, Springer-Verlag, 1989.
6. G. Huet. Résolution d'équations dans les langages d'ordre $1, 2, \ldots, \omega$. Thèse d'Etat, Université Paris VII, 1976.
7. J.-M. Hullot. Canonical Forms and Unification. In *Proc. 5th Conf. on Automated Deduction, Les Arcs*, Vol. 87 of *Lecture Notes in Computer Science*, Springer-Verlag, 1980.
8. C. Kirchner. Computing unification algorithms. In *Proceedings of the First Symposium on Logic in Computer Science*, Boston, 200-216, 1990.
9. C. Kirchner and H. Kirchner. *Rewriting, Solving, Proving.* `http://www.loria.fr/~ckirchne/`, 2000.
10. C. Kirchner and F. Klay. Syntactic Theories and Unification. In *LICS 5*, 270-277, 1990.
11. F. Klay. Undecidable Properties in Syntactic Theories. In *RTA 4*,ed. R. V. Book, LNCS vol. 488, 136-149, 1991.
12. D. E. Knuth and P. B. Bendix. Simple word problems in universal algebra. In *Computational Problems in Abstract Algebra*, ed. J. Leech, 263-297, Pergamon Press, 1970.
13. C. Lynch and B. Morawska. Goal Directed E-Unification. `http://www.clarkson.edu/~clynch/PAPERS/goal_long.ps/`, 2001.
14. C. Lynch and B. Morawska. Approximating E-Unification. `http://www.clarkson.edu/~clynch/PAPERS/approx.ps/`, 2001.
15. C. Lynch and B. Morawska. Decidability and Complexity of Finitely Closable Linear Equational Theories. `http://www.clarkson.edu/~clynch/PAPERS/linear.ps/`, 2001.
16. A. Middeldorp, S. Okui, T. Ida. Lazy Narrowing: Strong Completeness and Eager Variable Elimination. In *Theoretical Computer Science* 167(1,2), pp. 95-130, 1996.

The Unification Problem for Confluent Right-Ground Term Rewriting Systems

Michio Oyamaguchi and Yoshikatsu Ohta

Faculty of Engineering, Mie University
1515 kamihama-cho, Tsu-shi, 514-8507, Japan
{mo,ohta}@cs.info.mie-u.ac.jp

Abstract. The unification problem for term rewriting systems(TRSs) is the problem of deciding, for a given TRS R and two terms M and N, whether there exists a substitution θ such that $M\theta$ and $N\theta$ are congruent modulo R (i.e., $M\theta \leftrightarrow_R^* N\theta$). In this paper, the unification problem for confluent right-ground TRSs is shown to be decidable. To show this, the notion of minimal terms is introduced and a new unification algorithm of obtaining a substitution whose range is in minimal terms is proposed. Our result extends the decidability of unification for canonical (i.e., confluent and terminating) right-ground TRSs given by Hullot (1980) in the sense that the termination condition can be omitted. It is also exemplified that Hullot's narrowing technique does not work in this case. Our result is compared with the undecidability of the word (and also unification) problem for terminating right-ground TRSs.

1 Introduction

The unification problem for TRSs is the problem of deciding, for a TRS R and two terms M and N, whether M and N are unifiable modulo R, that is, whether there exists a substitution θ (called an R-unifier) such that $M\theta$ and $N\theta$ are congruent modulo R (i.e., $M\theta \leftrightarrow_R^* N\theta$). The unification problem is undecidable in general and even if we restrict to subclasses of TRSs, a lot of negative results have been shown, e.g., undecidability for canonical (i.e., terminating and confluent) TRSs [3] (having the decidable word problem), terminating and right-ground TRSs (since the word problem for this class is undecidable [10] and the word problem, $M \leftrightarrow_R^* N$, is a special case of the unification problem). On the other hand, several positive results have been obtained, e.g., unification is decidable for ground TRSs [2], left-linear and right-ground TRSs [8,2], canonical right-ground TRSs [4], shallow TRSs [1], linear standard TRSs [8] and semi-linear TRSs [5]. The narrowing (or paramodulation) technique [4,8] is strong and useful for showing the decidability of unification, in fact used for obtaining many of the above decidability results. But, this technique is difficult to apply to nonterminating TRSs. Thus, new techniques ensuring the decidability of unification for nonterminating TRSs are needed. This is a motivation of investigation of this paper.

A. Middeldorp (Ed.): RTA 2001, LNCS 2051, pp. 246–260, 2001.

In this paper, we consider the unification problem for confluent right-ground TRSs which may be nonterminating. This problem is a natural problem, since for extending the decidability of unification for canonical right-ground TRSs, we have two choices: one is to omit the termination condition and the other to omit the confluence or right-ground condition, but the latter is impossible by the undecidability results for terminating right-ground TRSs and canonical TRSs. In this paper, we show that the termination condition can be omitted, i.e., unification is decidable for confluent right-ground TRSs. This result can be also regarded as a solution to one of the open problems posed by Nieuwenhuis [8].

We first see that the narrowing technique does not work in this case. Let $R_1 = \{eq(x,x) \to t, eq(not(x),x) \to f, t \to not(f), f \to not(t)\}$ where only x is a variable. Note that R_1 is nonterminating, confluent [11] and right-ground. Let $M = eq(y, not(y))$ and $N = y$ where y is a variable. In this case, the narrowing technique does not work: since any nonvariable subterm of M(or N) and the left-hand-side of every rule are not $\emptyset$-unifiable, this technique can not decide whether M and N are R_1-unifiable. (Note that a substitution θ satisfying $y\theta = f$ is an R_1-unifier of M and N.)

So, we use a more general technique analogous to lazy narrowing [6,7] and RU [3, p. 284] each of which consists of more primitive operations which can simulate narrowing. But, the most crucial point is to transform such a technique into a new decision procedure which can decide whether a problem instance is unifiable or not. Up to our knowledge, such attempts were very few so far. To obtain our result, we introduce the notion of minimal terms for nonterminating right-ground TRSs, which play a role similar to irreducible (i.e., normal form) terms in terminating TRSs. Then, we construct a unification algorithm which takes as input a confluent right-ground TRS R and two terms M and N and produces an R-unifier θ of M and N such that $x\theta$ is minimal for each variable x iff M and N are unfiable modulo R. Such θ whose range is minimal is called locally minimal and a key idea for ensuring the correctness of our algorithm.

2 Preliminaries

We assume that the reader is familiar with standard definitions of rewrite systems (see [3]) and we just recall here the main definitions and notations used in the paper.

We use ε to denote the empty string and $\emptyset$ to denote the empty set. For a set S, let $\mathrm{Power}(S) = \{S' \mid S' \subseteq S\}$, i.e., the set of all the subsets of S, and let $|S|$ be the cardinality of S. Let X be a set of variables, let F be a finite set of operation symbols graded by an arity function $\mathit{arity} : F \to \{0, 1, 2, \cdots\}$, and let T be the set of terms constructed from X and F. A term M is *ground* if M has no variable. Let G be the set of ground terms. For a term M, we use $\mathcal{O}(M)$ to denote the set of positions of M, $M|_u$ to denote the subterm of M at position u, and $M[N]_u$ to denote the term obtained from M by replacing the subterm $M|_u$ by term N. For a sequence $(u_1, \cdots, u_n)$ of pairwise disjoint positions and terms $L_{u_1}, \cdots, L_{u_n}$, we use $M[L_{u_1}, \cdots, L_{u_n}]_{(u_1,\ldots,u_n)}$ to denote the term obtained from

M by replacing each subterm $M|_{u_i}$ by L_{u_i}, $(1 \le i \le n)$. Let $\mathcal{O}_x(M)$ be the set of positions of variable $x \in X$ in M, i.e., $\mathcal{O}_x(M) = \{u \in \mathcal{O}(M) \mid M|_u = x\}$. Let $\mathcal{O}_X(M) = \bigcup_{x \in X} \mathcal{O}_x(M)$ and $\mathcal{O}_F(M) = \mathcal{O}(M) \setminus \mathcal{O}_X(M)$. Let $V(M)$ be the set of variables occurring in M. We use $|M|$ to denote the size of M, i.e., the number of symbols in M. For a position u, we use $|u|$ to denote the length of u. The root symbol of M is denoted by root(M). $M|_u$ is a *leaf* symbol of M if $|M|_u| = 1$. Let $M[N/x]$ be the term obtained from M by replacing all occurrences of x by N. This notation is extended to sets of terms: for $\Gamma \subseteq T$, let $\Gamma[N/x] = \{M[N/x] \mid M \in \Gamma\}$. Let $\mathcal{O}_{\text{nv}}(M) = \{u \in \mathcal{O}(M) \mid \forall v \in \mathcal{O}_X(M).\ v|u\}$.

Let $\gamma : M_1 \overset{u_1}{\to} M_2 \cdots \overset{u_{n-1}}{\to} M_n$ or $M_1 \overset{u_1}{\leftrightarrow} M_2 \cdots \overset{u_{n-1}}{\leftrightarrow} M_n$ be a rewrite sequence. Then, $\mathcal{R}(\gamma) = \{u_1, \cdots, u_{n-1}\}$, i.e., the set of the redex positions of γ. If $\varepsilon \notin \mathcal{R}(\gamma)$, then γ is called ε-invariant (or ε-inv). A position u in a set of positions U is *minimal* if $v \not< u$ for any $v \in U$. Let Min(U) be the set of minimal positions of U.

Definition 1. *We use $M \approx N$ to denote a pair of terms M and N. $M \approx N$ is* unifiable modulo *a TRS R (or simply R-*unifiable*) if there exists a substitution θ and a rewrite sequence γ such that $\gamma : M\theta \leftrightarrow^* N\theta$. Such θ and γ are called an R-*unifier *and a* proof *of $M \approx N$, respectively. This notion is extended to sets of term pairs: for $\Gamma \subseteq T \times T$, θ is an R-unifier of Γ if θ is an R-unifier of every pair $M_i \approx N_i$ of Γ. In this case, Γ is R-unifiable. As a special case of R-unifiability, $M \approx N$ is $\emptyset$-unifiable if there exists a substitution θ such that $M\theta = N\theta$, i.e., $\emptyset$-unifiability coincides with usual unifiability.*

2.1 Standard Right-Ground TRS and Minimal Term

Definition 2. *A right-ground TRS R is said to be* standard *if $|\alpha| = 1$ or $|\beta| = 1$ for any rule $\alpha \to \beta \in R$.*

Let $R = \{\alpha_1 \to \beta_1, \cdots, \alpha_n \to \beta_n\}$ be a right-ground TRS. The corresponding standard TRS R' is constructed as follows. Let $c_1, \cdots, c_n \in F$ be new pairwise distinct constants which do not appear in R. Then, $R' = \{\alpha_i \to c_i, c_i \to \beta_i \mid 1 \le i \le n\}$ is standard.

We can show that R is confluent iff R' is confluent, and for any terms M, N which do not contain $c_1, \cdots, c_n$ and any substitution θ, $M\theta \downarrow_R N\theta$ iff $M\theta \downarrow_{R'} N\theta$. The proof is straightforward, so omitted. Thus, the R-unification problem for confluent right-ground TRS R reduces to that for the above corresponding standard R'. Hence, without loss of generality we can assume that a confluent right-ground TRS R is standard. Henceforth, we consider a fixed right-ground TRS R which is confluent and standard.

Definition 3. *Let $H(M) = Max\{|u| \mid u \in O(M)\}$, i.e., the height of M. We define $H_m(M)$ as $\{H(M|_u) \mid u \in \mathcal{O}(M)\}_m$. Here, we use $\{\cdots\}_m$ to denote a multiset and below we use $\sqcup$ to denote the multiset union. Let $\ll$ be the multiset extension of $<$ and let $\underline{\ll}$ be $\ll \cup =$. For a term M, let $L(M) = \{N \mid N \leftrightarrow^* M\}$ and $L_{min}(M) = \{N \in L(M) \mid \forall N' \in L(M).\ H_m(N) \underline{\ll} H_m(N')\}$. A term M is* minimal *iff $M \in L_{min}(M)$.*

Note that $L_{min}(M)$ is well-defined: $L_{min}(M) \neq \emptyset$. We have the following lemmata.

Lemma 1. *Let M be a minimal term and let $\gamma : M \leftrightarrow^* N$. Then for any $u \in \mathcal{R}(\gamma)$ there exists $v \in \mathcal{O}(M)$ such that $H(M|_v) = 0$ and $v \leq u$. (That is, only leaf symbols of M are rewritten in γ.)*

Proof. Note that since R is standard, $L \overset{\varepsilon}{\leftrightarrow} L'$ implies that $H(L) = 0$ or $H(L') = 0$ holds. Thus, minimality of M ensures this property, since there exists no $\delta : M|_v \leftrightarrow^* L$ satisfying that $H(M|_v) > 0$ and $H(L) = 0$. □

Lemma 2. *For any term M, $L_{min}(M)$ is finite and computable.*

Proof. We prove this lemma by induction on $H(M)$. First suppose that $H(M) = 0$. Obviously, $M \in L_{min}(M)$. For a term $N \neq M$, if $N \in L_{min}(M)$, then $H(N) = 0$ and $N \leftrightarrow^+ M$. Since R is right-ground and confluent, we have $N \downarrow M$ and $N, M \in F$. Since joinability of right-ground TRSs is decidable [9] and F is finite, $L_{min}(M)$ is finite and computable.

Next suppose that $H(M) > 0$. We first check whether there exists a rule $\alpha \to \beta \in R$ such that if $|\alpha| = 1$, then $M \leftrightarrow^* \alpha$, otherwise $M \leftrightarrow^* \beta$. This is also decidable by similar arguments as above. If so, then $\alpha \in L_{min}(M)$ and $|\alpha| = 1$ or $\beta \in L_{min}(M)$ and $|\beta| = 1$, since R is standard. Thus, $L_{min}(M) = L_{min}(\alpha)$ or $L_{min}(M) = L_{min}(\beta)$. It follows that $L_{min}(M)$ is finite and computable. Otherwise, $M \leftrightarrow^* N$ implies that $M \leftrightarrow^* N$ is ε-invariant for any term N, i.e., $\text{root}(M) = \text{root}(N)$. Let $f = \text{root}(M)$ and let $k = \text{arity}(f)$. Since $L_{min}(M|_i)$ is finite and computable for all $1 \leq i \leq k$ according to the induction hypothesis, so is $L_{min}(M) = \{ f(N_1, \cdots, N_k) \mid N_i \in L_{min}(M|_i) \text{ for } 1 \leq i \leq k\}$. □

2.2 Locally Minimal Unifier and New Pair of Terms

Definition 4. *Let $\Gamma \subseteq T \times T$. A substitution θ is a* locally minimal *R-unifier of Γ if θ is an R-unifier of Γ and $x\theta$ is minimal for any $x \in Dom(\theta)$.*

In this paper, we give a new unification algorithm which takes a pair of terms $M \approx N$ as input and produces a locally minimal unifier θ of $M \approx N$ iff $M \approx N$ is R-unifiable. For this purpose, we need pairs of terms having new types. $M \asymp_U N$ and $M \approx_{\text{vf}} N$, which are respectively called term pairs with type $\asymp_U$ and with type vf, are introduced where $M, N \in T$ and $U \subseteq \mathcal{O}(N)$ is a set of pairwise disjoint positions. Let $E_0 = \{M \approx N, M \asymp_U N, M \approx_{\text{vf}} N, \textbf{fail} \mid M, N \in T$ and $U \subseteq \mathcal{O}(N)$ is a set of pairwise disjoint positions$\}$. Here, **fail** is introduced as a special symbol and we assume that there exists no R-unifier of **fail**. R-unifiers of these new pairs are required to satisfy additional conditions derived from these types:

Definition 5. *A substitution θ is an R-unifier of $M \asymp_U N$ if θ is an R-unifier of $M \approx N$ and the following condition holds: if $U = \emptyset$ then $M\theta \to^* N\theta$, otherwise there exists $L = N[L_{u_1}, \cdots, L_{u_n}]_{(u_1, \cdots, u_n)}$ for some L_{u_i}, $1 \leq i \leq n$, where $U = \{u_1, \cdots, u_n\}$, such that $M\theta \to^* L$ and for any $u_i \in U$, $L_{u_i} \leftrightarrow^* (N|_{u_i})\theta$.*

*A substitution θ is an R-*unifier *of $M \approx_{\text{vf}} N$ if θ is an R-unifier of $M \approx N$ and there exists $\gamma : M\theta \leftrightarrow^* N\theta$ such that $\mathcal{O}_X(N)$ is a frontier in γ, i.e., $u|v$ or $v \leq u$ holds for any $u \in \mathcal{R}(\gamma)$ and $v \in \mathcal{O}_X(N)$.*

Note that if $U = \{\varepsilon\}$, then θ is an R-unifier of $M \asymp_{\{\varepsilon\}} N$ iff θ is an R-unifier of $M \approx N$ by definitions. So, $M \asymp_{\{\varepsilon\}} N$ is replaced by $M \approx N$ and excluded from E_0. Also, pairs of form $M \asymp_U x$, where $M \in T, x \in X, U \subseteq \mathcal{O}(x)$, are not used and excluded from E_0.

Example 1. Let R_1 be the TRS shown in Section 1.

1. $eq(not(x), x) \approx not(x)$ is R_1-unifiable, since any substitution θ satisfying $x\theta = t$ is an R_1-unifier: $eq(not(t), t) \to f \to not(t)$.
2. $eq(f, not(not(t))) \asymp_{\{1,21\}} eq(y, not(y))$ is R_1-unifiable, since any substitution θ satisfying $y\theta = f$ is an R_1-unifier: $eq(f, not(not(t))) \stackrel{21}{\leftarrow} eq(f, not(f))$.
3. $eq(t, not(t)) \approx_{\text{vf}} eq(not(f), y)$ is R_1-unifiable, since any substitution θ satisfying $y\theta = f$ is an R_1-unifier: $eq(t, not(t)) \stackrel{1}{\to} eq(not(f), not(t)) \stackrel{2}{\leftarrow} eq(not(f), f)$.

3 *R*-Unification Algorithm

We are ready to give our R-unification algorithm Φ. Our algorithm consists of a set of primitive operations analogous to those of Lazy Narrowing [6,7] and [3]. Each primitive operation takes a finite set of pairs $\Gamma \subseteq E_0$ and produces some $\Gamma' \subseteq E_0$, denoted by $\Gamma \Rightarrow_\Phi \Gamma'$. This operation is called a transformation. Such a transformation is made nondeterministically: $\Gamma \Rightarrow_\Phi \Gamma_1, \Gamma \Rightarrow_\Phi \Gamma_2, \cdots, \Gamma \Rightarrow_\Phi \Gamma_k$ are allowed for some $\Gamma_1, \cdots, \Gamma_k \subseteq E_0$. Let $\Rightarrow_\Phi^*$ be the reflexive transitive closure of $\Rightarrow_\Phi$. Our algorithm starts from $\Gamma_0 = \{M_0 \approx N_0\}$, where $M_0, N_0 \in T$, and makes primitive transformations repeatedly. We will prove that there exists a sequence $\Gamma_0 \Rightarrow_\Phi^* \Gamma$ such that Γ is $\emptyset$-unifiable iff Γ_0 is R-unifiable.

Our algorithm is divided into three stages.

3.1 Stage I

The transformation Φ_1 of Stage I takes as input a finite subset Γ of E_0 and has a finite number of nondeterministic choices $\Gamma \Rightarrow_{\Phi_1} \Gamma_1, \cdots, \Gamma \Rightarrow_{\Phi_1} \Gamma_k$ for some $\Gamma_1, \cdots, \Gamma_k \subseteq E_0$, i.e., Φ_1 is finite-branching. In this case, we write $\Phi_1(\Gamma) = \{\Gamma_1, \cdots, \Gamma_k\}$ by regarding Φ_1 as a function.

We begin with the initial $\Gamma = \{M_0 \approx N_0\}$ and repeatedly apply the transformation Φ_1 until the current Γ satisfies the **stop condition** of Stage I defined below. We consider all possibilities in order to ensure the correctness of the algorithm. If Γ satisfies this condition, then Γ becomes an input of the next stage. The **stop condition** of Stage I is as follows.

$$\Gamma \cap \{\textbf{fail}, M \approx_{\text{vf}} N \mid M, N \in T\} \neq \emptyset \text{ or } \Gamma \subseteq X \times X$$

To describe the transformations used in Stage I, we need the following auxiliary function:

$$\begin{aligned}\text{decompose}(M, N, U) = &\{M|_i \asymp_{U/i} N|_i \mid 1 \leq i \leq k \text{ and } U/i \neq \{\varepsilon\}\} \\ &\cup \{M|_i \approx N|_i \mid 1 \leq i \leq k \text{ and } U/i = \{\varepsilon\}\}\end{aligned}$$

where $k = \text{arity}(\text{root}(M))$ and $U/i = \{u \mid i \cdot u \in U\}$. (Note that $\emptyset/i = \emptyset$.)

In Stage I, we nondeterministically apply Conversion or choose an element p in $\Gamma \setminus X \times X$ and apply one of the following transformations (TT, $\mathrm{TL}_{\rightarrow}$, GG, VG, VT) to Γ according to the type of the chosen p. That is, for $p = M \approx N$,

$M \setminus N$	$T \setminus (G \cup X)$	G	X
$T \setminus (G \cup X)$	TT	TT	VT
G	TT	GG	VG
X	VT	VG	–

and for $p = M \asymp_U N$,

$M \setminus N$	$T \setminus (G \cup X)$	G
$T \setminus (G \cup X)$	$\mathrm{TL}_{\rightarrow}$	$\mathrm{TL}_{\rightarrow}$
G	$\mathrm{TL}_{\rightarrow}$	GG if $U = \emptyset$ $\mathrm{TL}_{\rightarrow}$ if $U \neq \emptyset$
X	VT	VG

If no transformation is possible, $\Gamma \Rightarrow_{\Phi_1} \{\mathbf{fail}\}$.

Let $\Gamma' = \Gamma \setminus \{p\}$. We write $p \simeq M \approx N$ if $p = M \approx N$ or $p = N \approx M$. In order to help understanding of the transformations, we assume that θ is a locally minimal unifier of p and we list the conditions that are assumed on a proof γ of $p\theta$. When applying the transformations we of course lack this information and so we just have to check that the conditions of the transformations are satisfied.

Conversion

If $\Gamma \subseteq \{x \approx L, L \approx x, x \asymp_U L \mid x \in X \text{ and } L \in T \setminus G\}$, then

$$\Gamma \Rightarrow_{\Phi_1} \mathrm{conv}(\Gamma)$$

where $\mathrm{conv}(\Gamma) = \{x \asymp_{\mathrm{vf}} P \mid x \in X \text{ and } (x \approx P \in \Gamma \text{ or } P \approx x \in \Gamma \text{ or } x \asymp_U P \in \Gamma)\}$. Note that $\mathrm{conv}(\Gamma)$ satisfies the **stop condition** of Stage I.

In the following examples, we use the TRS R_1 shown in Section 1.

Example 2. $\{eq(y, not(y)) \approx y\} \Rightarrow_{\Phi_1} \{y \approx_{\mathrm{vf}} eq(y, not(y))\}$

TT Transformation

If $p \simeq M \approx N$ with $M, N \notin X$ and either $M \in T \setminus (G \cup X)$ or $N \in T \setminus (G \cup X)$, we choose one of the following three cases. Let $k = \mathrm{arity}(\mathrm{root}(M))$. We guess that there exists a sequence $\gamma : M\theta \rightarrow^* \leftarrow^* N\theta$.

1. If $\mathrm{root}(M) = \mathrm{root}(N)$ then

$$\Gamma' \cup \{M \approx N\} \Rightarrow_{\Phi_1} \Gamma' \cup \{M|_i \approx N|_i \mid 1 \leq i \leq k\}$$

In this case, we guess that $\gamma : M\theta \rightarrow^* \leftarrow^* N\theta$ is ε-invariant.

2. If $M \notin G$ then we choose a rule $\alpha \to \beta \in R$ that satisfies $\mathrm{root}(M) = \mathrm{root}(\alpha)$ and

$$\Gamma' \cup \{M \approx N\} \Rightarrow_{\Phi_1} \Gamma' \cup \mathrm{decompose}(M, \alpha, \mathcal{O}_X(\alpha)) \cup \{\beta \approx N\}$$

We rename each variable occurring in α to a fresh variable if necessary. In this case, we guess the leftmost ε-reduction step in $\gamma : M\theta \to^* \alpha\sigma \to \beta \to^* \leftarrow^* N\theta$ (where the subsequence $M\theta \to^* \alpha\sigma$ is ε-invariant).
3. If $M \in G$ then we choose a rule $\alpha \to \beta \in R$ that satisfies $M \to^* \beta$ and

$$\Gamma' \cup \{M \approx N\} \Rightarrow_{\Phi_1} \Gamma' \cup \{\beta \approx N\}$$

and then do a single transformation on $\beta \approx N$ by case 1 or 2 of this TT transformation. [1] Note that it is decidable whether or not $M \to^* \beta$ [9]. In this case, we guess the rightmost $\overset{\varepsilon}{\to}$-reduction step in $\gamma : M\theta \to^* \alpha\sigma \to \beta \to^* \leftarrow^* N\theta$.

Example 3. In case 1 of the TT transformation,

$$\{eq(not(x), t) \approx eq(y, not(y))\} \Rightarrow_{\Phi_1} \{not(x) \approx y, t \approx not(y)\}$$

By choosing rule $eq(not(x), x) \to f$ in case 2,

$$\begin{aligned}
&\{eq(not(x), t) \approx eq(y, not(y))\} \Rightarrow_{\Phi_1} \\
&\qquad \mathrm{decompose}(eq(not(x), t), eq(not(x'), x'), \{11, 2\}) \cup \{f \approx eq(y, not(y))\} \\
&\qquad = \{not(x) \approx_{\{1\}} not(x'), t \approx x', f \approx eq(y, not(y))\}
\end{aligned}$$

TL$_\to$ Transformation

If $p = M \asymp_U N$ with $M \notin X$ and if $M \in G$ then $N \notin G$ or $U \neq \emptyset$, we choose one of the following three cases. We assume that $U \neq \{\varepsilon\}$, since $M \asymp_{\{\varepsilon\}} N$ can be replaced by $M \approx N$. We guess that there exists a sequence $\gamma : M\theta \to^* L \leftrightarrow^* N\theta$ for some term L such that for the subsequence $\gamma' : L \leftrightarrow^* N\theta$ and for any $v \in \mathcal{R}(\gamma')$, there exists $u \in U$ such that $u \leq v$.

1. If $\mathrm{root}(M) = \mathrm{root}(N)$ then

$$\Gamma' \cup \{M \asymp_U N\} \Rightarrow_{\Phi_1} \Gamma' \cup \mathrm{decompose}(M, N, U)$$

and if $M \in G$, then apply the VG transformation described later to all $L \approx L' \in \mathrm{decompose}(M, N, U) \cap (G \times X)$. In this case, we guess that $\gamma : M\theta \to^* L \leftrightarrow^* N\theta$ is ε-invariant.

[1] To prove the termination of the algorithm, each transformation must decrease the "size" of Γ. There are some cases when making one transformation, the "size" of Γ does not decrease. Making two TT transformations successively, we can ensure the termination. By the same reason, we make a finite number of successive transformations in some cases of the TL$_\to$ and VT transformations.

2. If $M \notin G$ then we choose a rule $\alpha \to \beta \in R$ that satisfies $\text{root}(M) = \text{root}(\alpha)$ and

$$\Gamma' \cup \{M \asymp_U N\} \Rightarrow_{\Phi_1} \Gamma' \cup \text{decompose}(M, \alpha, \mathcal{O}_X(\alpha)) \cup \{\beta \asymp_U N\}$$

We rename each variable occurring in α to a fresh variable if necessary. In this case, we guess the leftmost ε-reduction step in $\gamma : M\theta \to^* \alpha\sigma \to \beta \to^* L \leftrightarrow^* N\theta$ (where the subsequence $M\theta \to^* \alpha\sigma$ is ε-invariant).

3. If $M \in G$ then we choose a rule $\alpha \to \beta \in R$ that satisfies $M \xrightarrow{*} \beta$ and

$$\Gamma' \cup \{M \asymp_U N\} \Rightarrow_{\Phi_1} \Gamma' \cup \{\beta \asymp_U N\}$$

and then transform $\beta \asymp_U N$ by case 1 of the $\text{TL}_\to$ transformation. Note that it is decidable whether or not $M \to^* \beta$ [9]. In this case, we guess the rightmost $\xrightarrow{\varepsilon}$-reduction step in $\gamma : M\theta \to^* \alpha\sigma \to \beta \to^* L \leftrightarrow^* N\theta$.

Example 4. In case 1 of the $\text{TL}_\to$ transformation,

$$\{eq(not(x), t) \asymp_{\{1,21\}} eq(y, not(y))\} \Rightarrow_{\Phi_1} \{not(x) \approx y, t \asymp_{\{1\}} not(y)\}$$

By choosing rule $eq(not(x), x) \to f$ in case 2,

$$\begin{aligned}\{eq(not(x), t) \asymp_{\{1,21\}} eq(y, not(y))\} &\Rightarrow_{\Phi_1} \\ \{not(x) \asymp_{\{1\}} not(x'), t \approx x'\} &\cup \{f \asymp_{\{1,21\}} eq(y, not(y))\}\end{aligned}$$

GG Transformation

1. If $p = M \approx N$ with $M, N \in G$ and $M \downarrow N$ then

$$\Gamma' \cup \{M \approx N\} \Rightarrow_{\Phi_1} \Gamma'$$

Note that it is decidable whether or not $M \downarrow N$ [9].

2. If $p = M \asymp_\emptyset N$ with $M, N \in G$ and $M \to^* N$ then

$$\Gamma' \cup \{M \asymp_\emptyset N\} \Rightarrow_{\Phi_1} \Gamma'$$

Note that it is decidable whether or not $M \to^* N$ [9].

Example 5.

$$\{eq(f, not(f)) \approx f\} \Rightarrow_{\Phi_1} \emptyset$$

Note that $eq(f, not(f)) \downarrow f$ holds, e.g., $eq(f, not(f)) \to eq(not(t), not(f)) \to eq(not(not(f)), not(f)) \to f$.

VG Transformation

1. If $p \simeq x \approx M$ with $x \in X$ and $M \in G$, we choose an element M' in $L_{min}(M)$.

$$\Gamma' \cup \{x \approx M\} \Rightarrow_{\Phi_1} \Gamma'[M'/x]$$

2. If $p = x \asymp_U M$ with $x \in X$ and $M \in G$, we choose an element M' in $L_{min}(M)$.

$$\Gamma' \cup \{x \asymp_U M\} \Rightarrow_{\Phi_1} \Gamma'[M'/x] \cup \{M' \asymp_U M\}$$

Example 6. By choosing $p = (y, f)$ and $M' = f$ (note that $f \in L_{min}(f)$),

$$\{y \approx f, eq(y, not(y)) \approx f\} \Rightarrow_{\Phi_1} \{eq(f, not(f)) \approx f\}$$

VT Transformation

1. If $p \simeq (x, M)$ with $x \in X$ and $M \in T\backslash(G \cup X)$, we choose a rule $\alpha \to \beta \in R$ and a position $v \in \mathcal{O}(M)$ such that $M|_v \notin G \cup X$.

$$\Gamma' \cup \{x \approx M\} \Rightarrow_{\Phi_1} \Gamma' \cup \{x \approx M[\beta]_v, M|_v \approx \beta\}$$

   and if $v = \varepsilon$, then apply the VG transformation to $x \approx \beta$. In this case, we guess the sequence $\gamma : x\theta \leftrightarrow^* M\theta[\alpha\sigma]_v \stackrel{v}{\to} M\theta[\beta]_v \leftrightarrow^* M\theta$ (or $x\theta \leftrightarrow^* M\theta[\beta]_v \stackrel{v}{\leftarrow} M\theta[\alpha\sigma]_v \leftrightarrow^* M\theta$) for some σ and $v \in \mathrm{Min}(\mathcal{R}(\gamma))$.
2. If $p = x \asymp_U M$ with $x \in X$ and $M \in T\backslash(G \cup X)$, we choose a rule $\alpha \to \beta \in R$ and a position $v \in \mathcal{O}(M)$ such that $M|_v \notin G \cup X$.

$$\Gamma' \cup \{x \asymp_U M\} \Rightarrow_{\Phi_1} \Gamma' \cup \{x \asymp_{U'} M[\beta]_v, \beta \asymp_{U/v} M|_v\}$$

   where $U' = \{u \in U \mid u|v\}$, and if $v = \varepsilon$, then apply the VG transformation to $x \asymp_\emptyset \beta$. Here, we assume that $\gamma : x\theta \to^* x\theta[\alpha\sigma]_v \to x\theta[\beta]_v \leftrightarrow^* M\theta[\beta]_v \leftrightarrow^* M\theta$ for some σ and $v \in \mathrm{Min}(\mathcal{R}(\gamma))$ where $x\theta[\alpha\sigma]_v \to x\theta[\beta]_v$ is the rightmost v-reduction and there is no $u \in U$ such that $u \leq v$.

Example 7. By choosing $v = \varepsilon$ and rule $eq(not(x), x) \to f$,

$$\{eq(y, not(y)) \approx y\} \Rightarrow_{\Phi_1} \{y \approx f, eq(y, not(y)) \approx f\}$$

After that we apply the VG transformation for $p = (y, f)$.

3.2 Stage II

Below we define the one step transformation Φ_2 of Stage II. We write $\Gamma \Rightarrow_{\Phi_2} \Gamma'$ if $\Phi_2(\Gamma) \ni \Gamma'$.

We begin with Γ which is the output of Stage I. Then, we repeatedly apply the transformation Φ_2 until the current Γ satisfies the **stop condition** of Stage II defined below. We consider all possibilities in order to ensure the correctness of the algorithm. If Γ satisfies this condition, then we check the $\emptyset$-unifiability of Γ in the Final Stage.

Definition 6. *Let* $\Gamma_X = \{x \approx y \mid x, y \in X \text{ and } x \approx_{\mathrm{vf}} y \in \Gamma\}$ *and* $\Gamma_T = \{P \approx_{\mathrm{vf}} Q \in \Gamma \mid P \notin X \text{ or } Q \notin X\}$. *We do not distinguish* Γ_X *and* $\Gamma \setminus \Gamma_T$, *since* $x \approx y \in \Gamma_X$ *iff* $x \approx_{\mathrm{vf}} y \in \Gamma \setminus \Gamma_T$. *Let* $\sim_{\Gamma_X}$ *be the equivalence relation derived from* Γ_X, *i.e., the reflexive transitive and symmetric closure of* Γ_X. *Let* $[x]_{\sim_{\Gamma_X}}$ *be the equivalence class of* $x \in X$.

Definition 7. *([12])* Γ *is in* solved form *if for any* $x \approx_{\mathrm{vf}} P, y \approx_{\mathrm{vf}} Q \in \Gamma_T$, $(x, y \in X)$ *and* $(x \sim_{\Gamma_X} y \Rightarrow P = Q)$ *hold.*

The **stop condition** of Stage II is that Γ satisfies one of the following two conditions.

(1) For any $P \approx_{\mathrm{vf}} Q \in \Gamma_T$, we have $P \in X$ and $Q \in T$, and Γ is in solved form.
(2) $\Gamma = \{\mathbf{fail}\}$.

(Note. $\Gamma = \emptyset$ satisfies condition (1).)

To describe the transformations used in Stage II, we need the following definitions.

Definition 8. *For a term* M *we define* $\#_0(M)$ *by*

$$\#_0(M) = \begin{cases} \{(|\mathcal{O}_X(M)|, H(M))\}_m & \text{if } V(M) \neq \emptyset \\ \emptyset & \text{otherwise} \end{cases}$$

For $(i, j), (i', j') \in \mathcal{N} \times \mathcal{N}$ *where* $\mathcal{N}$ *is the set of nonnegative integers, we use a lexicographic ordering* $>$. *This measure is defined to give the number of variable positions of term* M *the highest priority and will be used in Section 4 to define* $\mathrm{size}(\Gamma)$ *for* $\Gamma \subseteq E_0$.

Definition 9. *For* $P, Q \notin X$, *we define function* $\mathrm{common}(P, Q)$ *as follows. Let* $U = \mathrm{Min}(\mathcal{O}_X(P) \cup \mathcal{O}_X(Q))$ *and let* $V = \mathrm{Min}\{v \in \mathcal{O}_F(P) \cup \mathcal{O}_F(Q) | \forall u \in U.\ u | v\}$. *If* $P|_v \downarrow Q|_v$ *holds for any* $v \in V$ *and* $P[c, \cdots, c]_{(v_1, \cdots, v_n)} = Q[c, \cdots, c]_{(v_1, \cdots, v_n)}$, *where* c *is a constant in* G *and* $V \cup U = \{v_1, \cdots, v_n\}$, *then* $\mathrm{common}(P, Q) = \mathbf{true}$ *otherwise* **false**. *Note that it is decidable whether* $P|_v \downarrow Q|_v$ *[9]. For example, let*

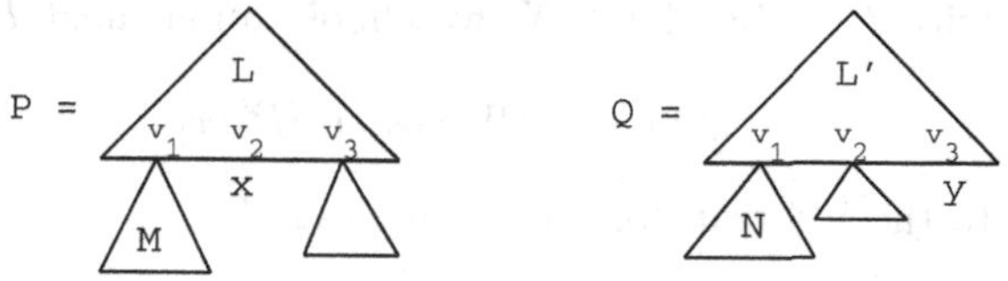

where $M, N \in G$. *If* $M \downarrow N$ *and* $P[c, c, c]_{(v_1, v_2, v_3)} = Q[c, c, c]_{(v_1, v_2, v_3)}$ *then* $\mathrm{common}(P, Q) = \mathbf{true}$.

In Stage II, we first choose an element p in $\Gamma \setminus X \times X$ nondeterministically and then apply one of the following transformations to Γ according to the type of the chosen p. If no transformation is possible, $\Gamma \Rightarrow_{\Phi_2} \{\mathbf{fail}\}$. Let $\Gamma' = \Gamma \setminus \{p\}$.

Decomposition

If $p = x \approx_{\rm vf} P$ with $x \in X$ and $P \in T \setminus G$ and there exists a pair $q = y \approx_{\rm vf} Q \in \Gamma_T$ such that $x \sim_{\Gamma_X} y$ and $P \neq Q$ and $Q \notin G$ and common(P, Q) then,

$$\begin{aligned}\Gamma'' \cup \{x \approx_{\rm vf} P, y \approx_{\rm vf} Q\} \Rightarrow_{\Phi_2} \Gamma'' &\cup \{y \approx_{\rm vf} Q\}\\ &\cup \{P|_u \approx_{\rm vf} Q|_u \mid u \in U \text{ and } P|_u \in X\}\\ &\cup \{Q|_u \approx_{\rm vf} P|_u \mid u \in U \text{ and } P|_u \notin X\}\end{aligned}$$

where $\Gamma'' = \Gamma' \setminus \{q\}$ and $U = \mathrm{Min}(\mathcal{O}_X(P) \cup \mathcal{O}_X(Q))$. Here, we assume that $\#_{\rm O}(P) \gtrsim \#_{\rm O}(Q)$.

In this case, if there exists a locally minimal R-unifier θ of Γ, then there exists a rewrite sequence $\delta : P\theta \leftrightarrow^* x\theta \leftrightarrow^* y\theta \leftrightarrow^* Q\theta$. Since $x\theta$ and $y\theta$ are minimal, the sequence δ has no reduction above the leaves of $x\theta$ and $y\theta$ by Lemma 1. For any reduction position v of the subsequences $P\theta \leftrightarrow^* x\theta$ and $y\theta \leftrightarrow^* Q\theta$, we have $v \not< u$ for any $u \in \mathrm{Min}(\mathcal{O}_X(P) \cup \mathcal{O}_X(Q))$. So, we can decompose subgoals $x \approx_{\rm vf} P$ and $y \approx_{\rm vf} Q$ into $P|_u \approx_{\rm vf} Q|_u$ or $Q|_u \approx_{\rm vf} P|_u$. For the termination and validity of the algorithm, we leave $y \approx_{\rm vf} Q$, whose size is not greater than the size of $x \approx_{\rm vf} P$, in Γ.

Example 8. Let $\Gamma = \{p, q, x \approx_{\rm vf} y\}$ with $p = x \approx_{\rm vf} eq(not(w), t)$ and $q = y \approx_{\rm vf} eq(z, not(f))$. Then, common$(eq(not(w), t), eq(z, not(f)))$ is **true** because $t \to not(f)$ and $eq(not(w), t)[c, c]_{(1,2)} = eq(c, c) = eq(z, not(f))[c, c]_{(1,2)}$ hold. $\#_{\rm O}(eq(not(w), t)) = \#_{\rm O}(eq(z, not(f))) = \{(1, 2)\}_m$. So, we can make the following Decomposition transformation:

$$\begin{aligned}&\{x \approx_{\rm vf} eq(not(w), t), y \approx_{\rm vf} eq(z, not(f)), x \approx_{\rm vf} y\} \Rightarrow_{\Phi_2}\\ &\qquad \{x \approx_{\rm vf} eq(not(w), t), z \approx_{\rm vf} not(w), x \approx_{\rm vf} y\}\end{aligned}$$

GT Transformation

If $p = P \approx_{\rm vf} Q$ with $P \in G$ and $Q \in T \setminus (G \cup X)$ and common(P, Q) then

$$\Gamma' \cup \{P \approx_{\rm vf} Q\} \Rightarrow_{\Phi_2} \Gamma' \cup \{P|_u \approx_{\rm vf} Q|_u \mid u \in \mathcal{O}_X(Q)\}$$

VG Transformation

If $p \simeq x \approx_{\rm vf} P$ with $P \in G$ and $x \in X$, we choose an element $P' \in L_{min}(P)$.

$$\Gamma' \cup \{x \approx_{\rm vf} P\} \Rightarrow_{\Phi_2} \Gamma'[P'/x]$$

This is similar to the VG transformation at Stage I.

GG Transformation

If $p = P \approx_{\rm vf} Q$ with $P, Q \in G$ and $P \downarrow Q$ then

$$\Gamma' \cup \{P \approx_{\rm vf} Q\} \Rightarrow_{\Phi_2} \Gamma'$$

Note that it is decidable whether or not $P \downarrow Q$ [9].

3.3 Final Stage

Let Γ_f be the output of Stage II. If Γ_f is $\emptyset$-unifiable, then our algorithm answers 'R-unifiable', otherwise fail, i.e., no answer.

Note that since $\emptyset$-unifiability is equal to usual unifiability, any unification algorithm can be used [3,12]. In fact, since Γ_f satisfies the **stop condition** of Stage II, Γ_f is in solved form, so that it is known that Γ_f is unifiable iff Γ_f is not cyclic [12]. The definition that Γ_f is cyclic is given as follows.

Definition 10. *A relation $\mapsto$ over X is defined as follows: $x \mapsto y$ iff there exist $x' \sim_{\Gamma_{fX}} x, y' \sim_{\Gamma_{fX}} y, P \in T \setminus (G \cup X)$ such that $x' \approx_{\mathrm{vf}} P \in \Gamma_{fT}$ and $y' \in V(P)$ hold. Let $\mapsto^+$ be the transitive closure of $\mapsto$. Then, Γ is* cyclic *if there exists $x \in X$ such that $x \mapsto^+ x$.*

We will prove later that Γ_f is not cyclic if there exists a locally minimal R-unifier of Γ_f.

Correctness condition of Φ:

(1) $\Rightarrow^*_{\Phi_1} \cdot \Rightarrow^*_{\Phi_2}$ is terminating and finite-branching, and
(2) $\Gamma_0 = \{M_0 \approx N_0\}$ is R-unifiable iff there exist Γ_1 and Γ_f such that $\Gamma_0 \Rightarrow^*_{\Phi_1} \Gamma_1 \Rightarrow^*_{\Phi_2} \Gamma_f$, Γ_1 satisfies the **stop conditions** of Stage I, Γ_f satisfies the one of Stage II, and Γ_f is $\emptyset$-unifiable (i.e., not cyclic and $\Gamma_f \neq \{\mathbf{fail}\}$).

Note that since Φ is a nondeterministic algorithm, we need an exhaustive search of all the computation $\Rightarrow^*_{\Phi_1} \cdot \Rightarrow^*_{\Phi_2}$ from Γ_0, but it is ensured that we can decide whether Γ_0 is R-unifiable or not within finite time by (1) and (2) above.

Our algorithm can be easily transformed into one which produces a locally minimal unifier of Γ_0 iff Γ_0 is R-unifiable, since the information can be obtained when VG transformations are made.

3.4 Example

We consider $R_1 = \{\ eq(x,x) \to t,\ eq(not(x),x) \to f,\ t \to not(f),\ f \to not(t)\}$ given in Section 1. For $\Gamma_0 = \{eq(y, not(y)) \approx y\}$, our algorithm Φ can make the following computation:

$$\begin{aligned} \{eq(y,not(y)) \approx y)\} &\Rightarrow_{VT} \{y \approx f, eq(y,not(y)) \approx f\} \\ &\Rightarrow_{VG} \{eq(f,not(f)) \approx f\} \\ &\Rightarrow_{GG} \emptyset \end{aligned}$$

Obviously, $\emptyset$ satisfies the **stop conditions** of Stages I and II and is $\emptyset$-unifiable. Hence, our algorithm decides that Γ_0 is R-unifiable. In fact, θ satisfying $y\theta = f$ is an R-unifier which can be computed by our algorithm.

Note that $\{eq(y, not(y)) \approx y\}$ is transformed into $\{y \approx_{\mathrm{vf}} eq(y, not(y))\}$ by Conversion which satisfies the **stop condition** of Stages I and II. But $\{y \approx_{\mathrm{vf}} eq(y, not(y))\}$ is cyclic, so this computation sequence fails in the final stage.

4 Correctness of Algorithm Φ

In this section, we give the lemmata needed to conclude the correctness of Algorithm Φ and the main theorem. We only outline the proof of part (1) on the correctness condition, i.e., termination of Φ. The reader is referred to the full version of this paper [13] for the complete proofs.

Definition 11. *For $\Gamma \subseteq E_0$, let* $\mathrm{core}(\Gamma) = \{M \approx N \mid M \approx N \in \Gamma$ or $M \asymp_U N \in \Gamma$ or $M \approx_{\mathrm{vf}} N \in \Gamma\}$.

Definition 12. *Substitutions θ and θ' are* consistent *if $x\theta = x\theta'$ for any $x \in Dom(\theta) \cap Dom(\theta')$.*

Definition 13. *Let Φ : $\mathrm{Power}(E_0) \rightarrow \mathrm{Power}(\mathrm{Power}(E_0))$ be a transformation. Then, Φ is valid iff the following validity conditions (V1) and (V2) hold. For any $\Gamma \subseteq E_0$, let $\Phi(\Gamma) = \{\Gamma_1, \cdots, \Gamma_n\}$.*

(V1) If θ is a locally minimal R-unifier of Γ, then there exist i $(1 \leq i \leq n)$ and a substitution θ' such that θ' is consistent with θ and θ' is a locally minimal R-unifier of Γ_i.

(V2) If there exists i $(1 \leq i \leq n)$ such that $\mathrm{core}(\Gamma_i)$ *is R-unifiable then* $\mathrm{core}(\Gamma)$ *is R-unifiable.*

4.1 Correctness of Stage I

Lemma 3. *Stage I is terminating and finite-branching.*

Proof. For $\Gamma \subseteq E_0$, we define $\mathrm{size}(\Gamma)$ as $(\#_1(\Gamma), \#_2(\Gamma), \#_3(\Gamma), \#_4(\Gamma))$. Here

$$\#_1(\Gamma) = \sqcup_{P \approx Q \in \Gamma}(\#_0(P) \sqcup \#_0(Q)) \sqcup (\sqcup_{P \asymp_U Q \in \Gamma} \#_0(P))$$
$$\#_2(\Gamma) = \sqcup_{P \asymp_U Q \in \Gamma} \#_0(Q)$$
$$\#_3(\Gamma) = \sqcup_{P \asymp_U Q \in \Gamma}\{|u| \mid u \in U\}_m$$
$$\#_4(\Gamma) = |\Gamma|$$

We use a lexicographic ordering $>$ to compare $\mathrm{size}(\Gamma)$ and $\mathrm{size}(\Gamma')$ for all $\Gamma, \Gamma' \in E_0$. For every transformation $\Phi_1(\Gamma) = \{\Gamma_1, \cdots, \Gamma_k\}$ in Stage I, we can prove that $\mathrm{size}(\Gamma) > \mathrm{size}(\Gamma_i)$ for every $1 \leq i \leq k$ according to showing the following tables.

	$\#_1$	$\#_2$	$\#_3$	$\#_4$
TT	$\gg$			
case 1 of $\mathrm{TL}_{\rightarrow}$	$\geqq$	$\geqq$	$\gg$	
case 2 of $\mathrm{TL}_{\rightarrow}$	$\gg$			
case 3 of $\mathrm{TL}_{\rightarrow}$	$\geqq$	$\geqq$	$\gg$	

	$\#_1$	$\#_2$	$\#_3$	$\#_4$
GG	$=$	$=$	$=$	$>$
VG	$\gg$			
case 1 of VT	$\gg$			
case 2 of VT	$\geqq$	$\gg$		

Moreover, if Γ is a finite set, then k is finite, i.e., Stage I is finite-branching. Thus, this lemma holds. □

Lemma 4. *Stage I is valid [13].*

4.2 Correctness of Stage II

Lemma 5. *Stage II is terminating and finite-branching.*

Proof. For $\Gamma \subseteq \{M \approx_{\mathrm{vf}} N \mid M, N \in T\}$, we define the size($\Gamma$) as $(\$_1(\Gamma), \$_2(\Gamma))$. Here

$$\$_1(\Gamma) = \sqcup_{P \approx_{\mathrm{vf}} Q \in \Gamma}(\#_0(P) \sqcup \#_0(Q))$$
$$\$_2(\Gamma) = |\Gamma|$$

We use a lexicographic ordering $>$ to compare size(Γ) and size(Γ') for all $\Gamma, \Gamma' \in E_0$. For every transformation $\Phi_2(\Gamma) = \{\Gamma_1, \cdots, \Gamma_k\}$ in Stage II, we can show that size(Γ) $>$ size(Γ_i) for every $1 \leq i \leq k$. Moreover, if Γ is a finite set, then k is finite. Thus, this lemma holds. □

Lemma 6. *Stage II is valid [13].*

4.3 Correctness of Final Stage

Lemma 7. *Assume that Γ_f satisfies the* **stop condition** *of Stage II. Then, Γ_f is not cyclic if there exists a locally minimal R-unifier θ of Γ_f.*

Proof. Let θ be a locally minimal R-unifier of Γ_f. Note that for any $x, x' \in X$ if $x \sim_{\Gamma_{f}X} x'$ then $x\theta \leftrightarrow^* x'\theta$ holds, so that $H(x\theta) = H(x'\theta)$ holds by local minimality of θ. Now, we show that for any $x \approx_{\mathrm{vf}} P \in \Gamma_f$ and $y \in V(P)$, if $P \notin X$, then $H(x\theta) > H(y\theta)$ holds. Let $y = P|_u$ for some $u \neq \varepsilon$. Then $x\theta|_u \leftrightarrow^* y\theta$ holds, since θ is an R-unifier of Γ_f. Local minimality of θ ensures that $H(x\theta|_u) \geq H(y\theta)$. Hence, $H(x\theta) > H(y\theta)$. It follows that for any $x, y \in X$ if $x \mapsto y$, then $H(x\theta) > H(y\theta)$ holds. Therefore, it is impossible that we have $x \mapsto^+ x$. That is, Γ_f is not cyclic. □

Lemma 8. *Assume that Γ_f satisfies the* **stop condition** *of Stage II. Then, if there exists a locally minimal R-unifier of Γ_f then Γ_f is $\emptyset$-unifiable.*

Proof. Obviously, $\Gamma_f \neq \{\textbf{fail}\}$, so that Γ_f is in solved form. By Lemma 7, Γ_f is not cyclic, so that Γ_f is $\emptyset$-unifiable. □

(Note. The converse of Lemma 8 does not necessarily hold.)

5 Conclusion

Now, we can deduce our main theorem.

Theorem 1. *The unification problem for confluent right-ground term rewriting systems is decidable.*

Proof. By Lemmata 3 and 5, part (1) of the correctness condition of Φ holds and by Lemmata 4 and 6, Stages I and II are valid, so that if $\Gamma_0 = \{M_0 \approx N_0\}$ is R-unifiable, then there exist Γ_1 and Γ_f such that $\Gamma_0 \Rightarrow^*_{\Phi_1} \Gamma_1 \Rightarrow^*_{\Phi_2} \Gamma_f$, Γ_1 satisfies the **stop conditions** of Stage I, Γ_f satisfies the one of Stage II, and there exists a locally minimal R-unifier of Γ_f. Hence, by Lemma 8, the only-if-part of part (2) of the correctness condition of Φ holds. Conversely, the if-part is ensured by validity of the transformations of Φ_1 and Φ_2. Thus, part (2) of the correctness condition of Φ holds. Therefore, decidability of $\emptyset$-unifiability ensures this theorem. □

Acknowledgements. We would like to thank Prof. A. Middeldorp and all the anonymous referees of this paper for their detailed helpful comments, which have enhanced the quality of the paper. This work was supported in part by Grand-in-Aid for Scientific Research 12680344 from Japan Society for the Promotion of Science.

References

1. H. Comon, M. Haberstrau and J.-P. Jouannaud, *Syntacticness, Cycle-Syntacticness and Shallow Theories*, Information and Computation, 111 (1), pp. 154–191, 1994.
2. M. Dauchet and S. Tison, *The Theory of Ground Rewriting Systems is Decidable*, Proc. 5th IEEE Symp. Logic in Computer Science, pp. 242–248, 1990.
3. N. Dershowitz and J.-P. Jouannaud, *Rewrite Systems*, Handbook of Theoretical Computer Science, Vol. B, ed. J. van Leeuwen, pp. 243–320, North-Holland, Amsterdam, 1990.
4. J.-M. Hullot, *Canonical Forms and Unification*, Proc. 5th CADE, LNCS 87, pp. 318–334, 1980.
5. F. Jacquemard, C. Meyer and C. Weidenbach, *Unification in Extensions of Shallow Equational Theories*, In Proc. 9th Rewriting Techniques and Applications, LNCS 1379, pp. 76–90, 1998.
6. A. Middeldorp and S. Okui, *A Deterministic Lazy Narrowing Calculus*, Journal of Symbolic Computation 25 (6), pp. 733–757, 1998.
7. A. Middeldorp, S. Okui and T. Ida, *Lazy Narrowing: Strong Completeness and Eager Variable Elimination*, Theoret. Comput. Sci. 167, pp. 95–130, 1996.
8. R. Nieuwenhuis, *Basic Paramodulation and Decidable Theories*, Proc. 11th Annual IEEE Symp. Logic in Computer Science, pp. 473–482, 1996.
9. M. Oyamaguchi, *The Reachability and Joinability Problems for Right-Ground Term-Rewriting Systems*, Journal of Information Processing, Vol. 13, No. 3, pp. 347–354, 1990.
10. M. Oyamaguchi, *On the Word Problem for Right-Ground Term-Rewriting Systems*, Trans. IEICE, Vol. E73, No. 5, pp. 718–723, 1990.
11. M. Oyamaguchi and Y. Toyama, *On the Church-Rosser Property of E-overlapping and Simple-Right-Linear TRS's*, Research Reports of the Faculty of Engineering, Mie Univ., 20, pp. 99–118, 1995.
12. A. Martelli and G. Rossi, *Efficient Unification with Infinite Terms in Logic Programming*, Proc. Fifth Generation Computer Systems, pp. 202–209, 1984.
13. `http://www.cs.info.mie-u.ac.jp/~mo/rta01/`.

On Termination of Higher-Order Rewriting

Femke van Raamsdonk

Division of Mathematics and Computer Science
Faculty of Sciences, Vrije Universiteit
De Boelelaan 1081a, 1081 HV Amsterdam
The Netherlands
femke@cs.vu.nl

CWI
P.O. Box 94079, 1090 GB Amsterdam

Abstract. We discuss the termination methods using the higher-order recursive path ordering and the general scheme for higher-order rewriting systems and combinatory reduction systems.

1 Introduction

A rewriting system is said to be terminating if all rewrite sequences are finite. Many methods to prove termination of first-order term rewriting have been studied. For higher-order rewriting, where bound variables may be present, there are so far significantly fewer results available. What makes this situation even worse is that there are several brands of higher-order rewriting, and it is often not immediately clear how to apply or adapt a result obtained in one framework to another one. We distinguish here three variants of higher-order rewriting. First there are the higher-order rewriting systems (HRSs) introduced by Nipkow [14]. Here rewriting is defined modulo $\beta\eta$ of simply typed λ-calculus. Second there are the combinatory reduction systems (CRSs) introduced by Klop [8]. Third there are the algebraic-functional systems (AFSs) introduced by Jouannaud and Okada [4]. Here the reduction relation of interest is the union of β-reduction and the reduction relation induced by the algebraic rewrite rules (which may be higher-order). Matching in an AFS is syntactic (not modulo β).

An important method to prove termination of a first-order term rewriting system is the one using the recursive path ordering (rpo) due to Dershowitz [3]. Jouannaud and Rubio [6] present a generalization of the recursive path ordering to the higher-order case, in the framework of AFSs. The crucial idea is to show well-foundedness of the ordering using the notion of computability from the proof of termination of typed λ-calculi due to Tait and Girard. The usual proof of well-foundedness of rpo, and also the proofs of well-foundedness of several earlier orderings designed to prove termination of higher-order rewriting ([10,11,5]) rely instead on Kruskal's tree theorem. Because so far no sufficiently expressive higher-order variant of Kruskal's tree theorem seems to be known, those higher-order term orderings don't have the full power of rpo.

A. Middeldorp (Ed.): RTA 2001, LNCS 2051, pp. 261–275, 2001.

The main purpose of this paper is to make the termination method using the higher-order version of the recursive path ordering (horpo) more widely available by presenting it for HRSs and CRSs. The fact that horpo can be adapted to prove termination of HRSs is already remarked in [6], and worked out in [7]. Here we take a different approach, as explained in Section 4.

Another method to prove termination of AFSs is due to Jouannaud and Okada [4], and makes use of the notion of general scheme. The general scheme is designed to make the proof of termination of typed λ-calculus due to Tait and Girard adaptable to the case of the particular AFS. Blanqui [2] studies versions of the general scheme for higher-order rewriting with a CRS-like syntax and for HRSs. Here we consider a simpler form of the general scheme, closer to the one considered in [6]. If we consider pure horpo and the pure general scheme, then the two methods to prove termination are incomparable, as shown by examples. The general scheme can be used to upgrade horpo, as done in [4]. In this way the power of both methods is combined.

Finally, we remark that for HRSs there is a semantical method to prove termination due to Van De Pol [16], using an interpretation (to be given by the user) of the function symbols as functionals.

2 Higher-Order Rewriting

In this section we briefly recall the syntax higher-order rewriting systems (HRSs) as introduced by Nipkow [14] and combinatory reduction systems (CRSs) as introduced by Klop [8]. For more detailed accounts we refer to [14,12,17,8,9]. Examples of higher-order rewriting systems in HRS, CRS, and AFS format are available at `http://www.cs.vu.nl/~femke/papers.html`.

2.1 Higher-Order Rewriting Systems

In a HRS we work modulo the $\beta\eta$-relation of simply typed λ-calculus. *Types* are built from a non-empty set of base types and the binary type constructor $\rightarrow$ as usual. For every type we assume a countably infinite set of *variables* of that type, written as $x, y, z, \ldots$. A *signature* is a non-empty set of typed function symbols. The set of *preterms* of type A over a signature Σ consists exactly of the expressions s for which we can derive $s : A$ using the following rules:

1. $x : A$ for a variable x of type A,
2. $f : A$ for a function symbol f of type A in Σ,
3. if $A = A' \rightarrow A''$, and $x : A'$ and $s : A''$, then $(x. s) : A$,
4. if $s : A' \rightarrow A$ and $t : A'$, then $(s\,t) : A$.

The abstraction operator $\_.\_$ binds variables, so occurrences of x in s in the preterm $x. s$ are bound. We work modulo type-preserving α-conversion and assume that bound variables are renamed whenever necessary in order to avoid unintended capturing of free variables. Parentheses may be omitted according to the usual conventions. We make use of the usual notions of *substitution* of a preterm t for the free occurrences of a variable x in a preterm s, notation $s[x := t]$, and *replacement in a context*, notation $C[t]$. We write $s \supseteq s'$ if s' is a subpreterm of s, and use $\supset$ for the strict subpreterm relation.

The *β-reduction relation*, notation $\rightarrow_\beta$, is the smallest relation on preterms that is compatible with formation of preterms and that satisfies the following:

$$(x.\, s)\, t \rightarrow_\beta s[x := t]$$

The *restricted η-expansion relation*, notation $\rightarrow_{\overline{\eta}}$, is defined as follows. We have

$$C[s] \rightarrow_{\overline{\eta}} C[x.\,(s\, x)]$$

if $s : A \rightarrow B$, and $x : A$ is a fresh variable, and no β-redex is created (hence the terminology *restricted* η-expansion). The latter condition is satisfied if s is not an abstraction (so not of the form $z.\, s'$), and doesn't occur in $C[s]$ as the left part of an application (so doesn't occur in a sub-preterm of the form $(s\, s')$).

In the sequel we employ only preterms in $\overline{\eta}$-normal form, where every subpreterm has the right number of arguments. Instead of $s_0 s_1 \ldots s_m$ we often write $s_0(s_1, \ldots, s_m)$. A preterm is then of the form $x_1 \ldots x_n.\, s_0(s_1, \ldots, s_m)$ with $s_0(s_1, \ldots, s_m)$ of base type and all s_i in $\overline{\eta}$-normal form.

A *term* is a preterm in β-normal form. It is also in $\overline{\eta}$-normal form because $\overline{\eta}$-normal forms are closed under β-reduction. A term is of the form $x_1 \ldots x_n.\, a(s_1, \ldots, s_m)$ with a a function symbol or a variable. Because the $\beta\overline{\eta}$-reduction relation is confluent and terminating on the set of preterms, every $\beta\overline{\eta}$-equivalence class of preterms contains a unique term, which is taken as the representative of that class.

Because in the discussion we will often use preterms, we use here the notation s^σ for the replacement of variables according to the substitution σ (*without* reduction to β-normal from), and write explicitly $s^\sigma\!\downarrow_\beta$ for its β-normal form. This is in contrast with the usual notations for HRSs.

A *rewrite rule* is a pair of terms (l, r), written as $l \rightarrow r$, satisfying the following requirements:

1. l and r are of the same base type,
2. l is of the form $f(l_1, \ldots, l_n)$,
3. all free variables in r occur also in l,
4. a free variable x in l occurs in the form $x(y_1, \ldots, y_n)$ with y_i η-equivalent to different bound variables.

The last requirement guarantees that the rewrite relation is decidable because unification of patterns is decidable [13]. The rewrite rules induce a rewrite relation $\rightarrow$ on the set of terms which is defined by the following rules:

1. if $s \rightarrow t$ then $x(\ldots, s, \ldots) \rightarrow x(\ldots, t, \ldots)$,
2. if $s \rightarrow t$ then $f(\ldots, s, \ldots) \rightarrow f(\ldots, t, \ldots)$,
3. if $s \rightarrow t$ then $x.\, s \rightarrow x.\, t$,
4. if $l \rightarrow r$ is a rewrite rule and σ is a substitution then $l^\sigma\!\downarrow_\beta \rightarrow r^\sigma\!\downarrow_\beta$.

The last clause in this definition shows that HRSs use higher-order pattern matching, unlike AFSs, where matching is syntactic.

2.2 Combinatory Reduction Systems

We assume a countably infinite set of *variables*, written as $x, y, z, \ldots$. We make use of the notion of *arity* which is a natural number indicating how many arguments a symbol is supposed to get. For every arity a countably infinite set of *metavariables*, written as $X, Y, Z, \ldots$ of that arity is assumed. A *signature* is a non-empty set of function symbols, each with a fixed arity. The set of *metaterms* over a signature Σ is defined by the following clauses:

1. a variable x is a metaterm of arity 0
2. if f is a function symbol in Σ of arity m and $s_1, \ldots, s_m$ are metaterms, then $f(s_1, \ldots, s_m)$ is a metaterm of arity 0,
3. if s is a metaterm of arity m, then $[x]s$ is a metaterm of arity $m + 1$,
4. a metavariable Z of arity n is a metaterm of arity n
5. if s_0 is a metaterm of arity m and $s_1, \ldots, s_m$ are metaterms of arity 0, then $s_0(s_1, \ldots, s_m)$ is a metaterm of arity 0 (a meta-application).

This definition of metaterm differs from the usual one: a metaterm can be a metavariable applied to metaterms, but also an abstraction applied to one or more metaterms. In this way the metaterms contain what are usually called the substitutes. Another difference is that here metaterms have an arity. The abstraction operator $[\_]\_$ binds variables, so occurrences of x in s in the metaterm $[x]s$ are bound. We write $s \supseteq s'$ if s' is a submetaterm of s and use $\supset$ for the strict submetaterm relation. For every $m \geq 1$ we have a b-reduction rule:

$$([x_1 \ldots x_m]s_0)(s_1, \ldots, s_m) \rightarrow_b s_0[x_1 := s_1 \ldots x_m := s_m].$$

Because a variable x doesn't occur in a submetaterm of the form $x(s_1, \ldots, s_m)$, an application of the b-reduction rule doesn't create new b-redexes. The relation $\rightarrow_b$ is like a development; it is confluent and terminating on the set of metaterms.

A *term* is a metaterm without metavariables or meta-application. A *rewrite rule* of a CRS is a pair of terms (l, r), written as $l \rightarrow r$, satisfying the following:

1. l and r are closed metaterms of arity 0,
2. l is of the form $f(l_1, \ldots, l_n)$,
3. all metavariables in r occur also in l,
4. all metavariables in l occur in the form $Z(x_1, \ldots, x_m)$ with $x_1, \ldots, x_m$ different bound variables.

The restriction concerning the arity in the first clause makes that some rewrite rules that fit in the usual definition of a CRS are not allowed here. An example is $a \rightarrow [x]a$. We use the notation s^σ for s where all metavariables are replaced according to the definition of the substitution σ. Such a substitution assigns terms of arity n to metavariables of arity n. The rewrite rules induce a rewrite relation $\rightarrow$ on the set of terms which is defined by the following rules:

1. if $s \rightarrow t$ then $f(\ldots, s, \ldots) \rightarrow f(\ldots, t, \ldots)$,
2. if $s \rightarrow t$ then $[x]s \rightarrow [x]t$,
3. if $l \rightarrow r$ is a rewrite rule and σ is a substitution, then $l^\sigma{\downarrow_b} \rightarrow r^\sigma{\downarrow_b}$.

3 Computability

In the following sections we will make use of the notion of computability due to Tait and Girard with respect to a relation $\gg$ on terms, preterms, or metaterms. Here we give the definition for both the typed HRS case and the untyped CRS case where we use the arity of a metaterm. The definition and the properties we use later are the well-known ones, also used in [6].

Definition 1. *The expressions of type A (of some arity) that are* computable *with respect to $\gg$ are defined by induction on the structure of A as follows:*

1. *If $s : B$ with B a base type (with s of arity 0) then s is computable with respect to $\gg$ if t is well-founded with respect to $\gg$ for all t such that $s \gg t$.*
2. *If s is of type $A_1 \to \ldots \to A_n \to B$ with B a base type (of arity n), then s is computable with respect to $\gg$ if for all computable $u_1, \ldots, u_n$ of type $A_1, \ldots, A_n$ we have that $s(u_1, \ldots, u_n)$ is computable with respect to $\gg$.*

The following lemma concerns computability with respect to some relation $\gg$.

Lemma 1.

1. *If s is computable then s is well-founded with respect to $\gg$.*
2. *If s is computable and $s \gg t$ then t is computable.*
3. *If $s : B$ with B a base type (of arity 0) and s' is computable for every s' such that $s \gg s'$ then s is computable.*
4. *(HRS case) If $s[x := u]$ is computable for every computable u of the right type, and $(x.\,s)\,(u) \gg s[x := u]$, then $x.\,s$ is computable.*
   (CRS case) If $s[x := u]$ is computable for every computable u and we have $([x]s)\,(u) \gg s[x := u]$, then $[x]s$ is computable.

4 The Higher-Order Recursive Path Ordering

This section is concerned with the higher-order version of the recursive path ordering (horpo). Jouannaud and Rubio [6] define horpo for what we call here AFSs. Here we present a method to prove termination of HRSs and a method to prove termination of CRSs. Both methods use an adaption of horpo as in [6].

4.1 Horpo for HRSs

We assume well-founded precedence $\rhd$ on the set function symbols. We write $\equiv$ for the equivalence relation on types induced by identifying all base types.

Definition 2. *We have $s \succ t$ for preterms $s : A$ and $t : A'$ if $A \equiv A'$ and one of the following clauses holds:*

1. $s = f(s_1, \ldots, s_m)$
   $t = g(t_1, \ldots, t_n)$
   $f \rhd g$
   for all $i \in \{1, \ldots, n\}$: either $s \succ t_i$ or $s_j \succeq t_i$ for some j

2. $s = f(s_1, \dots, s_m)$
   $t = f(t_1, \dots, t_m)$
   $(s_1, \dots, s_m) \succ (t_1, \dots, t_m)$
   for all $i \in \{1, \dots, n\}$*: either* $s \succ t_i$ *or* $s_j \succeq t_i$ *for some* j
3. $s = f(s_1, \dots, s_m)$
   $s_i \succeq t$ *for some* s_i
4. $s = f(s_1, \dots, s_m)$
   $t = t_0(t_1, \dots, t_n)$
   for all $i \in \{0, \dots, n\}$*: either* $s \succ t_i$ *or* $s_j \succeq t_i$ *for some* j
5. $s = x(s_1, \dots, s_m)$
   $t = x(t_1, \dots, t_m)$
   $s_i \succ t_i$ *for some* i
   $s_i \succeq t_i$ *for all* i
6. $s = s_0(s_1, \dots, s_m)$
   $t = t_0(t_1, \dots, t_m)$
   $s_i \succ t_i$ *for some* i
   $s_i \succeq t_i$ *for all* i
7. $s = x. s_0$
   $t = x. t_0$
   $s_0 \succ t_0$

The first three clauses are the same as for the first-order case. The difference is that here we need to take care of the types: in order to derive $s \succ t$ we need that the types of s and t are equivalent. If for instance $s : A$ and $t : A \to A$ then it is not possible to compare s and t using $\succ$. The condition 'for all i either $s \succ t_i$ or $s_j \succeq t_i$' in the clauses 1, 2, and 4 is to be understood as follows: if $t_i : A'$ with $A \equiv A'$ then $s \succ t$, otherwise we have $t_i : B'$ and then $s_j \succeq t_i$ for some $s_j : B$ with $B \equiv B'$. Note further that in clause 2 we use the notation $\succ$ also for the multiset or lexicographic (depending on the function symbol f) extension of the relation $\succ$ on preterms. The clause 4 takes care of substitution. For instance, we have $f(x. z(x), a) \succ (x. z(x))\, a$ because $x. z(x) \succeq x. z(x)$ and $f(x. z(x), a) \succ a$. In clause 6 it is assumed that s_0 is not (η-equivalent to) a function symbol or a variable and that $m \geq 1$. The clauses 2, 5, 6, and 7 make that $\succ$ is compatible with the structure of preterms.

Note that horpo is defined on preterms, not on terms. It is not the case that $s \succ t$ implies $s{\downarrow_\beta} \succ t{\downarrow_\beta}$. For instance $(x. a)\, b \succ (x. a)\, c$ if $b \rhd c$ but not $a \succ a$.

The termination method using horpo is as follows: *a HRS is terminating if for every rewrite rule* $l \to r$ *there exists a preterm* r' *such that* $l \succ r' \twoheadrightarrow_\beta r$. We call this the horpo criterion (for HRSs).

Example 1.

1. Consider for example the beta-reduction rule of untyped λ-calculus:

$$\mathsf{app}(\mathsf{abs}(x. Z(x)), Z') \to Z(Z')$$

   Clause 3 does *not* yield that $\mathsf{abs}(x. Z(x)) \succ x. Z(x)$ because $\mathsf{abs}(x. Z(x))$ has type T and $x. Z(x)$ has type $\mathsf{T} \to \mathsf{T}$.

2. The following rewrite rule can be shown to be terminating using horpo:
$$\mathsf{map}(x.\,F(x), \mathsf{cons}(h,t)) \to \mathsf{cons}(F(h), \mathsf{map}(x.\,F(x), t)).$$
We take $r' = \mathsf{cons}((x.\,F(x))(h), \mathsf{map}(x.\,F(x), t))$. First, we have $x.\,F(x) \succeq x.\,F(x)$, and $\mathsf{map}(x.\,F(x), \mathsf{cons}(h,t)) \succ h$ by clause 3, and hence we have $l \succ (x.\,F(x))\,(h)$ by clause 4. Further, $\mathsf{cons}(h,t) \succ t$, so $l \succ \mathsf{map}(x.\,F(x), t)$ by clause 2. Using the precedence $\mathsf{map} \rhd \mathsf{cons}$ we conclude $l \succ r'$ by clause 1.
3. Horpo cannot be used to show termination of the rewrite rule
$$f(a) \to g(x.\,a)$$
with $f : A \to A$ and $g : (A \to A) \to A$ because there is no subterm in the left-hand side to deal with the subterm $x.\,a$ of functional type.

In the remainder of this section we show that the condition $l \succ r' \twoheadrightarrow_\beta r$ for every rewrite rule $l \to r$ indeed guarantees termination of the HRS. We make use of the notion of computability with respect to $\succ \cup \to_\beta$. The following lemma follows immediately from the definition of $\succ$.

Lemma 2. *If $s \succ t$ for preterms s and t then $C[s] \succ C[t]$.*

The proof of the following lemma makes use of induction on a triple as in [6].

Lemma 3. *If $s_1, \ldots, s_m$ are computable, then $f(s_1, \ldots, s_m)$ is computable.*

Proof. The proof proceeds by induction on triples of the form $(f, (s_1, \ldots, s_m), n)$ with f a function symbol, $s_1, \ldots, s_m$ computable preterms, and n a natural number, ordered by $(\rhd, \succ \cup \to_\beta, >)$. This ordering is also written as $>$.

Let $s_1, \ldots, s_m$ be computable preterms. Suppose that $f(s_1, \ldots, s_m) \succ t$ or $f(s_1, \ldots, s_m) \to_\beta t$. We show that t is computable. The following cases are distinguished.

1. $t = g(t_1, \ldots, t_n)$ with $f \rhd g$ and for all $i \in \{1, \ldots, n\}$: either $s \succ t_i$ or $s_j \succeq t_i$ for some j.
If $s \succ t_i$, then because $(f, (s_1, \ldots, s_m), |t|) > (f, (s_1, \ldots, s_m), |t_i|)$ we can apply the induction hypothesis and conclude that t_i is computable. If $s_j \succeq t_i$ for some j, we have that t_i is computable because computability is closed under $\succ$ and s_j is by assumption computable. So all the t_i are computable. Suppose that $g(t_1, \ldots, t_n) \succ u$ or $g(t_1, \ldots, t_n) \to_\beta u$. Because we have $(f, (s_1, \ldots, s_m), |t|) > (g, (t_1, \ldots, t_n), |u|)$, the preterm u is computable by the induction hypothesis. Hence t is computable.
2. $t = f(t_1, \ldots, t_m)$ with $(s_1, \ldots, s_m) \succ (t_1, \ldots, t_m)$ and for all $i \in \{1, \ldots, m\}$: either $s \succ t_i$ or $s_j \succeq t_i$ for some j.
We can show as in the previous case that all t_i are computable. Suppose that $f(t_1, \ldots, t_m) \succ u$ or $f(t_1, \ldots, t_m) \to_\beta u$. The preterm u is computable because $(f, (s_1, \ldots, s_m), t) > (f, (t_1, \ldots, t_m), u)$, Hence t is computable.
3. $s_i \succeq t$ for some s_i.
Because s_i is computable with respect to $\succ \cup \to_\beta$ by assumption, and computability is closed under $\succ$, also t is computable.

4. $t = t_0(t_1, \ldots, t_n)$ with for all $i \in \{1, \ldots, n\}$: either $s \succ t_i$ or $s_j \succeq t_i$ for a j.
   As before, we can show that all t_i are computable. By the definition of computability we have that t is computable.
5. $t = f(s_1, \ldots, s'_i, \ldots, s_m)$ with $s_i \to_\beta s'_i$.
   Suppose that $f(\ldots, s'_i, \ldots) \succ u$ or $f(\ldots, s'_i, \ldots) \to_\beta u$. Because we have that $(f, (s_1, \ldots, s_m), t) > (f, (s_1, \ldots, s'_i, \ldots, s_m), u)$, it follows from the induction hypothesis that u is computable. This yields that t is computable.

Lemma 4. *If σ is a computable substitution, then s^σ is computable.*

Proof. By induction on the definition of preterms using Lemmas 3 and 1.

A consequence of this lemma is that all preterms (and hence all terms) are computable. That means that there is no infinite sequence of preterms $s_0 \succ \cup \to_\beta s_1 \succ \cup \to_\beta s_2 \succ \cup \to_\beta \ldots$ where every step is either $\succ$ or $\to_\beta$. Now the aim is to use this to show that there is no infinite sequence of terms $s_0 \to s_1 \to s_2 \to \ldots$ with $\to$ the rewrite relation of a HRS satisfying the horpo criterion.

Lemma 5. *Let $l \to r$ be a rewrite rule with $l \succ d \twoheadrightarrow_\beta r$ for some preterm d. Let σ be a substitution. Then there exists a preterm u such that $l^\sigma{\downarrow_\beta} \succ u \twoheadrightarrow_\beta r^\sigma{\downarrow_\beta}$.*

Proof. The proof proceeds by induction on $|l| + |d|$. We distinguish cases according to the definition of $\succ$. Let $l = f(l_1, \ldots, l_m)$.

1. $d = g(d_1, \ldots, d_n)$ with $f \rhd g$ and for all $i \in \{1, \ldots, n\}$: either $l \succ d_i$ or $l_j \succeq d_i$ for some j.
   We have $r = g(r_1, \ldots, r_n)$ with $r_i = d_i{\downarrow_\beta}$. If $l \succ d_i$ then by the induction hypothesis a preterm u_i with $l^\sigma{\downarrow_\beta} \succ u_i \twoheadrightarrow_\beta r_i^\sigma{\downarrow_\beta}$ exists. If $l_j \succeq r_i$ then by the induction hypothesis a preterm u_i with $l_j^\sigma{\downarrow_\beta} \succeq u_i \twoheadrightarrow_\beta r_i^\sigma{\downarrow_\beta}$ exists. Hence we have $l^\sigma{\downarrow_\beta} = f(l_1^\sigma{\downarrow_\beta}, \ldots, l_m^\sigma{\downarrow_\beta}) \succ g(u_1, \ldots, u_n) \twoheadrightarrow_\beta g(r_1^\sigma{\downarrow_\beta}, \ldots, r_n^\sigma{\downarrow_\beta}) = r^\sigma{\downarrow_\beta}$.
   So take $u = g(u_1, \ldots, u_n)$.
2. $d = f(d_1, \ldots, d_m)$ with $(l_1, \ldots, l_m) \succ (d_1, \ldots, d_m)$ and for all $i \in \{1, \ldots, n\}$: either $l \succ d_i$ or $l_j \succeq d_i$ for some j.
   We have $r = f(r_1, \ldots, r_m)$ with $r_i = d_i{\downarrow_\beta}$. For both the lexicographic and the multiset extension of $\succ$ the existence of suitable preterms u_i follows from the induction hypothesis. Then $l^\sigma{\downarrow_\beta} = f(l_1^\sigma{\downarrow_\beta}, \ldots, l_m^\sigma{\downarrow_\beta}) \succ f(u_1, \ldots, u_m) \twoheadrightarrow_\beta f(r_1^\sigma{\downarrow_\beta}, \ldots, r_m^\sigma{\downarrow_\beta}) = r^\sigma{\downarrow_\beta}$. So we take $u = f(u_1, \ldots, u_m)$.
3. $l_i \succeq d$ for some l_i.
   By the induction hypothesis there is a preterm u with $l_i^\sigma{\downarrow_\beta} \succeq u \twoheadrightarrow_\beta r^\sigma{\downarrow_\beta}$.
   Hence $l^\sigma{\downarrow_\beta} \succ u \twoheadrightarrow_\beta r^\sigma{\downarrow_\beta}$.
4. $d = d_0(d_1, \ldots, d_n)$ with for all $i \in \{1, \ldots, n\}$: either $l \succ d_i$ or $l_j \succeq d_i$ for some j.
   We have $r = d_0(d_1, \ldots, d_n){\downarrow_\beta}$. By the induction hypothesis there exist u_i's such that for every i we have either $l^\sigma{\downarrow_\beta} \succ u_i \twoheadrightarrow_\beta ((d_i{\downarrow_\beta})^\sigma){\downarrow_\beta}$ or $l_j^\sigma{\downarrow_\beta} \succeq u_i \twoheadrightarrow_\beta ((d_i{\downarrow_\beta})^\sigma){\downarrow_\beta}$. We have $l^\sigma{\downarrow_\beta} = f(l_1^\sigma{\downarrow_\beta}, \ldots, l_m^\sigma{\downarrow_\beta}) \succ u_0(u_1, \ldots, u_n) \twoheadrightarrow_\beta r^\sigma{\downarrow_\beta}$ because $r^\sigma{\downarrow_\beta} = ((d_0(d_1, \ldots, d_n)){\downarrow_\beta})^\sigma{\downarrow_\beta}$ which equals the β-normal form of $((d_0{\downarrow_\beta})^\sigma){\downarrow_\beta}\ (((d_1{\downarrow_\beta})^\sigma){\downarrow_\beta}, \ldots, ((d_n{\downarrow_\beta})^\sigma){\downarrow_\beta})$.

5. $l = x(l_1, \ldots, l_m)$ and $d = x(d_1, \ldots, d_m)$ with $l_i \succ d_i$ for some i and $l_i \succeq d_i$ for all i.
   Because all l_i are η-equivalent to different bound variables, we can only have $l \succeq d$ because $l = d$. Then also $l^\sigma\downarrow_\beta \succeq d^\sigma\downarrow_\beta$.
6. $l = x.\, l_0$ and $d = x.\, d_0$ with $l_0 \succ d_0$.
   We have $r = x.\, d_0\downarrow_\beta$. By the induction hypothesis a preterm u_0 exists such that $l_0^\sigma\downarrow_\beta \succ u_0 \twoheadrightarrow_\beta d_0^\sigma\downarrow_\beta$. Hence $l^\sigma\downarrow_\beta \succ x.\, u_0 \twoheadrightarrow_\beta r^\sigma\downarrow_\beta$.

The previous lemma doesn't hold if the left-hand side of a rewrite rule is not a pattern. Consider for example the would-be rewrite rule $f(z(a)) \to f(z(b))$. We have $f(z(a)) \succ f(z(b))$ using the precedence $a \rhd b$. However using the substitution $\sigma = \{z \mapsto x.\, a\}$ we do not have $f(z(a))^\sigma\downarrow_\beta = f(a) \succ f(a) = f(z(b))^\sigma\downarrow_\beta$.

Lemma 6. *If $s \to t$ in a HRS satisfying the horpo criterion then there exists a preterm u such that $s \succ u$ and $u \twoheadrightarrow_\beta t$.*

Proof. By induction on the definition of the rewrite relation using Lemma 5.

Theorem 1. *A HRS satisfying the horpo criterion is terminating.*

Proof. Suppose that we have an infinite rewrite sequence $s_0 \to s_1 \to s_2 \to \ldots$. By Lemma 6 $s_0 \succ u_0 \twoheadrightarrow_\beta s_1 \succ u_1 \twoheadrightarrow_\beta s_2 \ldots$. This contradicts Lemma 4.

A problem in proving termination for HRSs, also discussed in [16], is that a relation that is both monotonic and closed under β-reduction is reflexive and hence not well-founded. For instance if $b > c$, then monotonicity yields $(x.\, a)\, b > (x.\, a)\, c$, and closure under β-reduction yields $a > a$. There are different ways to deal with this problem.

In [7], the starting point is horpo for AFSs (here also written as $\succ$). Then a subrelation $>$ of $\succ$ is given that is β-stable. This means that $l > r$ implies $l^\sigma\downarrow_\beta \succ r^\sigma\downarrow_\beta$. Because $\succ$ is monotonic, this yields $C[l^\sigma\downarrow_\beta] \succ C[r^\sigma\downarrow_\beta]$. Since $\succ$ is well-founded, this yields termination of the rewriting relation. The method to prove termination is hence: show that $l > r$ for every rewrite rule $l \to r$. The relation $>$ is obtained from $\succ$ by restricting the clauses dealing with application, in order to make $>$ β-stable. A consequence of this restriction is that we do not have $\mathsf{map}(x.\, F(x), \mathsf{cons}(h, t)) > F(h)$. So it seems that this ordering cannot be used to prove termination of the HRS for map.

The approach taken here is different: it is shown that if $l \succ d \twoheadrightarrow_\beta r$, then $l^\sigma\downarrow_\beta \succ u \twoheadrightarrow_\beta r^\sigma\downarrow_\beta$. This implies that $C[l^\sigma\downarrow_\beta] \succ C[u] \twoheadrightarrow_\beta C[r^\sigma\downarrow_\beta]$. Because $\succ \cup \to_\beta$ is well-founded, this yields termination of the rewriting relation.

4.2 Horpo for CRSs

A well-founded precedence $\rhd$ on the set of function symbols is assumed.

Definition 3. *We have $s \succ t$ for metaterms s and t if s and t have the same arity and one of the following clauses holds:*

1. $s = f(s_1, \ldots, s_m)$
$t = g(t_1, \ldots, t_n)$
$f \rhd g$
for all $i \in \{1, \ldots, n\}$*: either* $s \succ t_i$ *or* $s_j \succeq t_i$ *for some* j

2. $s = f(s_1, \ldots, s_m)$
$t = f(t_1, \ldots, t_m)$
$(s_1, \ldots, s_m) \succ (t_1, \ldots, t_m)$
for all $i \in \{1, \ldots, n\}$*: either* $s \succ t_i$ *or* $s_j \succeq t_i$ *for some* j

3. $s = f(s_1, \ldots, s_m)$
$s_i \succ t$ *for some* i

4. $s = f(s_1, \ldots, s_m)$
$t = t_0(t_1, \ldots, t_n)$
for all $i \in \{0, \ldots, n\}$*: either* $s \succ t_i$ *or* $s_j \succeq t_i$ *for some* j

5. $s = Z(s_1, \ldots, s_m)$
$t = Z(t_1, \ldots, t_m)$
for some i*:* $s_i \succ t_i$
for all i *:* $s_i \succeq t_i$

6. $s = s_0(s_1, \ldots, s_m)$
$t = t_0(t_1, \ldots, t_m)$
for some i*:* $s_i \succ t_i$
for all i *:* $s_i \succeq t_i$

7. $s = [x]s_0$
$t = [x]t_0$
$s_0 \succ t_0$

The method to prove termination of a CRS is similar to the one for HRSs: *a CRS is terminating if for every rewrite rule* $l \to r$ *there exists a metaterm* r' *such that* $l \succ r' \twoheadrightarrow_b r$. We call this the horpo criterion (for CRSs).

Example 2.

1. Consider the beta-reduction rule of untyped λ-calculus:

$$\mathsf{app}(\mathsf{abs}([x]Z(x)), Z') \to Z(Z').$$

We do *not* have $\mathsf{abs}([x]Z(x)) \succ [x]Z(x)$ because $\mathsf{abs}([x]Z(x))$ has arity 0 and $[x]Z(x)$ has arity 1.

2. Consider the following rewrite rule from the CRS for map:

$$\mathsf{map}([x]F(x), \mathsf{cons}(H, T)) \to \mathsf{cons}(F(H), \mathsf{map}([x]F(x), T)).$$

We take $r' = \mathsf{cons}(([x]F(x))\,(H), \mathsf{map}([x]F(x), T))$. Since $[x]F(x) \succeq [x]F(x)$ and $l \succ H$, we have $l \succ ([x]F(x))(H)$. Further, $\mathsf{map}([x]F(x), \mathsf{cons}(H, T)) \succ T$ and hence using the precedence $\mathsf{map} \rhd \mathsf{cons}$ we have $l \succ r'$ by clause 1.

3. Horpo cannot be used to show termination of the rewrite rule

$$f(a) \to g([x]a)$$

because there is no subterm in $f(a)$ to deal with $[x]a$ which has arity 1.

Now we can show that the horpo criterion indeed guarantees termination of the CRS. We use computability with respect to $\succ \cup \rightarrow_b$. The proofs and auxiliary results are similar to the ones for the HRS case. That is, we show the following:

- If σ is computable, then s^σ is computable. Therefore all metaterms without metavariables (but possibly with meta-applications) are computable. The key step is again to show that a metaterm $f(s_1, \ldots, s_m)$ is computable if $s_1, \ldots, s_m$ are computable. This is shown by induction on a triple.
- If $s \rightarrow t$ then there exists a t' such that $s \succ t'$ and $t' \twoheadrightarrow_b t$.

This is used to prove the following result.

Theorem 2. *A CRS satisfying the horpo criterion is terminating.*

5 The General Scheme

The general scheme states conditions on the right-hand side of a rewrite rule that guarantee a termination proof à la Tait and Girard to work. There occur several incarnations of the general scheme in the literature. The first one is due to Jouannaud and Okada [4]. In many later works different versions of the general scheme (depending on the form of the AFS and its typing system) are considered. For instance termination of the calculus of constructions and algebraic rewriting, proved using the general scheme, is shown in [1]. In [2] the general scheme is used to prove termination of IDTSs, which are typed higher-order rewriting systems with a CRS-like syntax. Also a HRS version is given. Here we present two versions of the general scheme: one for HRSs and one for CRSs. They are simpler than the ones in [2]. Another difference is that here we consider CRSs and not IDTSs. The general schemes used here are close to the one presented for AFSs in [6]; the main difference is in the treatment of β-reduction.

5.1 The General Scheme for HRSs

We assume a well-founded ordering $\rhd$ on the set of function symbols.

Definition 4. *Let $s = f(s_1, \ldots, s_m)$ and let X be a set of variables not occurring free in s. We have $t \in \mathsf{C}(s, X)$ for a preterm t if one of the following clauses holds:*

1. $t = g(t_1, \ldots, t_n)$
   $f \rhd g$
   $t_i \in \mathsf{C}(s, X)$ *for all* i
2. $t = f(t_1, \ldots, t_m)$
   $(s_1, \ldots, s_m) \supset (t_1, \ldots, t_m)$
   $t_i \in \mathsf{C}(s, X)$ *for all* i
3. $t = s_i$ *for some* i
4. $t \subset s_i$ *for some i, with t of base type, and all variables of t occur free in s*
5. $t = x \in X$

6. $t = t_0(t_1, \ldots, t_n)$
   $t_i \in \mathsf{C}(s, X)$ *for all* i
7. $t = x.\, t_0$
   $t_0 \in \mathsf{C}(s, X \cup \{x\})$ *(x not free in s)*

In clause 2 we use $\sqsupset$ for the multiset- or lexicographic extension of $\sqsupset$. We write $\mathsf{C}(s)$ for $\mathsf{C}(s, \emptyset)$. Note that we do not include β-reduction.

The termination method using the general scheme works as follows: *a HRS is terminating if for every rewrite rule $l \to r$ there is a preterm r' such that $r' \in \mathsf{C}(l)$ and $r' \twoheadrightarrow_\beta r$.* We call this the general scheme criterion (for HRSs).

Example 3.

1. It is not possible to show that the beta-reduction rule of untyped λ-calculus
$$\mathsf{app}(\mathsf{abs}(x.\, Z(x)), Z') \to Z(Z')$$
is terminating. The preterm $Z(x)$ is of base-type but contains a free variable (x) that is not free in the left-hand side of the rewrite rule. So clause 4 does not yield that $Z(x) \in \mathsf{C}(\mathsf{app}(\mathsf{abs}(x.\, Z(x))), X)$ for any X. Further note that the variable Z (or its η-expanded form) is not in $\mathsf{C}(l)$ because of its type.
2. Using the general scheme we can show termination of the rewrite rule
$$\mathsf{map}(x.\, F(x), \mathsf{cons}(h, t)) \to \mathsf{cons}(F(h), \mathsf{map}(x.\, F(x), t)).$$
We take $r' = \mathsf{cons}((x.\, F(x))(h), \mathsf{map}(x.\, F(x), t))$. We have $x.\, F(x) \in \mathsf{C}(l)$ by clause 3 and $h \in \mathsf{C}(l)$ by clause 4, and hence $(x.\, F(x))\, h \in \mathsf{C}(l)$ by clause 6. Further, $t \in \mathsf{C}(l)$ by clause 4 and hence $\mathsf{map}(x.\, F(x), t) \in \mathsf{C}(l)$ by clause 2. Now we conclude by clause 1, using the precedence $\mathsf{map} \rhd \mathsf{cons}$.
3. The following rewrite rule cannot be shown to be terminating using horpo:
$$f(a) \to g(x.\, a).$$
It can be shown to be terminating using the general scheme. We have $a \in \mathsf{C}(f(a), \{x\})$ by clause 3 and hence $x.\, a \in \mathsf{C}(f(a))$ by clause 7. Then, using the precedence $f \rhd g$, we have $g(x.\, a) \in \mathsf{C}(f(a))$ by clause 1.
4. It is not possible to show termination of the rewrite rule
$$f(a) \to f(b)$$
using the general scheme. Note that clause 2 cannot be applied. If b is a constructor, termination follows using the general scheme as in [2]. However, if b is not a constructor (for instance if also the rule $b \to c$ is present) this version of the general scheme cannot be used anymore either.

Now we show that the general scheme criterion indeed guarantees termination of the HRS. We use computability with respect to $\rightsquigarrow \cup \to_\beta$. Here the relation $\rightsquigarrow$ is defined as the smallest one that is closed under preterm formation, and that contains $l^\sigma \downarrow_\beta \rightsquigarrow r^\sigma$ for every rewrite rule $l \to r$. Also in [8,15] such a decomposition of the rewrite step is used to get more grip on the rewrite relation. The aim is to show that all preterms are computable with respect to $\rightsquigarrow \cup \to_\beta$. The development is similar to the one in [6] and consists of the following steps:

- If l is the left-hand side of a rewrite rule and $t \in \mathsf{C}(l)$, then $t^\sigma \in \mathsf{C}(l^\sigma\downarrow_\beta)$.
- If $s_1, \ldots, s_m$ are computable, then $f(s_1, \ldots, s_m)$ are computable. We show that t is computable for every t such that $f(s_1, \ldots, s_m) \rightsquigarrow \cup \rightarrow_\beta t$. This is done by induction on a pair. In case the reduction takes place in an s_i, we use the induction hypothesis. In case the reduction takes place at the root, we use the information that $r \in \mathsf{C}(l)$ for every rewrite rule.
- We conclude that all preterms are computable with respect to $\rightsquigarrow \cup \rightarrow_\beta$.

This is used to prove the following result.

Theorem 3. *A HRS satisfying the general scheme criterion is terminating.*

Proof. A rewrite step $s \rightarrow t$ can be decomposed as $s \rightsquigarrow u \twoheadrightarrow_\beta t$. Termination follows because all preterms are computable with respect to $\rightsquigarrow \cup \rightarrow_\beta$.

5.2 The General Scheme for CRSs

We assume a well-founded precedence $\rhd$ on the set of function symbols.

Definition 5. *Let $s = f(s_1, \ldots, s_m)$ and let X be a set of variables not occurring in s. We have $t \in \mathsf{C}(s, X)$ if one of the following clauses hold:*

1. $t = g(t_1, \ldots, t_n)$
   $f \rhd g$
   $t_i \in \mathsf{C}(s, X)$ *for all* i
2. $t = f(t_1, \ldots, t_m)$
   $(s_1, \ldots, s_m) \sqsupset (t_1, \ldots, t_m)$
   $t_i \in \mathsf{C}(s, X)$ *for all* i
3. $t = s_i$ *for some* i
4. $t \subset s_i$ *for some* i*, all variables in* t *occur in* s*, and* t *is of arity* 0
5. $t = x \in X$
6. $t = t_0(t_1, \ldots, t_n)$
   $t_i \in \mathsf{C}(s, X)$ *for all* i
7. $t = [x]t_0$
   $t_0 \in \mathsf{C}(s, X \cup \{x\})$

In clause 2 we use $\sqsupset$ for the multiset- or lexicographic extension of $\sqsupset$. Again we write $\mathsf{C}(s)$ for $\mathsf{C}(s, \emptyset)$. Note that we do not include b-reduction.

The termination method using the general scheme is as follows: *A CRS is terminating if for every rewrite rule $l \rightarrow r$ there is a metaterm r' such that $r' \in \mathsf{C}(l)$ and $r' \twoheadrightarrow_b r$.* We call this the general scheme criterion (for CRSs).

Example 4.

1. Consider the beta-reduction rule for untyped λ-calculus:

   $$\mathsf{app}(\mathsf{abs}([x]Z(x)), Z') \rightarrow Z(Z').$$

   Clause 4 does not yield that $[x]Z(x) \in \mathsf{C}(\mathsf{app}(\mathsf{abs}([x]Z(x)), Z'))$ because $[x]Z(x)$ has arity 1. Also we do not have $Z \in \mathsf{C}(\mathsf{app}(\mathsf{abs}([x]Z(x)), Z'))$. Further, clause 4 does not yield that $Z(x) \in \mathsf{C}(\mathsf{app}(\mathsf{abs}([x]Z(x)), Z'))$ because the variable x is not free in $\mathsf{app}(\mathsf{abs}([x]Z(x)), Z')$.

2. Using the general scheme we can show termination of the rewrite rule
$$\mathsf{map}([x]F(x), \mathsf{cons}(H,T)) \to \mathsf{cons}(F(H), \mathsf{map}([x]F(x),T)).$$
We take $r' = \mathsf{cons}([x]F(x)(H), \mathsf{map}([x]F(x),T))$. Then we have $r' \in \mathsf{C}(l)$.
3. Using the general scheme we can show termination of the rewrite rule
$$f(z) \to g([x]a).$$
We have $a \in \mathsf{C}(f(a), \{x\})$ and hence $[x]a \in \mathsf{C}(f(a))$. Take $r' = g([x]a)$, then $r' \in \mathsf{C}(f(a))$. This rule cannot be shown to be terminating using horpo.
4. Using the general scheme we can show termination of the rewrite rule
$$f([x]Z(x)) \to Z([y]y).$$
We take $r' = ([x]Z(x))\,([y]y)$. Then $r' \in \mathsf{C}(f([x]Z(x)))$ by clause 6 because $[x]Z(x) \in \mathsf{C}(f([x]Z(x)))$ by clause 3 and $[y]y \in \mathsf{C}(f([x]Z(x)))$ by clause 7.

It can be shown that the general scheme criterion indeed guarantees termination of a CRS as in the previous subsection. We consider computability with respect to $\rightsquigarrow \cup \to_b$ where the $\rightsquigarrow$ is defined as the smallest relation that is closed under term formation and that satisfies: $l^\sigma{\downarrow_b}\rightsquigarrow r^\sigma$ for every rewrite rule $l \to r$.

Theorem 4. *A CRS satisfying the general scheme criterion is terminating.*

5.3 Horpo and the General Scheme

In [6], horpo is made into a stronger ordering by using also the general scheme. This can be done here as well. Then the last line of the conditions in the clauses 1, 2 and 4 becomes the following: *for all i: either $s \succ t_i$ or $s_j \succeq t_i$ for some j or $t_i \in \mathsf{C}(s)$.* This stronger ordering can for instance be used to prove termination of the rewrite rule $f(a) \to f(x.\,b)$ using the precedence $a \rhd b$. For this rewrite rule, neither pure horpo nor the pure general scheme can be used to prove termination. We leave a further study of this issue to future work.

Acknowledgements. I am grateful to the anonymous referees for their remarks.

References

1. F. Barbanera, M. Fernández, and H. Geuvers. Modularity of strong normalization in the algebraic lambda-cube. *Jounal of Functional Programming*, 7(6):613–660, 1997. An earlier version appears in the Proceedings of LICS '94.
2. F. Blanqui. Termination and confluence of higher-order rewrite systems. In L. Bachmair, editor, *Proceedings of the 10th International Conference on Rewriting Techniques and Applications (RTA '99)*, number 1833 in LNCS, pages 47–62, Norwich, UK, July 2000. Springer Verlag.

3. N. Dershowitz. Orderings for term rewriting systems. *Theoretical Computer Science*, 17(3):279–301, 1982.
4. J.-P. Jouannaud and M. Okada. A computation model for executable higher-order algebraic specification languages. In *Proceedings of the 6th annual IEEE Symposium on Logic in Computer Science (LICS '91)*, pages 350–361, Amsterdam, The Netherlands, July 1991.
5. J.-P. Jouannaud and A. Rubio. A recursive path ordering for higher-order terms in η-long β-normal form. In H. Gantzinger, editor, *Proceedings of the 7th International Conference on Rewriting Techniques and Applications (RTA '96)*, number 1103 in LNCS, pages 108–122, New Brunswick, USA, July 1996. Springer Verlag.
6. J.-P. Jouannaud and A. Rubio. The higher-order recursive path ordering. In *Proceedings of the 14th annual IEEE Symposium on Logic in Computer Science (LICS '99)*, pages 402–411, Trento, Italy, July 1999.
7. J.-P. Jouannaud and A. Rubio. Higher-order recursive path orderings à la carte. `http://www.lri.fr/~{}jouannau/biblio.html`, 2000.
8. J.W. Klop. *Combinatory Reduction Systems.* Number 127 in Mathematical Centre Tracts. CWI, Amsterdam, The Netherlands, 1980. PhD Thesis.
9. J.W. Klop, V. van Oostrom, and F. van Raamsdonk. Combinatory Reduction Systems: introduction and survey. *Theoretical Computer Science*, 121:279–308, 1993. Special issue in honour of Corrado Böhm.
10. C. Loria-Saenz and J. Steinbach. Termination of combined (rewrite and lambda-calculus) systems. In M. Rusinowitch and J.-L. Rémy, editors, *Proceedings of the 3rd International Workshop on Conditional Term Rewriting Systems (CTRS '92)*, number 656 in LNCS, pages 143–147, Pont-à-Mousson, France, April 1993. Springer Verlag.
11. O. Lysne and J. Piris. A termination ordering for higher order rewrite systems. In J. Hsiang, editor, *Proceedings of the 6th International Conference on Rewriting Techniques and Applications (RTA '95)*, number 914 in LNCS, pages 26–40, Kaiserslautern, Germany, April 1995. Springer Verlag.
12. R. Mayr and T. Nipkow. Higher-order rewrite systems and their confluence. *Theoretical Computer Science*, 192:3–29, 1998.
13. D. Miller. A logic programming language with lambda-abstraction, function variables, and simple unification. *Journal of Logic and Computation*, 1(4):497–536, 1991.
14. T. Nipkow. Higher-order critical pairs. In *Proceedings of the 6th annual IEEE Symposium on Logic in Computer Science (LICS '91)*, pages 342–349, Amsterdam, The Netherlands, July 1991.
15. V. van Oostrom. Higher-order families. In H. Gantzinger, editor, *Proceedings of the 7th International Conference on Rewriting Techniques and Applications (RTA '96)*, number 1103 in LNCS, pages 392–407, New Brunswick, USA, July 1996. Springer Verlag.
16. J.C. van de Pol. *Termination of higher-order rewrite systems.* PhD thesis, Utrecht University, Utrecht, The Netherlands, December 1996.
17. F. van Raamsdonk. Higher-order rewriting. In P. Narendran and M. Rusinowitch, editors, *Proceedings of the 9th International Conference on Rewriting Techniques and Applications (RTA '99)*, number 1631 in LNCS, pages 220–239, Trento, Italy, July 1999. Springer Verlag.

Matching with Free Function Symbols – A Simple Extension of Matching?

Christophe Ringeissen

LORIA — INRIA
615, rue du Jardin Botanique
BP 101, 54602 Villers-lès-Nancy Cedex France
ringeiss@loria.fr

Abstract. Matching is a solving process which is crucial in declarative (rule-based) programming languages. In order to apply rules, one has to match the left-hand side of a rule with the term to be rewritten. In several declarative programming languages, programs involve operators that may also satisfy some structural axioms. Therefore, their evaluation mechanism must implement powerful matching algorithms working modulo equational theories. In this paper, we show the existence of an equational theory where matching is decidable (resp. finitary) but matching in presence of additional (free) operators is undecidable (resp. infinitary). The interest of this result is to easily prove the existence of a frontier between matching and matching with free operators.

1 Introduction

Solving term equations is a ubiquitous process when performing any kind of deduction. For instance, even for performing a basic rewriting step, one has to solve an equation between the left-hand side of a rule and a term in which the instantiation of variables is forbidden. This is called a match-equation. Due to the increasing interest of rewriting [7] for specifying and programming in a declarative way, it is crucial to be able to design efficient and expressive matching algorithms. Many declarative programming languages use rewriting and its extensions as operational semantics. These languages enable the programmer to write rule-based specifications/programs involving operators which, for the sake of expressivity, may satisfy some structural axioms like associativity, associativity-commutativity, distributivity, etc... Consequently, the underlying matching algorithms should be able to work modulo equational theories. Moreover, since different equational theories must be implemented in the same framework, we naturally face the problem of combining equational theories.

The combination problem for the union of theories has been thoroughly studied during the last fifteen years and this has led to several combination algorithms for unification [13,14,11,2,1] and matching [8,9,5] in the union of signature-disjoint equational theories.

In the context of combining decision algorithms or solvers for a union of theories, some interesting and famous problems remain open. For instance, the combination algorithm given in [1] for unification needs as input decision algorithms

A. Middeldorp (Ed.): RTA 2001, LNCS 2051, pp. 276–290, 2001.

for unification with additional free function symbols. Combining an equational theory with the empty theory generated by some additional free function symbols is the simplest case of combination we can consider. Currently, it remains to be shown whether there exists an equational theory for which unification with free constants is decidable (finitary) but unification with free function symbols (ie. general unification) is not. Similarly, an interesting question arises for the problems of matching with free constants and matching with free function symbols (ie. general matching): is it possible that one is decidable (finitary) but not the other? Here, we give an answer to this question: Yes, definitely. Thus, the integration of a matching algorithm into the implementation of a declarative programming language or a deduction system can be jeopardized, since we often need to consider additional free function symbols possibly introduced by the programmer or the end-user of the system. We cannot expect a universal method for this integration.

The paper is organized as follows. Section 2 introduces the basic notations and concepts of general unification and general matching. Given a regular and collapse-free equational theory E, we present in Section 3 an equational theory T_E which is a conservative extension of E. A combined matching algorithm for T_E is described in Section 4, and we show that general T_E-matching is as difficult as E-unification. Based on different instances of T_E, we exhibit in Section 5 an equational theory where matching is finitary (resp. decidable) whilst general matching becomes infinitary (resp. undecidable). Finally, we conclude in Section 6 with some final remarks on the difference between unification (with free constants) and unification with free function symbols. By lack of space, some proofs are omitted in this version but can be found in [10].

2 General Unification and General Matching

In this section, we introduce the main definitions and concepts of interest for this paper, as well as already known results about these concepts. Our notations are compatible with the usual ones [4,6].

2.1 Definitions

A *first-order (finite) signature* Σ is a (finite) set of ranked function symbols. The rank of a function symbol f is an integer called *arity* and denoted by $ar(f)$. A function symbol c of arity 0 is called a constant. The denumerable set of variables is denoted by $\mathcal{X}$. The set of Σ-terms, denoted by $\mathcal{T}(\Sigma, \mathcal{X})$ is the smallest set containing $\mathcal{X}$ and constants in Σ, such that $f(t_1, \ldots, t_n)$ is in $\mathcal{T}(\Sigma, \mathcal{X})$ whenever $f \in \Sigma$, $ar(f) = n$ and $t_i \in \mathcal{T}(\Sigma, \mathcal{X})$ for $i = 1, \ldots, n$. The terms $t_{|\omega}$, $t[u]_\omega$ and $t[\omega \hookleftarrow u]$ denote respectively the subterm of t at the position ω, the term t such that $t_{|\omega} = u$, and the term obtained from t by replacing the subterm $t_{|\omega}$ by u. The symbol of t occurring at the position ω (resp. the top symbol of t) is written $t(\omega)$ (resp. $t(\epsilon)$). A Σ-rooted term is a term whose top symbol is in Σ. The set of variables of a term t is denoted by $\mathcal{V}ar(t)$. A term is *ground* if it contains no variable. A term is *linear* if each of its variables occurs just once. A Σ-substitution σ is an endomorphism of $\mathcal{T}(\Sigma, \mathcal{X})$ denoted by $\{x_1 \mapsto t_1, \ldots, x_n \mapsto t_n\}$ if there are only finitely many variables $x_1, \ldots, x_n$ not mapped to themselves. Application of a substitution σ to a term t (resp. a substitution ϕ) is written $t\sigma$ (resp. $\phi\sigma$). A substitution σ is *idempotent* if $\sigma = \sigma\sigma$. We call *domain* of the substitution σ the set of variables $\mathcal{D}om(\sigma) = \{x | x \in \mathcal{X} \text{ and } x\sigma \neq x\}$. Substitutions are denoted by letters $\sigma, \mu, \gamma, \phi, \ldots$

Given a first-order signature Σ, and a set E of Σ-axioms (i.e. pairs of Σ-terms, denoted by $l = r$), the *equational theory* $=_E$ is the congruence closure of E under the law of substitutivity. The equational theory is *regular* if $\mathcal{V}ar(l) = \mathcal{V}ar(r)$ for all $l = r$ in E, *linear* if l, r are linear for all $l = r$ in E and *collapse-free* if there is no axiom $l = x$ in E, where l is a non-variable Σ-term and x is a variable. Despite of a slight abuse of terminology, E will be often called an equational theory. The corresponding term algebra is denoted by $\mathcal{T}(\Sigma, \mathcal{X})/=_E$. We write $t \longleftrightarrow_E t'$ if $t = u[l\sigma]_\omega$ and $t' = u[r\sigma]_\omega$ for some term u, position ω, substitution σ, and equation $l = r$ in E (or $r = l$). We assume that Σ is the signature of E, that is the signature consisting of all function symbols occurring in E. This means that there are no *free* function symbols in the signature of E. Let $\mathcal{C}$ (resp. $\mathcal{F}$) be a denumerable set of additional constants (resp. function symbols) such that $\Sigma \cap \mathcal{C} = \emptyset$ (resp. $\Sigma \cap \mathcal{F} = \emptyset$). Function symbols in $\mathcal{C}$ and $\mathcal{F}$ are *free* with respect to E. The empty theories generated respectively by $\mathcal{C}$ and $\mathcal{F}$ are denoted by $\mathfrak{C}$ and $\mathfrak{F}$. In this paper, we are interested in studying and comparing the unions of equational theories $E \cup \mathfrak{C}$ and $E \cup \mathfrak{F}$.

A substitution ϕ is an E-instance on $V \subseteq \mathcal{X}$ of a substitution σ, written $\sigma \leq_E^V \phi$ (and read as σ is more general modulo E than ϕ on V), if there exists some substitution μ such that $\forall x \in V,\ x\phi =_E x\sigma\mu$.

An E-unification problem is a conjunction of equations $\varphi = \bigwedge_{k \in K} s_k =_E^? t_k$ such that s_k, t_k are Σ-terms. A substitution σ is an E-solution of φ if $\forall k \in K,\ s_k\sigma =_E t_k\sigma$. Given an idempotent substitution $\sigma = \{x_k \mapsto t_k\}_{k \in K}$, $\hat{\sigma}$ is the E-unification problem $\bigwedge_{k \in K} x_k =_E^? t_k$ called in *solved form*.

Two E-unification problems are *equivalent* if they have the same set of E-solutions. The set of E-solutions $SU_E(\varphi)$ may be schematized in the compact form provided by the notion of complete set of E-solutions, denoted by $CSU_E(\varphi)$ and based on the subsumption ordering $\leq_E^{\mathcal{V}ar(\varphi)}$ for comparing substitutions. A *complete set of most general E-solutions* is a complete set of solutions $\mu CSU_E(\varphi)$ whose elements are incomparable with $\leq_E^{\mathcal{V}ar(\varphi)}$. The set $\mu CSU_E(\varphi)$ needs not exist in general. Unification problems can be classified according to the existence and the cardinality of $\mu CSU_E(\varphi)$. The E-unification problem φ is (of type) nullary if $\mu CSU_E(\varphi)$ does not exist. The E-unification problem φ is (of type) unitary, (resp. finitary, infinitary) if $\mu CSU_E(\varphi)$ exists and is at most a singleton (resp. finite, infinite). A class of E-unification problems is unitary (resp. finitary) if each E-unification problem in the class is unitary (resp. finitary), and a class of E-unification problems is infinitary if there exists an E-unification problem in the class which is infinitary, and if no problem in the class is nullary. A class of E-unification problems is decidable if there exists a decision algorithm such that for each φ in this class it returns yes or no whether $CSU_E(\varphi)$ is non-empty or not. In this paper, existential variables may be introduced in unification problems, to represent new variables, also called *fresh* variables. These new variables behave like other ones, except that their solutions can be eventually omitted by using the following transformation rule (where $\tilde{x}$ stands for a sequence of variables):

$$\textbf{EQE} \quad \exists v, \tilde{x} : \varphi \wedge v =^? t \mapsto\!\!\!\to \exists \tilde{x} : \varphi \quad \text{if } v \notin \mathcal{V}ar(\varphi) \cup \mathcal{V}ar(t)$$

The set of (free) variables occurring in a unification problem φ is denoted by $\mathcal{V}ar(\varphi)$.

Definition 2.1. An *E-unification problem with free constants* is an $E \cup \mathfrak{C}$-unification problem. A *general E-unification problem* is an $E \cup \mathfrak{F}$-unification problem. An *E-matching problem* is an $E \cup \mathfrak{C}$-unification problem $\varphi = \bigwedge_{k \in K} s_k =^?_{E \cup \mathfrak{C}} t_k$ such that t_k is ground for each $k \in K$. A *general E-matching problem* is an $E \cup \mathfrak{F}$-unification problem $\varphi = \bigwedge_{k \in K} s_k =^?_{E \cup \mathfrak{F}} t_k$ such that t_k is ground for each $k \in K$. An *E-word problem* is an $E \cup \mathfrak{C}$-unification problem $\varphi = \bigwedge_{k \in K} s_k =^?_{E \cup \mathfrak{C}} t_k$ such that s_k, t_k are ground for each $k \in K$. A *general E-word problem* is an $E \cup \mathfrak{F}$-unification problem $\varphi = \bigwedge_{k \in K} s_k =^?_{E \cup \mathfrak{F}} t_k$ such that s_k, t_k are ground for each $k \in K$.
(General) E-matching problems are often written $\varphi = \bigwedge_{k \in K} s_k \leq^?_E t_k$. (General) E-word problems are often written $\varphi = \bigwedge_{k \in K} s_k \equiv^?_E t_k$.

Bürckert has shown in [3] the existence of a theory where unification does not remain decidable if we consider additional free constants. On the other hand, for regular and collapse-free theories, we know that a unification algorithm with free constants can be derived from a unification algorithm by using the combination algorithm designed for example in [14].

In the paper, we are interested in the possible difference between matching (with additional free constants) and matching with additional free symbols (not only constants).

2.2 The State of the Art

We are now ready to present the results already known about general unification and general matching modulo an equational theory.

The next proposition is a consequence of the combination algorithm known for the union of arbitrary disjoint equational theories.

Proposition 2.2. [1] For any (disjoint) equational theories E_1 and E_2, general E_1-unification and general E_2-unification are decidable iff general $E_1 \cup E_2$-unification is decidable.

Now, the problem is to find how to get a general unification algorithm starting from a unification algorithm with free constants.

Proposition 2.3. [11] If E is a regular equational theory, then an E-unification algorithm with free constants can be extended to a general E-unification algorithm.

A similar result is known for the matching problem, which is a specific form of unification problem with free constants.

Proposition 2.4. [8] If E is a regular equational theory, then an E-matching algorithm can be extended to a general E-matching algorithm.

For the matching problem, we may have a better result in the sense that decidability of matching is sufficient for some theories, and we do not need to know how to compute a complete set of solutions.

Proposition 2.5. [9] If E is a linear or (a regular and collapse-free) equational theory, then general E-matching is decidable iff E-matching is decidable.

Here are two examples of regular and collapse-free theories:

$$A = \{x + (y + z) = (x + y) + z\} \qquad DA = \begin{cases} x + (y + z) = (x + y) + z \\ x * (y + z) \ = x * y + x * z \\ (y + z) * x \ = y * x + z * x \end{cases}$$

A is linear but DA is not. A-unification is infinitary whilst there exists a finitary A-matching algorithm. In the same way, DA-unification is undecidable whilst there exists a finitary DA-matching algorithm, and so DA-matching is decidable.

Proposition 2.2 initially proved for unification does not hold anymore if unification is replaced by matching.

Proposition 2.6. There exist (disjoint) equational theories E_1, E_2 such that general E_1-matching and general E_2-matching are decidable (resp. finitary) but general $E_1 \cup E_2$-matching is undecidable (resp. infinitary).

Proof. Corollary of a counter-example given in [8]. Let us consider the decidability problem. Let E_1 be the theory DA and E_2 be the nilpotent theory $E_2 = \{x \oplus x = 0\}$. General E_1-matching is decidable since E_1 is regular and collapse-free and E_1-matching is decidable [9]. On the other hand, general E_2-matching is decidable since general E_2-unification is decidable by using well-known (basic) narrowing techniques for E_2-unification [6] as well as for E_2-constant elimination [11]. But general $E_1 \cup E_2$-matching is undecidable since DA-unification with free constants is undecidable [12] and any DA-equation $s =^?_{DA} t$ is equivalent to the general $E_1 \cup E_2$-match-equation $f(s) \oplus f(t) \leq^?_{E_1 \cup E_2} 0$, where f is an additional free function symbol.

The same kind of proof is possible if we consider the type of unification and A instead of DA. □

At this stage, a natural question arises: could we prove the existence of a theory where unification (resp. matching) with free constants is decidable/finitary whilst unification (resp. matching) with free symbols becomes undecidable/infinitary? The counter-example presented in [8] permits to prove Proposition 2.6. However, it does not give an answer to the previous question in the case of matching. As an answer to this question, we prove in this paper the existence of equational theories for which matching is finitary (resp. decidable) and general matching is not. These theories are rather simple extensions of A and DA, and have similarities with the counter-example presented in [8].

3 A Non-disjoint Union of Theories

We present an equational theory T which is of greatest relevance for proving the difference between matching and general matching. We study its combination with other equational theories E.

Definition 3.1. Let T be the equational theory defined as follows:

$$T = \begin{cases} h(i(x), i(x)) & = 0 \\ h(1, x) & = 0 \\ h(x, 1) & = 0 \\ i(h(x, y)) & = 1 \\ i(i(x)) & = 1 \\ i(0) & = 1 \\ i(1) & = 1 \end{cases}$$

Given an equational theory E such that $\Sigma_E \cap \Sigma_T = \emptyset$, T_E denotes the following equational theory:

$$T_E = T \cup E \cup \{i(f(x_1, \ldots, x_{ar(f)})) = 1\}_{f \in \Sigma_E}$$

where $x_1, \ldots, x_{ar(f)}$ are distinct variables in $\mathcal{X}$. In the following, Σ^+ denotes the union of the set of free constants $\mathcal{C}$ and the signature of T_E, whereas Σ denotes the signature of E. The set of i-terms is defined by:

$$\mathcal{IT}(\Sigma, \mathcal{X}) = \{i(t) \mid t \in T(\Sigma, \mathcal{X})\}$$

If E is represented by a confluent and terminating TRS R, then T_R denotes the TRS obtained from T_E by replacing E by R and by orienting axioms given above from the left to the right.

Assumption 1 *E is a regular and collapse-free equational theory represented by a confluent and terminating TRS R.*

Proposition 3.2. T_R is a confluent and terminating TRS.

We introduce the basic notations and technicalities to state some useful properties about the theory T_E seen as a union of theories. More precisely, we want to show how any T_E-equality between heterogeneous Σ^+-terms can be reduced to an E-equality between pure Σ-terms. To this aim, the notion of *variable abstraction* allows us to purify heterogeneous terms by replacing *alien subterms* with *abstraction variables.*

Definition 3.3. The set of abstraction variables is defined as follows:

$$\mathcal{AV} = \{\chi_{[t\downarrow_{T_R}]} \mid t \in \mathcal{T}(\Sigma^+, \mathcal{X}), t(\epsilon) \in \Sigma^+ \backslash \Sigma\}$$

where $\mathcal{AV}$ is a set of new variables disjoint from $\mathcal{X}$. Let $\mathcal{X}^+$ be $\mathcal{X} \cup \mathcal{AV}$. The mapping π, that associates to each term t the variable $\chi_{[t\downarrow_{T_R}]} \in \mathcal{AV}$, uniquely extends to a Σ-homomorphism from $\mathcal{T}(\Sigma^+, \mathcal{X})$ to $\mathcal{T}(\Sigma, \mathcal{X}^+)$ defined as follows:

- $f(t_1,\ldots,t_n)^\pi = f(t_1^\pi,\ldots,t_n^\pi)$ if $f \in \Sigma$,
- $f(t_1,\ldots,t_n)^\pi = \chi_{[f(t_1,\ldots,t_n)\downarrow_{T_R}]}$ if $f \in \Sigma^+\backslash\Sigma$,
- $x^\pi = x$ if $x \in \mathcal{X}$.

The image t^π of a term t by this Σ-homomorphism is called the *π-abstraction* of t. The set of abstraction variables in t^π is $\mathcal{AV}(t^\pi) = \mathcal{AV} \cap \mathcal{V}ar(t^\pi)$. Similarly, the *$\pi$-abstraction* φ^π of an equational formula φ is defined by replacing the terms occurring in φ by their π-abstractions. The set of abstraction variables occurring in φ^π is $\mathcal{AV}(\varphi^\pi) = \mathcal{AV} \cap \mathcal{V}ar(\varphi^\pi)$. The π-instantiation of variables in $\mathcal{AV}(\varphi^\pi)$ is the unique substitution π_φ^{-1} such that $\pi_\varphi^{-1} = \{\chi_{[t\downarrow_{T_R}]} \to t\downarrow_{T_R}\}_{\chi_{[t\downarrow_{T_R}]}\in\mathcal{AV}(\varphi^\pi)}$. An alien subterm of t is a $\Sigma^+\backslash\Sigma$-rooted term such that all superterms are Σ-rooted terms. *AlienPos*(t) denotes the set of positions of alien subterms in t.

We are now ready to state the result relating T_E-equational proofs and E-equational proofs.

Proposition 3.4. For any $t, u \in \mathcal{T}(\Sigma^+, \mathcal{X})$, we have

$$t =_{T_E} u \Leftrightarrow t^\pi =_E u^\pi.$$

Proof. The ($\Leftarrow$) direction is obvious since for any term t, we have $t =_{T_E} t^\pi\pi^{-1}_{(t=^?t)}$. Consider now the ($\Rightarrow$) direction.

- If t and u are Σ-rooted terms, $t =_{T_E} u$ implies that $t\downarrow_{T_R} = u\downarrow_{T_R}$ and so, by Lemma A.1 (Section A), we have

$$t^\pi \longrightarrow_R^* (t\downarrow_{T_R})^\pi = (u\downarrow_{T_R})^\pi \longleftarrow_R^* u^\pi$$

- If t and u are $(\Sigma^+\backslash\Sigma)$-rooted terms, then $t =_{T_E} u$ is equivalent to $t^\pi = u^\pi$ by definition of the π-abstraction.
- If t is Σ-rooted, then its normal form is a Σ-rooted term by Lemma A.1. If u is $(\Sigma^+\backslash\Sigma)$-rooted, then its normal form is a $(\Sigma^+\backslash\Sigma)$-rooted term by Lemma A.2 and Lemma A.3. If u is a variable, then u is in normal form. Therefore, t and u cannot be T_E-equal if u is $(\Sigma^+\backslash\Sigma)$-rooted or a variable. Moreover, in both cases, u^π is a variable which cannot be E-equal to the Σ-term t^π, otherwise it contradicts the fact that E is collapse-free.

Note that t^π and u^π are simply identical if t and u are $(\Sigma^+\backslash\Sigma)$-rooted terms. This result has two main consequences:

- First, an E-matching algorithm can be reused without loss of completeness for solving T_E-matching problems involving only Σ-pure terms, and more generally left-pure E-matching problems (see Section 4).
- Second, each alien subterm occurring in t is T_E-equal to an alien subterm of u since E is a regular and collapse-free equational theory. This explains why we can reuse a purification rule (**Purif**, see Section 4) borrowed from a combined matching algorithm in a disjoint union of regular and collapse-free theories [9].

4 A Combined Matching Algorithm

To solve a T_E-matching problem, the idea is to proceed in a modular way by splitting the input problem into conjunctions of two separate subproblems. The first subproblem consists of match-equations with non-ground Σ-terms, and it will be solved using an E-matching algorithm. The second subproblem containing the rest of match-equations will be handled by a rule-based matching algorithm which takes into account the axioms of T. Let us first detail the two classes of subproblems we are interested in.

Definition 4.1. A T_E-matching problem φ is a *left-pure E-matching problem* if left-hand sides of match-equations in φ are Σ-terms and for each match-equation, the right-hand side is Σ-rooted if the left-hand side is Σ-rooted. A T_E-matching problem γ is a *01-matching problem* if right-hand sides of match-equations in γ are in $\{0,1\}$.

Assumption 2 *Left-hand sides and right-hand sides of T_E-matching problems are normalized with respect to T_R.*

A rule-based algorithm for reducing T_E-matching problem into conjunctions of a left-pure E-matching problem and a 01-matching problem, called **LeftPurif**, is defined by the repeated application of rules given below, where **Purif** must be applied in all possible ways (*don't know nondeterminism*), whilst **Dec** and **Clash** are applied in one possible way (*don't care nondeterminism*).

Dec $\exists\tilde{x} : \varphi \wedge f(p_1,\ldots,p_n) \leq^? f(t_1,\ldots,t_n) \mapsto\!\!\!\!\rightarrow \exists\tilde{x} : \varphi \wedge p_1 \leq^? t_1 \wedge \cdots \wedge p_n \leq^? t_n$
if $f \in \{h, i\}$

Clash $\exists\tilde{x} : \varphi \wedge p \leq^? t \mapsto\!\!\!\!\rightarrow Fail$
if $(t(\epsilon) \in \{h,i\} \cup \mathcal{C}, p \notin \mathcal{X}, p(\epsilon) \neq t(\epsilon))$ or $(t(\epsilon) \in \Sigma, p \notin \mathcal{X}, p(\epsilon) \notin \Sigma)$

EqualC $\exists\tilde{x} : \varphi \wedge c \leq^? c \mapsto\!\!\!\!\rightarrow \exists\tilde{x} : \varphi$
if $c \in \mathcal{C}$

Purif $\exists\tilde{x} : \varphi \wedge p \leq^? t$

$\mapsto\!\!\!\!\rightarrow$

$$\bigvee_{\{(\omega,\omega') \mid \omega \in AlienPos(p),\ \omega' \in AlienPos(t)\}} \exists v, \tilde{x} : \varphi \wedge p[\omega \hookleftarrow v] \leq^? t \wedge p_{|\omega} \leq^? t_{|\omega'} \wedge v \leq^? t_{|\omega'}$$

if $t(\epsilon), p(\epsilon) \in \Sigma, AlienPos(p) \neq \emptyset$
where v is fresh.

Fig. 1. LeftPurif Rules

The correctness of **LeftPurif** is stated by the following proposition:

Proposition 4.2. **LeftPurif** applied on a T_E-matching problem φ terminates and leads to a finite set of normal forms, such that the normal forms different from $Fail$ are all the elements of

$$\bigcup_{i \in I} \{\varphi_i \wedge \gamma_i\}$$

such that

- φ_i is a left-pure E-matching problem, for $i \in I$.
- γ_i is a 01-matching problem[1], for $i \in I$.
- The union

$$\bigcup_{\sigma \in CSU_E(\varphi_i^{\pi})} \bigcup_{\phi \in CSU_{T_E}(\gamma_i \sigma \pi_{\varphi}^{-1})} \sigma\phi$$

 is a $CSU_{T_E}(\varphi)$.

Remark 4.3. A 01-matching problem remains a 01-matching problem after instantiation of its variables. Hence, the matching problem $\gamma_i \sigma \pi_{\varphi}^{-1}$ in Proposition 4.2 is a 01-matching problem.

Therefore, by using an E-matching algorithm, we are able to reduce arbitrary T_E-matching problems into T_E-matching problems where right-hand sides are in $\{0, 1\}$. These matching problems will be solved by applying repeatedly transformation rules in Figure 2 and in Figure 3.

Proposition 4.4. **Match01** applied on a 01-matching problem γ terminates and leads to a finite set of solved forms, such that the solved forms different from $Fail$ are all the elements of a $CSU_{T_E}(\gamma)$.

Proof. The proof is divided in three parts:

- Transformation rules in **Match01** are correct and complete.
  Rules **Equal**, **Diff**, **RepM** are known to be correct and complete in any equational theory. Correctness of rules **Match-*** can be directly proved by applying axioms of T_E. The completeness of remaining rules can be shown using Lemma A.6 and Lemma A.5 given in Section A.
- Solved forms are indeed normal forms with respect to the transformation rules in **Match01**.
  In [10], we prove that a transformation rule can be applied on any 01-matching problem (with T_R-normalized left-hand sides) which is not in solved form.
- The repeated application of rules in **Match01** always terminates.
  We can use as complexity measure a lexicographic combination of three noetherian orderings:

[1] excluding left-pure E-match-equations $x \leq^? t$ of φ, where $x \in \mathcal{X}, t \in \{0, 1\}$.

Equal $\exists\tilde{x} : \varphi \wedge p \leq^? t \mapsto\!\!\!\twoheadrightarrow \exists\tilde{x} : \varphi$
if $p \in \mathcal{T}(\Sigma^+)$ and $p = t$

Diff $\exists\tilde{x} : \varphi \wedge p \leq^? t \mapsto\!\!\!\twoheadrightarrow Fail$
if $p \in \mathcal{T}(\Sigma^+)$ and $p \neq t$

RepM $\exists\tilde{x} : \varphi \wedge x \leq^? t \mapsto\!\!\!\twoheadrightarrow \exists\tilde{x} : \varphi\{x \mapsto t\} \wedge x \leq^? t$
if $x \in \mathcal{V}(\varphi)$

ClashE $\exists\tilde{x} : \varphi \wedge p \leq^? t \mapsto\!\!\!\twoheadrightarrow Fail$
if $p(\epsilon) \in \Sigma$ and $t \in \{0, 1\}$

Fail0 $\exists\tilde{x} : \varphi \wedge i(u) \leq^? 0 \mapsto\!\!\!\twoheadrightarrow Fail$

Fail1 $\exists\tilde{x} : \varphi \wedge h(s,t) \leq^? 1 \mapsto\!\!\!\twoheadrightarrow Fail$

Fail-h $\exists\tilde{x} : \varphi \wedge h(s,t) \leq^? 0 \mapsto\!\!\!\twoheadrightarrow Fail$
if $s \notin \mathcal{X} \cup \mathcal{IT}(\mathcal{C}, \mathcal{X}), t \notin \mathcal{X} \cup \mathcal{IT}(\mathcal{C}, \mathcal{X})$

Fig. 2. Match01 Rules

1. The first component is $NME(\gamma)$, the number of match-equations in γ.
2. The second component is $\neg SV(\gamma) = \mathcal{V}(\gamma) \backslash SV(\gamma)$, where $SV(\gamma)$ is the set of solved variables in γ.
3. The third component is $MSL(\gamma)$, the multiset of sizes of left-hand sides of match-equations in γ.

Rules	$NME(\gamma)$	$\neg SV(\gamma)$	$MSL(\gamma)$
Equal	$\downarrow$		
Diff	$\downarrow$		
RepM	$=$	$\downarrow$	
ClashE	$\downarrow$		
Fail0	$\downarrow$		
Fail1	$\downarrow$		
Fail-h	$\downarrow$		
Match-hxy(1)	$\downarrow$		
Match-hxy(2-3)	$=$	$\downarrow=$	$\downarrow$
Match-hxi/hix(1)	$\downarrow$		
Match-hxi/hix(2-3)	$=$	$\downarrow=$	$\downarrow$
Match-hti/hit	$=$	$=$	$\downarrow$
Match-htx/hxt	$=$	$\downarrow=$	$\downarrow$
Match-hivic/hiciv	$=$	$\downarrow=$	$\downarrow$
Match-hiviv(1)	$\downarrow$		
Match-hiviv(2-3)	$=$	$=$	$\downarrow$
Match-i	$\downarrow$		

Match-hxy $\exists \tilde{x} : \varphi \wedge h(x, y) \leq^? 0 \twoheadrightarrow \begin{array}{l} \exists z, \tilde{x} : \begin{array}{l}(\ \varphi\{x \mapsto i(z), y \mapsto i(z)\} \\ \wedge\ x =^? i(z) \wedge y =^? i(z))\end{array} \\ \vee\ \exists \tilde{x} : \varphi \wedge x \leq^? 1 \\ \vee\ \exists \tilde{x} : \varphi \wedge y \leq^? 1 \end{array}$

if $x, y, z \in \mathcal{X}$
where z is fresh.

Match-hxi $\exists \tilde{x} : \varphi \wedge h(x, i(vc)) \leq^? 0 \twoheadrightarrow \begin{array}{l} \exists \tilde{x} : \varphi\{x \mapsto i(vc)\} \wedge x =^? i(vc) \\ \vee\ \exists \tilde{x} : \varphi \wedge x \leq^? 1 \\ \vee\ \exists \tilde{x} : \varphi \wedge i(vc) \leq^? 1 \end{array}$

if $x \in \mathcal{X}, vc \in \mathcal{X} \cup \mathcal{C}$

Match-hix $\exists \tilde{x} : \varphi \wedge h(i(vc), x) \leq^? 0 \twoheadrightarrow \begin{array}{l} \exists \tilde{x} : \varphi\{x \mapsto i(vc)\} \wedge x =^? i(vc) \\ \vee\ \exists \tilde{x} : \varphi \wedge x \leq^? 1 \\ \vee\ \exists \tilde{x} : \varphi \wedge i(vc) \leq^? 1 \end{array}$

if $x \in \mathcal{X}, vc \in \mathcal{X} \cup \mathcal{C}$

Match-hti $\exists \tilde{x} : \varphi \wedge h(t, i(vc)) \leq^? 0 \twoheadrightarrow \exists \tilde{x} : \varphi \wedge i(vc) \leq^? 1$
if $t \notin \mathcal{X} \cup \mathcal{IT}(\mathcal{C}, \mathcal{X})$

Match-hit $\exists \tilde{x} : \varphi \wedge h(i(vc), t) \leq^? 0 \twoheadrightarrow \exists \tilde{x} : \varphi \wedge i(vc) \leq^? 1$
if $t \notin \mathcal{X} \cup \mathcal{IT}(\mathcal{C}, \mathcal{X})$

Match-htx $\exists \tilde{x} : \varphi \wedge h(t, x) \leq^? 0 \twoheadrightarrow \exists \tilde{x} : \varphi \wedge x \leq^? 1$
if $x \in \mathcal{X}, t \notin \mathcal{X} \cup \mathcal{IT}(\mathcal{C}, \mathcal{X})$

Match-hxt $\exists \tilde{x} : \varphi \wedge h(x, t) \leq^? 0 \twoheadrightarrow \exists \tilde{x} : \varphi \wedge x \leq^? 1$
if $x \in \mathcal{X}, t \notin \mathcal{X} \cup \mathcal{IT}(\mathcal{C}, \mathcal{X})$

Match-hivic $\exists \tilde{x} : \varphi \wedge h(i(v), i(c)) \leq^? 0 \twoheadrightarrow \begin{array}{l} \exists \tilde{x} : \varphi \wedge v \leq^? c \\ \vee\ \exists \tilde{x} : \varphi \wedge i(v) \leq^? 1 \end{array}$

if $v \in \mathcal{X}, c \in \mathcal{C}$

Match-hiciv $\exists \tilde{x} : \varphi \wedge h(i(c), i(v)) \leq^? 0 \twoheadrightarrow \begin{array}{l} \exists \tilde{x} : \varphi \wedge v \leq^? c \\ \vee\ \exists \tilde{x} : \varphi \wedge i(v) \leq^? 1 \end{array}$

if $v \in \mathcal{X}, c \in \mathcal{C}$

Match-hiviv $\exists \tilde{x} : \varphi \wedge h(i(v), i(v')) \leq^? 0 \twoheadrightarrow \begin{array}{l} \exists \tilde{x} : \varphi\{v' \mapsto v\} \wedge v =^? v' \\ \vee\ \exists \tilde{x} : \varphi \wedge i(v) \leq^? 1 \\ \vee\ \exists \tilde{x} : \varphi \wedge i(v') \leq^? 1 \end{array}$

if $v, v' \in \mathcal{X}$

Match-i $\exists \tilde{x} : \varphi \wedge i(v) \leq^? 1 \twoheadrightarrow \bigvee_{f \in \Sigma^+} (\exists \tilde{z}, \tilde{x} : \begin{array}{l} \varphi\{v \mapsto f(z_1, \ldots, z_{ar(f)})\} \\ \wedge\ v =^? f(z_1, \ldots, z_{ar(f)})) \end{array}$

if $v \in \mathcal{X}$
where $\tilde{z} = z_1, \ldots, z_{ar(f)}$ are distinct fresh variables.

Fig. 3. **Match01** Rules (continued)

Proposition 4.5. Let E be a regular and collapse-free equational theory such that E can be represented by a confluent and terminating TRS. Given an E-matching algorithm, it is possible to construct a T_E-matching algorithm.

Proposition 4.6. General T_E-matching is undecidable (resp. infinitary) if E-unification is undecidable (resp. infinitary).

Proof. Any T_E-equation $s =^?_{T_E} t$ is equivalent to the general T_E-match-equation $h(i(g(s)), i(g(t))) \leq^?_{T_E} 0$, where g is an additional free unary function symbol.

5 Matching versus General Matching

5.1 A Theory with Finitary Matching and Infinitary General Matching

Consider the equational theory T_A:

$$T_A = \begin{cases} h(i(x), i(x)) = 0 \\ h(1, x) = 0 \\ h(x, 1) = 0 \\ i(h(x, y)) = 1 \\ i(i(x)) = 1 \\ i(0) = 1 \\ i(1) = 1 \\ i(x + y) = 1 \\ x + (y + z) = (x + y) + z \end{cases}$$

Theorem 5.1. *There exists an equational theory for which matching is finitary and general matching is infinitary.*

Proof. An A-matching algorithm is known. By Proposition 4.5, it follows that T_A-matching is finitary. Moreover, A-unification is infinitary. According to Proposition 4.6, general T_A-matching is infinitary.

5.2 A Theory with Decidable Matching and Undecidable General Matching

Consider the equational theory T_{DA}:

$$T_{DA} = \begin{cases} h(i(x), i(x)) = 0 \\ h(1, x) = 0 \\ h(x, 1) = 0 \\ i(h(x, y)) = 1 \\ i(i(x)) = 1 \\ i(0) = 1 \\ i(1) = 1 \\ i(x + y) = 1 \\ i(x * y) = 1 \\ x + (y + z) = (x + y) + z \\ x * (y + z) = x * y + x * z \\ (y + z) * x = y * x + z * x \end{cases}$$

Theorem 5.2. *There exists an equational theory for which matching is decidable and general matching is undecidable.*

Proof. A DA-matching algorithm is known. By Proposition 4.5, it follows that T_{DA}-matching is decidable. Moreover, DA-unification is undecidable. According to Proposition 4.6, general T_{DA}-matching is undecidable.

6 Conclusion

One major problem in the context of combining decision/solving algorithms for the union of theories is to show that a decision/solving algorithm cannot be always extended in order to allow additional free function symbols. In this direction, we point out the existence of a gap between matching and matching with additional free function symbols. This new result is obtained thanks to an equational theory T_E defined as a non-disjoint union of theories involving a given equational theory T with non-linear and non-regular axioms plus an arbitrary regular and collapse-free equational theory E, with axioms like A (Associativity) and D (Distributivity). On the one hand, we present a combined matching algorithm for T_E. This matching algorithm requires an E-matching algorithm. On the other hand, we show that general T_E-matching is undecidable (resp. infinitary) if E-unification is undecidable (resp. infinitary). Then, to end the proof of our main result, it is sufficient to remark that there exist regular and collapse-free theories E having an E-matching algorithm but for which E-unification is either infinitary or undecidable. Our result suggests that the same separation between unification without and with free function symbols should hold. In fact, for the order-sorted case where we introduce a sort s_T for T_E and a subsort s_E for E, a separation can be established. In this order-sorted framework, any unification problem of sort s_T can be reduced to an equivalent matching problem. Therefore, the result stated for matching can be applied to prove the existence of an order-sorted equational theory for which solving unification problems of a given sort becomes undecidable when free function symbols are considered [10].

References

1. F. Baader and K. U. Schulz. Unification in the union of disjoint equational theories: Combining decision procedures. *JSC*, 21(2):211–243, February 1996.
2. A. Boudet. Combining unification algorithms. *JSC*, 16(6):597–626, 1993.
3. H.-J. Bürckert. Matching — A special case of unification? *JSC*, 8(5):523–536, 1989.
4. N. Dershowitz and J.-P. Jouannaud. Rewrite Systems. In J. van Leeuwen, editor, *Handbook of Theoretical Computer Science*, chapter 6, pages 244–320. Elsevier Science Publishers B. V. (North-Holland), 1990.
5. S. Eker. Fast matching in combination of regular equational theories. In J. Meseguer, editor, *First Intl. Workshop on Rewriting Logic and its Applications*, volume 4. Electronic Notes in Theoretical Computer Science, September 1996.

6. J.-P. Jouannaud and Claude Kirchner. Solving equations in abstract algebras: a rule-based survey of unification. In Jean-Louis Lassez and G. Plotkin, editors, *Computational Logic. Essays in honor of Alan Robinson*, chapter 8, pages 257–321. The MIT press, Cambridge (MA, USA), 1991.
7. C. Kirchner and H. Kirchner, editors. *Second Intl. Workshop on Rewriting Logic and its Applications*, Electronic Notes in Theoretical Computer Science, Pont-à-Mousson (France), September 1998. Elsevier.
8. T. Nipkow. Combining matching algorithms: The regular case. *JSC*, 12:633–653, 1991.
9. Ch. Ringeissen. Combining decision algorithms for matching in the union of disjoint equational theories. *Information and Computation*, 126(2):144–160, May 1996.
10. Ch. Ringeissen. Matching with Free Function Symbols — A Simple Extension of Matching? (Full Version), 2001. Available at: `http://www.loria.fr/~ringeiss`.
11. M. Schmidt-Schauß. Unification in a Combination of Arbitrary Disjoint Equational Theories. *JSC*, 8(1 & 2):51–99, 1989.
12. P. Szabó. *Unifikationstheorie erster Ordnung*. PhD thesis, Universität Karlsruhe, 1982.
13. E. Tidèn. *First-order unification in combinations of equational theories*. PhD thesis, The Royal Institute of Technology, Stockholm, 1986.
14. K. Yelick. Unification in combinations of collapse-free regular theories. *JSC*, 3(1 & 2):153–182, April 1987.

A Equational Properties of T_E

In this appendix, one can find the lemmas needed to state the results expressed by Proposition 3.4 and Proposition 4.4.

Lemma A.1. For any Σ-rooted term t, we have

$$t^{\pi} \longrightarrow_R^* (t\downarrow_{T_R})^{\pi}$$

where $t\downarrow_{T_R}$ is a Σ-rooted term.

Proof. Assume there exists t' such that $t \longrightarrow_R t'$. Since E is collapse-free, t' is necessarily a Σ-rooted term.

If the rewrite rule from R is applied at a position greater than (or equal to) a position in $AlienPos(t)$, then $t'^{\pi} = t^{\pi}$ since a one-step rewriting of a $(\Sigma^+\backslash\Sigma)$-term with a rewrite rule of R does not change its top-symbol.

Otherwise, the same rewrite rule can be applied on t^{π} and we get $t^{\pi} \longrightarrow_R t'^{\pi}$.

Then, the result directly follows from an induction on the length of the rewrite derivation.

Lemma A.2. For any terms $s, t \in \mathcal{T}(\Sigma^+, \mathcal{X})$,

$$h(s,t)\downarrow_{T_R} = h(s\downarrow_{T_R}, t\downarrow_{T_R}) \vee h(s,t)\downarrow_{T_R} = 0$$

Proof. A rewrite step is always applied at a position different from ϵ, except if it is the last step. This last step leads to the normal form 0 if the top-symbol is h and a rule is applicable at the top position.

Lemma A.3. For any term $t \in \mathcal{T}(\Sigma^+, \mathcal{X})$,

$$i(t)\downarrow_{T_R} = i(t\downarrow_{T_R}) \vee i(t)\downarrow_{T_R} = 1$$

Proof. Similar to the proof of Lemma A.2. The normal form is 1 if the top-symbol is i and a rule is applicable at the top position.

Lemma A.4. For any term $t \in \mathcal{IT}(\mathcal{C}, \mathcal{X})$, t is in normal form wrt. T_R, and there is no other term T_E-equal to t.

Proof. Let $t \in \mathcal{IT}(\mathcal{C}, \mathcal{X})$. It is easy to check that no rule in T_R can be applied on t. Furthermore, t cannot be equal to another different term as shown next:

- First, it cannot be equal to 0 and 1 which are in normal form.
- It cannot be equal to a term with h as top-symbol, by Lemma A.2.
- It cannot be equal to another term in $\mathcal{IT}(\mathcal{C}, \mathcal{X})$, since all these terms are in normal form.
- It cannot be equal to a term in $\mathcal{IT}(\Sigma^+, \mathcal{X})\backslash\mathcal{IT}(\mathcal{C}, \mathcal{X})$, since the normal form of the latter is necessarily 1.
- It cannot be equal to a Σ-rooted term, by Proposition 3.4. Otherwise, it contradicts the fact that E is collapse-free.

Lemma A.5. For any term $u \in \mathcal{T}(\Sigma^+, \mathcal{X})\backslash\{1\}$,

$$u =_{T_E} 1 \Longleftrightarrow u \in \mathcal{IT}(\Sigma^+, \mathcal{X})\backslash\mathcal{IT}(\mathcal{C}, \mathcal{X})$$

Proof. By applying Lemma A.2, Lemma A.3, and the fact that normal forms of Σ-rooted terms are Σ-rooted, 1 can only be the normal form of a term in $\mathcal{IT}(\Sigma^+, \mathcal{X})$ (or the normal form of itself). Then, we can remark that terms in $\mathcal{IT}(\Sigma^+, \mathcal{X})\backslash\mathcal{IT}(\mathcal{C}, \mathcal{X})$ can be reduced to 1.

Lemma A.6. For any term $u \in \mathcal{T}(\Sigma^+, \mathcal{X})\backslash\{0\}$,

$$u =_{T_E} 0 \Longleftrightarrow \exists s, t : u = h(s,t) \wedge \begin{pmatrix} s, t \in \mathcal{IT}(\mathcal{C}, \mathcal{X}) \wedge s = t \\ \vee\, (\neg(s, t \in \mathcal{IT}(\mathcal{C}, \mathcal{X})) \wedge (s =_{T_E} 1 \vee t =_{T_E} 1)) \end{pmatrix}$$

Proof. By applying Lemma A.2, Lemma A.3, and the fact that normal forms of Σ-rooted terms are Σ-rooted, 0 can only be the normal form of a term u with h as top-symbol, or the normal form of 0. Since $u \neq 0$, there exist s, t such that $u = h(s,t)$, and there is necessarily a rewrite proof of the form

$$u = h(s,t) \longrightarrow^*_{T_R} h(s\downarrow_{T_R}, t\downarrow_{T_R}) \rightarrow_{T_R} 0$$

Let us consider the last rewrite step:

- If the applied rule is $h(i(x), i(x)) \rightarrow 0$, then $s\downarrow_{T_R}, t\downarrow_{T_R} \in \mathcal{IT}(\mathcal{C}, \mathcal{X})$ and $s\downarrow_{T_R} = t\downarrow_{T_R}$. By Lemma A.4, we have $s = s\downarrow_{T_R} = t\downarrow_{T_R} = t$.
- If the applied rule is $h(1, i(x)) \rightarrow 0$, then $s\downarrow_{T_R} = 1$. By Lemma A.5, we have $s \in \mathcal{IT}(\Sigma^+, \mathcal{X})\backslash\mathcal{IT}(\mathcal{C}, \mathcal{X})$.
- In a symmetric way, if the applied rule is $h(i(x), 1) \rightarrow 0$, then $t\downarrow_{T_R} = 1$. By Lemma A.5, we have $t \in \mathcal{IT}(\Sigma^+, \mathcal{X})\backslash\mathcal{IT}(\mathcal{C}, \mathcal{X})$.

Deriving Focused Calculi for Transitive Relations

Georg Struth

Institut für Informatik, Albert-Ludwigs-Universität Freiburg
Georges-Köhler-Allee, D-79110 Freiburg i. Br., Germany
struth@informatik.uni-freiburg.de

Abstract. We propose a new method for deriving focused ordered resolution calculi, exemplified by chaining calculi for transitive relations. Previously, inference rules were postulated and *a posteriori* verified in semantic completeness proofs. We derive them from the theory axioms. Completeness of our calculi then follows from correctness of this synthesis. Our method clearly separates deductive and procedural aspects: relating ordered chaining to Knuth-Bendix completion for transitive relations provides the semantic background that drives the synthesis towards its goal. This yields a more restrictive and transparent chaining calculus. The method also supports the development of approximate focused calculi and a modular approach to theory hierarchies.

1 Introduction

The integration of theory-specific knowledge and the systematic development of focused calculi are indispensable for applying logic in computer science. Focused first-order reasoning about equations, transitive relations and orderings has been successfully achieved, for instance, with ordered chaining and superposition calculi via integration of term rewriting techniques. In presence of more mathematical structure, however, the standard techniques do not sufficiently support a systematic development: Inference rules must be postulated ad hoc and a posteriori justified in rather involved semantic completeness proofs.

We therefore propose an alternative solution: a method for systematically developing focused ordered resolution calculi by deriving their inference rules. We exemplify the method by a simple case, deriving ordered chaining calculi for transitive relations from the ordered resolution calculus with the transitivity axiom. More complex examples, including derived chaining and tableau calculi for various lattices, can be found in [14].

Chaining calculi [11,6,10,5,7] are instances of theory resolution. The transitivity axiom $x < y < z \longrightarrow x < z$ is replaced by a chaining rule

$$\frac{\Gamma \longrightarrow \Delta, x < y \quad \Gamma' \longrightarrow \Delta', y < z}{\Gamma, \Gamma' \longrightarrow \Delta, \Delta', x < z}$$

that is a derived rule

A. Middeldorp (Ed.): RTA 2001, LNCS 2051, pp. 291–305, 2001.

$$\dfrac{\dfrac{\Gamma \longrightarrow \Delta, x < y \quad x < y < z \longrightarrow x < z}{\Gamma, y < z \longrightarrow \Delta, x < z} \qquad \Gamma' \longrightarrow \Delta', y < z}{\Gamma, \Gamma' \longrightarrow \Delta, \Delta', x < z}$$

of the resolution calculus internalizing the transitivity axiom. Chaining prunes the search space, in particular in its ordered variant [2,3], where inferences are constrained by syntactic orderings on terms, atoms and clauses. Completeness of ordered chaining is usually proved by adapting the model construction of equational superposition [1]. This strongly couples the deductive process and the syntactic ordering. It leads to certain unavoidable inference rules of opaque procedural content.

Our alternative proof-theoretic method is based on a different idea: Retranslate a refutational derivation in the ordered chaining or any superposition calculus to a refutational derivation in the ordered resolution calculus with theory axioms. Then a repeating pattern of non-theory clauses successively resolving with one instance of an axiom appears. Theory axioms are therefore independent: their interaction is not needed in refutational derivations. So why not using the converse of this observation and derive rule-based calculi from ordered resolution instead of postulating them? In a first step, make theory axioms independent. In a second step, search for patterns in refutational ordered resolution derivations with the independent axioms and turn them into derived inference rules internalizing the theory axioms.

It turns out that the closures of the theory axioms under ordered resolution modulo redundancy elimination have exactly the properties required for independent sets. So the first step of our method is the construction of this closure. In the second step, the inference rules are derived from the interaction of non-theory clauses with the closure, constrained by refined syntactic orderings that encode the goal-specific information about the construction. Completeness of the calculi thus reduces to correctness of their derivation. Deductive and procedural aspects are now clearly separated. In the present example, the transitivity axiom is already closed. The derived chaining rules are more restrictive and transparent than previous ones. Also the completeness proof is conceptually clear and simple. Chaining rules and their ordering constraints can be naturally motivated by Knuth-Bendix procedures for non-symmetric transitive relations [14]. In particular, a restriction of the calculus *is* such a Knuth-Bendix procedure.

Briefly, the main contributions of this text are the following:

- We propose a new method for deriving focused calculi based on ordered resolution.
- We use our method for syntactic completeness proofs for superposition calculi, in which the deductive and the procedural content are clearly separated.
- We derive a more restrictive and transparent ordered chaining calculus for transitive relations.
- We propose our method for approximating focused calculi and modular constructions with hierarchical theories.

Here, some proofs can only be sketched. All details can be found in [14,13].

The remainder of the text is organized as follows. Section 2 introduces basic definitions and the ordered resolution calculus. Section 3 introduces the syntactic orderings for the chaining calculus, section 4 the calculus itself. Section 5 contains the first step of the derivation of the chaining rules: the computation of the resolution basis. Section 6 contains the second step: the derivation of the inference rules. Section 7 discusses several specializations of the calculus and its relation to the set-of-support strategy and to Knuth-Bendix completion for non-symmetric rewriting. Section 8 contains a conclusion.

2 Preliminaries

Let $T_\Sigma(X)$ be a set of terms with signature Σ and variables in X. Let the language also contain one single binary relation symbol $<$. The set A of *atoms* then consists of expressions $t_1 < t_2$ over terms t_1, t_2. A *clause* is an expression

$$\{\!\!\{\phi_1, \ldots, \phi_m\}\!\!\} \longrightarrow \{\!\!\{\psi_1, \ldots, \psi_n\}\!\!\}.$$

Its *antecedent* $\{\!\!\{\phi_1, \ldots, \phi_m\}\!\!\}$ and *succedent* $\{\!\!\{\psi_1, \ldots, \psi_n\}\!\!\}$ are finite multisets of atoms. Antecedents are schematically denoted by Γ, succedents by Δ. Brackets will usually be omitted. The above clause represents the closed universal formula

$$(\forall x_1 \ldots x_k)(\neg\phi_1 \vee \cdots \vee \neg\phi_m \vee \psi_1 \vee \cdots \vee \psi_n).$$

A *Horn clause* contains at most one atom in its succedent.

We assume that $<$ denotes a relation satisfying the *transitivity axiom* (trans)

$$x < y, y < z \longrightarrow x < z$$

for all $x, y, z \in X$. We will however not build in *monotonicity axioms*

$$x < y \longrightarrow f(\ldots x \ldots) < f(\ldots y \ldots)$$

for functions $f \in \Sigma$, for reasons that will be explained in section 7. We write $x_1 < x_2 < x_3 < \cdots < x_{k-1} < x_n$ instead of $x_1 < x_2, x_2 < x_3, \ldots, x_{k-1} < x_n$.

Definition 1 (Ordered Resolution Calculus). *Given a well-founded ordering $\prec$ on atoms that is total on ground terms, the* ordered resolution calculus OR *consists of the following deduction inference rules.*

$$\frac{\Gamma \longrightarrow \Delta, \phi \qquad \Gamma', \psi \longrightarrow \Delta'}{\Gamma\sigma, \Gamma'\sigma \longrightarrow \Delta\sigma, \Delta'\sigma}, \qquad \text{(Ordered Resolution)}$$

$$\frac{\Gamma \longrightarrow \Delta, \phi, \psi}{\Gamma\sigma \longrightarrow \Delta\sigma, \phi\sigma}, \qquad \text{(Ordered Factoring)}$$

where σ is a most general unifier of ϕ and ψ. In the Ordered Resolution rule, $\phi\sigma$ is strictly maximal according to $\prec$ in the σ-instance of the first and maximal in that of the second premise. In the Ordered Factoring rule, $\phi\sigma$ is maximal in the σ-instance of the premise.

In all inference rules, *side formulas* are the parts of clauses denoted by capital Greek letters. Atoms occurring explicitly in the premises are called *minor formulas*, those in the conclusion *principal formulas.*

Let S be a clause set. We define the closures $\mathrm{cl}_{\models}(S)$, denoting the set of (semantic clausal) consequences of S, and $\mathrm{cl}_{\mathsf{OR}}(S)$, denoting the set of clauses derivable in OR from S. A clause C is *$\prec$-redundant* or simply redundant in S, if $C \in \mathrm{cl}_{\models}(C_1, \ldots, C_k)$ for some $S \ni C_1, \ldots, C_k \prec C$. Elimination of redundant clauses preserves the set of consequences; it is usually eagerly applied in OR. We denote this (non-monotonic) operation of OR-*closure modulo* $\prec$-redundancy elimination by $\mathrm{cl}_{\mathsf{OR}}^{mod}$. Since $\mathrm{cl}_{\models}(S) = \mathrm{cl}_{\models}(\mathrm{cl}_{\mathsf{OR}}^{mod}(S))$, this operation induces a basis transformation. It need not terminate, since resolution is only a semi-decision procedure. However, fair OR-strategies are refutationally complete: for all finite inconsistent clause sets, they derive the empty clause within finitely many steps.

Proposition 1. *If S is inconsistent, then $\mathrm{cl}_{\mathsf{OR}}^{mod}(S)$ contains the empty clause.*

When S is consistent and $\mathrm{cl}_{\mathsf{OR}}^{mod}$-closed, it satisfies—in addition to being a basis—also an independence property, since by definition, all conclusions of *primary S-inferences*, that is OR-inferences with both premises from S, are redundant.

Proposition 2 ([1])**.** *Let S be consistent and closed under $\mathrm{cl}_{\mathsf{OR}}^{mod}$. Let $S \cup T$ be inconsistent for some clause set T. Then there exists a refutational* OR*-derivation without primary S-inferences.*

Thus $\mathrm{cl}_{\mathsf{OR}}^{mod}$ transforms a consistent clause set to an independent or irredundant basis which is also irreducible by OR-inferences: a *resolution basis.* By proposition 2, resolution bases allow resolution strategies similar to set of support, which is complete in the unordered case for arbitrary consistent clause sets. In the ordered case, $\mathrm{cl}_{\mathsf{OR}}$-closure is necessary for dispensing with primary theory inferences. The computation of the resolution basis shifts part of proof-search complexity from run time to compile time in a well-defined way. It constitutes the first step of our derivation of focused ordered resolution calculi. If the procedure does not terminate, at least finite approximations of resolution bases may be used for constructing efficient incomplete calculi. Moreover, a stratified or incremental computation of resolution bases for hierarchic theories is possible.

3 Syntactic Orderings for Ordered Chaining

The computation of a resolution basis, in particular its termination, crucially depends on the appropriate syntactic ordering $\prec$ on terms, atoms and clauses. A reasonable approach is to select the ordering in accordance with well-known decision procedures. We follow [3] and use the ordering for a variant of Knuth-Bendix completion for transitive relations. Atom orderings suffice to constrain the inferences of OR. Clause orderings are needed to handle redundancy.

Let $\prec$ be a total well-founded ordering on ground terms in T_Σ. Let $\mathbb{B}$ be the two-element boolean algebra with ordering $<_{\mathbb{B}}$. Let $M = T_\Sigma \times \mathbb{B} \times \mathbb{B} \times T_\Sigma$. Let A be a set of atoms occurring in some clause $C = \Gamma \longrightarrow \Delta$. The ordering $\prec_1 \subseteq$

$M \times M$ is the lexicographic combination of $\prec$ for the first and last component of M and $<_{\mathbb{B}}$ for the others. A ground *atom measure* (for clause C) is the mapping $\mu_C : A \longrightarrow M$ defined by $\mu_C : \phi \mapsto (t_\nu(\phi), p(\phi), s(\phi), t_\mu(\phi))$ for each (ground) atom $\phi \in A$ occurring in C. Hereby $t_\nu(\phi)$ $(t_\mu(\phi))$ denotes the maximal (minimal) term with respect to $\prec$ in ϕ. $p(\phi) = 1$ $(p(\phi) = 0)$, if ϕ occurs in Γ (in Δ). $s(\phi) = 1$ $(s(\phi) = 0)$, if $\phi = s < t$ and $s \succeq t$ $(s \prec t)$. The (ground) *atom ordering* $\prec_2 \subseteq A \times A$ is defined by $\phi \prec_2 \psi$ iff $\mu_C(\phi) \prec_1 \mu_C(\psi)$ for $\phi, \psi \in A$.

Hence $\prec_2$ is embedded in $\prec_1$ via the atom measure, as shown in the diagram

$$\begin{array}{ccc} A & \xrightarrow{\mu_C} & M \\ {\scriptstyle \prec_2}\downarrow & & {\scriptstyle \prec_1}\downarrow \\ A & \xrightarrow{\mu_C} & M \end{array}$$

The ordering $\prec_1$ is total and well-founded by construction. Via the embedding, $\prec_2$ inherits these properties. Intuitively, t_ν and s yield ordering constraints turning the Positive Chaining rule defined below into a clausal extension of a critical pair computation. p guides the derivation of inference rules in section 6 and in particular makes (trans) a resolution basis in section 5. t_μ disambiguates atoms for which the other components of the atom measure are identical.

As free variables are implicitly universally quantified, the orderings $\prec$, $\prec_1$ and $\prec_2$ are lifted to the non-ground case, defining the ordering $\prec' \subseteq T_\Sigma(X) \times T_\Sigma(X)$ by $s \prec' t$ iff $s\sigma \prec t\sigma$ for all ground substitutions σ. This criterion need not be decidable. Defining $\prec_1'$ and $\prec_2'$ is then obvious. These orderings are still well-founded, but need no longer be total. The following trivial consequence of lifting determines the ordering constraints of the non-ground ordered chaining calculus.

Lemma 1. *$s \not\succeq t$ if $t\sigma \succ s\sigma$ for some ground terms $s\sigma$ and $t\sigma$.*

Atom measure and ordering are extended to clauses, measuring clauses as multisets of their atoms and using the multiset extension of the atom orderings. The clause ordering on ground clauses inherits totality and well-foundedness from the atom ordering. Again, the non-ground extension need not be total. In unambiguous situations we will denote all orderings by $\prec$.

4 The Ordered Chaining Calculus

We now define the ordered chaining calculus for transitive relations. It is a more restrictive variant of [2]. Instead of their Transitivity Resolution rules we use Transitivity Factoring rules. We do not intend to build in monotonicity axioms, thus we restrict chaining to roots of terms. We only state one of two Negative Chaining and Transitivity Factoring rules. A simple inversion of the ordering gives the respective other one.

Definition 2. *Let $\prec$ be an atom ordering. The* ordered chaining calculus for transitive relations OC *consists of the deduction rules of the ordered resolution* *calculus* OR, *its associated* $\prec$*-redundancy elimination rules*[1] *and the following rules, where* σ *is a most general unifier of* s *and* s', $s\sigma \not\preceq r\sigma$ *and, when it occurs,* $s'\sigma \not\preceq t\sigma$.

$$\frac{\Gamma \longrightarrow \Delta, r < s \qquad \Gamma' \longrightarrow \Delta', s' < t}{\Gamma\sigma, \Gamma'\sigma \longrightarrow \Delta\sigma, \Delta'\sigma, r\sigma < t\sigma}, \qquad \text{(Positive Chaining)}$$

where the σ*-instances of the minor formulas is strictly greater than the instances of the respective side formulas.*

$$\frac{\Gamma, s < t \longrightarrow \Delta \qquad \Gamma' \longrightarrow \Delta', s' < r}{\Gamma\sigma, \Gamma'\sigma, r\sigma < t\sigma \longrightarrow \Delta\sigma, \Delta'\sigma}, \qquad \text{(Negative Chaining)}$$

where the σ*-instance of the first (second) minor formulas is (strictly) greater than the* σ*-instance of the respective side formula and* $s\sigma \neq t\sigma$.

$$\frac{\Gamma \longrightarrow \Delta, s < r, s' < r'}{\Gamma\sigma, r\sigma < r'\sigma \longrightarrow \Delta\sigma, s\sigma < r'\sigma} \qquad \text{(Transitivity Factoring)}$$

where the σ*-instance of the leftmost minor formula is strictly greater than the* σ*-instance of the premise. In addition, a second Negative Chaining rule and a second Transitivity Factoring rule are defined by inversion of* $<$.

Soundness and completeness of OC are the subject of section 5 and 6. The meaning of the rules and their ordering constraints is briefly discussed section 7.

The unordered variant of OC is an instance of theory resolution and therefore more focused than mere set of support with (trans). We will derive the OC-rules as an ordered variant of theory resolution. Again, they will be more focused than mere reasoning with resolution bases. But as a first step in the completeness proof, we must derive a resolution basis from (trans) according to lemma 2. This construction is the subject of the following section.

5 Constructing the Resolution Basis

We now perform the first step of the derivation of OC. Our theory is (trans). With the orderings of section 3, we compute its resolution basis; its OR-closure modulo redundancy elimination. Here, (trans) *is* the resolution basis.

Lemma 2. *For the atom ordering* $\prec$, *(trans) is a resolution basis of the theory of the transitive relation* $<$.

Proof. Order an arbitrary ground instance $a < b < c \longrightarrow a < c$ of (trans) by the possible term orderings $\prec$. Let $A \equiv a < b$, $B \equiv b < c$ and $C \equiv a < c$. There are six different orderings $\prec$ for a, b, c. We consider them as three different cases.

(case i) If a is maximal, then $t_\nu(A) = t_\nu(C) \succ t_\nu(B)$ and $p(A) >_{\mathbb{B}} p(C)$. Hence $A \succ C \succ B$.

[1] Section 2 only defines a semantic *notion* of redundancy. Every set of inference rules implementing this notion is admitted.

(case ii) If b is maximal, then $t_\nu(A) = t_\nu(B) \succ t_\nu(C)$, $p(A) = p(B)$ and $s(B) \succ s(A)$. Hence $B \succ A \succ C$.

(case iii) If c is maximal, then $t_\nu(B) = t_\nu(C) \succ t_\nu(A)$ and $p(B) >_{\mathbb{B}} p(C)$. Hence $B \succ C \succ A$.

Thus for all atom orderings $\prec$ and ground instances of (trans), the antecedent is greater than the succedent. By definition of non-ground atom and clause orderings, the result can immediately be lifted. Thus there are no primary theory inferences of (trans). Factoring is inapplicable, since (trans) is a Horn clause. Concluding, (trans) is a resolution basis for the theory of transitive relations. □

Proposition 2 and lemma 2 immediately imply the following facts, which are essential for the arguments in the following section.

Corollary 1. *(i) For every inconsistent clause set containing (trans) there exists a refutational proof without primary theory resolution inferences.*

(ii) For every term ordering, every refutational proof using (trans) is free of primary theory resolution inferences.

(ii) strengthens (i). It holds in this particular case, because primary theory inferences are not only redundant, but a priori prohibited by the ordering constraints. Lemma 2 and corollary 1 express a trivial resolution basis computation, thereby establishing an ordered counterpart of set of support with (trans). The following section derives inference rules from the resolution basis in the spirit of theory resolution. In section 7 we will then compare the performance of OC with that of resolution basis proofs and a further intermediate system.

6 Deriving the Chaining Rules

We now derive the inference rules of OC from OR-derivations with (trans). Our main assumptions are refutational completeness of OC (theorem 1) and the fact that our ordering constraints rule out primary theory inferences of (trans) (corollary 1). The derivation proceeds in several steps. First, we show that certain permutations of inferences, which introduce inferences that violate the ordering constraints, have no critical impact on the structure of refutational proofs. Second, we show that corollary 1 can be further strengthened to the OR-rules as macro inferences consisting of two consecutive resolution inferences in which two non-theory clauses and one instance of (trans) participate. The combination of these two properties yields a partition of arbitrary refutational OR derivations into pattern corresponding to the OC-rules. Since all instances of (trans) appears within such macro inferences, they can be internalized. This generates the rules of the ordered chaining calculus. We first consider the ground case, where the constraints on inference rules are simplified.

The Negative Chaining rule, for instance, becomes

$$\frac{\Gamma \longrightarrow \Delta, a < b \qquad \Gamma', a < c \longrightarrow \Delta'}{\Gamma, \Gamma', b < c \longrightarrow \Delta, \Delta'}, \tag{1}$$

where a is the strictly maximal term in both premises and the minor formulas are maximal in the first and strictly maximal in the second clause. In general, by lemma 1, ordering constraints of the form $t \not\succ s$ are replaced by those of the form $t \prec s$. (1) can be derived in OR as

$$\dfrac{\dfrac{\Gamma \longrightarrow \Delta, \boxed{a} < b \quad \boxed{a} < b < c \longrightarrow a < c}{\Gamma, b < c \longrightarrow \Delta, \boxed{a} < c} \qquad \Gamma', \boxed{a} < c \longrightarrow \Delta'}{\Gamma, \Gamma', b < c \longrightarrow \Delta, \Delta'}$$

We put maximal terms in boxes, where they occur in minor formulas. There are two obstacles arising for the derivation of the Negative Chaining rule in (1). First, it may happen that $\Gamma', a < c \longrightarrow \Delta'$ is a second instance of (trans), $a < c < d \longrightarrow a < d$, say. We call such a situation a *secondary theory inference.* Second, it may happen that Δ contains an atom bigger than $a < c$. Then the second inference in (1) violates the ordering constraints and a refutational OR-derivation using the first inference of (1) must continue with an inference on a bigger atom. We call such a situation a *blocking inference.* Secondary theory inferences are problematic because they prevent us from making (trans) implicit. Blocking inference prevent us from deriving instances of chaining rules. We call a OR-derivation *regular*, when it contains neither primary and secondary theory inferences nor blocking inferences.

Lemma 3. *For every inconsistent clause set there exists a regular refutational derivation in* OR *(possibly violating the ordering constraints).*

Proof. We only give a sketch. A concise proof can be found in [14,13]. We proceed in three steps. We show that a refutational derivations exist without (i) secondary theory inferences, (ii) blocking inferences and (iii) both kinds of inferences.

(ad i) By induction on the size of clauses. We inspect secondary theory inferences in a refutational OR-derivation. Consider again, for instance, inference (1) with right-hand premise $a < c < d \longrightarrow a < d$ and conclusion $\Gamma, b < c < d \longrightarrow \Delta, a < d$. We replace this derivation by the inference

$$\frac{\Gamma \longrightarrow \Delta, \boxed{a} < b \qquad \boxed{a} < b < d \longrightarrow a < d}{\Gamma, b < d \longrightarrow \Delta, a < d}, \tag{2}$$

with $a < b < d \longrightarrow a < d$. The conclusions of (1) and (2) only differ at the expressions $b < c < d$ and $b < d$ in the antecedents. We argue that every refutational derivation with (1) can be replaced by a refutational derivation with (2). Every refutational derivation with (1) must contain clauses $\Gamma' \longrightarrow \Delta', b < c$ and $\Gamma'' \longrightarrow \Delta'', c < d$ eliminating the respective atoms from the conclusion. These clauses are also available in the derivation with (2). There we need only smaller instances of (trans)—as required by the ordering constraints on b, c and d—to eliminate $b < d$ from the conclusion of (2), too. By the induction hypothesis, these new instances do not introduce secondary theory inferences. Other cases with secondary theory inferences are similar.

(ad ii) By induction on the size of clauses. Consider, for instance, a derivation

$$\dfrac{\dfrac{\Gamma \longrightarrow \Delta, \boxed{a} < b, a < c' \quad \boxed{a} < b < c \longrightarrow a < c}{\Gamma, b < c \longrightarrow \Delta, a < c, \boxed{a} < c'} \qquad \Gamma'', \boxed{a} < c' \longrightarrow \Delta''}{\Gamma, \Gamma'', b < c \longrightarrow \Delta, \Delta'', a < c}$$

The second inference is blocking, since it disables a derivation of the Negative Chaining rule in (1) with the smaller clause $\Gamma', a < c \longrightarrow \Delta'$. We permute these two inferences and violate the ordering constraints. It can be shown that under this permutation we still obtain a refutational proof. Since the side formulas Γ' and Δ' are small, the change in the derivation remains local in the proof, up to some copying of proof trees and factoring inferences. This procedure is iterated bottom up on all blocking inferences in a refutational derivation, simply disregarding previous violations of the ordering constraints.

(ad iii) By induction on the size of clauses. Primary theory inferences are ruled out *ab initio* by the ordering constraints. We inspect the proof bottom up. We first transform secondary theory inferences (if they exist) up to the first blocking inference. This does not introduce new blocking inferences, by the induction hypothesis. We then transform the first blocking inference. This permutation introduces at most one secondary theory inference at the top level. We then transform all secondary theory inferences up to the second blocking inference. Whenever we copy a proof tree in the transformation, we simultaneously transform all copies. Therefore the procedure terminates after finitely many steps for each proof and yields a regular derivation. □

We are now prepared for our main theorem.

Theorem 1. *The ground ordered chaining calculus* OC *is refutationally complete: For every inconsistent ground clause set containing (trans) there exists a refutational derivation in* OC.

Proof. Consider a regular derivation to the empty clause. Such a derivation exists by lemma 3. Hence in all inferences either both premises are non-theory clauses (with respect to the theory of transitivity) or one premise is a non-theory clause and the other an instance of (trans). The former inferences are handled by Ordered Resolution and Ordered Factoring. The latter yield the inference rules of the ordered chaining calculus, as the following argument shows. Consider the ordered resolution and ordered factoring steps of non-theory clauses with (trans). By the proof of lemma 2, for every ordering $\prec$, the antecedent of (trans) is greater than the succedent. Therefore all non-theory clauses that are resolved with an instance of (trans) have their minor formula in the succedent. We put terms at which the chaining takes place in boxes.

We consider the possible ordering constraints on a ground instance $a < b < c \longrightarrow a < c$ of (trans) in interaction with non-theory clauses.

(case i) Let a be maximal in the axiom and assume that there is a clause $\Gamma \longrightarrow \Delta, a < b$ in which the atom $a < b$ is strictly maximal. So in particular a does not occur in Γ. Then a possible resolution inference is

$$\frac{\Gamma \longrightarrow \Delta, \boxed{a} < b \qquad \boxed{a} < b < c \longrightarrow a < c}{\Gamma, b < c \longrightarrow \Delta, a < c}, \tag{3}$$

In the conclusion, the maximal element a may occur more than once, but only in Δ. There are three subcases to consider.

(case i a) $a < c$ is strictly maximal in the conclusion, thus cannot occur in Δ by definition. Assume there exists a non-theory clause $\Gamma', \boxed{a} < c \longrightarrow \Delta'$ and a resolution inference

$$\frac{\Gamma', \boxed{a} < c \longrightarrow \Delta' \qquad \Gamma, b < c \longrightarrow \Delta, \boxed{a} < c}{\Gamma, \Gamma', b < c \longrightarrow \Delta, \Delta'}, \tag{4}$$

thus deriving a ground instance of the first Negative Chaining rule. In inference 4, the ordering constraints of ordered resolution requires the minor formula of the left-hand premise to be maximal, but not strictly maximal[2]. This gives the ordering constraints of the ground variant of this Negative Chaining rules.

(case i b) Under the assumption of (case i a), but the second premise is another instance $a < c < d \longrightarrow a < d$ of (trans). An inference with this instance would be a secondary theory inference contradicting the assumption of regularity.

(case i c) Under the assumption of (case i a), but the second premise is another instance $d < a, \boxed{a} < c \longrightarrow d < c$ of (trans). An inference with this instance would again be a secondary theory inference contradicting our assumption of regularity.

(case i d) a occurs in $\Delta = \bar{\Delta}, a < c'$ and $c = c'$. Then the derivation must continue with factoring as

$$\frac{\Gamma, b < c \longrightarrow \bar{\Delta}, a < c, a < c}{\Gamma, b < c \longrightarrow \bar{\Delta}, a < c}, \tag{5}$$

thus deriving a ground instance of one of the Transitivity Factoring rules.

(case i e) a occurs in $\Delta = \bar{\Delta}, a < c'$ and $c' \succ c$. Then the non-theory clause in inference 3 has the form $\Gamma \longrightarrow \bar{\Delta}, a < c', a < b$ with $a < c'$ greater than $a < c$. Then the next inference would be a blocking inference contradicting our assumption of regularity.

(case ii) Let b be maximal in the axiom. Then, by (case ii) of the proof of lemma 2, the atom $b < c$ is strictly maximal in the axiom and there is a clause $\Gamma \longrightarrow \Delta, \boxed{b} < c$ with $b \succ c$. Then a possible inference with an instance of the transitivity axiom is

$$\frac{\Gamma \longrightarrow \Delta, \boxed{b} < c \qquad a < b, \boxed{b} < c \longrightarrow a < c}{\Gamma, a < b \longrightarrow \Delta, a < c} \tag{6}$$

Since moreover $b \succ a$ an inference of a non-theory clause with the resolvent of 6 must have the form

$$\frac{\Gamma' \longrightarrow \Delta', a < \boxed{b} \qquad \Gamma, a < \boxed{b} \longrightarrow \Delta, a < c}{\Gamma, \Gamma' \longrightarrow \Delta, \Delta', a < c} \tag{7}$$

such that a ground instance of the Positive Chaining rule is derived. Also, the ordering constraints of ordered resolution require both minor formulas in the non-theory clauses to be strictly maximal, as for the Positive Chaining rule.

[2] This is because ordered factoring is only defined for succedents.

Here the ordering constraints and lemma 2 rule out that the second inference is a secondary theory inference or a blocking inference.

(case iii) Let c be maximal in the axiom. This case is analogous to (case i) and yields the second Negative Chaining rule. Assume that there is a clause $\Gamma \longrightarrow \Delta, b < c$ in which the atom $b < c$ is strictly maximal. So in particular c does not occur in Γ and Δ does not contain an atom of the form $c < d$. Then a possible resolution inference is

$$\frac{\Gamma \longrightarrow \Delta, b < \boxed{c} \qquad a < b < \boxed{c} \longrightarrow a < c}{\Gamma, a < b \longrightarrow \Delta, a < c}, \tag{8}$$

In the conclusion, the maximal term a may occur more than once, but only in Δ. There are three subcases to consider.

(case iii a) $a < c$ is strictly maximal in the conclusion and does not occur in Δ. Assume there exists a non-theory clause $\Gamma', a < \boxed{c} \longrightarrow \Delta'$ and a resolution inference

$$\frac{\Gamma', a < \boxed{c} \longrightarrow \Delta' \qquad \Gamma, a < b \longrightarrow \Delta, a < \boxed{c}}{\Gamma, \Gamma', a < c \longrightarrow \Delta, \Delta'}, \tag{9}$$

thus deriving a ground instance of the second Negative Chaining rule. In inference 9, the ordering constraints of ordered resolution again require the minor formula of the left-hand premise to be maximal, but not strictly maximal. This gives the ordering constraints of the ground variant of the second Negative Chaining rule.

(case iii b) Under the assumption of (case i a), but the second premise is another instance $d < a < \boxed{c} \longrightarrow d < c$ of (trans). An inference with this instance would be a secondary theory inference contradicting our assumption of regularity.

(case iii c) Under the assumption of (case i a), but the second premise is another instance $b < c < d \longrightarrow b < d$ of (trans). This inference would however violate the ordering constraints such that the argument of regularity need not be applied. $c < d$ and not $b < c$ is strictly maximal in that instance, since $s(c < d) = 1 >_{\mathbb{B}} 0 = s(b < c)$.

(case iii d) c occurs in $\Delta = \bar{\Delta}, a' < c$ and $a = a'$. Then the proof must continue with factoring as

$$\frac{\Gamma, a < b \longrightarrow \bar{\Delta}, a < c, a < c}{\Gamma, a < b \longrightarrow \bar{\Delta}, a < c}, \tag{10}$$

thus deriving a ground instance of the second Transitivity Factoring rule.

(case iii e) a occurs in $\Delta = \bar{\Delta}, a < c'$ and $c' \succ c$. Then the the non-theory clause in inference 8 has the form $\Gamma \longrightarrow \bar{\Delta}, a < c', a < b$ with $a < c'$ greater than $a < c$. Then the next inference would be a blocking inference contradicting our assumption of regularity, in analogy to (case i e). □

Theorem 1 can be extended to the non-ground case either by lifting or directly by considering non-ground non-theory clauses.

Theorem 2. *The ordered chaining calculus* OC *is refutationally complete.*

Proof. Revisit the cases of theorem 1. The unification constraints and the ordering constraints on atoms follow from those of the ordered resolution calculus. The ordering constraints on terms follow from lifting of ground term orderings, as expressed lemma 1.

Soundness of OC is trivial from the completeness proof, since all rules have been derived from OR with the transitivity axiom.

7 Discussion

Inspection of the proofs of theorem 1 and 2 immediately yields an ordered chaining calculus for Horn clauses, simply omitting Ordered Factoring and Transitivity Factoring. In particular, blocking inferences do not occur in Horn derivations.

The completeness results further specialize to decidability of certain subcases. To this end, the clause ordering $\prec$ must be extended to make all inference rules monotonic: their conclusions must be smaller than their maximal premises. Then closure computations lead to smaller and smaller new clauses, but like with Knuth-Bendix completion, this does not yet mean termination. $\prec$ is in general transfinite such that for each object greater than some limit ordinal, infinitely many smaller objects exist and there is no limit on the number of instances of each clause that can be taken. At least the closure computation of OC from a finite set of *ground* non-theory clauses terminates, since no new terms are generated by the computation. Therefore the term ordering $\prec$ can be finitely enumerated and the maximal clause in the input set is a finite upper bound of the clauses in the closure.

Avoidance of eager generation of fresh variables by primary theory inferences is a main advantage of OC. Our discussion of secondary theory inferences, however, shows that even a set-of-support-like strategy with the resolution basis cannot avoid an accumulation of undesired variables. This problem can be circumvented by using the resolution basis together with additional bookkeeping to avoid secondary theory inferences. But still, every resolution inference between a non-theory clause $\Gamma \longrightarrow \Delta, a < b$ and (trans) leads to intermediate results, like $\Gamma, b < x \longrightarrow \Delta, a < x$, that must be stored, whereas they are left implicit in OC. From a theoretical point of view, this increase of proof-length may seem insignificant, it also has no impact on the complexity of proof search, but practically the ordered chaining calculi seem superior to strategies directly using the resolution basis.

Our derivation method of ordered chaining calculi is a *synthetic* approach. So far, we have derived the inference rules, but not motivated the ordering constraints, which include the procedural aspects of the calculi and integrate theory-specific knowledge. In fact, these constraints find a natural explanation via concepts of rewriting and the Knuth-Bendix procedure for non-symmetric transitive relations, as specified in [14][3]. In a real synthesis situation, the se-

[3] Similar ideas appear in [2]. An approach to rewriting with quasi-orderings has first been proposed by [8].

mantic information is needed *a priori* to encode the desired properties of the inference rules via the ordering. Here, a first step in the development process would be to model the Positive Chaining rule as an extension of the critical pair computation rule of the Knuth-Bendix procedure for transitive relations. The ordering can then be further modified for other "meaningful" rules.

Consider a presentation of a transitive relation $<$ on a set of ground terms in presence of a syntactic ordering $\prec$, which is well-founded and total on ground terms, atoms and $<$-chains. The presentation can be partitioned into a decreasing part $R = (< \cap \succ)$, an increasing part $S = (< \cap \prec)$ and the reflexive part $\Delta_<$ of $<$, which is neither increasing nor decreasing. In analogy to the Church-Rosser theorem of equational rewriting one can ask for conditions to replace $(R \cup S)$-chains by elements of $R^+S^* \cup R^*S^+$, that is by chains that are monotonically decreasing from their initial and final elements. The following statement generalizes Newman's lemma to non-symmetric rewriting.

Lemma 4. *Let $(R \cup S^{-1})$ be well-founded. Then $SR \subseteq R^+S^* \cup R^*S^+$ implies $(R \cup S)^+ \subseteq R^+S^* \cup R^*S^+$.*

If this replacement property holds, then $s < t$ can be tested by first checking the reflexive part of $<$ or else spanning the R-digraph from s and the inverse S-digraph from t and searching for a common vertex. By well-foundedness of $\prec$, both digraphs are acyclic. This test is a decision procedure when both digraphs have finite out-degree.

To enforce this replacement property, the Knuth-Bendix procedure for non-symmetric transitive relations iteratively adds critical pairs in SR to an initial presentation. In the ground case, this is precisely the role of the restriction

$$\frac{\longrightarrow a < b \qquad \longrightarrow b < c}{\longrightarrow a < c}$$

of the ground Positive Chaining rule to positive atoms, where b must be strictly maximal according to the ordering constraints[4]. The measure and ordering needed for this computation are essentially the atom measure and ordering without the second component. Therefore the theory of rewriting and completion for non-symmetric transitive relations may serve as a starting point for developing OC, extending the ordering constraints to the clausal level. In the non-ground case, OC extends a variant of ordered Knuth-Bendix completion (c.f. [9] for its equational counterpart).

Non-symmetric rewriting also explains why we did not build in monotonicity, that is why chainings have been restricted to roots of terms. In equational rewriting, critical pairs arise only when one rewrite rule applies to another rule at a position that is not labeled by a variable. In case of transitive relations however, also critical pairs corresponding to certain variable positions must be considered. Then it is impossible to bound the position in an instance of a variable where the rule is applied and these critical pairs cannot be finitely represented within first-order logic. Since the ordered Positive Chaining rule generalizes critical pair

[4] Note that no ground critical pair rule appears for equational completion.

computations, monotonicity causes complications in connection with this rule. In particular, the complications depend on decidability of second-order context unification (c.f. [8]), which to our knowledge is open. But even without monotonicity, chaining inferences involving an atom $s < x$ or $x < t$ cannot completely be avoided, when Γ and Δ are non-empty (c.f. [2]).

For a more detailed discussion of these issues and a comparison with the model construction and calculi of [2] consider [14,13].

8 Conclusion and Related Work

We proposed a new method for deriving focused calculi based on ordered resolution. We used our method for syntactic completeness proofs for superposition calculi, in which the deductive and the procedural content are clearly separated. Here, we derived a more restrictive and transparent ordered chaining calculus for transitive relations; applications to various lattice calculi can be found in [14]. Our method can also be used for approximating focused calculi and modular constructions with hierarchical theories.

Focusing means integrating theory-specific deductive and procedural knowledge into inference rules. The rules are derived from an axiomatic representation via ordered resolution. This proof-theoretic approach is in contrast to previous semantic approaches, where inferences had to be guessed and *a posteriori* verified through model constructions. The derivation takes two steps. First, the theory axioms are transformed to a resolution basis, which has the property that no resolution inferences between its members must be considered in refutation derivations. If this transformation does not terminate, one can at least extract a resolution basis per hand or use a finite approximation for an incomplete calculus. Second, the interaction of this resolution basis with non-theory clauses in refutational ordered resolution derivations is considered. Again one can either extract inference rules of a complete calculus that exhaustively internalizes the theory axioms or search for sound but incomplete approximations.

In case of transitive relations our method leads to a more restrictive and transparent calculus and allows a more fine-grained consideration of the relation of ordered chaining to ordered resolution, of the ordering constraints and of the relation to rewriting techniques. It supports a concise evaluation and comparison of the proof search complexity of chaining calculi with related methods.

For unordered resolution, the set of support strategy has been developed in [15], a scenario for theory resolution in [12]. Both methods are based on semantic considerations. Ordered chaining calculi have been proposed in [2,3] and later without the opaque Transitivity Resolution rule in [4], based on additional coding. Although such coding is beyond our simple and natural approach, for which the Transitivity Factoring rule is procedurally transparent, it seems interesting to reconsider our calculus in this direction in the future. Besides our lattice calculi, a consideration of equational theories also appears very interesting for empirically demonstrating the power and applicability of our method for deriving focused theory-specific calculi based on ordered resolution.

References

1. L. Bachmair and H. Ganzinger. Rewrite-based equational theorem proving with selection and simplification. *J. Logic and Computation*, 4(3):217–247, 1994.
2. L. Bachmair and H. Ganzinger. Rewrite techniques for transitive relations. In *Ninth Annual IEEE Symposium on Logic in Computer Science*, pages 384–393. IEEE Computer Society Press, 1994.
3. L. Bachmair and H. Ganzinger. Ordered chaining calculi for first-order theories of transitive relations. *Journal of the ACM*, 45(6):1007–1049, 1998.
4. L. Bachmair and H. Ganzinger. Strict basic superposition. In *15th International Conference on Automated Deduction*, volume 1421 of *LNAI*, pages 160–174. Springer-Verlag, 1998.
5. W. Bledsoe, K. Kunen, and R. Shostak. Completeness results for inequality provers. *Artificial Intelligence*, 27:255–288, 1985.
6. W. W. Bledsoe and L. M. Hines. Variable elimination and chaining in a resolution-based prover for inequalities. In W. Bibel and R. Kowalski, editors, *5th Conference on Automated Deduction*, volume 87 of *LNCS*, pages 70–87. Springer-Verlag, 1980.
7. L. M. Hines. Completeness of a prover for dense linear logics. *J. Automated Reasoning*, 8:45–75, 1992.
8. J. Levy and J. Agustí. Bi-rewrite systems. *J. Symbolic Computation*, 22:279–314, 1996.
9. U. Martin and T. Nipkow. Ordered rewriting and confluence. In M. Stickel, editor, *10th International Conference on Automated Deduction*, volume 449 of *LNCS*, pages 366–380. Springer-Verlag, 1990.
10. M. M. Richter. Some reordering properties for inequality proof trees. In E. Börger, G. Hasenjaeger, and D. Rödding, editors, *Logic and Machines: Decision Problems and Complexity, Proc. Symposium "Rekursive Kombinatorik"*, volume 171 of *LNCS*, pages 183–197. Springer-Verlag, 1983.
11. J. R. Slagle. Automatic theorem proving with built-in theories including equality, partial ordering, and sets. *Journal of the ACM*, 19(1):120–135, 1972.
12. M. Stickel. Automated deduction by theory resolution. In A. Joshi, editor, *9th International Joint Conference on Artificial Intelligence*, pages 1181–1186. Morgan Kaufmann, 1985.
13. G. Struth. Deriving focused calculi for transitive relations (extended version). http://www.informatik.uni-freiburg.de/~struth/papers/focus.ps.gz.
14. G. Struth. *Canonical Transformations in Algebra, Universal Algebra and Logic*. PhD thesis, Institut für Informatik, Universität des Saarlandes, 1998.
15. L. Wos, G.A. Robinson, and D.F. Carson. Efficiency and completeness of the set of support strategy in theorem proving. *Journal of the ACM*, 12(4):536–541, 1965.

A Formalised First-Order Confluence Proof for the λ-Calculus Using One-Sorted Variable Names

(*Barendregt Was Right after all ... almost*)

René Vestergaard[1] and James Brotherston[2]

[1] CNRS-IML, Marseille, France, vester@iml.univ-mrs.fr*
[2] University of Edinburgh, Scotland, jjb@dcs.ed.ac.uk**

Abstract. We present the titular proof development which has been implemented in Isabelle/HOL. As a first, the proof is conducted exclusively by the primitive induction principles of the standard syntax and the considered reduction relations: the naive way, so to speak. Curiously, the Barendregt Variable Convention takes on a central technical role in the proof. We also show (i) that our presentation coincides with Curry's and Hindley's when terms are considered equal up-to α and (ii) that the confluence properties of all considered calculi are equivalent.

1 Introduction

The λ-calculus is a higher-order language: terms can be abstracted over terms. It is intended to formalise the concept of a *function*. The terms of the λ-calculus are typically generated inductively thus: $\Lambda^{\text{var}} ::= x \mid \Lambda^{\text{var}}\Lambda^{\text{var}} \mid \lambda x.\Lambda^{\text{var}}$

A λ-term, $e \in \Lambda^{\text{var}}$, is hence finite and is either a variable, an application of one term to another, or the functional abstraction (aka binding) of a variable over a term, respectively. On top of the terms, we define *reduction relations*, as we shall see shortly. Intuitively, we will also want to consider terms that only differ in the particular names used to express abstraction to be equal. However, this is a slightly tricky construction as far as the algebra of the syntax goes and we will only undertake it after mature consideration.

It is common, informal practice to take the variables to belong to a single infinite set of names, $\mathcal{VN}$, with a decidable equality relation, =, and that is indeed what we will do. Recent research [8,17] has shown that there can be formalist advantages to employing a certain amount of ingenuity on the issue of variable names. Still, we make a point of following the naive approach. In fact, the main contribution of this paper is to show that it is not only possible but also feasible and even instructive to use this, the naive set-up, for formal purposes. This is relevant both from a foundational and a practical perspective.

* Supported under EU TMR grant # ERBFMRXCT-980170: LINEAR. Work done in part while visiting LFCS, University of Edinburgh from Heriot-Watt University.
** Supported by a grant from LFCS, University of Edinburgh.

A. Middeldorp (Ed.): RTA 2001, LNCS 2051, pp. 306–321, 2001.

The latter more-so as we, as a first, give a rational reconstruction of the widely used and very helpful Barendregt Variable Convention (BVC) [1].

We stress that Λ^{var} is first-order abstract syntax (FOAS) and therefore comes equipped with a primitive (first-order) principle of *structural induction* [2]:

$$\frac{\forall x.P(x) \quad \forall e_1, e_2.P(e_1) \wedge P(e_2) \rightarrow P(e_1 e_2) \quad \forall x, e.P(e) \rightarrow P(\lambda x.e)}{\forall e.P(e)}$$

Similarly, the syntax also comes equipped with a primitive recursion principle so we can define auxiliary notions (e.g., free variables) by case-splitting.

The Issues. In the set-up of FOAS defined over one-sorted variable names ($\mathrm{FOAS}_{\mathcal{VN}}$), name-overlaps seem inevitable when computing. Traditionally, one therefore renames offending binders when appropriate. This has a two-fold negative impact: (i) the notion 'sub-term of' on which structural induction depends is typically broken,[1] and (ii) as a term can reduce in different directions, the resulting name for a given abstraction cannot be pre-determined. Consider, e.g., the following example taken from [11] — for precise definitions see Section 1.2:

$$(\lambda x.(\lambda y.\lambda x.xy)x)y \begin{array}{l} \overset{\beta^{C}}{\dashrightarrow} (\lambda y.\lambda x.xy)y \overset{\beta^{C}}{\dashrightarrow} \lambda x.xy \\ \underset{\beta^{C}}{\dashrightarrow} (\lambda x.\lambda z.zx)y \underset{\beta^{C}}{\dashrightarrow} \lambda z.zy \end{array}$$

Equational reasoning about $\mathrm{FOAS}_{\mathcal{VN}}$ can thus seemingly only be conducted up-to post-fixed "name-unification". Aside from any technical problems this might pose, the formal properties we establish require some interpretation.

The basic problems with $\mathrm{FOAS}_{\mathcal{VN}}$ has directly resulted in the inception of syntax formalisms (several of them recent) which overcome the issues by native means [4,5,6,7,8,12]. In general, they mark a conceptual and formal departure from the naive qualities of $\mathrm{FOAS}_{\mathcal{VN}}$. This is in part unfortunate because $\mathrm{FOAS}_{\mathcal{VN}}$ is the de facto standard in programming language theory where, as a result of the problems, it is customary to reason while "assuming the BVC" [1]:[2]

> "**2.1.12.** Terms that are α-[equivalent] are identified."
> "**2.1.13.** If $M_1, \ldots, M_n$ occur in a certain mathematical context, [their] bound variables are chosen to be different from the free variables."
> "**2.1.14.** Using 2.1.12/13 one can work with λ-terms the naive way."

Our Contribution. We

- show that it is possible and feasible to conduct formal equational proofs about higher-order languages by simple, first-order means
- show that this can be done over $\mathrm{FOAS}_{\mathcal{VN}}$, as done by hand

[1] Thanks to Regnier for observing that this need not happen with parallel substitution.

[2] We make reference to Barendregt because it is common practice to do so. Many other people have imposed hygiene conditions on variables.

- formally justify informal practices; in particular, the BVC [1,24]
- contribute to a much needed proof theoretical analysis of binding [8,9]
- introduce a quasi-complete range of positive and negative results about the preservation and reflection of confluence under a large class of mappings

1.1 Terminology and Conventions

We say that a term *reduces* to another if they are related by a reduction relation and we denote it by an infix arrow. The sub-term a reduction step acts upon is called the *redex* and it is said to be *contracted.* A reduction relation for which a redex remains so when occurring in any sub-term position is said to be *contextually closed.* We will distinguish *raw* and *real* calculi: inductive structures vs. the former factored by an equivalence. We use dashed respectively full-lined relational symbols for them. The first 5 of the following notions can be given by proper inductive constructions.

- The converse of a relation, $\rightarrow$, is written $(\rightarrow)^{-1}$.
- Composition is: $a \rightarrow_1 ; \rightarrow_2 c \Leftrightarrow^{\text{def}} \exists b \,.\, a \rightarrow_1 b \wedge b \rightarrow_2 c$.
- Given two reduction relations $\rightarrow_1$ and $\rightarrow_2$, we have: $\rightarrow_{1 \cup 2} =^{\text{def}} \rightarrow_1 \cup \rightarrow_2$.
- Transitive, reflexive closures: $(\rightarrow)^\star =^{\text{def}} \twoheadrightarrow =^{\text{``def''}} = \cup(\rightarrow; \twoheadrightarrow)$.
- Transitive, reflexive, and symmetric closures: $=_A =^{\text{def}} (\rightarrow_A \cup (\rightarrow_A)^{-1})^\star$.
- A relation which is functional will be written with a based arrow: $\mapsto$.
- A term reducing to two terms is called a *divergence.*
- Two diverging reduction steps, as defined above, are said to be *co-initial.*
- Two reduction steps that share their end-term are said to be *co-final.*
- A divergence is *resolvable* if there exist connecting co-final reduction steps.
- A relation has the *diamond property,* $\diamond$, if any divergence can be resolved.
- A relation, $\rightarrow$, is *confluent,* Confl, if $\diamond(\twoheadrightarrow)$.
- *Weak confluence* is transitive, reflexive resolution of any divergence.
- An abstract rewrite systems, *ARS*, is a relation on a set: $\rightarrow \subseteq A \times A$.
- *Residuals* are the descendants of terms under reduction.

1.2 Classic Presentations of the λ-Calculus

We will here review Curry's seminal formalist presentation of the λ-calculus [3]. We will also review Hindley [11] as, to the best of our knowledge, he is the first to give serious consideration to the problems with names in equational proofs. The process of term substitution epitomises the issues. The two Λ^{var}-terms: $\lambda x.y$ and $\lambda z.y$, e.g., have the same intuitive meaning. If we intend to substitute, say, the term x in for y, simple syntactic replacement would result in the intuitively different terms: $\lambda x.x$ and $\lambda z.x$. Some subtlety is therefore required.

Curry's Presentation. Curry [3] essentially defines the terms of the λ-calculus to be Λ^{var} with the proviso that variable names are ordered linearly. He defines substitution as follows — for *free variables*, FV(−), see Section 3, Figure 1:

$$y\langle x := e\rangle = \begin{cases} e & \text{if } x = y \\ y & \text{otherwise} \end{cases}$$

$$(e_1 e_2)\langle x := e\rangle = e_1\langle x := e\rangle e_2\langle x := e\rangle$$

$$(\lambda y.e')\langle x := e\rangle = \begin{cases} \lambda y.e' & \text{if } x = y \\ \lambda y.e'\langle x := e\rangle & \text{if } x \neq y \wedge (y \notin \mathrm{FV}(e) \vee x \notin \mathrm{FV}(e')) \\ \lambda z.e'\langle y := z\rangle\langle x := e\rangle & \text{o/w; } \textit{first } z \notin \{x\} \cup \mathrm{FV}(e) \cup \mathrm{FV}(e') \end{cases}$$

Curry is seminal in giving a precise definition of substitution which takes into account that scoping is static. He then defines the following reduction relations for λ which are closed contextually:

- $\lambda y.e\langle x := y\rangle \dashrightarrow_{\alpha^{\mathrm{C}}} \lambda x.e$, if $y \notin \mathrm{FV}(e)$
- $(\lambda x.e)e' \dashrightarrow_{\beta^{\mathrm{C}}} e\langle x := e'\rangle$

Unfortunately, following on from here, Curry makes no further mentioning of α in the proofs of the equational properties of the λ-calculus. Instead, all proofs are seemingly conducted implicitly on α-equivalence classes although these are not formally introduced.

Hindley's Presentation. This situation, amongst others, was rectified by Hindley [11]. In order to address α-equivalence classes explicitly, Hindley introduced a restricted α-relation which we call α^{H}. The relation is given as the contextual closure of:

- $\lambda x.e \dashrightarrow_{\alpha^{\mathrm{H}}} \lambda y.e\langle x := y\rangle$, if $x \neq y$, $y \notin \mathrm{FV}(e) \cup \mathrm{BV}(e)$, and $x \notin \mathrm{BV}(e)$

The α^{H}-relation has the nice property that the renaming clause of $-\langle - := -\rangle$ is not invoked, cf. Lemma 9. Furthermore, a number of Hindley's results conspire to establish the following property:

Lemma 1 (From Lemma 4.7, Lemma 4.8, Corollary 4.8 [11])

$$=\!=_{\alpha^{\mathrm{C}}} = \dashrightarrow\!\!\!\!\rightarrow_{\alpha^{\mathrm{C}}} = \dashrightarrow\!\!\!\!\rightarrow_{\alpha^{\mathrm{H}}} = =\!=_{\alpha^{\mathrm{H}}}$$

Notation 2 *To have an axiomatisation-independent name for α-equivalence on Λ^{var}, we will also refer to the relation of the above lemma as $\aleph$ (read: aleph).*

With this result in place, Hindley undertakes a formal study of α-equivalence classes which leads to the definition of a further β-relation, this time on α^{H}-equivalence classes:

$$\lfloor e \rfloor_{\mathrm{H}} =^{\mathrm{def}} \{e' \mid e =\!=_{\alpha^{\mathrm{H}}} e'\}$$

$$\lfloor e_1 \rfloor_{\mathrm{H}} \rightarrow_{\beta^{\mathrm{H}}} \lfloor e_2 \rfloor_{\mathrm{H}} =^{\mathrm{def}} \exists e_1' \in \lfloor e_1 \rfloor_{\mathrm{H}}, e_2' \in \lfloor e_2 \rfloor_{\mathrm{H}}. e_1' \dashrightarrow_{\beta^{\mathrm{C}}} e_2'$$

It is this relation which Hindley proves confluent albeit with no formal considerations concerning the invoked proof principles. This puts Hindley's treatment of the λ-calculus firmly apart from the present article. Interestingly, Hindley also points out that the obtained (real) confluence result implies confluence of the combined α^{C}- and β^{C}-relation. We are able to formally substantiate this remark of Hindley, cf. Theorem 16.

1.3 Related Work

There has recently been a substantial amount of work on proof principles for syntax that seemingly are more advanced than the first-order principles we use [4,5,6,7,8,10,12]. These lines of work, in particular the continuations of [6,8,22], are all very interesting but orthogonal to the work we present here. We suggest that a study of the proof-theoretical strength of these different proof principles might be informative and we leave it as potential future work.

There are a number of formalisations of β-confluence [13,17,18,22,23]. Apart from [17] which uses two-sorted variable names to distinguish bound and free variables, they all use either higher-order abstract syntax or de Bruijn indexing.

We know of no formal proof developments, be it for equational properties or otherwise, that are based on $\text{FOAS}_{\mathcal{VN}}$. That said, Schroer [21] does undertake a hand-proof of confluence of the λ-calculus which explicitly pays attention to variable names but with its 700+ pages, it is perhaps not as approachable as could be desired. Besides, no particular attention is paid to the employed proof principles and no formalisation is undertaken.

Acknowledgements. The first author wishes to thank Olivier Danvy, Jean-Yves Girard, Stefan Kahrs, Don Sannella, Randy Pollack, and in particular Joe Wells for fruitful discussions. The second author wishes to thank James Margetson, Larry Paulson, and Markus Wenzel for help and advice on using Isabelle/HOL. Finally, both authors wish to thank LFCS and the anonymous referees.

1.4 A Word on Our Proofs

The Isabelle/HOL proof development underpinning the present article was undertaken mainly by the second author in the space of roughly 9 weeks. It is available from the first author's homepage. At the time of writing, the confluence properties for our λ^{var}-calculus (Section 3) and the λ-calculus proper have been established. The Isabelle proof development closely follows the presentation we give here. There are one or two differences which are exclusively related to the use of alternative but equivalent induction principles in certain situations.

We started from scratch and learned theorem proving and Isabelle as we went along. Our proofs are mainly brute-force in that Isabelle apparently had problems overcoming the factorial blow-up in search space arising from the heavily conditioned proof goals for our conditional rewrite rules. Presently, the size of our proof scripts is in the order of 4000 lines of code.

The second author's Honours project will contain more detailed information about the proof development itself and will focus in part on the automation issue. The first author's thesis will focus more generally on first-order equational reasoning about higher-order languages.

2 Abstract Proof Techniques for Confluence

We now present the (new and well-known) abstract rewriting methods we use.

2.1 Preservation and Reflection of Confluence

Surprisingly, the results in this section seems to be new. Although they are very basic and related to the areas of rewriting modulo and refinement theory, we have not found any comprehensive overlaps.[3] In any event, the presentation is novel and instructive for the present purposes. Before proceeding, we refer the reader to Appendix A for an explanation of our diagram notation.

Definition 3 (Ground ARS Morphism) *Assume two ARS: $\to_A \subseteq A \times A$ and $\to_B \subseteq B \times B$. A mapping, $\mathcal{M} : A \longrightarrow B$, will be said to be a* ground ARS morphism[4] *from $\to_A$ to $\to_B$ if it is total and onto on points and a homomorphism from $\to_A$ to $\to_B$:*

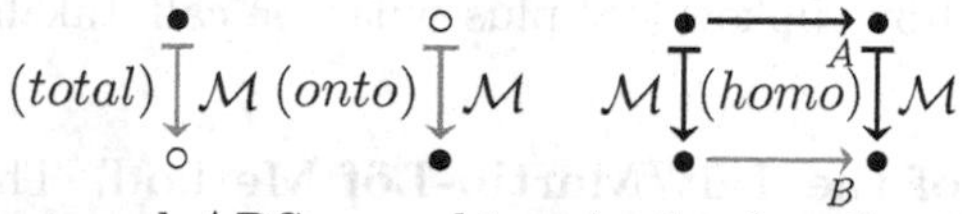

An example of a ground ARS morphism is the function that sends an object to its equivalence class relative to any equivalence relation (such as, α- or AC-equivalence): what one would call a "structural collapse". Notice that a ground ARS morphism prescribes surjectivity on objects but not on relations (and, as such, should not be called a "structural collapse" in itself). Instead, the following theorem analyses the various "degrees of relational surjectivity" relative to the confluence property.

Theorem 4 *Given a ground ARS morphism, $\mathcal{M}$, from $\to_A$ to $\to_B$, we have:*[5]

1.  $\Rightarrow \quad \diamond(\to_A) \not\Leftrightarrow \diamond(\to_B)$
2. 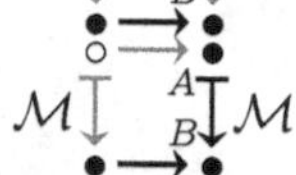 $\Rightarrow \quad \diamond(\to_A) \not\Leftrightarrow \diamond(\to_B)$
3. 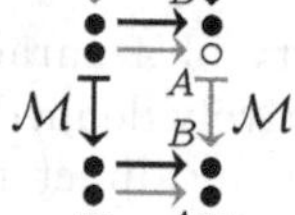 $\Rightarrow \quad \diamond(\to_A) \to \diamond(\to_B) \wedge \diamond(\to_A) \not\leftarrow \diamond(\to_B)$
4. 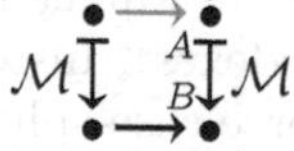 $\Rightarrow \quad \diamond(\to_A) \leftrightarrow \diamond(\to_B)$

Proof The positive results are straightforward to establish. The reflexive(!) versions of the following ARS provide counter-examples for all the negative results, left-to-right and right-to-left, respectively. Reflexivity is required to establish the ⋄ property in the first place.

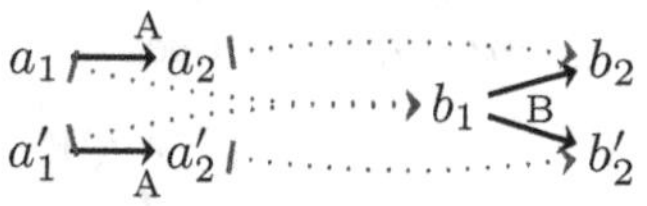

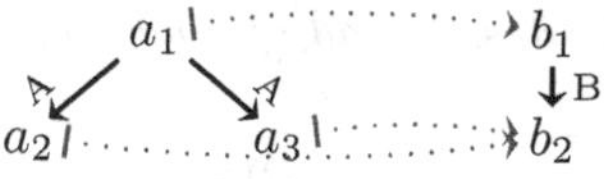

□

The asymmetry between cases 2 and 3 is due to the functionality of $\mathcal{M}$.

[3] A special case of Theorem 4, 4 is reported in [14] and we contradict a result in [19].
[4] The name is inspired from [20].
[5] In the theorem, the notation $\not\to$ ($\not\leftrightarrow$) means existence of counter-examples.

Implications. In order to preserve confluence under a "structural collapse" (i.e., a ground ARS morphism plus a premise from Theorem 4), we see from Theorem 4, cases 1 and 2 that it is insufficient to simply prove a raw diamond property which admit an initiality condition on well-formedness of raw terms. Observe that this is exactly what happens in the wider programming language community when using the BVC.

2.2 Resolving the Abstract Proof Burden of Confluence

We now sketch the abstract part of the Tait/Martin-Löf proof method for confluence as formalised by Nipkow [18] plus what we call Takahashi's Trick [24].

A Formalisation of the Tait/Martin-Löf Method. The Tait/Martin-Löf proof method uses a parallel relation that can contract any number of pre-existing redexes in one step, cf. Figure 4. The crucial step in applying the method is the following property of ARS.

Lemma 5 $(\exists \rightarrow_2 . \rightarrow_1 \subseteq \rightarrow_2 \subseteq \twoheadrightarrow_1 \ \wedge \ \diamond(\rightarrow_2)) \ \Rightarrow \ \mathrm{Confl}(\rightarrow_1)$

Proof A formalisation is provided in [18] and is re-used here. □

The point is that, since a parallel relation, $\rightarrow_2$ above, can contract an arbitrary number of redexes in parallel, only one reduction step is required to contract the unbounded copies of a particular redex that could have been created through duplication by a preceding reduction.

Takahashi's Trick. In order to prove the diamond property of a parallel β-relation, Takahashi [24] introduced the trick of using an inductively defined *complete development* relation, cf. Figure 5, rather than proceed by direct means (i.e., an involved case-splitting on the relative locations of redexes). Instead of resolving a parallel divergence "minimally" (i.e., by a brute-force case-splitting), Takahashi's idea is to go for "maximal" resolution: the term that has all pre-existing redexes contracted in one step is co-final for any parallel divergence. Abstractly, the following ARS result underpins Takahashi's idea up-to the guarding predicates which we have introduced.

Lemma 6 (Takahashi's Diamond Diagonalisation (Guarded)) *For any predicates, P and Q, and any relations, $\rightarrow_a$ and $\rightarrow_b$, we have*

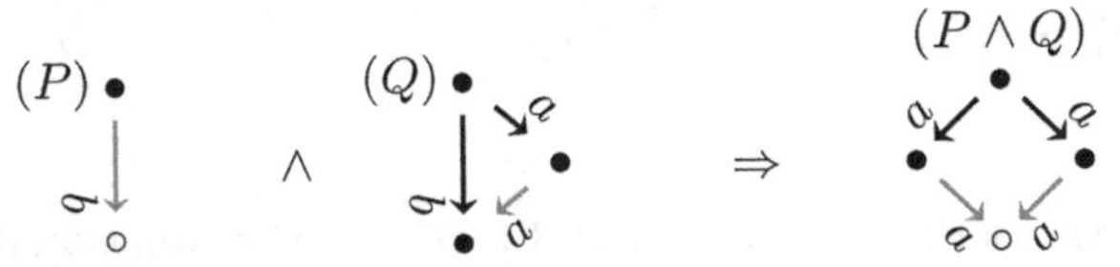

Proof Straightforward. □

The second premise is often called the triangle property when $\rightarrow_b$ is functional.

$$y[x := e] = \begin{cases} e \text{ if } x = y \\ y \text{ otherwise} \end{cases}$$
$$(e_1 e_2)[x := e] = e_1[x := e] e_2[x := e]$$
$$(\lambda y.e')[x := e] = \begin{cases} \lambda y.e'[x := e] \text{ if } x \neq y \wedge y \notin \mathrm{FV}(e) \\ \lambda y.e' \quad \text{otherwise} \end{cases}$$

$$\mathrm{FV}(y) = \{y\} \qquad \mathrm{Capt}_x(y) = \emptyset$$
$$\mathrm{FV}(e_1 e_2) = \mathrm{FV}(e_1) \cup \mathrm{FV}(e_2) \quad \mathrm{Capt}_x(e_1 e_2) = \mathrm{Capt}_x(e_1) \cup \mathrm{Capt}_x(e_2)$$
$$\mathrm{FV}(\lambda y.e) = \mathrm{FV}(e) \setminus \{y\} \qquad \mathrm{Capt}_x(\lambda y.e) = \begin{cases} \{y\} \cup \mathrm{Capt}_x(e) \text{ if } x \neq y \wedge x \in \mathrm{FV}(e) \\ \emptyset \quad \text{otherwise} \end{cases}$$

Fig. 1. Total but partially correct substitution, $-[- := -]$, free variables, $\mathrm{FV}(-)$, and variables capturing free occurrences of x, $\mathrm{Capt}_x(-)$, for Λ^{var}.

$$\frac{y \notin \mathrm{Capt}_x(e) \cup \mathrm{FV}(e)}{\lambda x.e \overset{y}{\dashrightarrow}_{i\alpha} \lambda y.e[x := y]}(\alpha) \qquad \frac{\mathrm{FV}(e_2) \cap \mathrm{Capt}_x(e_1) = \emptyset}{(\lambda x.e_1)e_2 \dashrightarrow_\beta e_1[x := e_2]}(\beta)$$

Fig. 2. Raw β- and indexed α-contraction — reduction is given by full contextual closure. By the premises no invoked substitution will result in free-variable capture.

$$\mathrm{BV}(x) = \emptyset \qquad \mathrm{UB}(x) = \mathrm{True}$$
$$\mathrm{BV}(e_1 e_2) = \mathrm{BV}(e_1) \cup \mathrm{BV}(e_2) \quad \mathrm{UB}(e_1 e_2) = \mathrm{UB}(e_1) \wedge \mathrm{UB}(e_2) \wedge \mathrm{BV}(e_1) \cap \mathrm{BV}(e_2) = \emptyset$$
$$\mathrm{BV}(\lambda x.e) = \mathrm{BV}(e) \cup \{x\} \qquad \mathrm{UB}(\lambda x.e) = \mathrm{UB}(e) \wedge x \notin \mathrm{BV}(e)$$

Fig. 3. The bound variables and the uniquely bound predicate for the terms of Λ^{var}.

3 The λ^{var}-Calculus

We will now formally define the λ^{var}-calculus and go on to show that its "structural collapse" under α is the λ-calculus proper as defined in Section 1.2.

Definition 7 (The λ^{var}-Calculus) *The terms of the λ^{var}-calculus are Λ^{var}, given on page 1.* Substitution, free variables *and* capturing variables *of raw terms are defined in Figure 1. The β- and indexed α-rewriting relations of λ^{var}: $\dashrightarrow_\beta$ and $\overset{-}{\dashrightarrow}_{i\alpha}$ are given inductively by contextual closure from Figure 2. Plain α-rewriting is given as: $e_1 \dashrightarrow_\alpha e_2 \Leftrightarrow^{\mathrm{def}} \exists z. e_1 \overset{z}{\dashrightarrow}_{i\alpha} e_2$*

The indexed α-rewriting relation will be used to conduct the ensuing proofs but is, as such, not needed for defining the λ^{var}-calculus. We stress that the (inductively defined) reduction relations also come equipped with first-order induction principles. We will typically refer to uses of these as *rule induction.* The main novelty in the above definition is the side-conditions on the contraction rules that makes binder-renaming unnecessary. The construct $\mathrm{Capt}_x(e)$ returns all the binding variables in e that have a free occurrence of x (relative to e) in their scope. It coincides with the all but forgotten notion of *not free for*. Substitution has been defined the way it has purely to enable us to prove certain "renaming sanity" properties for it, which we however will not present here.

Proposition 8 $\mathrm{FV}(e_2) \cap \mathrm{Capt}_x(e_1) = \emptyset \Rightarrow e_1[x := e_2] = e_1\langle x := e_2\rangle$

Proof By structural induction in e_1. The only non-trivial case is $e_1 \equiv \lambda y.e'_1$ which is handled by a tedious case-splitting on y. The main case is $y \neq x$ and $y \in \mathrm{FV}(e_2)$. Here, the premise of the proposition means that $y \notin \mathrm{Capt}_x(\lambda y.e')$ which immediately implies that $x \notin \mathrm{FV}(e')$ by $y \neq x$. We hence avoid $-\langle - := -\rangle$ performing a binder renaming. □

Lemma 9 $\dashrightarrow_{\alpha^{\mathrm{H}}} \subseteq \dashrightarrow_{\alpha} \subseteq (\dashrightarrow_{\alpha^{\mathrm{C}}})^{-1}$

Proof The first inclusion follows as the side-condition on $\dashrightarrow_{\alpha^{\mathrm{H}}}$ is subsumed by the side-condition on $\dashrightarrow_{\alpha}$. Any invoked substitutions thus coincide by Proposition 8 whose premise is established by the latter's side-condition. The reasoning for the second inclusion is analogous. □

Lemma 10 ($\dashrightarrow_{\alpha}$-Symmetry)

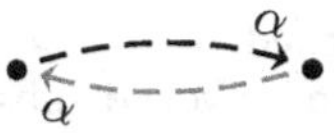

Lemma 11 $\aleph = \twoheadrightarrow_{\alpha} = =\!=_{\alpha}$

Proof From Lemmas 1 and 9 and Lemma 10, respectively. □

Lemma 12 $\dashrightarrow_{\beta} \subseteq \dashrightarrow_{\beta^{\mathrm{C}}} \subseteq \twoheadrightarrow_{\alpha}; \dashrightarrow_{\beta}$

Proof The first inclusion follows from Proposition 8. The second follows by observing that all the renamings required to perform the β^{C}-induced substitution preserve α^{C}-equivalence, i.e., $\aleph$-equivalence. By Lemma 11, they can thus be expressed by $\twoheadrightarrow_{\alpha}$. It suffices to observe that no renaming is performed following the "passing" of the substitution invoked by the β-rule. □

λ^{var} α-Collapses to the Real λ-Calculus. With these fundamental results in place, we have ensured the intuitive soundness of the following definition — which mimics Hindley's construction.

Definition 13 (The Real λ-Calculus)

- $\Lambda = \Lambda^{\mathrm{var}} / =\!=_{\alpha}$
- $\lfloor - \rfloor : \Lambda^{\mathrm{var}} \longrightarrow \Lambda$
  $e \mapsto \{e' \mid e =\!=_{\alpha} e'\}$
- $\lfloor e \rfloor \rightarrow_{\beta} \lfloor e' \rfloor \Leftrightarrow^{def} e =\!=_{\alpha}; \dashrightarrow_{\beta}; =\!=_{\alpha} e'$

Following on from the definition, we see that we have:

Proposition 14 $\lfloor e \rfloor \twoheadrightarrow_{\beta} \lfloor e' \rfloor \Leftrightarrow e\,(=\!=_{\alpha}; \dashrightarrow_{\beta}; =\!=_{\alpha})^{*}\, e' \vee e =\!=_{\alpha} e'$

$$\frac{}{x \dashrightarrow\!\!\!\!\!\!\!\!\!\;\;\mathrm{H}\;_{\beta}\; x} \qquad \frac{e \mathrel{-\!\!\mathrm{H}\!\!\rightarrow_\beta} e'}{\lambda x.e \mathrel{-\!\!\mathrm{H}\!\!\rightarrow_\beta} \lambda x.e'} \qquad \frac{e_1 \mathrel{-\!\!\mathrm{H}\!\!\rightarrow_\beta} e_1' \quad e_2 \mathrel{-\!\!\mathrm{H}\!\!\rightarrow_\beta} e_2'}{e_1 e_2 \mathrel{-\!\!\mathrm{H}\!\!\rightarrow_\beta} e_1' e_2'}$$

$$\frac{e_1 \mathrel{-\!\!\mathrm{H}\!\!\rightarrow_\beta} e_1' \quad e_2 \mathrel{-\!\!\mathrm{H}\!\!\rightarrow_\beta} e_2' \quad \mathrm{FV}(e_2') \cap \mathrm{Capt}_x(e_1') = \emptyset}{(\lambda x.e_1)e_2 \mathrel{-\!\!\mathrm{H}\!\!\rightarrow_\beta} e_1'[x := e_2']}$$

Fig. 4. The *parallel* β-relation: arbitrary, pre-existing β-redexes contracted in parallel.

Proof The left-most disjunct is the straightforward transitive version of our definition of real β. The right-most disjunct comes from the reflexive case, again by definition. □

We thus arrive at the following, rather appeasing, result.

Lemma 15 $\lfloor e \rfloor \twoheadrightarrow_\beta \lfloor e' \rfloor \Leftrightarrow e \dashrightarrow\!\!\!\!\rightarrow_{\alpha\cup\beta} e' \Leftrightarrow e \dashrightarrow\!\!\!\!\rightarrow_{\alpha^{\mathrm{C}}\cup\beta^{\mathrm{C}}} e' \Leftrightarrow \lfloor e \rfloor_{\mathrm{H}} \twoheadrightarrow_{\beta^{\mathrm{H}}} \lfloor e' \rfloor_{\mathrm{H}}$

Proof From Lemma 10, it is trivial to see that $(==_\alpha; \dashrightarrow_\beta; ==_\alpha)^\star \cup ==_\alpha = \dashrightarrow\!\!\!\!\rightarrow_{\alpha\cup\beta}$ and the first biimplication is established by Proposition 14. The second biimplication follows by Lemmas 9, 11, and 12. The last biimplication follows in an analogous manner. □

Equivalence of the Raw and the Real Calculi. The technical reason for calling the above result "appeasing" is that it allows us to prove the equational equivalence results for the raw and the real calculi we have made reference to. We consider the second result to be of particular interest.

Theorem 16

– $(\Lambda/=_\beta) = (\Lambda^{\mathrm{var}}/==_{\alpha\cup\beta}) = (\Lambda^{\mathrm{var}}/==_{\alpha^{\mathrm{C}}\cup\beta^{\mathrm{C}}}) = ((\Lambda^{\mathrm{var}}/==_{\alpha^{\mathrm{H}}})/=_{\beta^{\mathrm{H}}})$
– $\mathrm{Confl}(\rightarrow_\beta) \leftrightarrow \mathrm{Confl}(\dashrightarrow_{\alpha\cup\beta}) \leftrightarrow \mathrm{Confl}(\dashrightarrow_{\alpha^{\mathrm{C}}\cup\beta^{\mathrm{C}}}) \leftrightarrow \mathrm{Confl}(\rightarrow_{\beta^{\mathrm{H}}})$

Proof The first result is immediate following Lemma 15. As for the second result, the definitional totality and surjectivity of $\lfloor - \rfloor$ and $\lfloor - \rfloor_{\mathrm{H}}$ combined with Lemma 15 allow us to apply Theorem 4, case 4. □

Having thus formally convinced ourselves that we are about to solve the right problem, we will now present the details of the confluence proof.

4 An Equational λ^{var}-Property and λ-Confluence

As outlined in Sections 2 and 3, it suffices to find a raw relation over Λ^{var} which enjoys the diamond property in order to prove the confluence property for the λ-calculus. Taking the lead from the Tait/Martin-Löf method, this relation needs to contain a notion of parallel β-reduction.

$$\frac{}{x \dashrightarrow\!\!\!|_\beta\, x} \qquad \frac{e \dashrightarrow\!\!\!|_\beta\, e'}{\lambda x.e \dashrightarrow\!\!\!|_\beta\, \lambda x.e'} \qquad \frac{e \dashrightarrow\!\!\!|_\beta\, e'}{xe \dashrightarrow\!\!\!|_\beta\, xe'} \qquad \frac{e_1e_2 \dashrightarrow\!\!\!|_\beta\, e' \qquad e_3 \dashrightarrow\!\!\!|_\beta\, e_3'}{(e_1e_2)e_3 \dashrightarrow\!\!\!|_\beta\, e'e_3'}$$

$$\frac{e_1 \dashrightarrow\!\!\!|_\beta\, e_1' \qquad e_2 \dashrightarrow\!\!\!|_\beta\, e_2' \qquad \mathrm{FV}(e_2') \cap \mathrm{Capt}_x(e_1') = \emptyset}{(\lambda x.e_1)e_2 \dashrightarrow\!\!\!|_\beta\, e_1'[x := e_2']}$$

Fig. 5. The *complete development* β-relation: attempted contraction of all redexes.

Definition 17 Parallel *β-reduction, $-\!\!\Vdash\!\!\!\rightarrow_\beta$, is defined in Figure 4.*

The parallel β-relation admits the contraction of any number (including 0) of pre-existing β-redexes starting from within as long as no variable renaming is required. To give an impression of the level of detail of the formalisation, we can mention that the property which we need the most in the proof development is the following variable monotonicity result about the parallel β-relation:

Proposition 18 $e -\!\!\Vdash\!\!\!\rightarrow_\beta e' \Rightarrow \mathrm{FV}(e') \subseteq \mathrm{FV}(e) \wedge \mathrm{BV}(e') \subseteq \mathrm{BV}(e)$

In order to employ Takahashi's Trick, we need to ensure that any considered β-divergence can be resolved by a complete development step.

Definition 19 Complete β-development, $\dashrightarrow\!\!\!|_\beta$, *is defined in Figure 5.*

Observe, informally, that $\dashrightarrow\!\!\!|_\beta$ only is defined if all (pseudo-)redexes validate the side-condition on the β-rule. Or, more precisely, the relation is defined if it is possible to contract all (pseudo-)β-redexes starting from within — we will shortly show that this is indeed possible. For now, we merely present:

Lemma 20 $\dashrightarrow\!\!\!|_\beta \subseteq -\!\!\Vdash\!\!\!\rightarrow_\beta$

Proof Straightforward. □

The Overall Proof Structure. Having thus established the basics, we outline the proof of the diamond property of the following relation: $-\!-\!\!\twoheadrightarrow_\alpha; -\!\!\Vdash\!\!\!\rightarrow_\beta$, before supplying the actual details of the proof. The relation is inspired by the definitional reflection of the weak confluence property for the λ-calculus proper over the structural (α-)collapse of λ^{var}. In order to use the BVC in our proof, we first present it as a predicate on Λ^{var}, cf. Figure 3.

Definition 21 (Barendregt Conventional Form)

$$\mathrm{BCF}(e) = \mathrm{UB}(e) \wedge (\mathrm{BV}(e) \cap \mathrm{FV}(e) = \emptyset)$$

Lemma 22 $\diamond(-\!-\!\!\twoheadrightarrow_\alpha; -\!\!\Vdash\!\!\!\rightarrow_\beta)$

Proof
For the Ms given, we can construct the Ns in the divergence resolution on the right in order. The ensuing sections will detail the individual diagrams. The $\dashrightarrow_{\alpha_0}$-relation is introduced in Definition 25 as the fresh-naming restriction of $\dashrightarrow_{\alpha}$. It serves to facilitate the commutativity with β on either side of the diagram. We note that the result means that it suffices to address all naming issues before the combinatorially more complex β-divergence which can be addressed in isolation due to BCF-initiality.

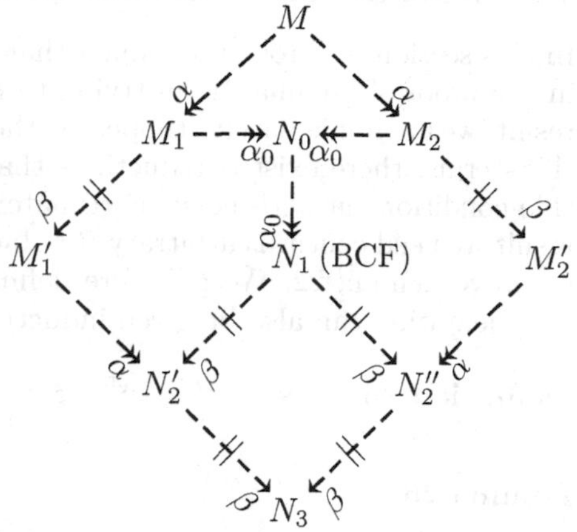

□

4.1 Substitutivity and Substitution Results

When proving a commutativity result about two relations, you typically proceed by rule induction over one of the relations. In what amounts to the non-trivial sub-cases of such a proof you therefore typically need to show that a substitution from the case-instantiated relation "distributes" over the other relation. Such results are called Substitutivity Lemmas. The non-trivial sub-cases of Substitutivity Lemmas, in turn, are called Substitution Lemmas. They establish commutativity of the substitutions from both the case-instantiations. Substitutivity and Substitution Lemmas are non-trivial to prove formally. For our present purposes we will merely display one of each to give an indication of the style. The key to understanding the following lemmas is the fact that $\mathrm{Capt}_x(e_1) \cap \mathrm{FV}(e_2) = \emptyset$ is the weakest predicate ensuring the correctness of substituting e_2 into e_1 for x.

Lemma 23 (Substitution)

$$
\begin{array}{l}
y \notin \mathrm{FV}(e_2) \wedge x \neq y \wedge (\mathrm{Capt}_x(e_3) \cap \mathrm{FV}(e_2) = \emptyset) \wedge (\mathrm{Capt}_y(e_1) \cap \mathrm{FV}(e_3) = \emptyset) \\
\wedge (\mathrm{Capt}_x(e_1) \cap \mathrm{FV}(e_2) = \emptyset) \wedge (\mathrm{Capt}_x(e_1[y := e_3]) \cap \mathrm{FV}(e_2) = \emptyset) \\
\Downarrow \\
e_1[y := e_3][x := e_2] = e_1[x := e_2][y := e_3[x := e_2]]
\end{array}
$$

Lemma 24 (Parallel β Substitutivity)

$$
\begin{array}{l}
e_1 \dashrightarrow\!\!\!\!\!\shortmid\!\!\shortmid_{\beta} e'_1 \wedge e_2 \dashrightarrow\!\!\!\!\!\shortmid\!\!\shortmid_{\beta} e'_2 \wedge (\mathrm{Capt}_x(e_1) \cap \mathrm{FV}(e_2) = \emptyset) \wedge (\mathrm{Capt}_x(e'_1) \cap \mathrm{FV}(e'_2) = \emptyset) \\
\Downarrow \\
e_1[x := e_2] \dashrightarrow\!\!\!\!\!\shortmid\!\!\shortmid_{\beta} e'_1[x := e'_2]
\end{array}
$$

We refer the interested reader to the complete Isabelle/HOL proof development at the homepage of the first author for full details.

4.2 Weak α- and β-Commutativity

In this section we prove the lemma that is needed on either side of the diagram in the proof of Lemma 22. In trying to prove a general α and β commutativity result, we are immediately stopped by the following naming issue: for virtually all Λ^{var}-terms, there exist α-reductions that can invalidate a previously validated side-condition on a β-redex. Fortunately, we can see that the commutativity result we need concerns arbitrary β-reductions but only α-reductions that suffice to prove Lemma 22. We therefore define a restricted, fresh-naming α-relation. The definition can also be given inductively.

Definition 25 $e \dashrightarrow_{\alpha_0} e' \Leftrightarrow^{\mathrm{def}} \exists z. e \overset{z}{\dashrightarrow}_{i\alpha} e' \wedge z \notin \mathrm{FV}(e) \cup \mathrm{BV}(e)$

Lemma 26

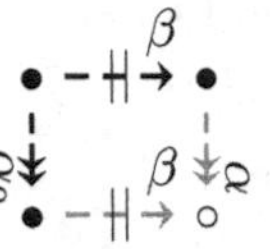

Proof By rule induction in $\dashrightarrow\!\!\!\rightarrow_{\alpha_0}$ with the induction step going through painlessly by freshness of the relevant z. □

4.3 The Diamond Property of Parallel β Up-to BVC-Initiality

We will now establish the lower part of the diagram in the proof of Lemma 22. It is proved using Takahashi's Trick, cf. Lemma 6. Initially, we thus need to establish the conditional existence of a non-renaming complete β-development.

Lemma 27 (BCF) $\bullet \dashrightarrow\!\!\!|_{\beta}\ \circ$

Proof By structural induction using Proposition 18 and Lemma 20. □

We stress that the proof is straightforward using the referenced variable monotonicity results as $\dashrightarrow\!\!\!|_{\beta}$ is inductively defined to contract from within. No complicated considerations concerning residuals are required. However, BCF-initiality is crucial for the property. The terms $(\lambda x.\lambda y.x)y$ and $\lambda y.(\lambda x.\lambda y.x)y$ fail to enjoy free/bound variable disjointness and unique binding, respectively, and neither completely develop. BCF-initiality is thus sufficient for the existence of a complete development but only necessary in a weak sense: breaking either conjunct of the BCF-predicate can prevent renaming-free complete development. Still, some non-BCFs completely develop, e.g., $(\lambda x.x)x$ and $\lambda x.(\lambda x.x)x$.

The second of the two required results for the application of Lemma 6 must establish that any parallel β-step always can "catch up" with a completely developing β-step by a parallel β-step, *with no renaming involved.*

Lemma 28

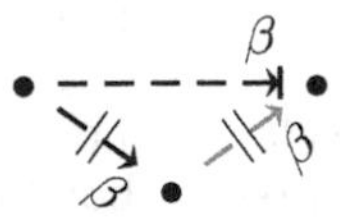

Proof By rule induction in $\dashrightarrow\!\!\!\!\!\!\shortmid\ _\beta$ using Lemma 24. □

It is interesting that the above property requires no initiality conditions, like the BCF-predicate, to be provable — except, that is, from well-definedness of $\dashrightarrow\!\!\!\!\!\!\shortmid\ _\beta$. This is mainly due to our use of the weakest possible side-condition on β-contraction to make β renaming free (i.e., $\mathrm{FV}(-) \cap \mathrm{Capt}\_(-) = \emptyset$). Had we instead required that the free variables of the argument were disjoint from the full set of bound variables in the body of the applied function (i.e., $\mathrm{FV}(-) \cap \mathrm{BV}(-) = \emptyset$), the property would not have been true. A counter-example is $(\lambda y.(\lambda x.y)z)\lambda z.z$. It takes advantage of complete developments contracting from within. Contracting the outermost redex first (e.g., by a parallel step) blocks the contraction of the residual of the innermost redex when the stronger side-condition is imposed: $(\lambda x.\lambda z.z)z$. No variable conflict is created between two residuals of the same term due to Hyland's Disjointness Property [15].[6]

Lemma 29

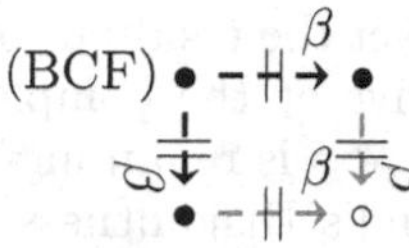

Proof From Lemmas 27 and 28 by using Takahashi's Trick, Lemma 6. □

4.4 Fresh-Naming α-Confluence with BVC-Finality

The last result we need for the proof of Lemma 22 is the top triangle with its leg. We prove it as two results (mainly out of formalisation considerations)—the first form suffices by Lemma 10:

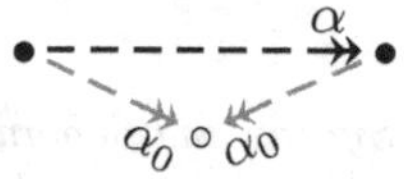

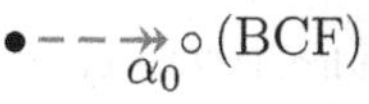

The proofs do not provide any insights and have been omitted.

4.5 Confluence

We have thus completed the proof of Lemma 22 and only one more lemma is needed before we can conclude our main result.

Lemma 30 $\dashrightarrow_{\alpha\cup\beta} \subseteq \dashrightarrow\!\!\!\!\rightarrow_\alpha$; $\dashrightarrow\!\!\!\!\!\!\shortmid\ _\beta \subseteq \dashrightarrow\!\!\!\!\rightarrow_{\alpha\cup\beta}$

Proof By rule induction observing that both $\dashrightarrow\!\!\!\!\rightarrow_\alpha$ and $\dashrightarrow\!\!\!\!\!\!\shortmid\ _\beta$ are reflexive. The proofs of the inclusions: $\dashrightarrow_\beta \subseteq \dashrightarrow\!\!\!\!\!\!\shortmid\ _\beta \subseteq \dashrightarrow\!\!\!\!\rightarrow_\beta$, go through straightforwardly. □

Theorem 31 (Confluence of the Raw and Real λ-Calculi)

$$\mathit{Confl}(\dashrightarrow_{\alpha\cup\beta}) \wedge \mathit{Confl}(\rightarrow_\beta) \wedge \mathit{Confl}(\dashrightarrow_{\alpha^{\mathrm{C}}\cup\beta^{\mathrm{C}}}) \wedge \mathit{Confl}(\rightarrow_{\beta^{\mathrm{H}}})$$

Proof By Lemmas 5, 22, and 30 and then Theorem 16. □

---

[6] "Any two residuals of some sub-term in a residual of the original term are disjoint".

5 Conclusion

We have completed a confluence proof applying to several raw and real λ-calculi. It has been done by using first-order induction principles over Λ^{var} and reduction relations, only. It is the first proof we know of which clearly makes the raw-/real-calculi distinction. It does so by introducing a new result about preservation/reflection of confluence. It is also the first formalised equational result about a higher-order language which conducts its inductive reasoning over $\text{FOAS}_{\mathcal{VN}}$, as you do informally by hand.

A Rational Reconstruction of the BVC. We proved two results about parallel and completely developing β-reduction, Lemmas 27 and 28, in order to apply Takahashi's Trick. In summary, they say that irrespective of which pre-existing β-redexes in a BCF-term you contract in parallel and without performing renaming, it is possible to contract the residuals of the rest in parallel and without performing renaming and arrive at the completely developed term. All in all, the residual theory of $\dashrightarrow_\beta$ in λ^{var} is renaming-free up-to BCF-initiality. This is partly a consequence of Hyland's Disjointness Property [15] and partly due to our careful use of substitution. Said differently, Barendregt's moral:

> "**2.1.14.** Using 2.1.12/13 one can work with λ-terms the naive way."

is formally justifiable and is, in fact, an entirely reasonable way to conduct equational proofs about the λ-calculus when due care is taken to clarify the raw vs. real status of the established property.

References

1. Barendregt: *The Lambda Calculus — Syntax and Semantics.* North-Holland, 1984.
2. Burstall: Proving properties of programs by struct. ind. *Comp.J.*, 12, 1967.
3. Curry, Feys: *Combinatory Logic.* North-Holland, 1958.
4. de Bruijn: Lambda calculus notation with nameless dummies, a tool for auto. formula manipulation, with appl. to the CR Theorem. *Indag. Math.*, 34, 1972.
5. Despeyroux, Hirschowitz: HOAS with ind. in COQ. *LPAR*, 1994. LNAI 822.
6. Despeyroux, Pfenning, Schürmann: Prim. rec. for HOAS. *TLCA*, 1997. LNCS 1210.
7. Fiore, Plotkin, Turi: Abstract syntax and variable binding. In Longo [16].
8. Gabbay, Pitts: A new approach to abstract syntax involving binders. In Longo [16].
9. Girard: From the rules of logic to the logic of rules. To appear in *MSCS.*
10. Gordon, Melham: Five axioms of alpha-conversion. *TPHOL*, 1996. LNCS 1125.
11. Hindley: *The CR Prop. and a Result in Comb. Logic.* PhD thesis, Newcastle, 1964.
12. Hofmann: Semantical analysis of HOAS. In Longo [16].
13. Huet: Residual theory in λ-calculus: A formal development. *JFP*, 4(3), 1994.
14. Jouannaud, Kirchner: Compl. of a set of rules mod. a set of eq. *SIAM*, 15, 1986.
15. Klop: *Combinatory Reduction Systems.* Mathematical Centre Tracts 127, 1980.
16. Longo (ed.): *LICS-14*, 1999. IEEE Computer Society Press.
17. McKinna, Pollack: Some lambda calculus and TT formalized. To appear in *JAR.*
18. Nipkow: More CR proofs (in Isabelle/HOL). *CADE-13*, 1996. LNCS 1104.
19. Rose: Explicit substitution – tutorial & survey. BRICS-LS-96-13, 1996.

20. Rutten: A calc. of transition systems (towards univ. coalg.). CWI-CS-R9503, 1995.
21. David E. Schroer. *The Church-Rosser theorem.* PhD thesis, Cornell, June 1965.
22. Schürmann: *Automating the Meta Theory of Ded. Syst.* PhD thesis, CMU, 2000.
23. Shankar: A mechanical proof of the Church-Rosser Theorem. *J. ACM*, 35(3), 1988.
24. Takahashi: Parallel reductions in λ-calculus. *I. and C.*, 118, 1995.

A Commutative Diagrams

Formally, a commutative diagram is a set of vertices and a set of directed edges between pairs of vertices. A vertex is written as either • or ∘. Informally, this denotes quantification modes over terms, universal respectively existential. A vertex may be guarded by a predicate. Edges are written as the relational symbol they pertain to and are either full-coloured (black) or half-coloured (gray). Informally, the colour indicates assumed and concluded relations, respectively. An edge connected to a ∘ must be half-coloured. A diagram must be type-correct on domains. A property is read off of a diagram thus:

1. write universal quantifications for all •s (over the relevant domains)
2. assume the full-coloured relations and the validation of any guard for a •
3. conclude the guarded existence of all ∘s and their relations

The following diagram and property correspond to each other (for $\rightarrow \subseteq A \times A$).

$$\begin{array}{ccc} (P)\,\bullet & \longrightarrow & \bullet \\ \downarrow & & \downarrow \\ \bullet & \longrightarrow & \circ\,(Q) \end{array} \qquad \begin{array}{c} \forall e_1, e_2, e_3 \in A \,.\; e_1 \rightarrow e_2 \,\wedge\, e_1 \rightarrow e_3 \,\wedge\, P(e_1) \\ \Downarrow \\ \exists e_4 \in A \,.\, e_2 \rightarrow e_4 \,\wedge\, e_3 \rightarrow e_4 \,\wedge\, Q(e_4) \end{array}$$

We will often leave quantification domains implicit and furthermore assume the standard disambiguating conventions for binding strength and associativity of connectives.

A Normal Form for Church-Rosser Language Systems

Jens R. Woinowski

TU Darmstadt
Fachbereich Informatik
`woinowski@iti.informatik.tu-darmstadt.de`

Abstract. In this paper the *context-splittable* normal form for rewritings systems defining Church-Rosser languages is introduced. Context-splittable rewriting rules look like rules of context-sensitive grammars with swapped sides. To be more precise, they have the form $uvw \to uxw$ with u, v, w being words, v being nonempty and x being a single letter or the empty word. It is proved that this normal form can be achieved for each Church-Rosser language and that the construction is effective.

1 Introduction

Church-Rosser languages (CRL's) basically are defined by a *length reducing* string rewriting system and a mechanism to handle the word ends [MNO88]. Here we call such a defining system a *Church-Rosser language system* (CRLS). CRL are a very interesting class of languages for three reasons: (i) Their word problem can be decided in deterministic linear time, although they are a strict superset of the *deterministic context-free languages* (DCFL)[MNO88]. (ii) Despite that fact, their definition is more intuitive than that of DCFL. (iii) They are the deterministic variant of the *growing context-sensitive languages* (GCSL), which was proved in [NO98] (for the definition of GCSL see [DW86]). Therefore, they fit into the Chomsky hierarchy very well [Cho59,McN99,BHNO00].

By assigning weights to single letters one can also define a *weight function* for words. This is the basis of the following important characterization result about Church-Rosser languages: Allowing *weight reduction* instead of length reduction does not improve the expressive power [NO98]. Given a rewriting system with weight reducing rules defining a Church-Rosser language it is possible to construct an equivalent one that only has length reducing rules.

In this paper, we use this fact to show the effective existence of a *context-splittable normal form* for every Church-Rosser language L. The defining (length reducing) rewriting system of L can be simulated by a weight reducing system that has rules of the form $uvw \to uxw$ with u, v, w being words, v being nonempty and x being a single letter or the empty word. We do not use the term 'context-sensitive' in order to stress the fact that the two forms are not fully corresponding to each other. Because context-splittable rule can also be deleting (i.e. of the form $v \to \square$), we do not always get a context-sensitive rule by swapping the sides.

A. Middeldorp (Ed.): RTA 2001, LNCS 2051, pp. 322–337, 2001.

One consequence of this normal form result is that the information flow during reductions is underlying stronger restrictions in a *context-splittable* CRLS (csCRLS): Any movement of a letter in either direction needs at least as many rule applications as the distance to be accomplished. Although this is only a refinement of the linear time bound for the reductions in CRLS's it might be handy for proofs.

This paper is organized in the following way: In the next section, we give the basic definitions. Section 3 contains the normal form theorem and the construction which is used to prove it. In order to enhance the readability of this paper we concentrate on the construction principles and do not give the proof in full detail. All details omitted are merely technical, they can be found in [Woi00a]. The last section of this paper contains some concluding remarks which also give an insight into possible further consequences of the normal form result, especially with respect to GCSL.

2 Basic Definitions

The reader is assumed to be familiar with definitions and notations of confluent string rewriting. For details, see [Jan88], [MNO88], and [BO98].

A *string-rewriting system (or simply rewriting system)* R on Σ is a subset of $\Sigma^* \times \Sigma^*$. For $(u, v) \in R$ we also write $(u \to v) \in R$ and call (u, v) a rule.

A *weight function* is a function $f : \Sigma \to \mathbb{N}$. It is recursively extended to a function on Σ^* by $f(wx) := f(w) + f(x)$ and $f(\Box) := 0$ (where $\Box$ is the empty word) with $w \in \Sigma^*$, $x \in \Sigma$. An example for a weight function is the *length function* with $f(x) := 1$ for all $x \in \Sigma$, then $f(w) = |w|$.

Throughout this article, a string-rewriting system R is called a *weight reducing system*, if there exists a weight function f such that $f(u) > f(v)$ for all $(u, v) \in R$.

Definition 1. *A* Church-Rosser language system (CRLS) *is a 6-tuple* $C = (\Gamma, \Sigma, R, k_l, k_r, y)$ *with finite alphabet* Γ, *terminal alphabet* $\Sigma \subset \Gamma$ ($\Gamma \setminus \Sigma$ *is the alphabet of nonterminals), finite confluent weight reducing system* $R \subseteq \Gamma^* \times \Gamma^*$, *left and right end marker words* $k_l, k_r \in (\Gamma \setminus \Sigma)^* \cap \textsc{Irr}(R)$, *and accepting letter* $y \in (\Gamma \setminus \Sigma) \cap \textsc{Irr}(R)$ *The* language defined by C *is defined as:* $L_C := \{w \in \Sigma^* | k_l \cdot w \cdot k_r \to^*_R y\}$

A language L *is called a* Church-Rosser language (CRL) *if there exists a* CRLS C *with* $L_C = L$.

The definition of Church-Rosser languages is due to McNaughton, Narendran, and Otto [MNO88]. The definition of Church-Rosser language systems given here is a convenient notation for their definition. Niemann and Otto proved in [NO98] that the expressive power of Church-Rosser languages is not enhanced by allowing arbitrary weight functions instead of the length function, so this fact is used, too.

The following definition of a syntactical restriction for CRLS will be proved to be a normal form in the main part of this paper.

Definition 2. *A* CRLS $C = (\Gamma, \Sigma, R, ¢, \$, y)$ *is* context-splittable (C is a csCRLS) *if* $¢, \$, y \in \textsc{Irr}(R) \cap \Gamma \setminus \Sigma$ *(let the* inner alphabet be $\Gamma_{inner} := \Gamma \setminus \{¢, \$, y\}$*) and for any rule* $r \in R$ *there exists a splitting* (u, v, w, x) *with:*

1. $r = (uvw, uxw)$
2. *v is non-empty.*
3. *uvw may contain at most one ¢ and if so at its beginning. Also it can have at most one \$ which only may appear at the end. All other letters of uvw have to be from the inner alphabet Γ_{inner}.*
4. *x is a single letter not equal to ¢ or \$ or it is the empty word.*
5. *If v contains ¢ or \$, then $x = y$, u and w are empty, and v is element of* $¢ \cdot \Gamma^*_{inner} \cdot \$$.
6. *If $x = y$, then u and w are empty, and v is element of* $¢ \cdot \Gamma^*_{inner} \cdot \$$.

The splitting (u, v, w, x) of a rule r allowed by this is called a context splitting, *u and w are called the* left and right context.

Example 1. These are some examples for the meaning of the definition (the splittings are marked by dots):

- $ab \cdot dea \cdot ab \to ab \cdot a \cdot ab$,
- $ab \cdot de \cdot aab \to ab \cdot \Box \cdot aab$ is another context-splitting of the same rule,
- $¢ \cdot ab \cdot \$ \to ¢ \cdot \Box \cdot \$$,
- $¢ \cdot ab\$ \cdot \Box \to ¢ \cdot \$ \cdot \Box$ is *not a valid* splitting,
- $abc \to \Box$ is a *deleting* context-splittable rule, and
- $abcd \to dcba$ is a rule that is not context-splittable.

Remark 1. We want to distinguish the notion of context-splittable CRLS's from that of context-sensitive grammars, because the former uses reductions and the latter productions for defining languages. Especially deleting rules of the form $v \to \Box$ have no counterpart in context-sensitive grammars. Therefore we do not use the term context-sensitive. Note that all other deleting rules with u or w being nonempty can be splitted in a different way such that x is not the empty word, just as in the example given above.

3 The Normal Form Theorem

Theorem 1. *Let* $C = (\Gamma, \Sigma, R, k_l, k_r, y)$ *be a* CRLS *with language* L_C*. Then there exists a* csCRLS C' *with* $L_{C'} = L_C$.

Proof. We will give an effective construction for such a new csCRLS C'.

Remark 2. Although we will not need a formal definition of the automaton model for the language family CRL, a variant of two pushdown automata, it will be helpful to understand parts of the construction. Informally, these automata work as follows (c.f. [Boo82]): They have two stacks and a scanning head. The initial configuration is a left stack with the left end marker word[1] in it, its leftmost letter at the bottom of the stack. The right stack contains the input, the leftmost letter of the input being on top, and below it the right end marker word. Now, letters from the right stack are shifted to the left, until a suffix of the left stack is the left side of a reduction rule of the CRLS. This suffix is deleted from the left stack and the right side of the rule is pushed onto the right stack. Because of the confluence of the underlying string-rewriting system we can make this automaton deterministic by choosing only one rule for every left side and by using the rule with the shortest left side if two or more different left sides apply. The procedure is repeated until the right stack is empty and the left stack is irreducible. If the left stack only contains the accepting letter of the CRLS the input is accepted. Note that since after a reduction operation the left stack will always be irreducible, one can directly combine one shift with each reduce operation, as in [MNO88].

3.1 The Construction Principles

Without restriction of generality we may assume that R is length reducing [NO98]. In order to construct a csCRLS C', we will make use of the following four principles:

1. Analogous to the automata model our new system will have the property that during the whole reduction process there is always exactly one place in the word where the next reduction rule can be applied.
2. We will use a compression alphabet which can store more than one letter of the input (resp. the derivated words) in one letter. This information will be represented by subscripts of the compression letters.
3. These compression letters will be enriched by surplus letters in their subscripts in order to spread necessary weight reductions over more than one letter.
4. Rules of the original system will, in most cases, be simulated by three or four rules in the new system.

The confluent weight reducing system will be built of five parts R_1 to R_5.

Definition 3. *With Γ being the alphabet of C which consists of all terminal and nonterminal letters, let $\overline{\Gamma}$ be a new alphabet, which is a disjoint copy of Γ. Then $\overline{}$ is the bijective morphism that maps Γ into $\overline{\Gamma}$, e.g. a to $\overline{a}$. Let $\natural$, ɋ, and \$ be new symbols. Let $\Gamma_\natural := \Gamma \cup \{\natural\}$ and $\overline{\Gamma}_\natural := \overline{\Gamma} \cup \{\natural\}$. Define*

$$W_\natural := \overline{\Gamma}_\natural^* \cdot \Gamma_\natural^* \cap ((\natural^{\leq 2} \cdot ((\overline{\Gamma} \cup \Gamma) \cdot \natural\natural)^* \cdot (\overline{\Gamma} \cup \Gamma) \cdot \natural^{\leq 2}) \cup \natural^{\leq 2}),$$

where $\natural^{\leq^2}$ is a shorthand for $\{\square, \natural, \natural\natural\}$.

[1] Note that both end marker words are irreducible.

As can be seen, this regular language consists of words, where between letters of Γ or $\overline{\Gamma}$ there are always exactly two $\sharp$'s. This language will not only be used to define the compression alphabet, but also to make some definitions during the construction process easier. The purpose of the $\sharp$'s will be explained later.

Definition 4. *Let* $\mu_l = \max\{|u| \mid (u,v) \in R\}$ *and* $\mu_r = \max\{|v| \mid (u,v) \in R\}$ *be the maximum length of the respective rule sides. Because* R *is length reducing* $\mu_l > \mu_r$ *holds. Let* $\mu := \max\{\mu_l, |k_l|, |k_r|\}$. *The* compression alphabet Γ_1 *is defined as:*

$$\Gamma_1 := \{\xi_w \mid w \in W_\sharp \wedge 1 \leq |w| \leq 3\mu + 5\}.$$

The elements of Γ_1 *are called* compression letters. *For distinction, letters in the index of a compression letter will be called* index letters.

Remark 3. At least $\mu + 1$ letters of the original alphabet Γ can be stored in one compression letter.

Definition 5. *Sometimes it is necessary to extract the information of the subscripts in a word from* Γ_1^*. *Let* $\check{}$ *be the morphism* $\check{} : (\Gamma_1 \cup \overline{\Gamma} \cup \Gamma \cup \{¢, \$\})^* \to (\overline{\Gamma}_\sharp \cup \Gamma_\sharp)^*$ *defined by:*

$$\check{x} := \begin{cases} w & x = \xi_w \in \Gamma_1 \\ x & \text{else} \end{cases}.$$

We assume, without loss of generality, that brackets are not in our alphabets so far and use them in the following for better readability.

3.2 Translating the Input

The first step in the simulation of C is to translate the input into the compression alphabet. At the same time, we will take care of k_l and k_r. The new end marker letters(!) of C' will be $k'_l := ¢$ and $k'_r := \$$.

Short words $w \in L_C, |w| \leq 2$ will be handled separately with a a set R_1 of rules: $R_1 := \{(¢w\$, y) \mid w \in L_C \wedge |w| \leq 2\}$

Obviously, R_1 can be computed easily.

For translating the input a set of rules R_2 will be used, which works as follows: Decompose k_l and k_r into single letters in the following way: $k_l = a_1 a_2 \cdots a_{|k_l|}$ and $k_r = c_1 c_2 \cdots c_{|k_r|}$. R_2 will be designed to be a suitable confluent and weight reducing rewriting system such that for every $w \in \Sigma^{>2}$ with $w = b_1 b_2 \cdots b_i \cdots b_{|w|}, b_i \in \Sigma (1 \leq i \leq |w|)$ we can make the following reduction with R_2:

$$¢w\$ \to^*_{R_2} ¢\xi_{(\overline{a}_1\sharp\sharp)(\overline{a}_2\sharp\sharp)\cdots(\overline{a}_{|k_l|}\sharp\sharp)(\overline{b}_1\sharp\sharp)} \xi_{b_2\sharp\sharp} \xi_{b_3\sharp\sharp} \cdots \xi_{b_{|w|-1}\sharp\sharp} \xi_{(b_{|w|}\sharp\sharp)(c_1\sharp\sharp)(c_2\sharp\sharp)\cdots(c_{|k_r|}\sharp\sharp)} \$.$$

Furthermore we require that the first translated letter to the right of the symbol ¢ is always produced by the last reduction step. This is necessary to give a precise moment in the reduction after which the rule sets defined in the following parts of the construction can begin to work.

Example 2. Let $k_l = dd, k_r = c$, and $w = abad$. Then R_2 translates the input to:

$$¢\xi_{\overline{d}\sharp\sharp\overline{d}\sharp\sharp\overline{a}\sharp\sharp}\xi_{b\sharp\sharp}\xi_{a\sharp\sharp}\xi_{d\sharp\sharp c\sharp\sharp}\$$$

It is easily verified that it is possible to construct R_2 such that $R_1 \cup R_2$ is confluent and, with a suitable weight function, weight reducing.

3.3 Simulating Shift Operations

The next step is similar to the shift operations of automata for CRL's. Sometimes it is necessary to *move right* (that is, shift) the position of a possible next reduction.

The right-most overlined letter marks the position at which the simulation of C will work with the following rule sets. In general, the right-most overlined index letter of a compressed word can be identified with the head position of the automaton described above.

The simulation of shifting is done with a further set of rules which is called R_3. A simulated shift has to take place whenever the overlined letters within the compression letters form an irreducible string w.r.t. R when the overlining is removed. We ommit the details of R_3, shifting corresponds to overlining the first letter in the indices which is in Γ. If a shift is necessary but no next such letter without overlining exists no next reduction is possible. Then the simulated system also would have come to an irreducible word.

The matter of weight reduction will be discussed later, at the moment simply assume that overlined index letters add slightly less to the weight than non-overlined ones.

Example 3. Assume that bba is irreducible and no suffix of a left-hand side of a rule in R. Then a simulated shift is necessary whenever $\overline{b}\sharp\sharp\overline{b}\sharp\sharp\overline{a}\sharp\sharp x$ $(x \in \Gamma)$ appears in the indices. For example, one of the necessary rules could be:

$$(\xi_{\overline{c}\sharp\sharp\overline{b}\sharp}\xi_{\sharp\overline{b}\sharp\sharp\overline{a}\sharp}\xi_{\sharp c\sharp}, \xi_{\overline{c}\sharp\sharp\overline{b}\sharp}\xi_{\sharp\overline{b}\sharp\sharp\overline{a}\sharp}\xi_{\sharp\overline{c}\sharp}) \in R_3$$

Again, we do not give the full details of R_3, because one can easily see that it is possible to construct it such that $R_1 \cup R_2 \cup R_3$ is confluent and weight reducing.

3.4 The "Weight Spreading" Strategy

Now we come to the core of the construction, which will be called *weight spreading.* The main idea is to simulate rules of R piecewise. In order to achieve a weight reducing system a second idea is used: the simulation will reduce the length of the subscripts of the compression letters.

First, we provide an example that shows the correct construction at work.

Example 4. Assume we have a rewriting system R with $(aaaa, bbb) \in R$. Then one case in the simulation is the following:

$$\begin{array}{ll}
\xi_{\overline{a}\sharp\sharp\overline{a}}\xi_{\sharp\sharp\overline{a}\sharp\sharp}\xi_{\overline{a}\sharp\sharp a} & \\
\quad\longrightarrow & \text{1. "lock" with new nonterminal} \\
\xi_{\overline{a}\sharp\sharp\overline{a}}\xi_{\sharp\sharp\overline{a}\sharp\sharp}\xi_t & \\
\quad\longrightarrow & \text{2. change first letter} \\
\xi_{\overline{b}\sharp\sharp}\xi_{\sharp\sharp\overline{a}\sharp\sharp}\xi_t & \\
\quad\longrightarrow & \text{3. change middle letter} \\
\xi_{\overline{b}\sharp\sharp}\xi_{b\sharp\sharp b}\xi_t & \\
\quad\longrightarrow & \text{4. change last letter, remove lock} \\
\xi_{\overline{b}\sharp\sharp}\xi_{b\sharp\sharp b}\xi_{\sharp\sharp a} &
\end{array}$$

The $\sharp$'s are used to spread the length reduction of the original rule over the compression letters in the simulation. Observe that (especially) in the last step of the example the result has three compression letters. More than three letters are not necessary during any rule simulation, because a right-hand side of a rule and the added $\sharp$'s fit into one compression letter (to be exact, sometimes some *unchanged context* appears on the right, making four letters necessary). Basically, the weight of a compression letter will be computed from the number of its index letters (overlined letters add *slightly* less to the weight). Therefore in the worst case—when the original rule has a length reduction of one—we reduce the number of index letters by three and spread this reduction over the resulting three compression letters. This is the cause for adding two $\sharp$'s after each index letter of $\Gamma \cup \overline{\Gamma}$ during the translation with R_2.

In generalising this example, a lot of cases have to be handled. The main idea is to identify all possibilities how the left-hand side of an original rule can be split over one or more letters of the compression alphabet. Also, sometimes it is necessary to allow some unchanged context in the first compression letter. Furthermore, the question of finding the right place for the reduction to work has to be handled.

3.5 Case Distinctions

For lack of space, we will not give the full construction. Instead, the necessary distinction of cases and an example how to handle one of these cases are given. With this example the reader should be able to understand the principles of the construction. The complete construction can be found in the technical report [Woi00a].

Definition 6. *The following notation for the set of compression letters whose subscript begins with a letter from Γ will be used:*

$$\xi_{\Gamma \cdot \Gamma_\sharp^*} := \{\xi_w \mid \xi_w \in \Gamma_1 \wedge w \in \Gamma \cdot \Gamma_\sharp^*\}$$

In some cases, the simulation is very easy. Especially, when the complete reduction can be simulated within one letter of the compression alphabet. We handle this first case with the following set:

$$
\begin{aligned}
T_1 := \{(u, v, \xi_{w_1w_2w_3}x) \mid\ & (u, v) \in R \\
& \wedge w_1w_2w_3 \in W_\sharp \\
& \wedge w_1 \in \overline{\Gamma}_\sharp^* \\
& \wedge \hat{w}_2 = \overline{u} \wedge w_2 \in \overline{\Gamma} \cdot \overline{\Gamma}_\sharp^* \cdot \overline{\Gamma} \cdot \{\sharp\sharp\} \\
& \wedge ((w_3 \in \Gamma\Gamma_\sharp^* \wedge x = \Box) \vee (w_3 = \Box \wedge x \in \{\$\} \cup \xi_{\Gamma \cdot \Gamma_\sharp^*})) \\
& \wedge (x \neq \$ \Rightarrow w_1w_2w_3\check{x} \in W_\sharp)\}.
\end{aligned}
$$

Now, for each element of $t \in T_1$ assume $u = a_1 \cdots a_{|u|}, a_i \in \Gamma$ $(1 \leq i \leq |u|)$ and $v = b_1b_2 \cdots b_{|v|}, b_i \in \Gamma$ $(1 \leq i \leq |v|)$. For each t add a rule r_t to a new system R_4 :

$$r_t := (\xi_{w_1w_2w_3}x, \xi_{w_1(\overline{b}_1\sharp\sharp)(b_2\sharp\sharp)\cdots(b_{|v|}\sharp\sharp)w_3}x)$$

If $w_1 = w_3 = v = \Box$, this would produce a letter $\xi_\Box$, which is not in the compression alphabet. In these cases identify $\xi_\Box \equiv \Box$. Then the rule will be simply deleting one symbol.

Now we come to the difficult part of the construction. What has to be done, if the left side of the original rule is not in one letter but distributed over the indices of several compression letters? Again we use a set which contains all cases of possible rule applications that are not covered by the above set T_1:

$$
\begin{aligned}
T_2 := \{(u, v, \xi_{w_1w_2}\xi_{w_3}\xi_{w_4} \cdots \xi_{w_{n-2}}\xi_{w_{n-1}w_n}x) \mid\ & (u, v) \in R \\
& \wedge 4 \leq n \\
& \wedge w_1 \cdots w_n \in W_\sharp \\
& \wedge w_1 \in \overline{\Gamma}_\sharp^* \\
& \wedge w_2w_3 \cdots w_{n-1} \in \overline{\Gamma} \cdot \overline{\Gamma}_\sharp^* \cdot \overline{\Gamma} \cdot \{\sharp\sharp\} \\
& \wedge w_2 \neq \Box \\
& \wedge \hat{w}_2\hat{w}_3 \cdots \hat{w}_{n-1} = \overline{u} \\
& \wedge ((w_n \in \Gamma \cdot \Gamma_\sharp^*) \vee (w_n = \Box \wedge x \in \{\$\} \cup \xi_{\Gamma \cdot \Gamma_\sharp^*})) \\
& \wedge (x \neq \$ \Rightarrow w_1 \cdots w_n\check{x} \in W_\sharp)\}
\end{aligned}
$$

For each $t \in T_2$ (T_2 is finite) a set of rules is added to R_4. This also needs some further nonterminals. These will be collected in the set Γ_2. Again, we assume $u = a_1 \cdots a_{|u|}, a_i \in \Gamma$ $(1 \leq i \leq |u|)$ and $v = b_1b_2 \cdots b_{|v|}, b_i \in \Gamma$ $(1 \leq i \leq |v|)$. The following cases have to be dealt with. We do not give all details, but with the example after the list of cases the construction should be clear enough. Whenever we speak of three or four rules these are a simulation of an original rule in as many steps.

Let $t = (u, v, \xi_{w_1w_2}\xi_{w_3}\xi_{w_4} \cdots \xi_{w_{n-2}}\xi_{w_{n-1}w_n}x) \in T_2$:

1. $n = 4, w_3 = \Box$: already covered by the rules for T_1.
2. $n = 4, v = \Box, w_3 \neq \Box$, all deleting rules need some extra care, four subcases:

2.1. $w_1 = \square, w_4 = \square$, one rule which simply deletes two letters: $r_t := (\xi_{w_1 w_2} \xi_{w_3 w_4} x, x)$.
2.2. $w_1 \neq \square, w_4 = \square$ use a new nonterminal ξ_t with $\psi(\xi_t) = \psi(\xi_{w_3 w_4}) - 1$ and three rules.
2.3. $w_1 = \square, w_4 \neq \square$ use only one rule: $r_t := (\xi_{w_1 w_2} \xi_{w_3 w_4} x, \xi_{w_4} x)$
2.4. $w_1 \neq \square, w_4 \neq \square$ use a new nonterminal ξ_t and three rules.

3. $n = 4, v \neq \square, w_3 \in \{\sharp, \sharp\sharp\}$, similar to the rules for T_1, no subcases, one rule: $r_t := (\xi_{w_1 w_2} \xi_{w_3 w_4} x, \xi_{w_1 \overline{b}_1 (\sharp\sharp b_2) \cdot (\sharp\sharp b_{|v|}) w_2''} \xi_{w_3 w_4} x)$
4. $n = 4, v \neq \square, |w_3| > 2$, (then w_3 contains at least one letter from $\overline{\Gamma}$), the reduction must be split over two nonterminals. For w_2 and w_3 there exist splittings such that $w_2 = w_2' w_2''$ and $w_3 = w_3' w_3''$ with $w_2'' w_3' = \sharp\sharp$. Since both w_2 and w_3 are nonempty, we know that there exists an $i, 1 \leq i < |u|$ with $w_2' = \overline{a}_1 \sharp\sharp \cdots \overline{a}_i$ and $w_3'' = \overline{a}_{i+1} \sharp\sharp \cdots \overline{a}_{|u|} \sharp\sharp$. Now there are five subcases, depending on the length of w_1, w_2'', and i, which we identify using the following notation:
Let $k := |v| - |u| + i$. If $k > 0$ this will be used to calculate a split point for the compressed word which is to be substituted. In some cases we need a new nonterminal ξ_t, which will be added to Γ_2. Let the weight of ξ_t be $\psi(\xi_t) := \psi(\xi_{w_3 w_4}) - 1$.
4.1. If $k \leq 0$ and $w_1 = \square$ add the rule $r_t := (\xi_{w_1 w_2} \xi_{w_3 w_4} x, \xi_{(\overline{b}_1 \sharp\sharp)(b_2 \sharp\sharp) \cdots (b_{|v|} \sharp\sharp) w_4} x)$.
4.2. $k \leq 0$, $w_2'' = \square$, and $w_1 \neq \square$: one new nonterminal and three rules.
4.3. $k > 0$ and $w_2'' = \square$: one new nonterminal and three rules.
4.4. $k \leq 0$, $w_2'' \neq \square$, and $w_1 \neq \square$: one new nonterminal and three rules.
4.5. $k > 0$ and $w_2'' \neq \square$: one new nonterminal and three rules.
5. $n = 5, v \neq \square, w_4 = \square$: already covered by 3. and 4.
6. $n = 5, v = \square, w_4 \neq \square$ deleting rules, we have four subcases:
6.1. $w_1 = \square, w_5 = \square$, add one rule which simply deletes three letters: $r_t := (\xi_{w_1 w_2} \xi_{w_3} \xi_{w_4 w_5} x, x)$
6.2. $w_1 \neq \square, w_5 = \square$: we use a new nonterminal ξ_t with $\psi(\xi_t) = \psi(\xi_{w_4 w_5}) - 1$ and three rules.
6.3. $w_1 = \square, w_5 \neq \square$ only one rule: $r_t := (\xi_{w_1 w_2} \xi_{w_3} \xi_{w_4 w_5} x, \xi_{w_5} x)$
6.4. $w_1 \neq \square, w_5 \neq \square$: one new nonterminal and three rules.
7. $n = 5, v \neq \square, w_4 \in \{\sharp, \sharp\sharp\}$, the complete reduction takes place in the first two nonterminals. This is similar to $n = 4$, $v \neq \square$,. These are the subcases:
7.1. $w_3 = \sharp$, then we can make a one rule reduction without new nonterminal: $(\xi_{w_1 w_2} \xi_{w_3} \xi w_4 w_5 x, \xi_{w_1 \overline{b}_1 (\sharp\sharp b_2) \cdots (\sharp\sharp b_{|v|}) w_3} \xi_{w_4 w_5} x)$
7.2. $w_3 \neq \sharp$, then w_3 contains at least one letter from $\overline{\Gamma}$, since $w_3 = \sharp\sharp$ would imply $w_4 \notin \{\sharp, \sharp\sharp\}$.
For w_2, w_3 and w_4 there exist splittings such that $w_2 = w_2' w_2''$, $w_3 = w_3' w_3'' w_3'''$ and $w_4 = w_4' w_4''$ with $w_2'' w_3' = \sharp\sharp$ and $w_3''' w_4' = \sharp\sharp$. Since both w_2 and w_3 are nonempty, we know that there exists an $i, 1 \leq i < |u|$ with $w_2' = \overline{a}_1 \sharp\sharp \cdots \overline{a}_i$ and $w_3'' = \overline{a}_{i+1} \sharp\sharp \cdots \overline{a}_{|u|}$. Now there are five subcases, depending on the length of w_1, w_2'', and i.

Let $k := |v| - |u| + i$. If $k > 0$ this will be used to calculate a split point for the compressed word which is to be substituted. In all cases with more than one rule we use a new nonterminal ξ_t, which will be added to Γ_2.

7.2.1. $k \leq 0$ and $w_1 = \Box$ (one rule),
7.2.2. $k \leq 0$, $w_2'' = \Box$, and $w_1 \neq \Box$ (three rules),
7.2.3. $k > 0$ and $w_2'' = \Box$ (three rules),
7.2.4. $k \leq 0$, $w_2'' \neq \Box$, and $w_1 \neq \Box$ (three rules), and
7.2.5. If $k > 0$ and $w_2'' \neq \Box$ (three rules).

8. $n = 5, v \neq \Box, w_3 \in \{\sharp, \sharp\sharp\}, |w_4| > 2$, then w_4 contains at least one letter from $\overline{\Gamma}$, since $|w_4| > 2$. Compute k as above, we have three subcases:
   8.1. $k \leq 0$ and $w_1 = \Box$ (one rule),
   8.2. $k \leq 0$ and $w_1 \neq \Box$ (three rules), and
   8.3. $k > 0$ (three rules).
9. $n = 5, v \neq \Box, |w_3| > 2, |w_4| > 2$, ($w_3$ cannot be empty), and $w_4 \notin \{\Box, \sharp, \sharp\sharp\}$ collection of subcases. In this case w_2,w_3, and w_4 contain at least one letter from $\overline{\Gamma}$.

   For w_2, w_3 and w_4 there exist splittings such that $w_2 = w_2'w_2''$, $w_3 = w_3'w_3''w_3'''$ and $w_4 = w_4'w_4''$ with $w_2''w_3' = \sharp\sharp$ and $w_3'''w_4' = \sharp\sharp$. Since both w_2, w_3 and w_4 are nonempty, we know that there exist an $i, j, 1 \leq i < j < |u|$ with $w_2' = \overline{a}_1\sharp\sharp\cdots\overline{a}_i$, $w_3'' = \overline{a}_{i+1}\sharp\sharp\cdots\overline{a}_j\sharp\sharp$, and $w_4'' = \overline{a}_{j+1}\sharp\sharp\cdots\overline{a}_{|u|}\sharp\sharp$.

   Let $k := |v| - |u| + i$. If $k > 0$ this will be used to calculate a split point for the compressed word which is to be substituted. Similarly, we will use $l := |v| - |u| + j$, note that $l > 0$ implies $|v| \geq 2$, we get the following subcases:

   9.1. $k \leq 0$, $l \leq 0$, and $w_1 = \Box$ (one rule),
   9.2. $k \leq 0$, $l > 0$, $w_1 = \Box$, and $w_3''' = \Box$ (three rules),
   9.3. $k \leq 0$, $l > 0$, $w_1 = \Box$, and $w_3''' \neq \Box$ (three rules),
   9.4. $k \leq 0$, $l \leq 0$, and $w_1 \neq \Box$ (three rules),
   9.5. $k \leq 0$, $l > 0$, $w_1 \neq \Box$, $w_2'' = \Box$, and $w_3''' = \Box$ (four rules),
   9.6. $k \leq 0$, $l > 0$, $w_1 \neq \Box$, $w_2'' = \sharp$, and $w_3''' = \Box$ (four rules),
   9.7. $k \leq 0$, $l > 0$, $w_1 \neq \Box$, $w_2'' = \sharp\sharp$, and $w_3''' = \Box$ (four rules),
   9.8. $k \leq 0$, $l > 0$, $w_1 \neq \Box$, $w_2'' = \Box$, and $w_3''' = \sharp$ (four rules),
   9.9. $k \leq 0$, $l > 0$, $w_1 \neq \Box$, $w_2'' = \sharp$, and $w_3''' = \sharp$ (four rules),
   9.10. $k \leq 0$, $l > 0$, $w_1 \neq \Box$, $w_2'' = \sharp\sharp$, and $w_3''' = \sharp$ (four rules),
   9.11. $k \leq 0$, $l > 0$, $w_1 \neq \Box$, $w_2'' = \Box$, and $w_3''' = \sharp\sharp$ (four rules),
   9.12. $k \leq 0$, $l > 0$, $w_1 \neq \Box$, $w_2'' = \sharp$, and $w_3''' = \sharp\sharp$ (four rules),
   9.13. $k \leq 0$, $l > 0$, $w_1 \neq \Box$, $w_2'' = \sharp\sharp$, and $w_3''' = \sharp\sharp$ (four rules),
   9.14. $k > 0$, $l > 0$, $w_2'' = \Box$, $w_3''' = \Box$ (four rules),
   9.15. $k > 0$, $l > 0$, $w_2'' = \Box$, $w_3''' = \sharp$ (four rules),
   9.16. $k > 0$, $l > 0$, $w_2'' = \Box$, $w_3''' = \sharp\sharp$ (four rules),
   9.17. $k > 0$, $l > 0$, $w_2'' = \sharp$, $w_3''' = \Box$ (four rules),
   9.18. $k > 0$, $l > 0$, $w_2'' = \sharp$, $w_3''' = \sharp$ (four rules),
   9.19. $k > 0$, $l > 0$, $w_2'' = \sharp$, $w_3''' = \sharp\sharp$ (four rules),
   9.20. $k > 0$, $l > 0$, $w_2'' = \sharp\sharp$, $w_3''' = \Box$, $w_3 \notin \overline{\Gamma}$, note that under these premises $l - k > 1$, (four rules),

9.21. $k > 0, l > 0, w_2'' = \sharp\sharp, w_3''' = \Box, w_3 \in \overline{\Gamma}$, note that in this case $k+1 = l$, (three rules),
9.22. $k > 0, l > 0, w_2'' = \sharp\sharp, w_3''' = \sharp$ (four rules), and
9.23. $k > 0, l > 0, w_2'' = \sharp\sharp, w_3''' = \sharp\sharp$ (four rules).

10. If $n \geq 6$ we only compress the information. In consequence, any reduction will take place by the rules of the cases for $n < 6$.
Consider all $t = (u, v, \xi_{w_1 w_2} \xi_{w_3} \xi_{w_4} \cdots \xi_{w_{n-2}} \xi_{w_{n-1} w_n} x) \in T_2$ with $n \geq 7$. Then $\xi_{w_{n-3} w_{n-2}} \in \Gamma_2$. So, add the rule:

$$r_t := (\xi_{w_{n-3}} \xi_{w_{n-2}} \xi_{w_{n-1} w_n} x, \cdots \xi_{w_{n-3} w_{n-2}} \xi_{w_{n-1} w_n} x).$$

3.6 An Example Case

In order to show how the subcases of the above main cases can be handled, one example for the case 9.16. is provided. For details refer to [Woi00a].

Example 5. Consider $t = (u, v, \xi_{w_1 w_2} \xi_{w_3} \xi_{w_4 w_5} x) \in T_2$, so $n = 5$. Assume $|w_3| > 2$ and $|w_4| > 2$. We know $\hat{w}_2 \hat{w}_3 \hat{w}_4 = \overline{u}$. Furthermore, decompose u and v into letters: $u = a_1 \cdots a_{|u|}, v = b_1 \cdots b_{|v|}$.

Then we know there exist $i, j > 0$ such that $j > i$ and $\hat{w}_2 = \overline{a}_1 \cdots \overline{a}_i$, $\hat{w}_3 = \overline{a}_{i+1} \cdots \overline{a}_j$, and $\hat{w}_4 = \overline{a}_{j+1} \cdots \overline{a}_{|u|}$.

We use i and j to determine how letters have to be spread over the subscripts of the new compression letters. Let $k := |v| - |u| + i$ and $l := |v| - |u| + j$. Obviously, $l > k$ always holds. In the following figure 1 the relation between i, j, k, and l is illustrated. As one can see, l and k allow to determine over how many compression letters the reduction result can be spread with respect to w_2, w_3, and w_4

$\hat{w}_2 = \overline{a}_1 \cdots \overline{a}_i$
$\hat{w}_3 = \overline{a}_{i+1} \cdots \overline{a}_j$
$\hat{w}_4 = \overline{a}_{j+1} \cdots \overline{a}_{|u|}$
$\overline{b}_1 \cdots \overline{b}_i$
$\overline{b}_{i+1} \cdots \overline{b}_j$
$\overline{b}_{j+1} \cdots \overline{b}_{|v|}$
$l > k > 0$ (3 comp. letters)
$\overline{b}_1 \cdots \overline{b}_i$
$\overline{b}_{i+1} \cdots \overline{b}_{|v|}$
$l > 0, k \leq 0$ (2 comp. letters)
$\overline{b}_1 \cdots \overline{b}_{|v|}$
$l, k \leq 0$ (1 comp. letter)

Fig. 1. Identifying subcases for rule simulation with i, j, k, and l.

In addition we have to consider the $\sharp$'s at the split points between w_2 and w_3, respectively between w_3 and w_4. It is clear that there exist w_2', w_2'', w_3', w_3'', w_3''', w_4', and w_4'' such that $w_2 = w_2' w_2''$, $w_3 = w_3' w_3'' w_3'''$, $w_4 = w_4' w_4''$, with $w_2'' w_3' = \sharp\sharp$ and $w_3''' w_4' = \sharp\sharp$.

Now the subcase 9.16. is identified by l and k as illustrated above, the length of w_2'', and the length of w_3'''. Example 4 above leads to $k = 1$, $l = 2$, $w_2'' = \square$, and $w_3''' = \sharp\sharp$. In general in this case we need a new nonterminal ξ_t which is added to Γ_2 and the following four rules:

$$
\begin{aligned}
r_{t,1} &:= (\xi_{w_1 w_2}\xi_{w_3}\xi_{w_4 w_5}x, \xi_{w_1 w_2}\xi_{w_3}\xi_t x)\\
r_{t,2} &:= (\xi_{w_1 w_2}\xi_{w_3}\xi_t x, \xi_{w_1 \overline{b}_1 (\sharp\sharp b_2)\cdots(\sharp\sharp b_k)\sharp\sharp}\xi_{w_3}\xi_t x)\\
r_{t,3} &:= (\xi_{w_1 \overline{b}_1 (\sharp\sharp b_2)\cdots(\sharp\sharp b_k)\sharp\sharp}\xi_{w_3}\xi_t x, \xi_{w_1 \overline{b}_1 (\sharp\sharp b_2)\cdots(\sharp\sharp b_k)\sharp\sharp}\xi_{b_{k+1}(\sharp\sharp b_{k+2})\cdots(\sharp\sharp b_{l+1})}\xi_t x)\\
r_{t,4} &:= (\xi_{w_1 \overline{b}_1 (\sharp\sharp b_2)\cdots(\sharp\sharp b_k)\sharp\sharp}\xi_{b_{k+1}(\sharp\sharp b_{k+2})\cdots(\sharp\sharp b_{l+1})}\xi_t x,\\
&\qquad \xi_{w_1 \overline{b}_1 (\sharp\sharp b_2)\cdots(\sharp\sharp b_k)\sharp\sharp}\xi_{b_{k+1}(\sharp\sharp b_{k+2})\cdots(\sharp\sharp b_{l+1})}\xi_{\sharp\sharp(b_{l+2}\sharp\sharp)\cdots(b_{|v|}\sharp\sharp)w_5}x)
\end{aligned}
$$

The following figure 2 illustrates the distribution of letters in the simulation, the length of the boxes is indicating their weight. Of the surplus letters $\sharp$ only the most important are explicitly shown.

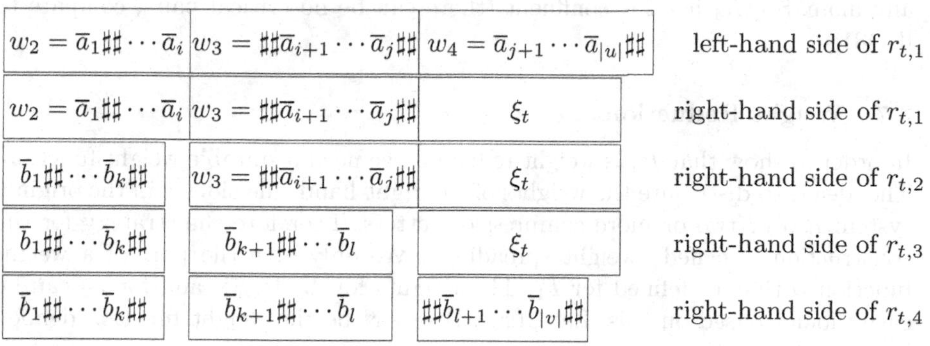

Fig. 2. Distribution of letters in the simulation.

We have seen how to handle case 9.16. In the other cases, the rules are derived in a similar way.

In order to assure confluence, the following changes to R_4 will be used. First of all, remember that the original rewriting system is confluent. That means, whenever two or more rules may be applied to one word, it does not matter which of them is chosen. Especially, with respect to acceptance or rejection of words the choice does not change the result. Now assume we had R_4 constructed as above. Then there may be rules that have overlapping left-hand sides. Note that these overlaps are always such that one left-hand side is a suffix of the other. This is due to the unique place of possible reductions obtained by the overlined index letters. Depending on the cases the possibly conflicting rules are derived from, we cannot control the resulting distribution of index letters over the compression letters. Therefore, R_4 might not be confluent. Instead, we drop some of the rules in three steps:

1. Omit all compressing rules (case 10. above), for which a left-hand side of a simulating rule is a suffix of or identical to the left-hand side of the compressing rule. Note that the compressing rule in that case is not necessary because the left-hand side of any simulating rule cannot be longer than that of a compressing rule.
2. Whenever two rules have the same left-hand side, chose one of them (arbitrarily). Because of the first step those rules always are either a single rule simulating a rule of R or the first of a sequence doing such a simulation. Therefore, if the rules dropped contain letters from Γ_2, we may also drop their successor rules.
3. Whenever one left-hand side of a rule is a suffix of the left-hand side of another rule, drop the rule with the longer left-hand side. Again, if we drop the start of a simulation sequence, the successors can be dropped, too.

It is important that dropping these rules does not change the accepted language, and that we may only do that because R is confluent.

After that process of omitting rules, R_4 has no overlaps of left-hand rule sides any more. So, R_4 itself is confluent (there can be no critical pairs, compare to [BO81]).

3.7 Weight Reduction

In order to show that R_4 is weight reducing, we need a suitable weight function. The idea is to distribute the weights of the right-hand rule sides v in the original system R over two or more compression letters. Therefore the strategy for the construction is called "weight spreading". We only give the part of a weight function ψ that is defined for Γ_1. The weights for Σ, $\{¢, \$\}$, and for Γ_2 can be easily found based on this. Let $\varphi(x) : W_{\sharp} \to \mathbb{N}$ be the weight function defined by:

$$\varphi(x) := \begin{cases} 2\mu_l + 2 & (x \in \Gamma) \\ 2\mu_l & \text{else} \end{cases}$$

Then $\psi(x) : \Gamma_1 \to \mathbb{N}$ is defined by $\psi(\xi_w) := \varphi(w) + 1$. The following property can be verified easily:

Claim. For all $\xi_v, \xi_w \in \Gamma_1$: $|v| > |w| \implies \psi(\xi_v) > \psi(\xi_w)$ and $\xi_{vw} \in \Gamma_1 \implies \psi(\xi_v) + \psi(\xi_w) > \psi(\xi_{vw})$.

So, ψ is the Γ_1 part of the required weight function. For Γ_2 the fact can be used that all Γ_1 weights are odd, so Γ_2 letters will have even weights just fitting "in between".

3.8 Final Rules

The last step is to define rules that accept the result of a reduction:

$$R_5 := \{(¢w\$, y) | w \in \Gamma_1^{\leq 3}, \check{w} = \overline{y}\sharp\sharp\}$$

3.9 Correctness of the Construction

Lemma 1. *Let* $\Gamma' := \Sigma \cup \{¢, \$, y\} \cup \Gamma_1 \cup \Gamma_2$, $\Sigma' := \Sigma$, $R' := R_1 \cup R_2 \cup R_3 \cup R_4 \cup R_5$. *Then* $C' := (\Gamma', \Sigma', R', ¢, \$, y)$ *is a* csCRLS *and* $L_{C'} = L_C$.

Proof.

1. To check if C' is well defined we only have to make sure that nowhere a $\xi_\Box$ would be necessary in R_4. By observing the relations between the subscript word lengths and n, i, j, k, and l of the subcases described above this can be shown to be true.
2. Checking the weight reduction of the rules is a rather tedious effort, but straight forward.
3. First, observe that the first three parts $R_1 \cup R_2 \cup R_3$ can be constructed to be confluent. Second, reduction rules of R_4 cannot overlap with those rules or with rules from R_5 (the latter because $y \in \text{IRR}(R)$). Overlaps between rules within R_4 cannot happen. (Note that as soon as a ξ_t is introduced there is always exactly one next rule that can be applied until the ξ_t is removed again. Especially, no such rule with ξ_t can be applied twice.) So, finally R' is confluent.
4. There exists a cover (comparable to the covers for context-free grammars discussed by Nijholt[2]) between accepting leftmost reductions that start with words in Σ^* in R and accepting reductions in R' on the same words. In consequence, both CRLS's accept the same languages.
5. Checking the context-splittability is the easiest part, it can be verified by simply looking at all rules.

With this lemma, the proof of the normal form theorem is complete. □

4 Conclusion and Further Questions

This paper shows that a normal form similar to context-sensitive grammars can be established for systems defining CRL's. The initial motivation of the author was to answer a question raised in the context of systems for prefix languages of CRL's [Woi00b]. There, a construction for systems defining prefix languages of CRL's was given which depends on the existence of the context-splittable normal form (it was called *prefix-splittable* at that time).

The normal form theorem is a very strong hint that the CRLS given by the construction in [Woi00b] cannot always be proved to be correct or false. This conjecture is due to the fact that CRL are a basis for the recursively enumerable languages. That means, given an alphabet Σ ($¢, \sharp \notin \Sigma$) and any r.e. language $L \subseteq \Sigma^*$ there is a CRL $L' \subseteq \Sigma^* \cdot \{¢\} \cdot \{\sharp\}^*$ such that deleting the letters ¢ and ♯ with a homomorphism h which leaves letters of Σ unchanged leads to $h(L') = L$ (see also [OKK97]).

[2] A. Nijholt. *Context-Free Grammars: Covers, Normal Forms, and Parsing.* Springer-Verlag, 1980.

Although the normal form theorem shows that *in principle* and *constructively* it is possible to find a prefix splittable system for every CRL, there can be conflicts with the prefix construction. This is a line of further research.

One can see that the piecewise simulation of rules leads to a rewriting system that has reducible right-hand sides. A rewriting system that has no such rules is called *interreduced*. The context-splittable normal form and the property of being interreduced seem to be dual to each other: We conjecture that there is a CRL that does not have an interreduced csCRLS. Besides, the construction of a csCRLS makes heavy use of weight reduction. This makes the existence of a length reducing context-splittable normal form doubtful: We conjecture that there is a CRL that has no length reducing csCRLS.

As mentioned above, the language classes CRL and the *deterministic growing context-sensitive languages (*DGCSL*)* are identical. The latter can be described by *shrinking deterministic two pushdown automata* sDTPDA whose definition is very similar to the automata mentioned above. The main differences are that they have bottom symbols and a state, that they can look at and replace both (single) top symbols of the stacks by arbitrarily long words (as long as the weight of the configuration shrinks), and the slightly different mode of accepting words. Niemann and Otto showed in [NO98] that this model is equivalent in its power of accepting languages to the definition of CRL's.

By dropping the condition of confluence we obtain the class of the *growing context-sensitive languages* (GCSL, defined in [DW86]) from the class CRL. We conjecture that a normal form corresponding to the one established here for CRL also holds for GCSL (thus justifying the use of the term *context-sensitive*). This would imply that the class of the *acyclic context-sensitive languages* (ACSL) coincides with GCSL (see [Bun96]).

Furthermore, by the construction given in this paper, it should be possible to require that the automata for CRL and GCSL are not allowed to replace the top symbols with arbitrarily long words. Instead, except for shifting operations, at most one letter per stack should suffice.

References

[BHNO00] M. Beaudry, M. Holzer, G. Niemann, and F. Otto. MacNaughton languages. Technical Report Mathematische Schriften Kassel, No. 26/00, Universität Kassel, November 2000.

[BO81] R. V. Book and C. O'Dunlaing. Testing for the Church-Rosser property. *Theoretical Computer Science*, 16:223–229, 1981.

[BO98] G. Buntrock and F. Otto. Growing context-sensitive languages and Church-Rosser languages. *Information and Computation*, 141:1–36, 1998.

[Boo82] R.V. Book. Confluent and other types of Thue systems. *JACM*, 29:171–182, 1982.

[Bun96] G. Buntrock. *Wachsende kontext-sensitive Sprachem.* Habilitationsschrift, Fakultät für Mathematik und Informatik, Universität Würzburg, Juli 1996.

[Cho59] N. Chomsky. On certain formal properties of grammars. *Information and Control*, 2(2):137–167, June 1959.

[DW86] E. Dahlhaus and M.K. Warmuth. Membership for growing context-sensitive grammars is polynomial. *Journal of Computer and System Sciences*, 33:456–472, 1986.

[Jan88] M. Jantzen. *Confluent String Rewriting*. Springer-Verlag, 1988.

[McN99] R. McNaughton. An insertion into the Chomsky hierarchy? In J. Karhumäki, H. A. Maurer, G. Paŭn, and G. Rozenberg, editors, *Jewels are Forever, Contributions on Theoretical Computer Science in Honour of Arto Salomaa*, pages 204–212. Springer-Verlag, 1999.

[MNO88] R. McNaughton, P. Narendran, and F. Otto. Church-Rosser Thue systems and formal languages. *Journal Association Computing Machinery*, 35:324–344, 1988.

[NO98] G. Niemann and F. Otto. The Church-Rosser languages are the deterministic variants of the growing context-sensitive languages. In M. Nivat, editor, *Foundations of Software Sscience and Computation Structures, Proceedings FoSSaCS'98*, volume 1378 of *LNCS*, pages 243–257, Berlin, 1998. Springer-Verlag.

[OKK97] F. Otto, M. Katsura, and Y. Kobayashi. Cross-sections for finitely presented monoids with decidable word problems. In H. Comon, editor, *Rewriting Techniques and Applications*, volume 1232 of *LNCS*, pages 53–67, Berlin, 1997. Springer-Verlag.

[Woi00a] J. R. Woinowski. A normal form for Church-Rosser language systems. Report, TU-Darmstadt, www.iti.informatik.tu-darmstadt.de/~woinowsk/, June 2000.

[Woi00b] J. R. Woinowski. Prefix languages of Church-Rosser languages. In *Proceedings of FST-TCS 2000 (Foundations of Software Technology/Theoretical Computer Science)*, Lecture Notes in Computer Science. Springer-Verlag, 2000.

Confluence and Termination of Simply Typed Term Rewriting Systems

Toshiyuki Yamada

Institute of Information Sciences and Electronics
University of Tsukuba, Tsukuba 305–8573, Japan

toshi@score.is.tsukuba.ac.jp

Abstract. We propose simply typed term rewriting systems (STTRSs), which extend first-order rewriting by allowing higher-order functions. We study a simple proof method for confluence which employs a characterization of the diamond property of a parallel reduction. By an application of the proof method, we obtain a new confluence result for orthogonal conditional STTRSs. We also discuss a semantic method for proving termination of STTRSs based on monotone interpretation.

1 Introduction

Higher-order function is one of the useful features in functional programming. The well-known higher-order function *map* takes a function as an argument and applies it to all elements of a list:

$$\begin{array}{lcl} map\ f\ [\,] & = & [\,] \\ map\ f\ (x : xs) & = & f\ x\ :\ map\ f\ xs \end{array}$$

It is not possible to directly express this definition by using a first-order term rewriting system, because the variable f is also used as a function. In order to deal with higher-order functions, one can use higher-order rewriting (e.g. Combinatory Reduction Systems [Klo80], Higher-Order Rewrite Systems [NP98]). Higher-order rewriting is a computation model which deals with higher-order terms. Higher-order functions and bound variables are usually used for constructing the set of higher-order terms. The use of bound variables enriches the descriptive power of higher-order rewrite systems. However, it makes the theory complicated. The aim of this paper is to give a simple definition of rewrite systems which conservatively extend the ordinary (first-order) term rewriting systems in such a way that they can naturally express equational specifications containing higher-order functions. We propose higher-order rewrite systems which are close to the format of functional programming languages with pattern matching. Based on the new definition of higher-order rewrite systems (given in the next section) we investigate the confluence property of the rewrite systems. We first study a method for proving confluence using parallel reduction (in Section 3). Based on the proof method introduced, we prove confluence of

A. Middeldorp (Ed.): RTA 2001, LNCS 2051, pp. 338–352, 2001.

orthogonal higher-order rewrite systems (in Section 4). This confluence result is further extended to the case of conditional higher-order rewriting (in Section 5). We also discuss a semantic method for proving termination of the newly proposed rewrite systems (in Section 6).

2 Simply Typed Term Rewriting Systems

We assume the reader is familiar with abstract rewrite systems (ARSs) and (first-order) term rewrite systems (TRSs). In this section we propose a simple extension of first-order rewrite systems (cf. [DJ90], [Klo92], [BN98]) to the higher-order case. The basic idea behind the following construction of terms is the same as typed combinatory logic (see e.g. [HS86]): variables may take arguments, and types are introduced in order to apply a function to its arguments correctly.

Definition 1 (simply typed term, position, substitution)
The set of *simple types* is the smallest set ST which contains the *base type* o and satisfies the property that $\tau_1 \times \cdots \times \tau_n \to \tau_0 \in \mathsf{ST}$ whenever $\tau_0, \tau_1, \ldots, \tau_n \in \mathsf{ST}$ with $n \geq 1$. We call a non-base type a *function type*. Let V^τ be a set of *variable symbols* of type τ and C^τ be a set of *constant symbols* of type τ, for every type τ. The set $\mathsf{T}(V,C)^\tau$ of *(simply typed) terms* of type τ is the smallest set satisfying the following two properties: (1) if $t \in V^\tau \cup C^\tau$ then $t \in \mathsf{T}(V,C)^\tau$, and (2) if $n \geq 1$, $t_0 \in \mathsf{T}(V,C)^{\tau_1 \times \cdots \times \tau_n \to \tau}$, and $t_i \in \mathsf{T}(V,C)^{\tau_i}$ for $i = 1, \ldots, n$, then $(t_0\, t_1 \cdots t_n) \in \mathsf{T}(V,C)^\tau$. Note that τ is not necessarily the base type. We also define $V = \bigcup_{\tau \in \mathsf{ST}} V^\tau, C = \bigcup_{\tau \in \mathsf{ST}} C^\tau$, and $\mathsf{T}(V,C) = \bigcup_{\tau \in \mathsf{ST}} \mathsf{T}(V,C)^\tau$. In order to be consistent with the standard definition of first-order terms, we use $t_0(t_1, \ldots, t_n)$ as an alternative notation for $(t_0\, t_1 \cdots t_n)$. The outermost parentheses of a term can be omitted. To enhance readability, infix notation is allowed. We use the notation t^τ to make the type τ of a term t explicit.

We do not confuse non-variable symbols with function symbols as in the first-order case, because a variable symbol of non-base type expresses a function. The term t_0 in a term of the form $(t_0\, t_1 \cdots t_n)$ expresses a function which is applied to the arguments $t_1, \cdots, t_n$. The construction of the simply typed terms allows arbitrary terms, including variables, at the position of t_0, while only constant symbols are allowed in the first-order case. Thus a set of first-order terms is obtained as a subset of simply typed terms which satisfies both $V^\tau = \varnothing$ whenever $\tau \neq o$, and $C^{\tau_1 \times \cdots \times \tau_n \to \tau} = \varnothing$ whenever $\tau_i \neq o$ for some i or $\tau \neq o$.

Let t be a term in $\mathsf{T}(V,C)$. The head symbol $\mathsf{head}(t)$ of t is defined as follows: (1) $\mathsf{head}(t) = t$ if $t \in V \cup C$, and (2) $\mathsf{head}(t) = t_0$ if $t = (t_0\, t_1 \cdots t_n)$. The set of variable symbols occurring in t is denoted by $\mathsf{Var}(t)$.

A *position* is a sequence of natural numbers. The empty sequence ϵ is called the *root position*. Positions are partially ordered by $\leq$ as follows: $p \leq q$ if there exists a position r such that $pr = q$. The set of positions in a term t is denoted by $\mathsf{Pos}(t)$. The *subterm* $t_{|p}$ of t at position p is defined as follows: (1) $t_{|p} = t$ if $p = \epsilon$, and (2) $t_{|p} = t_{i|q}$ if $t = (t_0\, t_1 \cdots t_n)$ and $p = iq$. The term obtained from a term s by replacing its subterm at position p with a term t is denoted by $s[t]_p$. The set $\mathsf{Pos}(t)$ is divided into three parts. We say $p \in \mathsf{Pos}(t)$ is at a *variable position* of t if $t_{|p} \in V$, at a *constant position* if $t_{|p} \in C$, otherwise p is at an *application position*. We denote the set of variable positions, constant positions, and application positions in a term t by $\mathsf{Pos_v}(t)$, $\mathsf{Pos_c}(t)$, and $\mathsf{Pos_a}(t)$, respectively.

A *substitution* σ is a function from V to $\mathsf{T}(V,C)$ such that its *domain*, defined as the set $\{\, x \in V \mid \sigma(x) \neq x \,\}$, is finite and $\sigma(x) \in \mathsf{T}(V,C)^\tau$ whenever $x \in V^\tau$. A substitution $\sigma : V \to \mathsf{T}(V,C)$ is extended to the function $\overline{\sigma} : \mathsf{T}(V,C) \to \mathsf{T}(V,C)$ as follows: (1) $\overline{\sigma}(t) = \sigma(t)$ if $t \in V$, (2) $\overline{\sigma}(t) = t$ if $t \in C$, and (3) $\overline{\sigma}(t) = (\overline{\sigma}(t_0)\,\overline{\sigma}(t_1) \cdots \overline{\sigma}(t_n))$ if $t = (t_0\,t_1 \cdots t_n)$. We will write $t\sigma$ instead of $\overline{\sigma}(t)$. A *renaming* is a bijective substitution from V to V. Two terms t and s are *unifiable* by a *unifier* σ if $s\sigma = t\sigma$.

Definition 2 (rewrite rule and rewrite relation)
Let $\mathsf{T}(V,C)$ be a set of simply typed terms. A *rewrite rule* is a pair of terms, written as $l \to r$, such that $\mathsf{Var}(r) \subseteq \mathsf{Var}(l)$, $\mathsf{head}(l) \in C$, and l and r are of the same type. The terms l and r are called the *left-hand side* and the *right-hand side* of the rewrite rule, respectively. Let R be a set of rewrite rules. We call $\mathcal{R} = (R, V, C)$ a *simply typed term rewriting system* (STTRS for short).

Let $\mathcal{R} = (R, V, C)$ be an STTRS. We say a term s *rewrites to* t, and write $s \to_{\mathcal{R}} t$, if there exists a rewrite rule $l \to r \in R$, a position $p \in \mathsf{Pos}(s)$ and a substitution σ such that $s_{|p} = l\sigma$ and $t = s[r\sigma]_p$. We call the subterm $s_{|p}$ a *redex* of s. In order to make the position p of a rewrite step explicit, we also use the notation $s \xrightarrow{p}_{\mathcal{R}} t$. Especially, a rewrite step at root position is denoted by $\xrightarrow{\epsilon}_{\mathcal{R}}$ and a rewrite step at non-root position is denoted by $\xrightarrow{>\epsilon}_{\mathcal{R}}$. When the underlying STTRS $\mathcal{R}$ is clear from the context, we may omit the subscript $\mathcal{R}$ in $\to_{\mathcal{R}}$.

The definition of STTRS is close to the algebraic functional systems defined by Jouannaud and Okada [JO91]. The main difference is that we dispenses with bound variables since our focus is on higher-order functions and not on quantification. It is easy to see that every variable symbol is in normal form because of the second restriction imposed on the rewrite rules. The following example shows that a simple functional programming language with pattern matching can be modeled by STTRS.

Example 3 (functional programming by STTRS)
Let C be the set which consists of constant symbols 0°, $\mathsf{S}^{o \to o}$, $[]^o$, $:^{o \times o \to o}$, $\mathsf{map}^{(o \to o) \times o \to o}$, $\circ^{(o \to o) \times (o \to o) \to (o \to o)}$, and $\mathsf{twice}^{(o \to o) \to (o \to o)}$, where $:$ and $\circ$ are infix symbols, and the set of variable symbols V contains x^o, xs^o, $F^{o \to o}$, and $G^{o \to o}$. We define the set of rewrite rules R as follows:

$$R = \left\{ \begin{array}{ll} \mathsf{map}\ F\ [] & \to [] \\ \mathsf{map}\ F\ (x : xs) & \to F\,x\ :\ \mathsf{map}\ F\ xs \\ (F \circ G)\ x & \to F\ (G\ x) \\ \mathsf{twice}\ F & \to F \circ F \end{array} \right\}.$$

Examples of rewrite sequences of the TRS $\mathcal{R} = (R, V, C)$ are:

$$\begin{array}{rl}
\mathsf{map}\ \underline{(\mathsf{twice\ S})}\ (0 : []) \to_{\mathcal{R}} & \underline{\mathsf{map}\ (\mathsf{S} \circ \mathsf{S})\ (0 : [])} \\
\to_{\mathcal{R}} & \underline{(\mathsf{S} \circ \mathsf{S})\ 0}\ :\ \mathsf{map}\ (\mathsf{S} \circ \mathsf{S})\ [] \\
\to_{\mathcal{R}} & \mathsf{S(S0)}\ :\ \underline{\mathsf{map}\ (\mathsf{S} \circ \mathsf{S})\ []} \\
\to_{\mathcal{R}} & \mathsf{S(S0)}\ :\ [] \\
\underline{\mathsf{map}\ (\mathsf{twice\ S})\ (0 : [])} \to_{\mathcal{R}} & \underline{(\mathsf{twice\ S})\ 0}\ :\ \mathsf{map}\ (\mathsf{twice\ S})\ [] \\
\to_{\mathcal{R}} & \underline{(\mathsf{S} \circ \mathsf{S})\ 0}\ :\ \mathsf{map}\ (\mathsf{twice\ S})\ [] \\
\to_{\mathcal{R}} & \mathsf{S(S0)}\ :\ \underline{\mathsf{map}\ (\mathsf{twice\ S})\ []} \\
\to_{\mathcal{R}} & \mathsf{S(S0)}\ :\ []
\end{array}$$

where the underlined redexes are rewritten.

3 Confluence by Parallel Moves

In this section, we develop a method for proving confluence using parallel reduction. We assume the reader is familiar with the basic notions of abstract rewrite systems (ARSs). For more detailed descriptions on abstract rewriting, see, for example, Klop's survey [Klo92].

Definition 4 (properties of an ARS)
Let $\mathcal{A} = (A, \to)$ be an ARS. We use the following abbreviations:

property	definition	abbreviation
$\mathcal{A}$ has the *diamond property*	$\leftarrow \cdot \to \subseteq \to \cdot \leftarrow$	$\diamond(\to)$
$\mathcal{A}$ is *confluent*	${}^*\!\leftarrow \cdot \to^* \subseteq \to^* \cdot {}^*\!\leftarrow$	$\mathrm{CR}(\to)$

Lemma 5 (confluence by simultaneous reduction)
Let $(A, \to)$ and $(A, \hookrightarrow)$ be ARSs.
(1) $\diamond(\to) \implies \mathrm{CR}(\to)$.
(2) If $\hookrightarrow^* = \to^*$ then $\mathrm{CR}(\hookrightarrow) \iff \mathrm{CR}(\to)$.

Proof. Straightforward. □

Definition 6 (parallel reduction)
Let $\mathcal{R} = (R, V, C)$ be an STTRS. The *parallel reduction relation* induced by $\mathcal{R}$ is the smallest relation $\nrightarrow\!\!\!\!\!\!\to_{\mathcal{R}}$ such that
(1) if $t \in V \cup C$ then $t \nrightarrow\!\!\!\!\!\!\to_{\mathcal{R}} t$,
(2) if $s \overset{\epsilon}{\to}_{\mathcal{R}} t$ then $s \nrightarrow\!\!\!\!\!\!\to_{\mathcal{R}} t$, and
(3) if $n \geq 1$ and $s_i \nrightarrow\!\!\!\!\!\!\to_{\mathcal{R}} t_i$ for $i = 0, \ldots, n$, then $(s_0\, s_1 \cdots s_n) \nrightarrow\!\!\!\!\!\!\to_{\mathcal{R}} (t_0\, t_1 \cdots t_n)$.
We may omit the underlying TRS $\mathcal{R}$ in $\nrightarrow\!\!\!\!\!\!\to_{\mathcal{R}}$ if it is not important.

One can easily verify that every parallel reduction relation is reflexive. Note also that $\to \subseteq \nrightarrow\!\!\!\!\!\!\to \subseteq \to^*$, hence $\nrightarrow\!\!\!\!\!\!\to^* = \to^*$. From Lemma 5 we know that the diamond property of a parallel reduction relation is a sufficient condition for the confluence of the underlying TRS: $\diamond(\nrightarrow\!\!\!\!\!\!\to) \implies \mathrm{CR}(\nrightarrow\!\!\!\!\!\!\to) \iff \mathrm{CR}(\to)$. The following lemma gives a characterization of the diamond property of a parallel reduction, which is inspired by Gramlich's characterization of the strong confluence of a parallel reduction [Gra96].

Lemma 7 (parallel moves)
We have $\mathrel{\longleftarrow\mathllap{\parallel\,}} \cdot \stackrel{\epsilon}{\rightarrow} \subseteq \mathrel{\mathrlap{\,\parallel}\longrightarrow} \cdot \mathrel{\longleftarrow\mathllap{\parallel\,}} \iff \mathrel{\longleftarrow\mathllap{\parallel\,}} \cdot \mathrel{\mathrlap{\,\parallel}\longrightarrow} \subseteq \mathrel{\mathrlap{\,\parallel}\longrightarrow} \cdot \mathrel{\longleftarrow\mathllap{\parallel\,}}$.

Proof. The implication from right to left is obvious by $\stackrel{\epsilon}{\rightarrow} \subseteq \mathrel{\mathrlap{\,\parallel}\longrightarrow}$. For the reverse implication, suppose $\mathrel{\longleftarrow\mathllap{\parallel\,}} \cdot \stackrel{\epsilon}{\rightarrow} \subseteq \mathrel{\mathrlap{\,\parallel}\longrightarrow} \cdot \mathrel{\longleftarrow\mathllap{\parallel\,}}$. We show that if $t \mathrel{\longleftarrow\mathllap{\parallel\,}} s \mathrel{\mathrlap{\,\parallel}\longrightarrow} u$ then there exists a term v such that $t \mathrel{\mathrlap{\,\parallel}\longrightarrow} v \mathrel{\longleftarrow\mathllap{\parallel\,}} u$, for all terms s, t, and u. The proof is by induction on the structure of s. We distinguish three cases according to the parallel reduction $s \mathrel{\mathrlap{\,\parallel}\longrightarrow} t$. If $s = t \in V \cup C$ then $t \mathrel{\mathrlap{\,\parallel}\longrightarrow} v = u$ by taking $v = u$. If $s \stackrel{\epsilon}{\rightarrow} t$ or $s \stackrel{\epsilon}{\rightarrow} u$ then we use the assumption. Otherwise, we have $s = (s_0\, s_1 \cdots s_n)$, $t = (t_0\, t_1 \cdots t_n)$, and $u = (u_0\, u_1 \ldots u_n)$ for some $n \geq 1$ with $t_i \mathrel{\longleftarrow\mathllap{\parallel\,}} s_i \mathrel{\mathrlap{\,\parallel}\longrightarrow} u_i\, (i = 0, \ldots, n)$. By the induction hypothesis, we know the existence of terms v_i $(i = 0, \ldots, n)$ such that $t_i \mathrel{\mathrlap{\,\parallel}\longrightarrow} v_i \mathrel{\longleftarrow\mathllap{\parallel\,}} u_i$. Let $v = (v_0\, v_1 \ldots v_n)$. Then we have $t \mathrel{\mathrlap{\,\parallel}\longrightarrow} v \mathrel{\longleftarrow\mathllap{\parallel\,}} u$ by definition. □

This lemma allows us to partially localize the test for the diamond property of a parallel reduction, though the complete localization is impossible as shown in the following example.

Example 8 (complete localization of parallel moves)
In this example we show that the implication $\leftarrow \cdot \stackrel{\epsilon}{\rightarrow} \subseteq \mathrel{\mathrlap{\,\parallel}\longrightarrow} \cdot \mathrel{\longleftarrow\mathllap{\parallel\,}} \implies \mathrel{\longleftarrow\mathllap{\parallel\,}} \cdot \stackrel{\epsilon}{\rightarrow} \subseteq \mathrel{\mathrlap{\,\parallel}\longrightarrow} \cdot \mathrel{\longleftarrow\mathllap{\parallel\,}}$ does not hold in general. Let $V = \varnothing$ and C be the set consisting of the constant symbol f of type $o \times o \to o$ and constant symbols $\mathsf{a}, \mathsf{b}, \mathsf{c}, \mathsf{d}, \mathsf{e}$ of type o. Consider the set R of rewrite rules defined by

$$R = \left\{ \begin{array}{ll} \mathsf{f\,a\,a} \to \mathsf{c} & \mathsf{a} \to \mathsf{b} \\ \mathsf{f\,a\,b} \to \mathsf{d} & \mathsf{c} \to \mathsf{d} \\ \mathsf{f\,b\,a} \to \mathsf{d} & \mathsf{d} \to \mathsf{e} \\ \mathsf{f\,b\,b} \to \mathsf{e} & \end{array} \right\}.$$

It is easy to see that the inclusion $\leftarrow \cdot \stackrel{\epsilon}{\rightarrow} \subseteq \mathrel{\mathrlap{\,\parallel}\longrightarrow} \cdot \mathrel{\longleftarrow\mathllap{\parallel\,}}$ holds. We have $\mathsf{f\,b\,b} \mathrel{\longleftarrow\mathllap{\parallel\,}} \mathsf{f\,a\,a} \stackrel{\epsilon}{\rightarrow} \mathsf{c}$ but $\mathsf{f\,b\,b} \mathrel{\mathrlap{\,\parallel}\longrightarrow} \cdot \mathrel{\longleftarrow\mathllap{\parallel\,}} \mathsf{c}$ is not satisfied.

Definition 9 (parallel moves property)
We say an STTRS satisfies the *parallel moves property*, and write $\mathrm{PM}(\rightarrow)$, if the inclusion $\mathrel{\longleftarrow\mathllap{\parallel\,}} \cdot \stackrel{\epsilon}{\rightarrow} \subseteq \mathrel{\mathrlap{\,\parallel}\longrightarrow} \cdot \mathrel{\longleftarrow\mathllap{\parallel\,}}$ holds.

Lemma 7 states that the parallel moves property is equivalent to the diamond property of a parallel reduction. The parallel moves property is a useful sufficient condition for proving the confluence of orthogonal rewrite systems.

Lemma 10 (confluence by parallel moves)
Every STTRS with the parallel moves property is confluent.
Proof. We have $\mathrm{PM}(\rightarrow) \iff \diamond(\mathrel{\mathrlap{\,\parallel}\longrightarrow}) \implies \mathrm{CR}(\rightarrow)$ by Lemmata 7 and 5 with $\mathrel{\mathrlap{\,\parallel}\longrightarrow}^* = \rightarrow^*$, see Fig.1. □

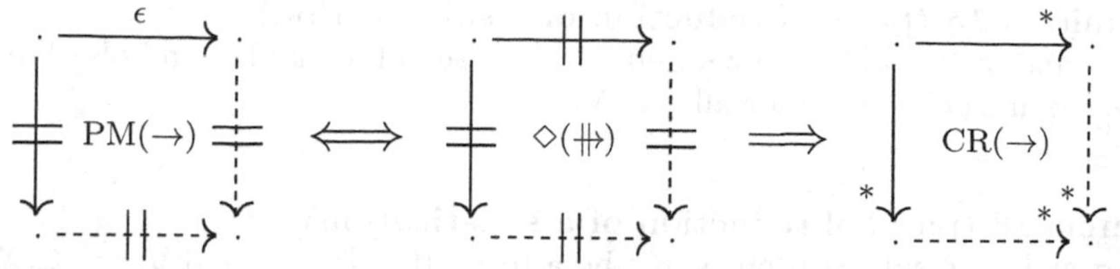

Fig. 1. Confluence by parallel moves

4 Confluence of STTRSs

In this section, we give a simple proof of the confluence of orthogonal STTRSs based on the parallel moves property. We are interested in orthogonal STTRSs, which are of practical importance, although there are confluence results obtained by weakening orthogonality, see, for example, [Hue80]. For the definition of orthogonality, we need the notions of left-linearity and overlap.

Definition 11 (left-linearity)
An STTRS is *left-linear* if none of left-hand sides of rewrite rules contain multiple occurrences of a variable symbol.

Definition 12 (overlap)
An STTRS is *overlapping* if there exist rewrite rules $l \to r$ and $l' \to r'$ without common variable symbols (after renaming) and a non-variable position $p \in \mathsf{Pos_c}(l) \cup \mathsf{Pos_a}(l)$ such that
- $l_{|p}$ and l' are unifiable, and
- if $p = \epsilon$ then $l \to r$ is not obtained from $l' \to r'$ by renaming variable symbols.

Example 13 (overlapping STTRS)
Let C be the set consisting of constant symbols $\mathsf{f}, \mathsf{g}, \mathsf{h}$ of type $o \to o$ and a, b of type o. Let V contain variable symbols F of type $o \to o$ and x of type o. We define

$$R = \left\{ \begin{array}{c} \mathsf{f}\ (F\ x) \to F\ \mathsf{b} \\ \mathsf{g}\ \mathsf{a} \to \mathsf{h}\ \mathsf{b} \end{array} \right\}.$$

The STTRS $\mathcal{R} = (R, V, C)$ is overlapping because the left-hand sides of the rewrite rules are unifiable at an application-position. In this STTRS the term $\mathsf{f}\ (\mathsf{g}\ \mathsf{a})$ has two different normal forms: $\mathsf{g}\ \mathsf{b}\ {}_{\mathcal{R}}\!\leftarrow \mathsf{f}\ (\mathsf{g}\ \mathsf{a}) \to_{\mathcal{R}} \mathsf{f}\ (\mathsf{h}\ \mathsf{b}) \to_{\mathcal{R}} \mathsf{h}\ \mathsf{b}$. Note that the redex $(\mathsf{g}\ \mathsf{a})$ in the initial term is destroyed by the application of the first rewrite rule.

Definition 14 (orthogonality)
An STTRS is *orthogonal* if it is left-linear and non-overlapping.

Now we are ready to give a proof that orthogonal STTRSs are confluent. We first extend the use of parallel reduction to substitutions.

Definition 15 (parallel reduction of a substitution)
Let σ and τ be substitutions and X be a set of variable symbols. We write $\sigma \mathrel{+\!\!\!\!\!\rightarrow}_{[X]} \tau$ if $\sigma(x) \mathrel{+\!\!\!\!\!\rightarrow} \tau(x)$ for all $x \in X$.

Lemma 16 (parallel reduction of a substitution)
Let σ and τ be substitutions and t be a term. If $\sigma \mathrel{+\!\!\!\!\!\rightarrow}_{[X]} \tau$ and $\mathsf{Var}(t) \subseteq X$ then $t\sigma \mathrel{+\!\!\!\!\!\rightarrow} t\tau$.

Proof. An easy induction on the structure of t. □

Lemma 17 (key properties for confluence)
Let $\mathcal{R} = (R, V, C)$ be an orthogonal STTRS.
(1) $\overset{\epsilon}{\leftarrow} \cdot \overset{\epsilon}{\rightarrow} \subseteq =$.
(2) For all rewrite rules $l \rightarrow r \in R$, substitutions σ, and terms t, if $t \mathrel{\leftarrow\!\!\!\!\!+} l\sigma \overset{\epsilon}{\rightarrow} r\sigma$ and not $l\sigma \overset{\epsilon}{\rightarrow} t$ then there exists a substitution τ such that $t = l\tau$ with $\sigma \mathrel{+\!\!\!\!\!\rightarrow}_{[\mathsf{Var}(l)]} \tau$.

Proof.
(1) Since $\mathcal{R}$ is non-overlapping, we can only use (two renamed versions of) the same rewrite rule in R for rewriting a term at the same position. Hence we always obtain the same term.
(2) Since $\mathcal{R}$ is non-overlapping, there is no term s' with $l_{|p}\sigma \overset{\epsilon}{\rightarrow} s'$, for all non-variable position p in l. Hence we have $l_{|p}\sigma \mathrel{+\!\!\!\!\!\rightarrow} t_{|p}$ for all variable positions p in l. Define the substitution τ by $\tau(x) = t_{|p}$ if $l_{|p} = x$ and $\tau(x) = x$ otherwise. This substitution is well-defined because there are no multiple occurrences of x in l by left-linearity. It is easy to see that $\sigma \mathrel{+\!\!\!\!\!\rightarrow}_{[\mathsf{Var}(l)]} \tau$ and $t = l\tau$ by construction.

□

Lemma 18 (Parallel Moves Lemma)
Every orthogonal TRS $\mathcal{R} = (R, V, C)$ has the parallel moves property, i.e., $\mathrel{\leftarrow\!\!\!\!\!+} \cdot \overset{\epsilon}{\rightarrow} \subseteq \mathrel{+\!\!\!\!\!\rightarrow} \cdot \mathrel{\leftarrow\!\!\!\!\!+}$.

Proof. Suppose $t \mathrel{\leftarrow\!\!\!\!\!+} s \overset{\epsilon}{\rightarrow} u$. We show $t \mathrel{+\!\!\!\!\!\rightarrow} \cdot \mathrel{\leftarrow\!\!\!\!\!+} u$. If $s \overset{\epsilon}{\rightarrow} t$ then the desired result follows from Lemma 17(1). Otherwise, we do not have $s \overset{\epsilon}{\rightarrow} t$. Since there exists a rewrite rule $l \rightarrow r \in R$ and a substitution σ such that $s = l\sigma$ and $u = r\sigma$, we know the existence of a substitution τ such that $t = l\tau$ with $\sigma \mathrel{+\!\!\!\!\!\rightarrow}_{[\mathsf{Var}(l)]} \tau$ by Lemma 17(2). Therefore, $t = l\tau \overset{\epsilon}{\rightarrow} r\tau \mathrel{\leftarrow\!\!\!\!\!+} r\sigma = u$ by Lemma 16 and $\mathsf{Var}(r) \subseteq \mathsf{Var}(l)$. Note that the case $s = t \in V$ is impossible because every variable symbol is in normal form and that the case $s = t \in C$ is contained in the case $s \overset{\epsilon}{\rightarrow} u$. □

Note that the Parallel Moves Lemma does not hold for orthogonal higher-order rewrite systems with bound variables as observed in the literature, see [vO97] and [vR99]. A proof of confluence in such rewrite systems can be found in [MN98]. Now we conclude this section by the main result of this section and an example of its application.

Theorem 19 (confluence by orthogonality)
Every orthogonal TRS is confluent.

Proof. By Lemma 10 and the Parallel Moves Lemma (Lemma 18). □

Example 20 (confluence by orthogonality)
The example STTRS given in Example 3 is confluent because it is orthogonal.

5 Confluence of Conditional STTRSs

In this section, we generalize the confluence result presented in the previous section to the case of conditional rewriting. Bergstra and Klop proved the confluence of first-order orthogonal CTRSs in [BK86]. Their proof depends on the notion of development and the fact that every development is finite. Our result in this section generalizes their result to STTRSs and also simplifies their confluence proof, based on the parallel moves property.

Definition 21 (conditional rewrite rule)
Let $\mathsf{T}(V,C)$ be a set of terms. A *conditional rewrite rule* $l \to r \Leftarrow c$ consists of a rewrite rule $l \to r$ and the *conditional part* c. Here c is a possibly empty finite sequence $c = l_1 \approx r_1, \dots, l_n \approx r_n$ of equations such that every pair of terms l_i and r_i are of the same type and $\mathsf{Var}(r) \subseteq \mathsf{Var}(c)$. If the conditional part is empty, we may simply write $l \to r$. Let R be a set of conditional rewrite rules. We call $\mathcal{R} = (R, V, C)$ a *conditional* STTRS.

A variable in the right-hand side or in the conditional part of a rewrite rule which does not appear in the corresponding left-hand side is called an extra variable. In this paper, we allow extra variables only in the conditional part but not in the right-hand sides of rewrite rules.

Definition 22 (rewrite relation)
The rewrite relation $\to_{\mathcal{R}}$ of a conditional STTRS $\mathcal{R} = (R, V, C)$ is defined as follows: $s \to_{\mathcal{R}} t$ if and only if $s \to_{\mathcal{R}_k} t$ for some $k \geq 0$. The minimum such k is called the *level* of the rewrite step. Here the relations $\to_{\mathcal{R}_k}$ are inductively defined:

$$\begin{aligned} \to_{\mathcal{R}_0} &= \varnothing, \\ \to_{\mathcal{R}_{k+1}} &= \{\, (t[l\sigma]_p, t[r\sigma]_p) \mid l \to r \Leftarrow c \in R,\ c\sigma \subseteq \to^*_{\mathcal{R}_k} \}. \end{aligned}$$

Here $c\sigma$ denotes the set $\{\, l'\sigma \approx r'\sigma \mid l' \approx r'$ belongs to $c\}$. Therefore $c\sigma \subseteq \to^*_{\mathcal{R}_k}$ with $c = l_1 \approx r_1, \dots, l_n \approx r_n$ is a shorthand for $l_1 \to^*_{\mathcal{R}_k} r_1, \dots, l_n \to^*_{\mathcal{R}_k} r_n$. We may abbreviate $\to_{\mathcal{R}_k}$ to $\to_k$ if there is no need to make the underlying conditional STTRS explicit.

Properties of conditional STTRSs are often proved by induction on the level of a rewrite step. So, it is useful for proving confluence of orthogonal conditional STTRSs to introduce the parallel reduction relations which are indexed by levels.

Definition 23 (parallel reduction relations indexed by levels)
Let $\mathcal{R}$ be a conditional STTRS. We define $\mathrel{\#\!\!\!\!\rightarrow}_{\mathcal{R}_k}$ as the smallest relation such that
(1) $t \mathrel{\#\!\!\!\!\rightarrow}_{\mathcal{R}_k} t$ for all terms t,
(2) if $s \overset{\epsilon}{\rightarrow}_{\mathcal{R}_k} t$ then $s \mathrel{\#\!\!\!\!\rightarrow}_{\mathcal{R}_k} t$, and
(3) if $k \geq j_i$ and $s_i \mathrel{\#\!\!\!\!\rightarrow}_{\mathcal{R}_{j_i}} t_i$ for $i = 0, \ldots, n$, then $(s_0\, s_1 \cdots s_n) \mathrel{\#\!\!\!\!\rightarrow}_{\mathcal{R}_k} (t_0\, t_1 \cdots t_n)$.
We may abbreviate $\mathrel{\#\!\!\!\!\rightarrow}_{\mathcal{R}_k}$ to $\mathrel{\#\!\!\!\!\rightarrow}_k$ when no confusion can arise.

Observe that $s \mathrel{\#\!\!\!\!\rightarrow}_{\mathcal{R}} t$ if and only if $s \mathrel{\#\!\!\!\!\rightarrow}_{\mathcal{R}_k} t$ for some level $k \geq 0$. It is also easy to verify that $\rightarrow_{\mathcal{R}_k} \subseteq \mathrel{\#\!\!\!\!\rightarrow}_{\mathcal{R}_k} \subseteq \rightarrow^*_{\mathcal{R}_k}$ for all levels $k \geq 0$.

Definition 24 (properties of an ARS with indexes)
Let $\mathcal{A} = (A, \bigcup_{i \in I} \rightarrow_i)$ be an ARS whose rewrite relations are indexed. We use the following abbreviations:

definition	abbreviation
${}_j\!\leftarrow \cdot \rightarrow_k \subseteq \rightarrow_k \cdot {}_j\!\leftarrow$	$\diamond^j_k(\rightarrow)$
${}^*_j\!\leftarrow \cdot \rightarrow^*_k \subseteq \rightarrow^*_k \cdot {}^*_j\!\leftarrow$	$\mathrm{CR}^j_k(\rightarrow)$

Lemma 25 (confluence by simultaneous reduction)
Let $(A, \bigcup_{i \in I} \rightarrow_i)$ and $(A, \bigcup_{i \in I} \hookrightarrow_i)$ be ARSs such that $\hookrightarrow^*_i = \rightarrow^*_i$ for all $i \in I$. We have $\diamond^j_k(\hookrightarrow) \implies {}_j\!\hookleftarrow \cdot \rightarrow^*_k \subseteq \rightarrow^*_k \cdot {}_j\!\hookleftarrow \implies \mathrm{CR}^j_k(\rightarrow)$.

Proof. Straightforward. □

Definition 26 (parallel moves property for conditional STTRSs)
We say a conditional STTRS satisfies the *parallel moves property* with respect to levels j and k, and write $\mathrm{PM}^j_k(\rightarrow)$, if the inclusion ${}_j\!\mathrel{\leftarrow\!\!\!\!\#} \cdot \overset{\epsilon}{\rightarrow}_k \subseteq \mathrel{\#\!\!\!\!\rightarrow}_k \cdot {}_j\!\mathrel{\leftarrow\!\!\!\!\#}$ holds.

Lemma 27 (parallel moves for conditional STTRSs)
The following two statements are equivalent, for all $m \geq 0$.
(1) $\mathrm{PM}^j_k(\rightarrow)$ for all j, k with $j + k \leq m$.
(2) $\diamond^j_k(\mathrel{\#\!\!\!\!\rightarrow})$ for all j, k with $j + k \leq m$.

Proof. The implication (2) ⇒ (1) is obvious because $\overset{\epsilon}{\rightarrow}_k \subseteq \mathrel{\#\!\!\!\!\rightarrow}_k$ by definition. For the proof of the implication (1) ⇒ (2), suppose statement (1) holds. We show that if $t \;{}_j\!\mathrel{\leftarrow\!\!\!\!\#}\; s \mathrel{\#\!\!\!\!\rightarrow}_k u$ and $j + k \leq m$ then there exists a term v such that $t \mathrel{\#\!\!\!\!\rightarrow}_k v \;{}_j\!\mathrel{\leftarrow\!\!\!\!\#}\; u$, for all terms s, t, u and levels j, k. The proof is by induction on the structure of s. We distinguish three cases according to the parallel reduction $s \mathrel{\#\!\!\!\!\rightarrow}_j t$. If $s = t$ then $t \mathrel{\#\!\!\!\!\rightarrow}_k v = u$ by taking $v = u$. If $s \overset{\epsilon}{\rightarrow}_j t$ or $s \overset{\epsilon}{\rightarrow}_k u$ then we can use the assumption (1) in both cases because $j + k \leq m$. Otherwise, we have $s = (s_0\, s_1 \cdots s_n)$, $t = (t_0\, t_1 \cdots t_n)$, and $u = (u_0\, u_1 \ldots u_n)$ for some $n \geq 1$ with $t_i \;{}_{j_i}\!\mathrel{\leftarrow\!\!\!\!\#}\; s_i \mathrel{\#\!\!\!\!\rightarrow}_{k_i} u_i (i = 0, \ldots, n)$. Since $j_i + k_i \leq j + k \leq m$, the induction hypothesis yields the existence of terms v_i such that $t_i \mathrel{\#\!\!\!\!\rightarrow}_{k_i} v_i \;{}_{j_i}\!\mathrel{\leftarrow\!\!\!\!\#}\; u_i$, for $i = 0, \ldots, n$. Let $v = (v_0\, v_1 \ldots v_n)$. Then we have $t \mathrel{\#\!\!\!\!\rightarrow}_k v \;{}_j\!\mathrel{\leftarrow\!\!\!\!\#}\; u$ by definition. □

The following lemma gives a sufficient condition for the confluence of CTRSs.

Lemma 28 (confluence by parallel moves)
If a CTRS satisfies $\mathrm{PM}^j_k(\rightarrow)$ for all levels j and k then $\mathrm{CR}^j_k(\rightarrow)$ holds for all j and k, hence it is confluent.

Proof. By Lemmata 27 and 25 with $\twoheadrightarrow\!\!\!\!\!\!+_i^* = \rightarrow_i^*$ for all levels i, see Fig.2. □

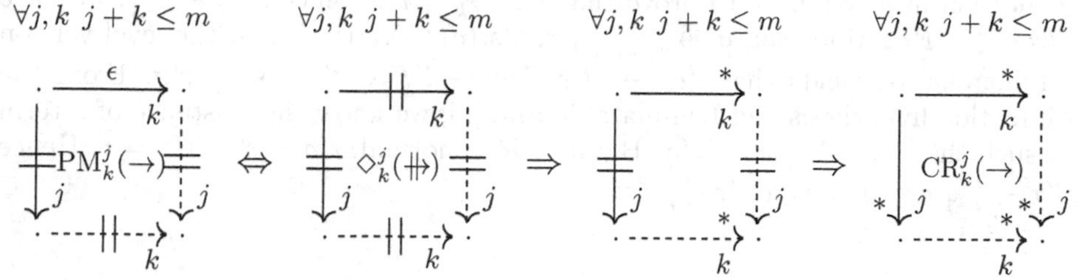

Fig. 2. Confluence of CTRSs by parallel moves

Imposing restrictions on reducibility of the right-hand sides of the conditions is important for ensuring the confluence of conditional STTRSs.

Definition 29 (normal conditional STTRS)
Let $\mathcal{R}$ be a conditional STTRS. A term t is called *normal* if it contains no variables and is a normal form with respect to the unconditional version of $\mathcal{R}$. Here the unconditional version is obtained from $\mathcal{R}$ by dropping all conditions. A conditional STTRS is called *normal* if every right-hand side of an equation in the conditional part of a rewrite rule is normal.

We extend the indexed version of a parallel reduction $\twoheadrightarrow\!\!\!\!\!\!+_k$ to the relation $\twoheadrightarrow\!\!\!\!\!\!+_{k[X]}$ on substitutions as in the unconditional case (Definition 15). It is easy to verify that the level version of Lemma 16 also holds. Now we are ready for proving the Parallel Moves Lemma for conditional STTRSs. In the conditional case, we must confirm that the conditions are satisfied after the change of the substitution.

Lemma 30 (Parallel Moves Lemma for conditional STTRSs)
Every orthogonal normal conditional STTRS $\mathcal{R} = (R, V, C)$ satisfies $\mathrm{PM}^j_k(\rightarrow)$, i.e., ${}_j\!\twoheadleftarrow\!\!\!\!\!\!+ \cdot \xrightarrow{\epsilon}_k \subseteq \twoheadrightarrow\!\!\!\!\!\!+_k \cdot {}_j\!\twoheadleftarrow\!\!\!\!\!\!+$, for all levels j and k.

Proof. We show that if $t \;{}_j\!\twoheadleftarrow\!\!\!\!\!\!+\; s \xrightarrow{\epsilon}_k u$ then there exists a term v such that $t \twoheadrightarrow\!\!\!\!\!\!+_k v \;{}_j\!\twoheadleftarrow\!\!\!\!\!\!+\; u$. The proof is by induction on $j + k$. The case $j + k = 0$ is trivial because $\xrightarrow{\epsilon}_0 = \varnothing$. Suppose $j + k > 0$. We distinguish two cases. If $s \xrightarrow{\epsilon}_j t$ then we have $s = u$ because $\mathcal{R}$ is non-overlapping and has no extra variable in the right hand sides of R. Hence we can take $v = s = u$. Consider the case that $s \xrightarrow{\epsilon}_j t$ does not hold. From $s \xrightarrow{\epsilon}_k u$ we know that there exists a conditional rewrite rule $l \to r \Leftarrow c \in R$ and a substitution σ such that $s = l\sigma$, $u = r\sigma$, and $c\sigma \subseteq \to^*_{k-1}$. Since $\mathcal{R}$ is non-overlapping, there is no term s' with $l_{|p}\sigma \xrightarrow{\epsilon}_j s'$ for all non-variable positions p in l. Define the substitution τ by $\tau(x) = t_{|p}$ if $l_{|p} = x$ and $\tau(x) = \sigma(x)$ otherwise. This substitution is well-defined by the left-linearity of $\mathcal{R}$. We have $\sigma \twoheadrightarrow\!\!\!\!\!\!\,\|\,_{j[\mathsf{Var}(l,c)]} \tau$ and $t = l\tau$ by the definition of τ. From $\mathsf{Var}(r) \subseteq \mathsf{Var}(l)$ and the level version of Lemma 16 we obtain $r\sigma \twoheadrightarrow\!\!\!\!\!\!\,\|\,_j r\tau$. It remains to show that $t = l\tau \twoheadrightarrow\!\!\!\!\!\!\,\|\,_k r\tau$. So, we will prove $c\tau \subseteq \to^*_{k-1}$. Let $l' \approx r'$ be an arbitrary condition in c. We have to prove that $l'\tau \to^*_{k-1} r'\tau$. Since $c\sigma \subseteq \to^*_{k-1}$, we have $l'\sigma \to^*_{k-1} r'\sigma$. Moreover, $\sigma \twoheadrightarrow\!\!\!\!\!\!\,\|\,_{j[\mathsf{Var}(l,c)]} \tau$, $\mathsf{Var}(c) \subseteq \mathsf{Var}(l,c)$, and the level version of Lemma 16 yields that $l'\sigma \twoheadrightarrow\!\!\!\!\!\!\,\|\,_j l'\tau$. Hence $l'\tau \,{}_j\!\twoheadleftarrow\!\!\!\!\!\!\,\|\, l'\sigma \to^*_{k-1} r'\sigma$. From the induction hypothesis and Lemmata 30 and 25, we know the existence of a term v such that $l'\tau \to^*_{k-1} v \,{}_j\!\twoheadleftarrow\!\!\!\!\!\!\,\|\, r'\sigma$. Because $\mathcal{R}$ is normal, $r'\sigma = r' = r'\tau = v$. Hence $l'\tau \to^*_{k-1} r'\tau$. Therefore $l\tau \twoheadrightarrow\!\!\!\!\!\!\,\|\,_k r\tau$. □

Theorem 31 (confluence by orthogonality)
Every orthogonal normal conditional STTRS is confluent.

Proof. By Lemma 28 and the Parallel Moves Lemma for conditional STTRSs (Lemma 30). □

6 Termination of STTRSs by Interpretation

In this section, we discuss how to prove termination of STTRSs by semantic method. We adapt the semantic proof technique both in the first-order case [Zan94] and in the higher-order case [vdP94]. In a semantic method, terms are interpreted as an element in a well-founded ordered set.

Definition 32 (monotone domain)
Given a non-empty set D_o, called a *base domain*, and a strict partial order (irreflexive transitive relation) $\succ^D_o$ on D_o, we recursively define the set D_τ and the relation $\succ^D_\tau$ on D_τ, for every function type $\tau = \tau_1 \times \cdots \tau_n \to \tau_0$, as follows:

$$D_\tau = \{\, \varphi : D_{\tau_1} \times \cdots \times D_{\tau_n} \to D_{\tau_0} \mid \varphi \text{ is monotone} \,\}$$

where a function $\varphi \in D_\tau$ is called *monotone* if

$$\varphi(x_1, \ldots, x_i, \ldots, x_n) \succ^D_{\tau_0} \varphi(x_1, \ldots, y, \ldots, x_n)$$

for all $i \in \{1, \ldots, n\}$ and $x_1 \in D_{\tau_1}, \ldots, x_n \in D_{\tau_n}, y \in D_{\tau_i}$ with $x_i \succ^D_{\tau_i} y$. We write $\varphi \succ^D_\tau \psi$ if $\varphi, \psi \in D_\tau$ and

$$\varphi(x_1, \ldots, x_n) \succ^D_{\tau_0} \psi(x_1, \ldots, x_n)$$

for all $x_1 \in D_{\tau_1}, \ldots, x_n \in D_{\tau_n}$.

Lemma 33 (properties of $\succ^D_\tau$)
Let $\succ^D_o$ be a strict partial order on a base domain D_o.
(1) $\succ^D_\tau$ also is a strict partial order for all types τ.
(2) If $\succ^D_o$ is well-founded, then $\succ^D_\tau$ also is well-founded for all types τ.

Proof. By induction on τ. □

We interpret every constant symbol by an algebra operation on a well-founded ordered set. A valuation gives an interpretation to variables.

Definition 34 (monotone interpretation)
Let $\succ^D_o$ be a strict partial order on a base domain D_o. A *monotone interpretation* I associates every constant symbol c of type τ with its interpretation $c_I \in D_\tau$. A *valuation* is a function which maps a variable symbol of type τ to an element in D_τ.

Let I be a monotone interpretation and α be a valuation. A term t of type τ is interpreted as an element of D_τ as follows: If t is a variable symbol, then $[\![t]\!]_\alpha = \alpha(t)$. If t is a constant symbol, then $[\![t]\!]_\alpha = t_I$. If $t = (t_0\, t_1 \cdots t_n)$ then $[\![t]\!]_\alpha = [\![t_0]\!]_\alpha([\![t_1]\!]_\alpha, \cdots, [\![t_n]\!]_\alpha)$. We define the relation $\succ_I$ on the set of terms as follows: $s \succ_I t$ if and only if s and t are of the same type, say τ, and $[\![s]\!]_\alpha \succ^D_\tau [\![t]\!]_\alpha$ for all valuations α.

The order $\succ_I$ compares two terms by interpretation. If a well-founded order $\succ$ on terms is closed under contexts and substitutions, then $l \succ r$ for all rewrite rules $l \to r \in \mathcal{R}$ suffices to show the termination of $\mathcal{R}$. It turns out that $\succ_I$ is closed under contexts and substitutions.

Lemma 35 (properties of $\succ_I$)
Let $\succ^D_o$ be a strict partial order on a base domain D_o and I be a monotone interpretation.
(1) $\succ_I$ also is a strict partial order.
(2) If $\succ^D_o$ is well-founded, then $\succ_I$ also is well-founded.

Proof. Easy consequences of Lemma 33. □

Lemma 36 (closure under contexts of $\succ_I$)
Let $\succ^D_o$ be a strict partial order on a base domain D_o and I be a monotone interpretation. Then, $\succ_I$ is closed under contexts, i.e., if $s \succ_I t$ then $u[s]_p \succ_I u[t]_p$ for all possible terms u and positions p in u.

Proof. The proof is by induction on p. The case $p = \epsilon$ is trivial. If $p = iq$, then u should have the form $(u_0\, u_1 \cdots u_n)$ and u_0 has function type, say $\tau = \tau_1 \times \cdots \times \tau_n \to \tau_0$. From the induction hypothesis, we know $u_i[s]_q \succ_I u_i[t]_q$. We distinguish two cases. If $i = 0$, then $[\![u_0[s]_q]\!]_\alpha \succ^D_\tau [\![u_0[t]_q]\!]_\alpha$ for all valuations α, by the induction hypothesis. Hence $[\![u[s]_p]\!]_\alpha = [\![(u_0[s]_q\, u_1 \cdots u_n)]\!]_\alpha = [\![u_0[s]_q]\!]_\alpha([\![u_1]\!]_\alpha, \cdots, [\![u_n]\!]_\alpha) \succ^D_{\tau_0} [\![u_0[t]_q]\!]_\alpha([\![u_1]\!]_\alpha, \cdots, [\![u_n]\!]_\alpha) = [\![u_0[t]_q\, u_1 \cdots u_n]\!]_\alpha = [\![u[t]_p]\!]_\alpha$ for all α. Therefore $u[s]_p \succ_I u[t]_p$. Consider the case $i > 0$. By the induction hypothesis, we have $[\![u_i[s]_q]\!]_\alpha \succ^D_\tau [\![u_i[t]_q]\!]_\alpha$ for all valuations α. Hence, $[\![u[s]_p]\!]_\alpha = [\![(u_0 u_1 \cdots u_i[s]_q \cdots u_n)]\!]_\alpha = [\![u_0]\!]_\alpha([\![u_1]\!]_\alpha, \cdots, [\![u_i[s]_q]\!]_\alpha, \cdots [\![u_n]\!]_\alpha) \succ^D_{\tau_0}$

$[\![u_0]\!]_\alpha([\![u_1]\!]_\alpha, \cdots, [\![u_i[t]_q]\!]_\alpha, \cdots, [\![u_n]\!]_\alpha) = [\![(u_0 u_1 \cdots u_i[t]_q \cdots u_n)]\!]_\alpha = [\![u[t]_p]\!]_\alpha$, for all valuations α, by monotonicity. Therefore $u[s]_p \succ_I u[t]_p$. □

Lemma 37 (closure under substitutions of $\succ_I$)
Let $\succ_o^D$ be a strict partial order on a base domain D_o and I be a monotone interpretation.

(1) $[\![t\sigma]\!]_\alpha = [\![t]\!]_{\alpha*\sigma}$ for all terms t, substitutions σ, and valuations α. Here the valuation $\alpha * \sigma$ is defined by $(\alpha * \sigma)(x) = [\![\sigma(x)]\!]_\alpha$.

(2) $\succ_I$ is closed under substitutions, i.e., if $s \succ_I t$ then $s\sigma \succ_I t\sigma$ for all substitutions σ.

Proof. (1) is by structural induction on t. (2) is an easy consequence of (1). □

The following theorem provides a semantic proof method for termination of STTRSs.

Theorem 38 (termination of STTRSs by interpretation)
Let $\mathcal{R} = (R, V, C)$ be an STTRS. If there exists a base domain D_o, a well-founded strict partial order $\succ_o^D$ on D_o, and a monotone interpretation I, such that $[\![l]\!]_\alpha \succ_I [\![r]\!]_\alpha$ for all rewrite rules $l \to r \in R$ and valuations α, then $\mathcal{R}$ is terminating.

Proof. It is easy to see that $\to_\mathcal{R} \subseteq \succ_I$ because we have $[\![l]\!]_\alpha \succ_I [\![r]\!]_\alpha$ for all rewrite rules $l \to r$ and valuations α by assumption, and the relation $\succ_I$ is closed under contexts and substitutions by Lemmata 36 and 37. For the proof by contradiction, suppose $\to_\mathcal{R}$ is not well-founded. So, there is an infinite rewrite sequence $t_0 \to_\mathcal{R} t_1 \to_\mathcal{R} \cdots$. From this sequence and the inclusion $\to_\mathcal{R} \subseteq \succ_I$, we obtain an infinite descending sequence $t_0 \succ_I t_1 \succ_I \cdots$. This contradicts the well-foundedness of the relation $\succ_I$, which follows from Lemma 35. □

Example 39 (termination of STTRSs by interpretation)
Consider again the STTRS $\mathcal{R}$ of Example 3. In order to prove the termination of $\mathcal{R}$, define the base domain as $D_o = \mathbb{N} - \{0, 1\}$, and the order $\succ_o^D$ by using the standard order $>$ on $\mathbb{N}$ as follows: $m \succ_o^D n$ if and only if $m > n$. Constant symbols are interpreted as follows:

$$\begin{aligned}
[\,]_I &= 2 \\
:_I(m, n) &= m + n \\
\mathsf{map}_I(\varphi, n) &= n \times \varphi(n) \\
\circ_I(\varphi, \psi)(n) &= \varphi(\psi(n)) + 1 \\
\mathsf{twice}_I(\varphi)(n) &= \varphi(\varphi(n)) + 2
\end{aligned}$$

Note that $[\,]_I \in D_o$, $:_I \in D_{o \times o \to o}$, $\mathsf{map}_I \in D_{(o \to o) \times o \to o}$, $\circ_I \in D_{(o \to o) \times (o \to o) \to (o \to o)}$, and $\mathsf{twice}_I \in D_{(o \to o) \to (o \to o)}$ are satisfied. It is easy to see that $[\![l]\!]_\alpha \succ_I [\![r]\!]_\alpha$ for all rewrite rules $l \to r \in R$ and valuations α. Therefore $\mathcal{R}$ is terminating.

The converse of Theorem 38 holds for the first order case [Zan94]. However, for STTRSs, the converse does not hold as shown in the following example.

Example 40 (incompleteness of the semantic proof method)
Let $C = \{0^o, 1^o, \mathsf{g}^{o \to o}\}$, the set of variable symbols V contain $F^{o \to o}$, and $R = \{\mathsf{g}\ (F\ 0\ 1) \to \mathsf{g}\ (F\ 1\ 0)\}$. It is not difficult to see that the STTRS (R, V, C) is terminating.

Suppose there is a base domain D_o, a well-founded strict partial order $\succ_o^D$ on D_o, and a monotone interpretation I, such that $[\![l]\!]_\alpha \succ_I [\![r]\!]_\alpha$ for all rewrite rules $l \to r \in R$ and valuations α. We have $\mathsf{g}_I(f(0_I, 1_I)) \succ_o^D \mathsf{g}_I(f(1_I, 0_I))$ for all monotone functions $f \in D_{o \times o \to o}$. Let h be an arbitrary monotone function in $D_{o \times o \to o}$. Then, the function h' defined by $h'(x, y) = h(y, x)$ is also monotone and hence $h' \in D_{o \times o \to o}$. Therefore, $\mathsf{g}_I(h(0_I, 1_I)) \succ_o^D \mathsf{g}_I(h(1_I, 0_I)) = \mathsf{g}_I(h'(0_I, 1_I)) \succ_o^D \mathsf{g}_I(h'(1_I, 0_I)) = \mathsf{g}_I(h(0_I, 1_I))$. This contradicts the irreflexivity of $\succ_o^D$.

At present, it is not known whether there is a complete proof method for termination of STTRSs based on monotone interpretation.

7 Concluding Remarks

We have proposed simply typed term rewriting systems (STTRSs), which is close to the format of functional programming languages with pattern matching. For proving the confluence of orthogonal rewrite systems, we introduced the parallel moves property, which is a useful sufficient condition obtained by localizing the test for the diamond property of a parallel reduction. We proved the confluence of orthogonal STTRSs and orthogonal normal conditional STTRSs. We also provided a semantic method for proving termination of STTRSs based on monotone interpretation.

Since the class of (conditional) STTRSs is a proper extension of the first-order case, all known results for the first-order TRSs can be applied to the subclass of our (conditional) STTRSs. We can also expect that many known results for the first-order TRSs can be lifted to the higher-order case without difficulty, because the behaviour of our higher-order extension is very close to that of the first-order (conditional) TRSs.

Suzuki et al. gave a sufficient condition for the confluence of orthogonal first-order conditional TRSs possibly with extra variables in the right-hand sides of the rewrite rules [SMI95]. The author conjectures that their result can be extended to the higher-order case.

Acknowledgments. I wish to thank Aart Middeldorp, Femke van Raamsdonk, and Fer-Jan de Vries for their comments on the preliminary version of this paper. I benefitted from a discussion with Jaco van de Pol and Hans Zantema, which resulted in Example 40. My thanks are also due to the anonymous referees for their helpful comments.

References

[BK86] J.A. Bergstra and J.W. Klop. Conditional rewrite rules: Confluence and termination. *Journal of Computer and System Science*, 32:323–362, 1986.

[BN98] F. Baader and T. Nipkow. *Term Rewriting and All That.* Cambridge University Press, 1998.

[DJ90] N. Dershowitz and J.-P. Jouannaud. Rewrite systems. In J. van Leeuwen, editor, *Handbook of Theoretical Computer Science*, volume B, chapter 6, pages 243–320. The MIT Press, 1990.

[Gra96] B. Gramlich. Confluence without termination via parallel critical pairs. In *Proceedings of the 21st International Colloquium on Trees in Algebra and Programming*, 1996. Lecture Notes in Computer Science 1059, pp. 211-225.

[HS86] J.R. Hindley and J.P. Seldin. *Introduction to Combinators and λ-Calculus.* Cambridge University Press, 1986.

[Hue80] G. Huet. Confluent reductions: Abstract properties and applications to term rewriting systems. *Journal of the Association for Computing Machinery*, 27(4):797–821, 1980.

[JO91] J.-P. Jouannaud and M. Okada. Executable higher-order algebraic specification languages. In *Proceedings of the 6th IEEE Symposium on Logic in Computer Science*, pages 350–361, 1991.

[Klo80] J.W. Klop. *Combinatory Reduction Systems.* PhD thesis, Rijks-universiteit, Utrecht, 1980.

[Klo92] J.W. Klop. Term rewriting systems. In S. Abramsky, D. Gabbay, and T. Maibaum, editors, *Handbook of Logic in Computer Science*, volume 2, chapter 1, pages 1–116. Oxford University Press, 1992.

[MN98] R. Mayr and T. Nipkow. Higher-order rewrite systems and their confluence. *Theoretical Computer Science*, 192:3–29, 1998.

[NP98] T. Nipkow and C. Prehofer. *Higher-Order Rewriting and Equational Reasoning*, volume I, pages 399–430. Kluwer, 1998.

[SMI95] T. Suzuki, A. Middeldorp, and T. Ida. Level-confluence of conditional rewrite systems with extra variables in right-hand sides. In *Proceedings of the 6th International Conference on Rewriting Techniques and Applications*, 1995. Lecture Notes in Computer Science 914, pp. 179–193.

[vdP94] J. van de Pol. Termination proofs for higher-order rewrite systems. In *Proceedings of the 1st International Workshop on Higher-Order Algebra, Logic and Term Rewriting*, 1994. Lecture Notes in Computer Science 816, pp. 305–325.

[vO97] V. van Oostrom. Developing developments. *Theoretical Computer Science*, 175:159–181, 1997.

[vR99] F. van Raamsdonk. Higher-order rewriting. In *Proceedings of the 10th International Conference on Rewriting Techniques and Applications*, 1999. Lecture Notes in Computer Science 1631, pp. 220–239.

[Zan94] H. Zantema. Termination of term rewriting: Interpretation and type elimination. *Journal of Symbolic Computation*, 17:23–50, 1994.

Parallel Evaluation of Interaction Nets with MPINE

Jorge Sousa Pinto*

Departamento de Informática
Universidade do Minho
Campus de Gualtar, 4710-057 Braga, Portugal
jsp@di.uminho.pt

Abstract. We describe the MPINE tool, a multi-threaded evaluator for Interaction Nets. The evaluator is an implementation of the present author's Abstract Machine for Interaction Nets [5] and uses POSIX threads to achieve concurrent execution. When running on a multi-processor machine (say an SMP architecture), parallel execution is achieved effortlessly, allowing for desktop parallelism on commonly available machines.

Interaction Nets

Interaction Nets [3] are a graph-rewriting formalism where the rewriting rules are such that only pairs of nodes, connected in a specific way, may be rewritten. Because of this restriction, the formalism enjoys strong local confluence. Although the system has been introduced as a visual, simple, and inherently parallel programming language, translations have been given of other formalisms into Interaction Nets, specifically term-rewriting systems [1] and the λ-calculus [2, 4]. When used as an intermediate implementation language for these systems, Interaction Nets allow to keep a close control on the sharing of reductions.

Interaction Nets have always seemed to be particularly adequate for being implemented in parallel, since there can never be interference between the reduction of two distinct redexes.

Moreover there are no global time or synchronization constraints. A parallel reducer for Interaction Nets provides a reducer for any of the formalisms that can be translated into these nets, without additional effort.

We present here MPINE (for Multi-Processing Interaction Net Evaluator), a parallel reducer for Interaction Nets which runs on generic shared-memory multiprocessors, based on the **POSIX** threads library. The system runs notably on the widely available SMP architecture machines running the Unix operating system.

A Concurrent Abstract Machine for Interaction Nets

In [5] the author has proposed an abstract machine for Interaction Net reduction, providing a decomposition of interaction steps into finer-grained operations. The

* Research done whilst staying at Laboratoire d'Informatique (CNRS UMR 7650), École Polytechnique. Partially supported by PRAXIS XXI grant BD/11261/97.

A. Middeldorp (Ed.): RTA 2001, LNCS 2051, pp. 353–356, 2001.

multi-threaded version of this machine is a device for the concurrent implementation of Interaction Nets on shared-memory architectures, based on a generalized version of the producer-consumers model: basic machine tasks are kept on a shared queue, from which a number of threads take tasks for processing. While doing so, new tasks may be generated, which will be enqueued.

Besides allowing for finer-grained parallelism than Interaction Nets themselves, the decomposition of interaction also allows for improvements concerning the synchronization requirements of the implementation. In particular, it solves a basic deadlock situation which arises when one naively implements Interaction Nets in shared-memory architectures, and it lightens the overheads of synchronization – the number of mutexes required by each parallel operation is smaller than that required by a full interaction step, which may be quite significant.

We direct the reader to [6] for details on these issues.

MPINE

The abstract machine may be implemented on any platform offering support for multi-threaded computation. MPINE, which uses **POSIX** threads, is, to the best of our knowledge, the first available parallel reducer for Interaction Nets.

The program has a text-based interface. The user provides as input an interaction net written in a language similar to that of [3], and the number of threads to be launched (which should in general be equal to the number of available processors in the target machine). The output of the program (if the input net has a normal form) is a description of the reduced net in the same language.

We give a very simple example in which we declare three agents (Zero, Successor, Addition) that will allow us to implement the sum of Natural numbers by rewriting with Interaction Nets, using the usual inductive definition.

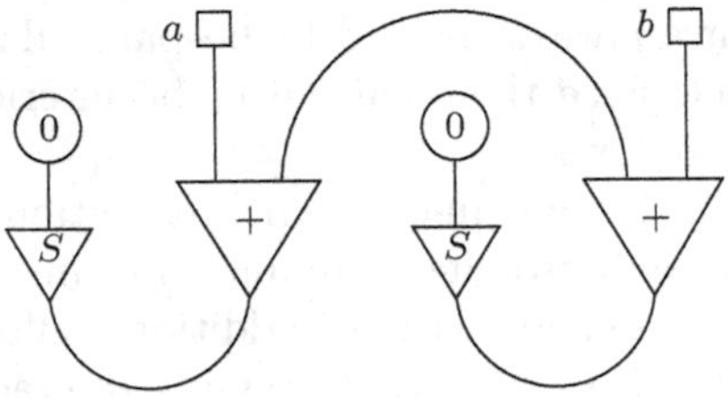

This net represents the two equations $S(0) + a = x$ and $S(0) + x = b$, or simply $S(0) + (S(0) + a) = b$. Each cell has a unique principal port. For + cells this represents the first argument of the sum. The file `example.net` contains a description of the net together with the interaction system in which it is defined:

```
agents
        Z       0;
        S       1;
        A       2;
rules
```

```
        A(x,x) >< Z;
        A(x,S(y)) >< S(A(x,y));
net
        S(Z) = A(a,x);
        S(Z) = A(x,b);
interface
        a;
        b;
end
```

The **agents** section contains declarations of agents, with their arity; then the **rules** and the interaction **net** are given as sequences of active pairs, written as equations. The file ends with the **interface** of the net, a sequence of terms. Interaction rules rewrite *active pairs* of cells connected through their principal ports. An example is given below, corresponding to the (term-rewriting) rule $S(y)+x \longrightarrow S(y+x)$. Interaction rules are written as pairs of terms by connecting together the corresponding free ports in both sides of each rule, as shown:

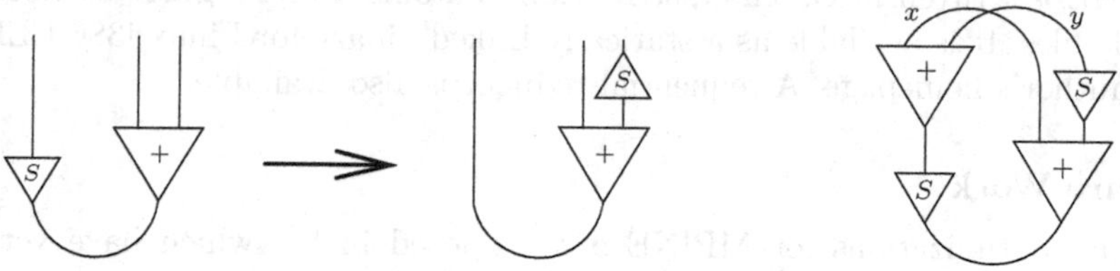

The following is an invocation of the reducer with 4 threads, with the above file as input, followed by the result produced:

```
> mpine -4 example.net
==============Initial Net===============
Displaying net equalities...
        S(Z) = A(x0,x1)
        S(Z) = A(x2,x0)
Displaying observable interface...
        x2
        x1
===============Reduced Net==============
Displaying observable interface...
        x0
        S(S(x0))
```

The input net is printed followed by the reduced net. Since this is a net with no cycles, there are no equations left (if there were they would be displayed). The system may also be instructed to print some statistics, including the number of tasks performed by each individual thread. The reader is referred to the user's guide for more information on additional features of the system.

Some Benchmark Results

We show a set of benchmark results for the reduction of nets obtained from λ-terms using the YALE translation of [4]. The terms are Church numerals, which we have selected simply because they generate abundant computations.

term	N.int	seq.red	par_2red	par_2/seq %
2232II	37272	5.02	5.54	110
423II	105911	29.8	15.7	52.7
333II	473034	552	310	56.2
2233II	1417653	7096	5381	75.8

For each term we show the number of interactions and the time taken to reduce the corresponding net, both by a sequential reducer (free of the synchronization overheads) and by MPINE running on a 2 processor machine. We also show the ratio of the two. These preliminary results are promising: in two of the above nets, the ideal goal of reducing by half the execution time is practically attained.

Availability

MPINE is written in C. The distribution contains a user's guide and some example files. It is available as a statically linked binary for Linux i386 ELF from the author's homepage. A sequential reducer is also available.

Future Work

Further optimizations for MPINE are proposed in [6], which have yet to be incorporated in the implementation. It also remains to test the implementation with a large set of terms, notably in machines with more than two processors.

References

1. Maribel Fernández and Ian Mackie. Interaction nets and term rewriting systems. *Theoretical Computer Science*, 190(1):3–39, January 1998.
2. Georges Gonthier, Martín Abadi, and Jean-Jacques Lévy. The geometry of optimal lambda reduction. In *Proceedings of the 19th ACM Symposium on Principles of Programming Languages (POPL'92)*, pages 15–26. ACM Press, January 1992.
3. Yves Lafont. Interaction nets. In *Proceedings of the 17th ACM Symposium on Principles of Programming Languages (POPL'90)*, pages 95–108. ACM Press, January 1990.
4. Ian Mackie. YALE: Yet another lambda evaluator based on interaction nets. In *Proceedings of the 3rd ACM SIGPLAN International Conference on Functional Programming (ICFP'98)*, pages 117–128. ACM Press, September 1998.
5. Jorge Sousa Pinto. Sequential and concurrent abstract machines for interaction nets. In Jerzy Tiuryn, editor, *Proceedings of Foundations of Software Science and Computation Structures (FOSSACS)*, number 1784 in Lecture Notes in Computer Science, pages 267–282. Springer-Verlag, 2000.
6. Jorge Sousa Pinto. *Parallel Implementation with Linear Logic (Applications of Interaction Nets and of the Geometry of Interaction).* PhD thesis, École Polytechnique, 2001.

Stratego: A Language for Program Transformation Based on Rewriting Strategies

System Description of Stratego 0.5

Eelco Visser

Institute of Information and Computing Sciences, Universiteit Utrecht,
P.O. Box 80089, 3508 TB Utrecht, The Netherlands
visser@acm.org, http://www.cs.uu.nl/~visser

1 Introduction

Program transformation is used in many areas of software engineering. Examples include compilation, optimization, synthesis, refactoring, migration, normalization and improvement [15]. Rewrite rules are a natural formalism for expressing single program transformations. However, using a standard strategy for normalizing a program with a set of rewrite rules is not adequate for implementing program transformation systems. It may be necessary to apply a rule only in some phase of a transformation, to apply rules in some order, or to apply a rule only to part of a program. These restrictions may be necessary to avoid non-termination or to choose a specific path in a non-confluent rewrite system.

Stratego is a language for the specification of program transformation systems based on the paradigm of rewriting strategies. It supports the separation of strategies from transformation rules, thus allowing careful control over the application of these rules. As a result of this separation, transformation rules are reusable in multiple different transformations and generic strategies capturing patterns of control can be described independently of the transformation rules they apply. Such strategies can even be formulated independently of the object language by means of the generic term traversal capabilities of Stratego.

In this short paper I give a description of version 0.5 of the Stratego system, discussing the features of the language (Section 2), the library (Section 3), the compiler (Section 4) and some of the applications that have been built (Section 5). Stratego is available as free software under the GNU General Public License from `http://www.stratego-language.org`.

2 The Language

In the paradigm of program transformation with rewriting strategies [14] a specification of a program transformation consists of a signature, a set of rules and a strategy for applying the rules. The abstract syntax trees of programs are represented by means of first-order terms. A signature declares the constructors of such terms. *Labeled conditional rewrite rules* of the form `L: l -> r where s`,

A. Middeldorp (Ed.): RTA 2001, LNCS 2051, pp. 357–361, 2001.

```
module lambda-transform
imports lambda-sig lambda-vars iteration simple-traversal
rules
  Beta : App(Abs(x, e1), e2) -> <lsubs>([(x,e2)], e1)
strategies
  simplify = bottomup(try(Beta))
  eager = rec eval(try(App(eval, eval)); try(Beta; eval))
  whnf  = rec eval(try(App(eval, id)); try(Beta; eval))
```

Fig. 1. A Stratego module defining several strategies for transforming lambda expressions using beta reduction. Strategy `simplify` makes a bottom-up traversal over an expression trying beta reduction at each subexpression once, even under lambda abstractions. Strategy `eager` reduces the argument of a function before applying it, but does not reduce under abstractions. Strategy `whnf` reduces an expression to weak head-normal form, i.e., does not normalize under abstractions or in argument positions. Strategies `eager` and `whnf` use the congruence operator `App` to traverse terms of the form `App(e1,e2)`, while strategy `simplify` uses the generic traversal `bottomup`. The strategy `lsubs` is a strategy for substituting expressions for variables. It is implemented in module `lambda-vars` using a generic substitution strategy.

with `l` and `r` patterns, express basic transformations on terms. A rewriting strategy combines rules into a program that determines where and in what order the rules are applied to a term. An example specification is shown in Figure 1.

A *strategy* is an operation that transforms a term into another term or fails. Rules are basic strategies that perform the transformation specified by the rule or fail when either the subject term does not match the left-hand side or the condition fails. Strategies can be combined into more complex strategies by means of a language of strategy operators. These operators can be divided into operators for sequential programming and operators for term traversal. The sequential programming operators *identity* (`id`), *failure* (`fail`), *sequential composition* (`;`), *choice* (`+`), *negation* (`not`), *test*, and *recursive closure* (`rec x(s)`) combine strategies that apply to the root of a term. To achieve transformations throughout a term, a number of *term traversal primitives* are provided. For each constructor `C/n`, the corresponding *congruence* operator `C(s1,...,sn)` expresses the application of strategies to the direct sub-terms of a term constructed with `C`. Furthermore, a number of term traversal operators express *generic traversal* to the direct sub-terms of a term without reference to the constructor of the term. These constructs allow the generic definition of a wide range of traversals over terms. For example, the strategy `all(s)` applies `s` to each direct sub-term of a term. Using this operator one can define `bottomup(s) = rec x(all(x); s)`, which generically defines the notion of a post-order traversal that visits each sub-term applying the parameter strategy `s` to it.

A number of abstraction mechanisms are supported. A *strategy definition* of the form `f(x1,...,xn) = s` defines the new operator `f` with `n` parameters as an abstraction of the strategy `s`. An *overlay* of the form `C(x1,...,xn) = t` captures the pattern `t` in a new pseudo-constructor `C` [9]. Constructors and strategy operators can be overloaded on arity. Strategies implemented in a foreign language (e.g., for accessing the file system) can be called via the `prim` construct.

The distinction between rules and strategies is actually only idiomatic, that is, rules are abbreviations for strategies that are composed from the actual primitives of transformation: matching terms against patterns and building instantiations of patterns. Thus, a rule `L: l -> r where s` is just an abbreviation of the strategy `L = {x1,...,xn: ?l; where(s); !r}`, where the `xi` are the variables used in the rule. The construct `{xs: s}` delimits the scope of the variables `xs` to the strategy `s`. The strategy `?t` matches the subject term against the pattern `t` binding the variables in `t` to the corresponding sub-terms of the subject term. The strategy `!t` builds an instantiation of the term pattern `t` by replacing the variables in `t` by the terms to which they are bound. Decoupling pattern matching and term construction from rules and scopes, and making these constructs into first-class citizens, opens up a wide range of idioms such as contextual rules and recursive patterns [9]. In these idioms a pattern match is passed on to a local traversal strategy to match sub-terms at variable depth in the subject term.

Finally, specifications can be divided into modules that can import other modules. The above constructs of Stratego together with its module system make a powerful language that supports concise specification of program transformations. An operational semantics of System S, the core of the language, can be found in [13,14]. A limitation of the current language is that only a weak type system is implemented. Work is in progress to find a suitable type system that reconciles genericity with type safety.

3 The Library

The Stratego Library [10] is a collection of modules (≈45) with reusable rules (≈130) and strategies (≈300). Included in the library are strategies for sequential control, generic traversal, built-in data type manipulation (numbers and strings), standard data type manipulation (lists, tuples, optionals), generic language processing, and system interfacing (I/O, process control, association tables).

The generic traversal strategies include one-pass traversals (such as topdown, bottomup, oncetd, and spinetd), fixed point traversal (such as reduce, innermost, and outermost), and traversal with environments. The generic language processing algorithms cover free variable extraction, bound variable renaming, substitution, and syntactic unification [11]. These algorithms are parameterized with the pattern of the relevant object language constructs and use the generic traversal capabilities of Stratego to ignore all constructs not relevant for the operation. For example, bound variable renaming is parameterized with the shape of variables and the binding constructs of the language.

4 The Compiler

The Stratego Compiler translates specifications to C code. The run-time system is based on the ATerm library [4], which supports the ATerm Format, a representation for first-order terms with prefix application syntax. The library implements writing and reading ATerms to and from the external format, which is used to exchange terms between tools. This enables component-based development of transformation tools. For example, a Stratego program can transform abstract syntax trees produced by any parser as long as it produces an ATerm representation of the abstract syntax tree for a program.

The compiler has been bootstrapped, that is, all components except the parser are specified in Stratego itself. The compiler performs various optimizations, including extracting the definitions that are used in the main strategy, aggressive inlining to enable further optimizations and merging of matching patterns to avoid backtracking. A limitation of the current compiler is that it does not support separate compilation and that compilation of the generated code by `gcc` is rather slow, resulting in long compilation times (e.g., 3 minutes for a large compiler component). Overcoming this limitation is the focus of current work.

5 Applications

Stratego is intended for use in a wide range of language processing applications including source-to-source transformation, application generation, program optimization, compilation, and documentation generation. It is not intended for interactive program transformation or theorem proving.

Examples of applications that use Stratego are XT, CodeBoost, HSX and a Tiger compiler. XT is a bundle of program transformation tools [6] in which Stratego is included as the main language for implementing program transformations. The bundle comes with a collection of grammars for standard languages and many tools implemented in Stratego for generic syntax tree manipulation, grammar analysis and transformation, and derivation of tools from grammars. CodeBoost is a framework for the transformation of C++ programs [2] that is developed for domain-specific optimization of C++ programs for numerical applications. HSX is a framework for the transformation of core Haskell programs that has been developed for the implementation of the warm fusion algorithm for deforesting functional programs [8]. The Tiger compiler translates Tiger programs [1] to MIPS assembly code [12]. The compiler includes translation to intermediate representation, canonicalization of intermediate representation, instruction selection, and register allocation.

6 Related Work

The creation of Stratego was motivated by the limitations of a fixed (innermost) strategy for rewriting, in particular based on experience with the algebraic specification formalism ASF+SDF [7]. The design of the strategy operators was inspired by the strategy language of ELAN [3], a specification language based on the paradigm of rewriting logic [5]. For a comparison of Stratego with other systems see [13,14]. A survey of program transformation systems in general can be found in [15]. The contributions of Stratego include: generic traversal primitives that allow definition of generic strategies; break-down of rules into primitives match and build giving rise to first-class pattern matching; many programming idioms for strategic rewriting; bootstrapped compilation of strategies; a foreign function interface; component-based programming based on exchange of ATerms.

Acknowledgements. I would like to thank Bas Luttik, Andrew Tolmach, Zino Benaissa, Patricia Johann, Joost Visser, Merijn de Jonge, Otto Skrove Bagge, Dick Kieburtz, Karina Olmos, Hedzer Westra, Eelco Dolstra and Arne de Bruijn for their contributions to the design, implementation and application of Stratego.

References

1. A. W. Appel. *Modern Compiler Implementation in ML.* Cambridge University Press, 1998.
2. O. S. Bagge, M. Haveraaen, and E. Visser. CodeBoost: A framework for the transformation of C++ programs. Technical report, Universiteit Utrecht, 2000.
3. P. Borovanský, C. Kirchner, H. Kirchner, P.-E. Moreau, and M. Vittek. Elan: A logical framework based on computational systems. In J. Meseguer, editor, *ENTCS*, volume 4, 1996. Workshop on Rewriting Logic and Applications 1996.
4. M. G. J. van den Brand, H. A. de Jong, P. Klint, and P. A. Olivier. Efficient annotated terms. *Software—Practice & Experience*, 30:259–291, 2000.
5. M. Clavel and J. Meseguer. Reflection and strategies in rewriting logic. In J. Meseguer, editor, *ENTCS*, volume 4, 1996. Workshop on Rewriting Logic and its Applications 1996.
6. M. de Jonge, E. Visser, and J. Visser. XT: A bundle of program transformation tools. In *ENTCS*, 2001. Language Descriptions, Tools and Applications 2001.
7. A. Van Deursen, J. Heering, and P. Klint, editors. *Language Prototyping*, volume 5 of *AMAST Series in Computing.* World Scientific, Singapore, 1996.
8. P. Johann and E. Visser. Warm fusion in Stratego: A case study in the generation of program transformation systems. *Annals of Mathematics and Artificial Intelligence.* (To appear).
9. E. Visser. Strategic pattern matching. In P. Narendran and M. Rusinowitch, editors, *Rewriting Techniques and Applications (RTA'99)*, volume 1631 of *Lecture Notes in Computer Science*, pages 30–44, Trento, Italy, July 1999. Springer-Verlag.
10. E. Visser. *The Stratego Library.* Institute of Information and Computing Sciences, Universiteit Utrecht, Utrecht, The Netherlands, 1999.
11. E. Visser. Language independent traversals for program transformation. In J. Jeuring, editor, *Workshop on Generic Programming (WGP2000)*, Ponte de Lima, Portugal, July 6, 2000. Technical Report UU-CS-2000-19, Universiteit Utrecht.
12. E. Visser. Tiger in Stratego: An exercise in compilation by transformation. `http://www.stratego-language.org/tiger/`, 2000.
13. E. Visser and Z.-e.-A. Benaissa. A core language for rewriting. In C. Kirchner and H. Kirchner, editors, *ENTCS*, volume 15, September 1998. Rewriting Logic and its Applications 1998.
14. E. Visser, Z.-e.-A. Benaissa, and A. Tolmach. Building program optimizers with rewriting strategies. *ACM SIGPLAN Notices*, 34(1):13–26, January 1999. Proceedings of the International Conference on Functional Programming (ICFP'98).
15. E. Visser et al. The online survey of program transformation. `http://www.program-transformation.org/survey.html`.

Author Index

Lecture Notes in Computer Science

For information about Vols. 1–1964
please contact your bookseller or Springer-Verlag

Vol. 1958: S. Fischer-Hübner, IT-Security and Privacy. XIII, 351 pages. 2001.

Vol. 1965: Ç. K. Koç, C. Paar (Eds.), Cryptographic Hardware and Embedded Systems – CHES 2000. Proceedings, 2000. XI, 355 pages. 2000.

Vol. 1966: S. Bhalla (Ed.), Databases in Networked Information Systems. Proceedings, 2000. VIII, 247 pages. 2000.

Vol. 1967: S. Arikawa, S. Morishita (Eds.), Discovery Science. Proceedings, 2000. XII, 332 pages. 2000. (Subseries LNAI).

Vol. 1968: H. Arimura, S. Jain, A. Sharma (Eds.), Algorithmic Learning Theory. Proceedings, 2000. XI, 335 pages. 2000. (Subseries LNAI).

Vol. 1969: D.T. Lee, S.-H. Teng (Eds.), Algorithms and Computation. Proceedings, 2000. XIV, 578 pages. 2000.

Vol. 1970: M. Valero, V.K. Prasanna, S. Vajapeyam (Eds.), High Performance Computing – HiPC 2000. Proceedings, 2000. XVIII, 568 pages. 2000.

Vol. 1971: R. Buyya, M. Baker (Eds.), Grid Computing – GRID 2000. Proceedings, 2000. XIV, 229 pages. 2000.

Vol. 1972: A. Omicini, R. Tolksdorf, F. Zambonelli (Eds.), Engineering Societies in the Agents World. Proceedings, 2000. IX, 143 pages. 2000. (Subseries LNAI).

Vol. 1973: J. Van den Bussche, V. Vianu (Eds.), Database Theory – ICDT 2001. Proceedings, 2001. X, 451 pages. 2001.

Vol. 1974: S. Kapoor, S. Prasad (Eds.), FST TCS 2000: Foundations of Software Technology and Theoretical Computer Science. Proceedings, 2000. XIII, 532 pages. 2000.

Vol. 1975: J. Pieprzyk, E. Okamoto, J. Seberry (Eds.), Information Security. Proceedings, 2000. X, 323 pages. 2000.

Vol. 1976: T. Okamoto (Ed.), Advances in Cryptology – ASIACRYPT 2000. Proceedings, 2000. XII, 630 pages. 2000.

Vol. 1977: B. Roy, E. Okamoto (Eds.), Progress in Cryptology – INDOCRYPT 2000. Proceedings, 2000. X, 295 pages. 2000.

Vol. 1978: B. Schneier (Ed.), Fast Software Encryption. Proceedings, 2000. VIII, 315 pages. 2001.

Vol. 1979: S. Moss, P. Davidsson (Eds.), Multi-Agent-Based Simulation. Proceedings, 2000. VIII, 267 pages. 2001. (Subseries LNAI).

Vol. 1980: M. Agosti, F. Crestani, G. Pasi (Eds.), Lectures on Information Retrieval. Proceedings, 2000. XI, 311 pages. 2001.

Vol. 1981: J.M.L.M. Palma, J. Dongarra, V. Hernández (Eds.), Vector and Parallel Processing – VECPAR 2000. Proceedings, 2000. XVI, 580 pages. 2001.

Vol. 1983: K.S. Leung, L.-W. Chan, H. Meng (Eds.), Intelligent Data Engineering and Automated Learning – IDEAL 2000. Proceedings, 2000. XVI, 573 pages. 2000.

Vol. 1984: J. Marks (Ed.), Graph Drawing. Proceedings, 2001. XII, 419 pages. 2001.

Vol. 1985: J. Davidson, S.L. Min (Eds.), Languages, Compilers, and Tools for Embedded Systems. Proceedings, 2000. VIII, 221 pages. 2001.

Vol. 1987: K.-L. Tan, M.J. Franklin, J. C.-S. Lui (Eds.), Mobile Data Management. Proceedings, 2001. XIII, 289 pages. 2001.

Vol. 1988: L. Vulkov, J. Waśniewski, P. Yalamov (Eds.), Numerical Analysis and Its Applications. Proceedings, 2000. XIII, 782 pages. 2001.

Vol. 1989: M. Ajmone Marsan, A. Bianco (Eds.), Quality of Service in Multiservice IP Networks. Proceedings, 2001. XII, 440 pages. 2001.

Vol. 1990: I.V. Ramakrishnan (Ed.), Practical Aspects of Declarative Languages. Proceedings, 2001. VIII, 353 pages. 2001.

Vol. 1991: F. Dignum, C. Sierra (Eds.), Agent Mediated Electronic Commerce. VIII, 241 pages. 2001. (Subseries LNAI).

Vol. 1992: K. Kim (Ed.), Public Key Cryptography. Proceedings, 2001. XI, 423 pages. 2001.

Vol. 1993: E. Zitzler, K. Deb, L. Thiele, C.A.Coello Coello, D. Corne (Eds.), Evolutionary Multi-Criterion Optimization. Proceedings, 2001. XIII, 712 pages. 2001.

Vol. 1995: M. Sloman, J. Lobo, E.C. Lupu (Eds.), Policies for Distributed Systems and Networks. Proceedings, 2001. X, 263 pages. 2001.

Vol. 1997: D. Suciu, G. Vossen (Eds.), The World Wide Web and Databases. Proceedings, 2000. XII, 275 pages. 2001.

Vol. 1998: R. Klette, S. Peleg, G. Sommer (Eds.), Robot Vision. Proceedings, 2001. IX, 285 pages. 2001.

Vol. 1999: W. Emmerich, S. Tai (Eds.), Engineering Distributed Objects. Proceedings, 2000. VIII, 271 pages. 2001.

Vol. 2000: R. Wilhelm (Ed.), Informatics: 10 Years Back, 10 Years Ahead. IX, 369 pages. 2001.

Vol. 2001: G.A. Agha, F. De Cindio, G. Rozenberg (Eds.), Concurrent Object-Oriented Programming and Petri Nets. VIII, 539 pages. 2001.

Vol. 2002: H. Comon, C. Marché, R. Treinen (Eds.), Constraints in Computational Logics. Proceedings, 1999. XII, 309 pages. 2001.

Vol. 2003: F. Dignum, U. Cortés (Eds.), Agent Mediated Electronic Commerce III. XII, 193 pages. 2001. (Subseries LNAI).

Vol. 2004: A. Gelbukh (Ed.), Computational Linguistics and Intelligent Text Processing. Proceedings, 2001. XII, 528 pages. 2001.

Vol. 2006: R. Dunke, A. Abran (Eds.), New Approaches in Software Measurement. Proceedings, 2000. VIII, 245 pages. 2001.

Vol. 2007: J.F. Roddick, K. Hornsby (Eds.), Temporal, Spatial, and Spatio-Temporal Data Mining. Proceedings, 2000. VII, 165 pages. 2001. (Subseries LNAI).

Vol. 2009: H. Federrath (Ed.), Designing Privacy Enhancing Technologies. Proceedings, 2000. X, 231 pages. 2001.

Vol. 2010: A. Ferreira, H. Reichel (Eds.), STACS 2001. Proceedings, 2001. XV, 576 pages. 2001.

Vol. 2011: M. Mohnen, P. Koopman (Eds.), Implementation of Functional Languages. Proceedings, 2000. VIII, 267 pages. 2001.

Vol. 2012: D.R. Stinson, S. Tavares (Eds.), Selected Areas in Cryptography. Proceedings, 2000. IX, 339 pages. 2001.

Vol. 2013: S. Singh, N. Murshed, W. Kropatsch (Eds.), Advances in Pattern Recognition – ICAPR 2001. Proceedings, 2001. XIV, 476 pages. 2001.

Vol. 2015: D. Won (Ed.), Information Security and Cryptology – ICISC 2000. Proceedings, 2000. X, 261 pages. 2001.

Vol. 2016: S. Murugesan, Y. Deshpande (Eds.), Web Engineering. IX, 357 pages. 2001.

Vol. 2018: M. Pollefeys, L. Van Gool, A. Zisserman, A. Fitzgibbon (Eds.), 3D Structure from Images – SMILE 2000. Proceedings, 2000. X, 243 pages. 2001.

Vol. 2020: D. Naccache (Ed.), Topics in Cryptology – CT-RSA 2001. Proceedings, 2001. XII, 473 pages. 2001

Vol. 2021: J. N. Oliveira, P. Zave (Eds.), FME 2001: Formal Methods for Increasing Software Productivity. Proceedings, 2001. XIII, 629 pages. 2001.

Vol. 2022: A. Romanovsky, C. Dony, J. Lindskov Knudsen, A. Tripathi (Eds.), Advances in Exception Handling Techniques. XII, 289 pages. 2001

Vol. 2024: H. Kuchen, K. Ueda (Eds.), Functional and Logic Programming. Proceedings, 2001. X, 391 pages. 2001.

Vol. 2025: M. Kaufmann, D. Wagner (Eds.), Drawing Graphs. XIV, 312 pages. 2001.

Vol. 2026: F. Müller (Ed.), High-Level Parallel Programming Models and Supportive Environments. Proceedings, 2001. IX, 137 pages. 2001.

Vol. 2027: R. Wilhelm (Ed.), Compiler Construction. Proceedings, 2001. XI, 371 pages. 2001.

Vol. 2028: D. Sands (Ed.), Programming Languages and Systems. Proceedings, 2001. XIII, 433 pages. 2001.

Vol. 2029: H. Hussmann (Ed.), Fundamental Approaches to Software Engineering. Proceedings, 2001. XIII, 349 pages. 2001.

Vol. 2030: F. Honsell, M. Miculan (Eds.), Foundations of Software Science and Computation Structures. Proceedings, 2001. XII, 413 pages. 2001.

Vol. 2031: T. Margaria, W. Yi (Eds.), Tools and Algorithms for the Construction and Analysis of Systems. Proceedings, 2001. XIV, 588 pages. 2001.

Vol. 2032: R. Klette, T. Huang, G. Gimel'farb (Eds.), Multi-Image Analysis. Proceedings, 2000. VIII, 289 pages. 2001.

Vol. 2033: J. Liu, Y. Ye (Eds.), E-Commerce Agents. VI, 347 pages. 2001. (Subseries LNAI).

Vol. 2034: M.D. Di Benedetto, A. Sangiovanni-Vincentelli (Eds.), Hybrid Systems: Computation and Control. Proceedings, 2001. XIV, 516 pages. 2001.

Vol. 2035: D. Cheung, G.J. Williams, Q. Li (Eds.), Advances in Knowledge Discovery and Data Mining – PAKDD 2001. Proceedings, 2001. XVIII, 596 pages. 2001. (Subseries LNAI).

Vol. 2037: E.J.W. Boers et al. (Eds.), Applications of Evolutionary Computing. Proceedings, 2001. XIII, 516 pages. 2001.

Vol. 2038: J. Miller, M. Tomassini, P.L. Lanzi, C. Ryan, A.G.B. Tettamanzi, W.B. Langdon (Eds.), Genetic Programming. Proceedings, 2001. XI, 384 pages. 2001.

Vol. 2039: M. Schumacher, Objective Coordination in Multi-Agent System Engineering. XIV, 149 pages. 2001. (Subseries LNAI).

Vol. 2040: W. Kou, Y. Yesha, C.J. Tan (Eds.), Electronic Commerce Technologies. Proceedings, 2001. X, 187 pages. 2001.

Vol. 2042: K.-K. Lau (Ed.), Logic Based Program Synthesis and Transformation. Proceedings, 2000. VIII, 183 pages. 2001.

Vol. 2043: , D. Craeynest, A. Strohmeier (Eds.), Reliable Software Technologies – Ada-Europe 2001. Proceedings, 2001. XV, 405 pages. 2001.

Vol. 2044: S. Abramsky (Ed.), Typed Lambda Calculi and Applications. Proceedings, 2001. XI, 431 pages. 2001.

Vol. 2045: B. Pfitzmann (Ed.), Advances in Cryptology – EUROCRYPT 2001. Proceedings, 2001. XII, 545 pages. 2001.

Vol. 2048: J. Pauli, Learning Based Robot Vision. IX, 288 pages. 2001.

Vol. 2051: A. Middeldorp (Ed.), Rewriting Techniques and Applications. Proceedings, 2001. XII, 363 pages. 2001.

Vol. 2052: V.I. Gorodetski, V.A. Skormin, L.J. Popyack (Eds.), Information Assurance in Computer Networks. Proceedings, 2001. XIII, 313 pages. 2001.

Vol. 2053: O. Danvy, A. Filinski (Eds.), Programs as Data Objects. Proceedings, 2001. VIII, 279 pages. 2001.

Vol. 2054: A. Condon, G. Rozenberg (Eds.), DNA Computing. Proceedings, 2000. X, 271 pages. 2001.

Vol. 2055: M. Margenstern, Y. Rogozhin (Eds.), Machines, Computations, and Universality. Proceedings, 2001. VIII, 321 pages. 2001.

Vol. 2056: E. Stroulia, S. Matwin (Eds.), Advances in Artificial Intelligence. Proceedings, 2001. XII, 366 pages. 2001. (Subseries LNAI).

Vol. 2057: M. Dwyer (Ed.), Model Checking Software. Proceedings, 2001. X, 313 pages. 2001.

Vol. 2059: C. Arcelli, L.P. Cordella, G. Sanniti di Baja (Eds.), Visual Form 2001. Proceedings, 2001. XIV, 799 pages. 2001.